OCEANOGRAPHY
and
MARINE BIOLOGY

AN ANNUAL REVIEW

Volume 55

OCEANOGRAPHY
and
MARINE BIOLOGY

AN ANNUAL REVIEW

Volume 55

Editors

S. J. Hawkins
Ocean and Earth Science, University of Southampton,
National Oceanography Centre, UK
and
The Marine Biological Association of the UK, The Laboratory, Plymouth, UK

A. J. Evans
Ocean and Earth Science, University of Southampton,
National Oceanography Centre, UK
and
The Marine Biological Association of the UK, The Laboratory, Plymouth, UK

A. C. Dale
Scottish Association for Marine Science, Argyll, UK

L. B. Firth
School of Biological and Marine Sciences, Plymouth University, UK

D. J. Hughes
Scottish Association for Marine Science, Argyll, UK

I. P. Smith
School of Biological Sciences, University of Aberdeen, United Kingdom

CRC Press
Taylor & Francis Group
Boca Raton London New York

CRC Press is an imprint of the
Taylor & Francis Group, an **informa** business

International Standard Serial Number: 0078-3218

CRC Press
Taylor & Francis Group
6000 Broken Sound Parkway NW, Suite 300
Boca Raton, FL 33487-2742

Printed on acid-free paper
Version Date: 20170206

International Standard Book Number-13: 978-1-138-19786-2 (Hardback)

Visit the Taylor & Francis Web site at
http://www.taylorandfrancis.com

and the CRC Press Web site at
http://www.crcpress.com

Contents

Preface

The 55th volume of *Oceanography and Marine Biology: An Annual Review* (OMBAR) contains six reviews that cover a range of topics, reflecting the wide readership of the series. A brief introductory comment outlines the inspiration for review articles in this special issue that stemmed from contributions to and discussions at the 2015 Aquatic Biodiversity & Ecosystems conference in Liverpool, UK.

OMBAR welcomes suggestions from potential authors for topics that could form the basis of appropriate reviews. Contributions from physical, chemical and biological oceanographers that seek to inform both oceanographers and marine biologists are especially welcome. Because the annual publication schedule constrains the timetable for submission, evaluation and acceptance of manuscripts, potential contributors are advised to contact the editors at an early stage of manuscript preparation. Contact details are listed on the title page of this volume.

The editors gratefully acknowledge the willingness and speed with which authors complied with the editors' suggestions and requests and the efficiency of CRC Press, especially Jennifer Blaise, Marsha Hecht and John Sulzycki, in ensuring the timely appearance of this volume.

Introduction to the Special Issue

ALLY J. EVANS[1,2], LOUISE B. FIRTH[3], ANDREW DALE[4],
DAVID HUGHES[4], I. PHILIP SMITH[5] & STEPHEN J. HAWKINS[1,2]

[1]*Ocean and Earth Science, National Oceanography Centre
Southampton, Southampton, SO14 3ZH, UK*
[2]*The Marine Biological Association of the UK, The Laboratory,
Citadel Hill, Plymouth, PL1 2PB, UK*
[3]*School of Biological and Marine Sciences, Plymouth University, Plymouth, PL4 8AA, UK*
[4]*Scottish Association for Marine Science, Scottish Marine
Institute, Oban, Argyll, Scotland, PA37 1QA, UK*
[5]*School of Biological Sciences, University of Aberdeen, Aberdeen, Scotland, AB24 2TZ, UK*

In this 55th volume of *Oceanography and Marine Biology: An Annual Review* (OMBAR), we present a special issue stemming directly and indirectly from contributions and discussions at the 2015 *Aquatic Biodiversity & Ecosystems Conference* in Liverpool, UK, plus related work from research teams represented at that conference.

The *Aquatic Biodiversity & Ecosystems Conference* was organised by Dr Louise Firth at the University of Liverpool in August–September 2015. The overarching theme was "evolution, interactions and global change in aquatic ecosystems", with a number of subthemes, within which over 250 papers were presented to delegates from 30 different countries. The meeting partially followed up on the 1990 *Plant–Animal Interactions in the Marine Benthos Conference* hosted by Professor Stephen Hawkins 25 years previously, which itself was prompted by a 1983 OMBAR review on grazing of intertidal algae by marine invertebrates (Hawkins & Hartnoll 1983).

Through structured workshops and informal discussions at the 2015 conference, suggestions were garnered for review articles for this special OMBAR issue, and for chapters in the forthcoming book *Interactions in the Marine Benthos—A Regional and Habitat Perspective*, a follow-up to *Plant–Animal Interactions in the Marine Benthos* (John et al. 1992) (which was inspired by the 1990 conference), to be published by Cambridge University Press. The neglected fields of herbivory in starfish (Martinez et al. 2017) and intertidal boulder-fields (Chapman 2017) are synthesized to stimulate further interest in these topics. Ecological dominance along rocky shores is discussed, focussing on ascidians (Rius et al. 2017). Neo et al. (2017) review the ecology of giant clams. Johnston et al. (2017) consider the spread of non-indigenous species. Enge et al. (2017) take this theme further by discussing defences against herbivory of invasive seaweeds. The themes of the conference—evolution, interactions and global change—are all explored in these contributions.

We hope that this special issue will be of value to oceanographers and marine biologists alike, and will inspire further research where knowledge gaps have been identified. We look forward to working with new contributors and welcome suggestions for reviews for future volumes. We would especially value contributions from oceanographers for coming issues.

We sadly heard of the passing away of Professor Roger Hughes at the conference in August 2015. As the then Editor in Chief of OMBAR, this special issue was planned with Roger. We dedicate this volume to his memory as a superb scientist, excellent editor, fine and fanatical fisherman and great guitarist.

References

Chapman, M.G. 2017. Intertidal boulder-fields: a much neglected, but ecologically important, intertidal habitat. *Oceanography and Marine Biology: An Annual Review* **55**, 35–53.

Enge, S., Sagerman, J., Wikström, S.A. & Pavia, H. 2017. A review of herbivore effects on seaweed invasions. *Oceanography and Marine Biology: An Annual Review* **55**, 421–440.

Hawkins, S.J. & Hartnoll, R.G. 1983. Grazing of intertidal algae by marine invertebrates. *Oceanography and Marine Biology: An Annual Review* **21**, 195–282.

John, D.M., Hawkins, S.J. & Price, J.H. 1992. *Plant-Animal Interactions in the Marine Benthos*. Oxford: Clarendon Press on behalf of the Systematics Association.

Johnston, E.L., Dafforn, K.A., Clark, G.F., Rius, M. & Floerl, O. 2017. How anthropogenic activities affect the establishment and spread of non-indigenous species post-arrival. *Oceanography and Marine Biology: An Annual Review* **55**, 389–419.

Martinez, A.S., Byrne, M. & Coleman, R.A. 2017. Filling in the grazing puzzle: a synthesis of herbivory in starfish. *Oceanography and Marine Biology: An Annual Review* **55**, 1–34.

Neo, M.L., Wabnitz, C.C.C., Braley, R.D., Heslinga, G.A., Fauvelot, C., Van Wynsberge, S., Andréfouët, S., Waters, C., Tan, A.S.-H., Gomez, E.D., Costello, M.J. & Todd, P.A. 2017. Giant clams (Bivalvia: Cardiidae: Tridacninae): a comprehensive update of species and their distribution, current threats and conservation status. *Oceanography and Marine Biology: An Annual Review* **55**, 87–390.

Rius, M., Teske, P.R., Manríquez, P.H., Suárez-Jiménez, R., McQuaid, C.D. & Castilla, J.C. 2017. Ecological dominance along rocky shores, with a focus on intertidal ascidians. *Oceanography and Marine Biology: An Annual Review* **55**, 55–84.

Oceanography and Marine Biology: An Annual Review, 2017, **55**, 1-34
© S. J. Hawkins, D. J. Hughes, I. P. Smith, A. C. Dale, L. B. Firth, and A. J. Evans, Editors
Taylor & Francis

FILLING IN THE GRAZING PUZZLE: A SYNTHESIS OF HERBIVORY IN STARFISH

ALINE S. MARTINEZ[1], MARIA BYRNE[1] & ROSS A. COLEMAN[1]*

[1]*Coastal and Marine Ecosystems Group, Marine Ecology Laboratories (A11), School of Life and Environmental Sciences, The University of Sydney New South Wales 2006, Australia*
Corresponding author: Ross A. Coleman
e-mail: ross.coleman@sydney.edu.au

Herbivory is an important ecological process controlling community structure and function in almost all ecosystems. The effects of herbivores on algal assemblages depend primarily on consumer and algal traits, but the strength of this interaction is contingent on physical and biological processes. Marine herbivory is particularly intense, where grazers can remove around 70% of primary production. Present understanding of marine herbivory is largely based on well-studied groups including herbivorous fishes, gastropods, crustaceans and sea urchins. Herbivory in other marine taxa is poorly understood, but nonetheless important. For instance, grazing by starfish has the potential to strongly affect algal assemblages. Most starfish feed by extruding their stomach and digesting their food externally. This feeding mechanism is distinctive and complex, and evolutionarily advantageous as it allows individuals to explore many different food sources. Variation in the feeding habits of herbivorous starfish is intriguing because some species are very specialized whereas others are more generalist, and the reasons for those variations are not well understood. Some herbivorous starfish are obligate herbivores while others vary from herbivory to carnivory between life stages or between populations within the same species. The question that then arises is how well we are able to predict grazing pressure from complex feeding habits on benthic systems? This review provides a synthesis of herbivory in starfish showing that: the majority of species forage on microalgae and soft tissue macroalgae; fidelity to an algal diet appears to be related to the size of individuals; and, feeding habits are likely to change with variation in food availability. Directions for future studies on the biology and ecology of herbivorous starfish are suggested to better understand variation in species feeding behaviour. Elucidating the mechanisms that contribute to variation in the behaviour of herbivorous starfish is crucial to predict the effects that these species exert on the structure of marine benthic communities. The influence of omnivorous species also warrants more detailed study. Such investigations are important in the context of climate change, given the potential for species invasions associated with range expansions.

Introduction

Herbivores play a key role in the world's ecosystems by consuming primary production and altering habitat structure. Mammalian herbivory, including *inter alia* grazing and browsing, is well known to directly affect the distribution of plants in terrestrial systems (Hempson et al. 2015), whereas the most important herbivores in structuring benthic communities in freshwater aquatic systems are snails, insects, crustaceans and small vertebrates (Huntly 1991, Steinman 1996). In marine ecosystems, fish, molluscs, small crustaceans and sea urchins have been shown to dramatically affect the structure of marine benthic communities (Hawkins & Hartnoll 1983, Hawkins et al. 1992, Coleman et al. 2006, Poore et al. 2012, Poore et al. 2014). Thus far, the strongest known impact on a marine habitat, in terms of change in marine algal/macrophyte canopy in the absence of a herbivore, is for molluscs in rocky intertidal habitats (Poore et al. 2012).

The effects of herbivores on marine plants or microalgal assemblages depend primarily on the feeding apparatus of the consumer and the interactions of these with the susceptibility of the primary producer to grazing (Hawkins et al. 1992). The feeding mechanism determines what resources herbivores can exploit (Steneck & Watling 1982, Hawkins et al. 1989, Kennish & Williams 1997), whereas food traits (e.g. detectability, toughness, palatability and digestibility etc.) will affect the intake or intensity of consumption (Lubchenco & Cubit 1980, Lubchenco & Gaines 1981, Duffy & Hay 1990, Norton et al. 1990, Paul et al. 2001, Dolecal & Long 2013). Variation in grazing pressure is also associated with the size and mobility of herbivores. Small and/or sedentary herbivores (e.g. amphipods, limpets) are likely to have local impacts on the assemblage while bigger and/or more mobile species (e.g. fish, sea urchins) require a broader foraging range and can affect assemblages at larger scales (Lawrence 1975, Hay 1984, Norton et al. 1990, Duffy & Hay 1991, Vergés et al. 2009). Apart from size and mobility of herbivores, abundance and composition of herbivores will together play an important role in the magnitude of the effects of grazing, thereby affecting the spatial heterogeneity and diversity of algal assemblages (Lubchenco & Gaines 1981, Benedetti-Cecchi et al. 2005, Jenkins et al. 2008, Griffin et al. 2010).

Although there is a vast literature on the ecology of plant-animal interactions, knowledge from marine systems is concentrated on a few groups of herbivores, i.e. fish, sea urchins, crabs, amphipods and gastropod grazers (Poore et al. 2012). There are a few gaps that need to be addressed in order to clarify grazing pressure by different herbivores. This is the case for starfish species, which are abundant and distributed worldwide. Research on the feeding behaviour of starfish has largely focused on predatory species (e.g. Brun 1972, Paine 1974, 1976, Town 1980, Keesing & Lucas 1992, Keesing 1995, Himmelman et al. 2005, Scheibling & Lauzon-Guay 2007, Estes et al. 2011, Mueller et al. 2011, Menge & Sanford 2013). This is because numerous large and non-cryptic starfish species are carnivores and play an important role in controlling densities of prey (Chesher 1969, Porter 1972, Paine 1974, Gaymer & Himmelman 2008, Kenyon & Aeby 2009, Pratchett et al. 2009, Uthicke et al. 2009; Kayal et al. 2012, Baird et al. 2013, Pratchett et al. 2014). Several species of starfish have been reported to also feed on other resources such as detritus, dead animals, and macro- and microalgae (for a review, see Sloan 1980, Jangoux 1982b). These species have been considered to be omnivores, however their diet is poorly known because of the difficulty in identification of gut contents, especially because of their peculiar extra-oral feeding mechanism.

Most starfish feed by everting their cardiac stomach and directly secreting enzymes from the epithelium onto the food item and thereby digest their food externally. This feeding mechanism is rare among other animals and allows starfish to utilize many different food types. Despite using the same extra-oral feeding mechanism, diet varies among starfish species, ranging from carnivores feeding on specific prey items to omnivorous feeders and herbivores (for a review, see Sloan 1980, Jangoux 1982b).

The variation in feeding biology among starfish species is related to digestive tract anatomy and tube foot morphology (Jangoux 1982a). There is, however, poor knowledge of the causes of variation in feeding habits among starfish species and populations. This is a particular concern for our understanding of those starfish recorded as feeding on plant or microphyte resources because of the lack of information on their ecological role as herbivores. Understanding the mechanisms involved with a herbivorous diet in starfish is important to predict the potential impacts of grazing by these animals on the structure of benthic assemblages. Therefore the aim of this review is to identify the main findings and the fundamental gaps in knowledge on the ecology of herbivorous starfish, using a systematic review of the literature. We compile information of starfish species that forage on any source of plant or microphytes and present a synthesis on the different aspects of the ecology of herbivory in starfish. We describe the food source exploited by herbivorous starfish, the changes in time and space on the use of food resource, and discuss the possible mechanisms driving these variations. We also highlight the potential function of herbivorous starfish in controlling the distribution of algal assemblages and the limited studies on interactions between herbivorous starfish and other herbivores. We expect that this synthesis will encourage researchers to investigate in detail the functional ecology of herbivorous starfish in marine benthic ecosystems.

Data collection and definition of terminology

Literature on starfish that forage on algae and/or marine macrophytes was compiled by searching the ISI Web of Science database (Jan 1900–Jan 2015) using the following criteria: TOPIC: (starfish or "sea star" or seastar or asteroidea) *AND* TOPIC: (herb* or graz*) *AND* TOPIC: (macrophyte* or alga* or seagrass* or eelgrass* or seaweed*). We selected every article that cited or described algal or vascular plant material in the feeding or diet of starfish species. The compilation was supplemented by subsequently searching material cited within those articles including published journal articles and unpublished theses. We also gathered information from published reviews on starfish diet (Sloan 1980, Jangoux 1982b) and searched the references cited therein. The information extracted from the literature included species name, food resource exploited, foraging behaviour, distribution (habitat and region) and size of starfish. Any detailed or more relevant information on feeding habits of a starfish was noted where appropriate. Starfish scientific names are presented as the current accepted species name according to the world Asteroidea database (accessed at http://www.marinespecies.org/asteroidea on Jan 7th, 2015). Previous species names in some of the references cited in this review can be found in Appendix 1.

Hereafter true macrophytes are referred to as "plants". Macroalgae and microalgae are used to distinguish multi- and unicellular algae. We consider that microalgae correspond to any biofilm assemblage (largely prokaryotes and microalgae). Algae are also considered in functional groups according to the classification of Steneck & Dethier (1994). Finally, it is important to keep in mind that all data of starfish presented as foraging on macroalgae or plant does not necessarily mean that the starfish is feeding on the macroalgae or plant *per se*, but may be consuming associated epiphytic organisms.

Composition of diet

Identifying the target plant or algal food source exploited by starfish species was used to detect components of the benthos that are likely to be affected by asteroid grazing. We found 57 species whose diet consisted entirely or partially of algae across a variety of habitats and ecosystems (Table 1). The great majority of starfish (90%) that consume any source of plant/algae are found on hard-bottom ecosystems, where 30% are exclusive to coral reefs and 32% to rocky shores (mostly temperate). The remaining 28% of species are found in hard and soft bottom systems, including seagrass beds. Only 10% are exclusive to seagrass and sand/mud sediments. The information on the feeding habits of starfish within different habitats is often descriptive and lacked accurate evidence of diet especially for those species that are found on sediments. After compiling information from the literature, however, a clear pattern of diet emerged for starfish that forage on algal/plant resources.

The most frequently documented algal group exploited by starfish were those that comprise biofilms. Biofilms usually contain a mix of diatoms, cyanobacteria, bacteria, spores of macroalgae and other microorganisms embedded in a matrix of extra-polymer substances (Anderson 1995, Decho 2000). The majority of the species (77%, 44 out of 57 spp.) were reported feeding on epibenthic biofilms on different substrata (Table 1). Thus, starfish feeding on biofilm were identified

Table 1 Classification of starfish herbivory according to temporal and spatial fidelity of a species to a plant/algae resource

Species	Size			Food resource	Ecosystem	Habitat	References
	R	r	R/r				
Obligate herbivores							
Asterinidae							
Aquilonastra anomala (H.L. Clark, 1921)	8	4.4	1.8	Biofilm (substrate feeding on algae and microbes)	Coral reefs		(Yamaguchi 1975)
Asterina gibbosa (Pennant, 1777)	16	9.4	1.7	Biofilm (diatoms/bacterial film), macroalgae (*Ulva*) and decaying algae (*Laminaria*)	Rocky shores	Intertidal rock pools, under boulders	(Crump & Emson 1978, Sloan 1980, Jangoux 1982b, Crump & Emson 1983)
Asterina phylactica (Emson & Crump, 1979)	7	4.4	1.6	Biofilm, epilithic organisms on algae	Rocky shores	Intertidal rocky pools, under boulders	(Emson & Crump 1979, Crump & Emson 1983)
Cryptasterina hystera (Dartnall & Byrne, 2003)	12	7.5	1.6	Biofilm (bare rock)	Rocky shores	Mid intertidal under small rocks	(Dartnall 1971; Martinez, A.S., unpublished, Dartnall et al. 2003)
Cryptasterina pacifica (Hayashi, 1977)	12	9.2	1.3	Biofilm (bare rock)	Coral reefs	Intertidal rock pools, under coral rubble	(Dartnall 1971, Dartnall et al. 2003)
Cryptasterina pentagona (Muller & Troschel, 1842)	11	5.6	2.0	Biofilm on coral rubble	Coral reefs	High intertidal rock pools, under coral rubble	(Dartnall 1971; Martinez A.S., unpublished)
Parvulastra exigua (Lamarck, 1816)	10	6.7	1.5	Biofilm (diatoms, spores, detritus, bacteria), Macroalgae (*Ulva, Enteromorpha, Ceramium, Corallina officinalis, Hormosira*, encrusting calcareous), detritus, shells.	Rocky shores, seagrass beds, soft bottom	Intertidal rock pools, crevices, under boulders; seagrass and lagoonal sandflats	(Duyverman 1976, Branch & Branch 1980, Jangoux 1982b, Arrontes & Underwood 1991, Stevenson 1992, Jackson et al. 2009, Pillay et al. 2010, Shepherd 2013, Martinez et al. 2016)
Parvulastra parvivipara (Keough & Dartnall, 1978)	3	2.7	1.1	Biofilm	Rocky shores	Mid to low intertidal rock pools, under boulders and small rocks	(Keough & Dartnall 1978, Wilson 2004)

Continued

Table 1 (Continued) Classification of starfish herbivory according to temporal and spatial fidelity of a species to a plant/algae resource

Species	Size			Food resource	Ecosystem	Habitat	References
	R	r	R/r				
Parvulastra vivipara (Dartnall, 1969) nomen dubium[a]	15 —	12.0 —	1.3 —	Biofilm (diatoms, cyanobacteria) Macroalgae and detritus	Rocky shores Rocky shores	Intertidal rock pools and under boulders	(Jangoux 1982b, Prestedge 1998, Polanowski 2002) (Jangoux 1982b)
Echinasteridae							
Echinaster luzonicus (Gray, 1840)	40	4.5	8.9	Biofilm (substrate feeding on algae and microbes)	Coral reefs, seagrass beds, soft bottom	Corals, seagrass, mud/ sand bottom, pebble, shell grits	(Yamaguchi 1975, Jangoux 1982b)
Echinaster purpureus (Gray, 1840)	80	8.3	9.6	Biofilm (epibenthic film of the hard substrate of the coral built formations)	Coral reefs, seagrass beds, soft bottom	Corals, seagrass, sand	(Thomassin 1976, Jangoux 1982b)
Goniasteridae							
Neoferdina cumingi (Gray, 1840)	40	9.5	4.2	Biofilm (substrate feeding on algae and microbes)	Coral reefs	Seaward reef	(Yamaguchi 1975, Jangoux 1982b)
Ophidiasteridae							
Cistina columbiae (Gray, 1840)	40	6.8	5.9	Biofilm (substrate feeding on algae and microbes)	Coral reefs	Coral and rock reefs	(Yamaguchi 1975, Jangoux 1982b)
Dactylosaster cylindricus (Lamarck, 1816)	21	2.1	10.0	Biofilm (substrate feeding on algae and microbes)	Coral reefs	Coral and rock reefs	(Yamaguchi 1975, Jangoux 1982b)
Linckia guildingi (Gray, 1840)	150	15.0	10	Biofilm (substrate feeding on algae and microbes)	Coral reefs	Reef flat and lagoon, seaward reef	(Yamaguchi 1975, Jangoux 1982b)

Continued

5

Table 1 (Continued) Classification of starfish herbivory according to temporal and spatial fidelity of a species to a plant/algae resource

Species	Size			Food resource	Ecosystem	Habitat	References
	R	r	R/r				
Linckia laevigata (Linnaeus, 1758)	122	22.0	5.5	Biofilm (epibenthic film of seagrass and hard/dead coral substratum), macroalgae (calcareous encrusting)	Coral reefs	Reef flat and less common on coarse sand flat, seaward reef	(Laxton 1974, Yamaguchi 1975, Thomassin 1976, Sloan 1980, Jangoux 1982b)
Linckia multifora (Lamarck, 1816)	62	7.0	8.9	Biofilm (substrate feeding on algae and microbes)	Coral reefs		(Yamaguchi 1975, Jangoux 1982b)
Nardoa variolata (Retzius, 1805)	150*	26.8	5.6	Biofilm (epibenthic film of sea grass)	Coral reefs, seagrass beds	Coral reef flats and less commonly on coarse sandy flats	(Thomassin 1976, Jangoux 1982b)
Ophidiaster cribrarius (Lütken, 1871)	29	4.5	6.4	Biofilm (substrate feeding on algae and microbes)	Coral reefs	Reef flat and lagoon	(Yamaguchi 1975, Jangoux 1982b)
Ophidiaster granifer (Lütken, 1871)	41	11.4	3.6	Biofilm (substrate feeding on algae and microbes)	Coral reefs	Reef flats, under boulders and coral rubble	(Yamaguchi 1975, Jangoux 1982b)
Ophidiaster hemprichi (Müller & Troschel, 1842)	120	11.3	10.7	Biofilm (substrate feeding on algae and microbes)	Coral reefs	Reef flat and lagoon, seaward reef	(Yamaguchi 1975, Jangoux 1982b)
Phataria unifascialis (Gray, 1840)	70	10.3	6.8	Biofilm (film of algae and invertebrates on rock), macroalgae	Rocky shores	Intertidal and subtidal rocks	(Morgan & Cowles 1997)
Oreasteridae							
Pentaceraster cumingi (Gray, 1840)	100	45.5	2.2	Macroalgae (bits of algae), calcareous encrusting algae and biofilm (rock)	Rocky shores, coral reefs	Rocks and reef flats	(Jangoux 1982b, Dee et al. 2012;
Pentaceraster mammillatus (Audouin, 1826)	100	38.5	2.6	Biofilm (epibenthic felt of microorganisms)	Coral reefs, seagrass beds	Seagrass and reef flats	(Thomassin 1976, Jangoux 1982b)

Continued

Table 1 (Continued)   Classification of starfish herbivory according to temporal and spatial fidelity of a species to a plant/algae resource

Species	Size			Food resource	Ecosystem	Habitat	References
	R	r	R/r				
Pentaceraster regulus (Müller & Troschel, 1842)	100	28.6	3.5	Biofilm (epibenthic film; organic matter)	Soft bottom	Sand, mud and rubble	(Coleman 1977, Jangoux 1982b)
Protoreaster lincki (Blainville, 1830)	150*	62.5	2.4	Biofilm (substrate film-feeder, epibenthic felt of microorganisms), algae, detritus and dissolved organic material	Seagrass beds, soft bottom	Seagrass, sand and cobble	(Thomassin 1976, Coleman 1977, Jangoux 1982b)
Protoreaster nodosus (Linnaeus, 1758)	150	50.0	3.0	Biofilm (epibenthic felt of microorganisms, film on *Halimeda* and on calcareous encrusting), decomposing macroalgae, seagrass and sand meiobenthos	Seagrass beds, soft bottom	Seagrass, sand and cobble	(Thomassin 1976, Jangoux 1982b, Scheibling & Metaxas 2008)
Facultative herbivores (ontogenetic changes in diet)							
Acanthasteridae							
Acanthaster planci (Linnaeus, 1758)	20	16.7	1.2	Calcareous encrusting macroalgae (*Poroliton*) when juveniles; Other macroalgae as adults during outbreaks (calcareous turfs, filamentous, foliose)	Coral reefs, soft bottom	Reefs, cobble, sand	(Yamaguchi 1973, Yamaguchi 1974, 1975, Sloan 1980, De'ath & Moran 1998)
Echinasteridae							
Echinaster (Othilia) sentus (Say, 1825)	—	—	—	Biofilm (benthic diatoms)	Coral reefs, seagrass beds, soft bottom	Coral and rocky reefs, seagrass and sand	(Sloan 1980)
Henricia leviuscula (Stimpson, 1857)	5	1.7	2.9	Biofilm (polychaete tubes)	Soft bottom	Sand and mud	(Birkeland et al. 1971, Sloan 1980)

Continued

Table 1 (Continued) Classification of starfish herbivory according to temporal and spatial fidelity of a species to a plant/algae resource

Species	Size			Food resource	Ecosystem	Habitat	References
	R	r	R/r				
Goniasteridae							
Mediaster aequalis (Stimpson, 1857)	10	4.2	2.4	Biofilm (tubes of polychaete, diatoms mats on rock)	Rocky shores, soft bottom	Subtidal rocks, cobble, sand and mud	(Mauzey et al. 1968, Birkeland et al. 1971, Sloan 1980, Jangoux 1982b, Sloan & Robinson 1983)
Luidiidae							
Luidia foliolata (Grube, 1866)	5	1.5	3.4	Biofilm (on tubes of polychaetes), detritus and microorganisms	Soft bottom	Sand and mud	(Birkeland et al. 1971, Sloan 1980)
Oreasteridae							
Culcita novaeguineae (Müller & Troschel, 1842)	7	5.4	1.3	Calcareous encrusting algae	Coral reefs		(Yamaguchi 1973, Yamaguchi 1975, Yamaguchi 1977, Sloan 1980)
Pterasteridae							
Pteraster tesselatus (Ives, 1888)	5	4.5	1.1	Biofilm (polychaete tubes, films of detritus and microorganisms) and sessile invertebrates (sponges)	Rocky shores	Rocks in deep seas	(Sloan 1980)
Stichasteridae							
Stichaster australis (Verrill, 1871)	10	8.3	1.2	Calcareous encrusting macroalgae (*Mesophyllum insigne*)	Rocky shores	Subtidal rocks	(Barker 1977, 1979, Sloan 1980)
Solasteridae							
Crossaster papposus (Linnaeus, 1767)	5	2.8	1.8	Biofilm (polychaete tubes, films of detritus and microorganisms)	Soft bottom	Sand and mud	(Birkeland et al. 1971, Sloan 1980)
Solaster stimpsoni (Verrill, 1880)	5	2.0	2.5	Biofilm (polychaete tubes, films of detritus and microorganisms)	Soft bottom	Sand and mud	(Birkeland et al. 1971, Sloan 1980)
Solaster dawsoni (Verrill, 1880)	5	2.2	2.3	Biofilm (polychaete tubes, films of detritus and microorganisms)	Soft bottom	Sand and mud	(Birkeland et al. 1971, Sloan 1980)

Continued

Table 1 (Continued) Classification of starfish herbivory according to temporal and spatial fidelity of a species to a plant/algae resource

Species	Size			Food resource	Ecosystem	Habitat	References
	R	r	R/r				
Facultative herbivores (spatial and temporal shifts in diet)							
Asterinidae							
Asterina stellifera (Möbius, 1859)	70	28.0	2	Biofilm (diatoms), red and green macroalgae, invertebrates (gastropods, anemone, polychaetes, small crabs, sponges, ascidians), and moribund/dead invertebrates	Rocky shores	Subtidal boulders	(Jangoux 1982b, Farias et al. 2012)
Meridiastra calcar (Lamarck, 1816)	60*	37.5	1.6	Macroalgae (*Laurencia* sp. and *Gelidium* sp.), biofilm (bare rock), invertebrates (gastropods, pelecipods), detritus and moribund animals	Rocky shores	Intertidal rock pools, subtidal rocks	Shepherd1968; Stevenson1992; Shepherd 2013
Meridiastra gunnii (Gray, 1840)	56*	43.1	1.3	Red macroalgae (*Laurencia* sp. and *Spyridia* sp.), seagrass, sessile invertebrates (bryozoans, spirorbids ascidians, sponges), carrion, detritus and moribund animals	Rocky shore, soft bottom	Intertidal rock pools with boulders, subtidal rocks and sand bottom	(Shepherd 1968, Jangoux 1982b, Shepherd 2013)
Meridiastra medius (O'Loughlin, Waters & Roy, 2003)	38	26.6	1.4	Calcareous macroalgae, biofilm (bare rock) and sessile invertebrates (bryozoans and ascidians)	Rocky shore, soft bottom	Intertidal rock pools with boulders, subtidal rocks and sand bottom	(Shepherd 2013)
Patiria miniata (Brandt, 1835)	67	21.0	3.2	Microalgae, early life stages of kelps (*Macrocystis pyrifera*), drift algae and seagrass, macroalgae (*Egregia, Rhodymenia, Microcladia, Iridea, Porphyra, Prionitis, Gigartina, Agardhiella, Laminaria,* and Delesseriaceae), sessile invertebrates (ascidians, sponges and bryozoans) and dead animals.	Rocky shores, kelp beds	Low intertidal rocks, subtidal rocks and kelps	(Araki 1964, Gerard 1976, Jangoux 1982b, Dayton et al. 1984, Leonard 1994)

Continued

9

Table 1 (Continued)   Classification of starfish herbivory according to temporal and spatial fidelity of a species to a plant/algae resource

Species	Size			Food resource	Ecosystem	Habitat	References
	R	r	R/r				
Patiria pectinifera (Muller & Troschel, 1842)	50	31.3	1.6	Biofilm (periphyton films), macroalgae (*Ulva*), calcareous turf algae (*Corallina*), encrusting calcareous algae (*Lithophyllum yessoense*), seagrass, small invertebrates, dead animals and detritus	Rocky shores, seagrass beds, soft bottom	Subtidal rocks, seagrass, shell grits, cobble and sand	(Bak 1981, Jangoux 1982b, Levin et al. 1987, Fujita 1999, Kurihara 1999)
Patiriella regularis (Verrill, 1867)	24	16.0	1.5	Biofilm (bare rock and diatom mats), encrusting calcareous and non-calcareous macroalgae, calcareous turf, green filamentous algae, invertebrates (gastropods, ascidians and crabs), detritus and shell debris	Rocky shores, kelp beds, soft bottom	Intertidal under boulders, underwater rocks, cobble, shell grit, sand and kelps	(Grace 1967, Crump 1969, Martin 1970, Burgett 1982, Jangoux 1982b, Burgett 1988)
Pseudonepanthia troughtoni (Livingstone, 1934)	85	9.3	9.1	Calcareous encrusting macroalgae (*Lithothamnia*), calcareous turf and sessile invertebrates (bryozoans, ascidians and sponges)	Rocky shores, soft bottom	Subtidal rocks, under boulders and sand	(Shepherd 1968, 2013)
Asteropseidae							
Dermasterias imbricata (Grube, 1857)	85	29.3	2.9	Green algae (*Enteromorpha*), sponges, anemones, ascidians, hydroids, holothurians, detritus, moribund animals	Rocky shores, soft bottom	Intertidal and subtidal rocks, cobble and sand	(Mauzey et al. 1968, Jangoux 1982b, Sloan & Robinson 1983)
Goniasteridae							
Fromia hemiopla (Fisher, 1913)	43	9.3	4.6	Biofilm (substrate feeding on algae and microbes) and sessile invertebrates (sponges)	Coral reef		(Yamaguchi 1975, Jangoux 1982b)
Nectria ocellata (Perrier, 1875)	97	34.6	2.8	Encrusting macroalgae, detritus and sessile invertebrates (sponges, ascidians and bryozoans)	Rocky shores, soft bottom	Subtidal rocks, shell grit and sand	(Shepherd 1967, Coleman 1977, Jangoux 1982b, Shepherd 2013)

Continued

Table 1 (Continued)   Classification of starfish herbivory according to temporal and spatial fidelity of a species to a plant/algae resource

| Species | Size | | | Food resource | Ecosystem | Habitat | References |
	R	r	R/r				
Nectria saoria (Shepherd, 1967)	66	24.4	2.7	Red and calcareous macroalgae, and sessile organisms (ascidians and bryozoans)	Rocky shores	Subtidal rocks	(Shepherd 1967, Jangoux 1982b, Shepherd 2013)
Nectria multispina (H.L. Clark, 1928)	85	32.7	2.6	Macroalgae (red algae), rock encrusting organisms, sessile invertebrates (sponges, ascidians and bryozoans) and small bivalves	Rocky shores	Subtidal rocks	(Shepherd 1967, Jangoux 1982b)
Mediaster aequalis (Stimpson, 1857)	100*	33.3	3.0	Macroalgae (*Ulva* spp. and *Porphyra* spp.), biofilm (tubes of polychaete, diatom mats on rock) microalgae (planktonic diatoms), detritus, sessile invertebrates (ascidians, sponges, bryozoans) and moribund animals	Rocky shores, soft bottom	Subtidal rocks, cobble, sand and mud	(Mauzey et al. 1968, Birkeland et al. 1971, Sloan 1980, Jangoux 1982b, Sloan & Robinson 1983)
Ophidiasteridae							
Gomophia egyptiaca (Gray, 1840)	50	7.4	6.8	Biofilm (substrate feeding on algae and microbes) and sessile invertebrates (sponges, ascidians) and moribund animals	Coral reef	Reef flat and lagoon	(Yamaguchi 1975, Jangoux 1982b)
Oreasteridae							
Oreaster reticulatus (Linnaeus, 1758)	165	126.9	1.3	Biofilm, green algae (*Chaetomorpha*, *Enteromorpha*), epiphytes, seagrass, invertebrates (sponges, sea urchins)	Seagrass beds, soft bottoms	Subtidal seagrass, sand and rubble	(Scheibling 1979, Scheibling 1980b,d, 1981, Jangoux 1982b, Scheibling 1982, Martin et al. 2001, Scheibling 2013)
Culcita schmideliana (Retzius, 1805)	120	109.1	1.1	Seagrass, sandy substratum, sponges, corals, film growing on dead corals	Seagrass beds, soft bottoms	Intertidal seagrass, sand	(Thomassin 1976, Jangoux 1982b)

Continued

11

Table 1 (Continued) Classification of starfish herbivory according to temporal and spatial fidelity of a species to a plant/algae resource

Species	Size			Food resource	Ecosystem	Habitat	References
	R	r	R/r				
Stichasteridae							
Granaster nutrix (Studer, 1885)	70	25.0	2.8	Red macroalgae, small gastropods, ectoprots	Rocky shore	Underwater rocks	(Jangoux 1982b)

[a] An uncertain starfish species from the family Asterinidae, formerly *Marginaster littoralis* Dartnall, 1970.

Note: Starfish size, food resource exploited, and distribution among ecosystems and habitats are presented for each species. Detailed information of food resource and habitat are presented whenever available from the literature. The references cited in this table include the primary source of information for a given species and all other studies to date that recorded that species feeding on plant/algal resources. Starfish metrics are shown for average sizes in mm (R – arm size, r – disc size) with exception for some species (maximum size instead) whose average size was not found (indicated by an asterisk) and for those species classified with ontogenetic change in diet.

as species feeding on "bare rock", "diatoms", "diatom mats on rock", "films of dead corals, detritus, tubes of polychaetes, shells, macroalgae, seagrass", "microalgae", "epibenthic film", "bacterial film", "epilithic organisms on algae" and "epibenthic felt of microorganisms"; or characterized as "substratal film-feeder" and "substratal feeding on algae and microbes".

Biofilms are one of the main food resources that are also exploited by several grazing gastropods. One of the reasons that gastropods forage on this food resource is that their feeding abilities are limited by the physical characteristics of their radular feeding apparatus (Steneck & Watling 1982, Hawkins et al. 1989). Asteroids do not have any feeding apparatus such as radulae, Aristotle's lantern in sea urchins or teeth (e.g. fish) that can mechanically rasp, cut or bite the algae from the substratum. The algal food resource utilized by herbivorous starfish relies on the digestive capacity of their stomach. Feeding on biofilms could be a cost-efficient source of food for starfish, since diatoms, bacteria and algal sporelings are protected only by individual cell walls, which are easier to break down compared to the structures of multicellular macrophytes. Macrophytes not only have thick walls, but also exhibit complex chemical defences (Hay 2009).

The enzymes produced in the pyloric ceca of some starfish (e.g. *Oreaster reticulatus*, *Patiria miniata*, *Patiriella regularis*), are capable of digesting some oligosaccharides and polysaccharides that are the sugar reserves of plants and algae (e.g. sucrose, trehalose, amylose, and laminarinose), but they have a weak effect on, or failed to break down, the structural components of plant/algal cell walls such as cellulose, alginate, agar, and carrageenan (Araki 1964, Araki & Giese 1970, Martin 1970, Scheibling 1980c). This suggests that starfish are potentially capable of digesting macroalgae, but that feeding on these algae with thick cell walls might not be cost-effective. Still some starfish seem to feed on macroalgae, but on less structured thalli forms. Soft-structured macrophytes such as soft foliose (e.g. *Ulva* spp.) and filamentous algae (e.g. *Ceramium*, *Chaetomorpha*, etc.) appear to constitute the main group of macroalgae in the diet of some herbivorous starfish. These algae can be intensively grazed by several herbivores including snails, amphipods, isopods, crabs, and fish (e.g. Choat 1982, Hawkins & Hartnoll 1983, Arrontes 1990, Poore 1994, Kennish & Williams 1997). This is because soft filamentous and foliose algae are generally easy to digest, have rapid growth rates and energy intake, and do not offer strong resistance to herbivory (Littler & Littler 1980, Littler & Arnold 1982, Steneck & Dethier 1994). The cell walls of soft and thin thalli algae are generally structured in a simple uni- or multiseriate cell configuration, which are not differentiated and heavily corticated unlike those of thick leathery or calcareous or crustose algae (Steneck & Dethier 1994). Thus, starfish could benefit from feeding on those soft algae by disrupting their simple cell walls with their enzymes and accessing the digestible cell content (Kristensen 1972, Scheibling 1980a). This hypothesis needs to be addressed through investigation of the digestive capacity of herbivorous species.

The fact that starfish are not able to readily break down the cell walls of plants/macroalgae raises the question of whether asteroids are able to feed on these more complex-structured primary producers. The diet of *Asterina gibbosa*, *Patiria miniata* and *Protoreaster nodosus* includes macroalgal detritus (Gerard 1976, Crump & Emson 1978, Scheibling & Metaxas 2008), and it is reasonable that these starfish species can eat decaying algae as the cell walls are already damaged. Indeed, feeding on decomposing macroalgae by *Patiria miniata* enhanced the nutritional value of the food source compared to intact layers of the algae (Gerard 1976). Some starfish species, however, were described as feeding on calcareous turf algae, calcareous and non-calcareous encrusting algae, red corticated algae and seagrass (Table 1). There is no certainty whether those starfish were actually feeding on the macroalgae itself or on the epiphytes or biofilm on those macrophytes. Some authors argue that starfish probably eat the epilithic organisms that grow on the macroalgae, but many authors noted that the plant/algae area on which a starfish had their stomach extruded became discoloured (Araki 1964, Crump 1969, Yamaguchi 1973, Scheibling 1979, Scheibling 1982, Levin et al. 1987, Leonard 1994, Farias et al. 2012). This is notably visible in calcareous algae, which may look bright orange or white after starfish feeding (Figure 1A,B). The physical structure of

Figure 1 Discolouration of algae and scars caused by starfish herbivory: (A) Calcareous turf discoloured (bright orange) by *Meridiastra calcar* feeding (Photo: Aline Martinez); (B-D) Feeding scars of *Parvulastra exigua* on (B) encrusting calcareous algae (Photo: Ross Coleman), on (C) bare rock, in detail (Photo: Aline Martinez), and (D) around rock pools on open rock (Photo: Aline Martinez).

those macrophytes, however, seems to not be affected by the starfish feeding. Indeed, studies of the effects of starfish grazing on calcareous algae showed that the top layer of live tissue on the algae was removed after being grazed, but the structure of the cell walls remained intact (Barker 1979, Burgett 1982, Fujita 1999). Hence it is more likely that starfish are targeting epiphytic organisms growing on the macroalgae than the macroalgae itself, although Bak (1981) showed that *Patiria pectinifera* intensely consumed eelgrass.

In summary, it seems that the plant/algal resources exploited by starfish are limited by their digestive enzymes. These enzymes can break down unicellular components of the biofilm and might be able to disrupt the cell walls of soft filamentous and foliose algae to readily digest and assimilate material from these macrophytes. Moreover, it is likely that starfish foraging on more robust macroalgae are eating epiphytes and other organisms growing on the thalli. Whether or not the starfish benefits from the apparently digested (discoloured) area on these macrophytes is not known.

Categories of herbivory

The definition of trophic guilds (e.g. carnivores, herbivores, omnivores etc.) in asteroid echinoderms is often convoluted and not clear for species that are not strictly carnivores. Trying to distinguish between herbivorous and omnivorous starfish becomes complicated because the behaviour of herbivorous species can vary across time (amount or type of plant/algae consumed by individuals varies at different times) and space (individuals at different places have different diets). There are also species that are herbivorous only during a specific life stage, usually the juvenile stage (e.g. Yamaguchi 1973, Barker 1977, 1979, Sloan 1980; Kamya et al. 2016) whereas other species might forage for their entire life on algae (e.g. Laxton 1974, Arrontes & Underwood 1991, Prestedge 1998, Wilson 2004, Jackson et al. 2009). Differentiating herbivorous behaviour becomes more complicated when the same species displays different feeding habits between populations, locations or seasons (e.g. Araki 1964, Mauzey et al. 1968, Shepherd 1968, Leonard 1994, Scheibling 2013). Therefore, we approached this review by classifying starfish into different herbivore status depending on temporal and spatial fidelity of a species to an algal resource. Thus, starfish species were classified into obligate and facultative herbivores (Table 1).

Obligate herbivores

Starfish species that primarily feed on algae were classified as obligate herbivores and are considered herbivorous starfish *sensu stricto*. Some species were noted to feed on detritus in addition to algae and were also included in the obligate herbivorous category. While detritus may be a range of decaying material, it is unlikely to include dead animals because starfish foraging on dead or decomposing animals are clearly identified in the literature. Obligate herbivores also include some species that, while foraging primarily on algae, on very rare occasions are reported to feed on animal-derived food resources, which was the case for *Parvulastra exigua* and *Pentaceraster cumingi* (Branch & Branch 1980, Dee et al. 2012).

Most herbivorous starfish (70%) belong to the family Asterinidae and Ophidiasteridae. The Asterinidae includes, *Aquilonastra anomala, Asterina gibbosa, A. phylactica, Cryptasterina hystera, C. pentagona, Parvulastra exigua, P. parvivipara* and *P. vivipara* (Dartnall 1971, Yamaguchi 1975, Duyverman 1976, Crump & Emson 1978, Keough & Dartnall 1978, Emson & Crump 1979, Jangoux 1982b, Chen & Chen 1992, Prestedge 1998; Dartnall et al. 2003). These herbivores are small (average radius varies from 3 to 15 mm; see Figure 2A for example) and many occur in intertidal pools on temperate rocky shores. There are also tropical species that occur in intertidal pools of coral reefs (*Aquilonastra anomola* and *Cryptasterina pentagona*). These herbivorous asterinids feed on surficial biofilms on rocks and coral rubble. *Parvulastra exigua* and *Asterina gibbosa* may also feed on macroalgae (encrusting algae, soft filamentous and foliose algae, and decaying algae), but these contribute a small portion of their diet (Crump & Emson 1978, Branch & Branch 1980).

The species from the Ophidiasteridae are nearly 10 orders of magnitude bigger (average radius varies from 30 to 160 mm) than those in the Asterinidae. Herbivorous ophidiasterids include species from the genera *Cistina, Dactylosaster, Linckia, Nardoa, Ophidiaster* and *Phataria* and occur mostly in tropical coral reef flats and lagoons. Although they are bigger than the asterinids, the extruded stomach (ca. 5–20 mm radius) of the biggest ophidiasterids (i.e. *Linckia* spp.; Figure 2B) is similar in size to that of the small asterinids.

The other eight herbivorous starfish are distributed within the families Echinasteridae (2 spp. from the genus *Echinaster*), Goniasteridae (*Neoferdina cumingi*) and Oreasteridae (2 spp. from the genus *Pentaceraster* and 3 spp. from the genus *Protoreaster*). The *Echinaster* spp. and *Neoferdina cumingi* are as big and similar in shape (i.e. long arms and small central disk) as *Linckia* spp. These starfish inhabit coral reef flats and seagrass beds where they feed on the biofilms on hard substrata and on epibenthic films of the sediment. The oreasterids are big starfish (average radius

Figure 2 Aboral view of different categories of herbivorous starfish: Obligate herbivores. (A) *Parvulastra exigua* in a feeding mode shape (round; Photo: Aline Martinez), (B) *Linckia laevigata* (Maria Byrne); and facultative herbivores, (C) *Acanthaster planci* at juvenile stage (Photo: Phil Mercurio) and (D) *Oreaster reticulatus* (Photo: Marcela Rosa).

100–150 mm) and typically inhabit soft bottoms; foraging on seagrass, sand, and among cobbles. *Pentaceraster cumingi* is the only oreasterid species restricted to hard bottoms where it feeds mostly on biofilms and calcareous encrusting algae (Dee et al. 2012). Oreasterids exploit a broad range of plant/algal resources including biofilm, macroalgae, decomposing algae and seagrass.

It appears that eating plant/algal resources by obligate herbivores is associated with the lack of ability to catch or manipulate animal resources as predatory starfish do. *Pentaceraster cumingi* consumes small sea urchins (corresponding to 1% of their diet), but feeding was observed on tethered sea urchins (Dee et al. 2012). Whether the starfish actively captured live sea urchins or was attracted by a moribund sea urchin already attacked by other predators (e.g. fish) is not clear. Other herbivorous species such as *Protoreaster nodosus* might not primarily forage on animal-derived resources, but benefit from ingesting small invertebrates (i.e. meiofauna) when foraging on plant/algal food (Thomassin 1976, Scheibling 1982, Scheibling & Metaxas 2008). Complementing an algal diet with animal resource can benefit animal growth as demonstrated for herbivorous crabs (Wolcott & O'Connor 1992). An explanation for this is that animal food is much richer in essential amino acids, vitamins and sterols compared to plant/algal resource (Phillips 1984).

Interestingly, obligate herbivorous starfish do not forage, or rarely forage, on soft sessile invertebrates such as sponges, bryozoans or ascidians, even though there is no need to handle these organisms. Possible causes for this feeding behaviour include feeding preferences, chemical defences of sessile invertebrates, especially sponges and ascidians, physical prey defences and/or digestive capabilities of starfish. This is an unexplored subject in herbivorous starfish that needs to be addressed to better understand their feeding behaviour.

Facultative herbivores

Many starfish species that forage on algae also feed on animal-derived material and this is quite variable among and within species, individual lifespan, seasons and location of populations. These species are classified as facultative herbivores, as their algal/plant diet changes in time or space.

Ontogenetic changes in diet

Herbivory in starfish that display ontogenetic changes in diet is reported from the early life stages of carnivorous or omnivorous species (e.g. Yamaguchi 1973, Barker 1977, 1979, Barker & Nichols 1983). Only a few species have been described as herbivores when juvenile and carnivores when adult (Table 1). The lack of information on the feeding behaviour of juvenile starfish is due to the difficulty in finding newly-settled or post-metamorphosed starfish *in situ* (Yamaguchi 1975, 1977). Juveniles are small and often cryptic. Despite the lack of detailed studies, there is evidence that some predatory starfish are herbivores when young (for a review, see Sloan 1980), for example *Acanthaster planci* (Yamaguchi 1973, Yamaguchi 1975; Kamya et al. 2016), *Mediaster aequalis* (Birkeland et al. 1971) and *Stichaster australis* (Birkeland et al. 1971, Yamaguchi 1973, Barker 1977, Sloan 1980).

The feeding behaviour of juveniles of the predatory starfish *Acanthaster planci* is well documented. Encrusting calcareous and coral rubble substrata induce settlement and metamorphosis of *A. planci* (Johnson et al. 1991). After metamorphosis, juveniles feed on calcareous encrusting algae for around 18 weeks (radius ca. 20 mm; Figure 2C), after which, they start foraging on small coral polyps (Henderson & Lucas 1971, Yamaguchi 1973, Yamaguchi 1975). Similarly, the coral reef starfish *Culcita novaeguineae* settles on encrusting calcareous algae and spends at least 12 months on this substratum (radius reaches ca. 3.5 mm) (Yamaguchi 1973, Yamaguchi 1975), but whether further feeding and growth rely on such algae is not known. Adult *C. novaeguineae* forage on sessile invertebrates including coral, but still eat algae. Similar to these tropical species, the temperate starfish *Stichaster australis* uses the coralline alga *Mesophyllum insigne* as a nursery area. Juvenile *Stichaster australis* remain in this habitat and feed on the algae for 15 to 18 months (until ca. 6.5 mm radius) and then switch to juvenile mussels (*Perna canaliculus*) (Barker 1977, 1979). The larvae of *Mediaster aequalis* settle and feed on the biofilm that grows on tubes of the polychaete *Phyllochaetopterus prolifica*, until the juvenile reaches around 5 to 10 mm radius and then feeds on animal prey and algae (Birkeland et al. 1971). In a survey of the feeding behaviour of juvenile starfish, Birkeland et al. (1971) noted that individuals smaller than 5 mm radius (*Henricia leviuscula*, *Luidia foliolata*, *Pteraster tesselatus*, *Crossaster papposus*, *Solaster stimpsoni* and *S. dawsoni*) feed on the tubes of *Phyllochaetopterus prolifica*. It appears that a period of herbivory in the early juvenile stage may be a common feature of carnivorous starfish, although not all juveniles studied were observed feeding on biofilm. Juvenile *Asterias rubens* and *Marthasterias glacialis* eat small encrusting invertebrates as soon as metamorphosis occurs, although these starfish settle in a variety of substrata covered with biofilm (Barker & Nichols 1983).

The extent to which herbivory occurs in young starfish appears to be influenced by the growth patterns of a given species. Newly-settled starfish only have one to two pairs of tube feet and so feeding depends on extruding the stomach (Sloan 1980). Juveniles of predatory starfish pass through a phase during which young starfish start foraging on small animals as soon as they are big enough to catch prey. At this stage the juveniles stop feeding on algae. When adult, many predatory species use their disc-ending tube feet to hold prey and open shells (Hennebert et al. 2013). These starfish may also have extensive batteries of specialized pedicellariae (jaw-like structures) that the adults use to ensnare large prey, including fishes (e.g. *Stylasterias forreri*, Chia & Amerongen 1975). For predatory starfish, it appears that morphological changes are required as a juvenile develops into a competent predator. However, ecological–morphological studies of diet change through growth of juveniles are necessary to test this hypothesis.

In addition to the intrinsic biology of asteroids, external factors such as availability of resource and predation risk might influence diet shifts in starfish, as has been shown for crabs. Juvenile crabs inhabit macrophyte habitats where they are safe from predators and forage on the available food (i.e. plant/algal resources) until they achieve a certain size at which they are less vulnerable to predation and then move to open areas (Laughlin 1982, Alexander 1986, Orth & Vanmontfrans 1987, Williams et al. 1990). In a similar way, Yamaguchi (1975) argued that the cryptic behaviour of tropical juvenile starfish is associated with the risk of predation, which has been shown to be high in the early life stage (Keesing & Halford 1992). Foraging on encrusting algae exploits a highly abundant algal resource available within sheltered-cryptic habitat in coral reefs. When animals become big enough to escape predation, starfish would then start foraging on open areas on the reefs, as demonstrated for *Acanthaster planci* (Yamaguchi 1973, Yamaguchi 1975). These observations suggest that herbivorous behaviour may have evolved in these species through a predator–prey interaction with the juvenile stages remaining in cryptic algal dominated habitat until they grow to a size where they are less vulnerable to predation.

It is clear that the ontogenetic change in starfish diet is related to starfish size (Birkeland 1989). Thus, investigations of feeding habits associated with morphological changes in post-metamorphosed juvenile starfish, as well as feeding behaviour associated with available food resources and predation risk, would be of great interest to understand the shifts from a herbivorous to a carnivorous diet in asteroids. This may help in understanding the mechanisms that drive different feeding behaviours in starfish, a taxon where feeding is not constrained by possession of a complex feeding apparatus.

Spatial and seasonal changes in diet

Most asteroids are considered to be opportunistic with feeding habits reflecting the availability of resources (for a review, see Sloan 1980, Jangoux 1982b). A more detailed examination of the literature indicates that several species classified as omnivores or even carnivores, display herbivorous behaviours linked to the variability of food between locations and seasons. Species are referred to here as facultative herbivores. Eighteen species were included in this category belonging to the families Asterinidae (8 spp.), Goniasteridae (5 spp.), Oreasteridae (2 spp.), Asteropseidae (1 sp.), Ophidiasteridae (1 sp.) and Stichasteridae (1 sp.). The most evident aspect of the feeding behaviour of facultative herbivorous starfish is that the proportion of algae eaten by a population varies according to spatial and temporal fluctuations in density of food resources.

A well-studied species is the red cushion star *Oreaster reticulatus* (Figure 2D), a species that inhabits soft bottoms around reef areas such as sand, seagrass and cobble. Populations of this species vary in their food selection, being either microphagous grazers or macrophagous predators, and the proportion of the food sources utilized changes with respect to the availability of macrofaunal prey (Scheibling 2013). *Oreaster reticulatus* forages on biofilms (microalgae and other microscopic organisms), detritus, filamentous algae and seagrass in locations where animal prey (sea urchins and sponges) is absent (Scheibling 1980d, 1982, Wulff 1995). Similarly, feeding by

Dermasterias imbricata varies between carnivory and herbivory depending on local variation in food resource (Mauzey et al. 1968). As a predator, this species forages on anemones in rocky habitats, and on anemones, sponges and ascidians on cobbles and sandy bottoms. When their primary prey is absent, most of the *D. imbricata* population (ca. 70%) concentrated their feeding on filamentous green algae (Mauzey et al. 1968). Changes in feeding behaviour across habitats or locations were also observed in the bat star *Patiria miniata*, which can intensively forage on kelp sporelings in kelp beds (Leonard 1994), but also forages on other macroalgae or invertebrates in rocky shores (Araki 1964). In *Meridiastra medius* the percentage of algae consumed varies between locations (Shepherd 2013), but it is not clear whether this is associated with resource availability. Other species (e.g. *Meridiastra gunnii, M. calcar, Nectria* spp., *Culcita schmideliana, Pseudonepanthia troughtoni, Granaster nutrix*) have a highly variable diet within locations that include both algae and invertebrates (Yamaguchi 1975, Jangoux 1982b, Shepherd 2013). These observations suggest that facultative herbivores may behave as opportunistic species, where prey items consumed may be proportional to the amount of resource available. It is not possible, however, to clarify whether these observations are related to resource availability because the variability of food resources within habitats was not evaluated.

Whilst herbivory in many facultative herbivores seems to be associated with absence of animal prey, some species display a different behaviour in which algal resources appear to be the main target resource when available. For example, *Mediaster aequalis* feeds on a great range of food resources that vary across habitats and season (Mauzey et al. 1968, Sloan 1980, Sloan & Robinson 1983). This starfish forages on sponges, hydroids, bryozoans and diatoms on rocky shores, whereas on sandy bottoms it predominantly feeds on sea pens (i.e. 62% of its diet) and also captures drift algae (2% of its diet; mostly *Ulva* and *Porphyra* spp.) (Mauzey et al. 1968). Alternatively, *Mediaster aequalis* consumes detritus and biofilm on muddy bottoms. The proportion of the algal resource exploited by this species in sandy habitats varies seasonally. In spring, when *Ulva* forms dense mats on the seafloor, *Mediaster aequalis* switches to an algae-dominated diet (51%) with sea pens being less important (17%), despite their year-round abundance (Mauzey et al. 1968).

In summary, herbivory appears to be driven by the lack of animal food resources in facultative herbivores, where the proportion of food consumed will depend on variation in prey availability within a location, a habitat, or among seasons. Perhaps the fact that *Acanthaster planci* intensively forages on macroalgae at the end of an outbreak event when corals were scarce (De'ath & Moran 1998) is strong evidence that algae might be targeted whenever animal resources are not available. This hypothesis, however, is not conclusive since there is a lack of manipulative studies on the relationships between feeding behaviour, availability of food resource and food preference in starfish that forage on animal and plant/algal resources.

Herbivory and starfish body metrics

Unlike virtually all other invertebrate taxa, starfish lack a specialized feeding structure. Instead, they have evolved extra-oral digestion of food due to their unusual ability to extrude their stomach. The feeding ecology of starfish has evolved independently of acquisition of specialized feeding structures. It seems that trophic modes among starfish may be influenced by other more general traits such as body size (Figure 3). Most obligate herbivorous starfish have short arm length (R) or in species with long arms, have a small disc size (r) (Figure 2A,B). This is also the case for juvenile individuals of carnivorous species (Figure 2C). Adult carnivores are generally bigger than the adults of herbivorous species. Large animals with larger disc size increase the area on which a starfish can extrude its stomach (Lawrence 2012, Lawrence 2013), which allows the starfish to feed on larger food resources and could facilitate the capture of mobile prey. On the other hand, feeding by species with small disc size regardless of arm length is restricted to a small surface area (Jangoux 1982a, Lawrence 2012, Lawrence 2013).

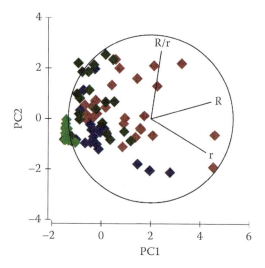

Figure 3 Exploratory analysis (PCA) of starfish body metrics: arm length (R), disc radius (r) and the ratio between arm length and disc radius (R/r). Arm length (R) is the distance between the centre of the starfish disc and the arm tip, and disc radius (r) is the distance between the centre of the starfish disc and the intercept between two arms or interradius. The different colours of diamond symbols represent the data of obligate herbivores (green), herbivores at juvenile stage (light green), facultative herbivores (blue) and random carnivorous species (red).

Another interesting observation is that small herbivorous species and some larger ones have a ratio of arm length and disc size (R/r) close to 1. The greater the ratio, the greater is the difference between arm length and the arm intercepts. Thus, ratios close to 1 may be indicative of lower arm flexibility compared with species that have greater ratios. A small R/r ratio may constrain the capacity to capture mobile or large prey. This hypothesis also supports the observation that carnivorous species have a distinct five-rayed or multi-rayed profile, so have a greater R/r (Figure 3). They are also larger and often have highly dexterous arms (e.g. *Asterias*, *Acanthaster*). The fact that some herbivorous and carnivorous species overlap in the R/r ratio is not explained, however. Regardless of the ratio, flexibility is contingent on the degree of connection and development of the arm ossicles (Blake 1989, Lawrence 2013). Thus, herbivorous starfish often have a pentagonal cushion-shape where the arms are not distinct from the disc or have arms that are not highly dexterous. In contrast, carnivorous species tend to have a stellate profile with flexible arms, with the exception perhaps of *Pisaster ochraceus*, which has stiff arms.

Other traits associated with starfish size may also influence starfish feeding habits. A possible explanation for starfish with similar metric sizes, but different feeding guilds, may be associated with the muscular strength of the arms and tube feet, as well as the ability of mutable connective tissue to "lock" the starfish in place (Eylers 1976, Hennebert et al. 2010). Some facultative herbivorous starfish, for example *Patiriella regularis*, do not appear to be capable of preying on bivalves, species that have strong adductor muscles capable of holding valves shut (Crump 1969). Indeed a few studies note the relative ease of removing herbivorous starfish by hand from the substratum compared to carnivorous species. Strength in starfish depends on the type and number of tube feet, which are used to adhere to the substratum, move and capture prey (Hennebert et al. 2013). Thus, it is likely that herbivorous starfish may have fewer tube feet that have less adhesion power than predatory species, which makes it difficult, if not impossible, to capture mobile or big prey.

Based on these observations, it is suggested that herbivory in starfish is more likely to occur in species with a low R/r or a small disc. The arms may be less dexterous than those of predatory species and their tube feet may not be as muscular. Studies of the relationship between feeding guild, starfish body profile (e.g. pentagonal vs. stellate), arm flexibility and strength would be needed to test these hypotheses.

Phylogeny and associated feeding habits

Asteroid phylogeny has been in a state of flux for decades with strong disagreement among morphologists (e.g. Blake 1989, Gale 2011) with many attempts to resolve species' and family relationships using molecular data (Mah & Foltz 2011a,b, Feuda & Smith 2015). Application of new molecular and statistical methods is starting to clarify the relationships between the different asteroid families and their morphological evolution (Mah & Blake 2012, Feuda & Smith 2015). Evolution of disc-ending tube feet and an eversible stomach are ancient features of the Asteroidea, with loss of both features in some soft sediment burrowing starfish (e.g. *Astropecten*; Feuda & Smith 2015). The pattern of extra-oral digestion using an eversible stomach is a near-uniform characteristic of starfish. Variations of the digestive system among starfish taxa, however, could be informative with respect to herbivorous habits. Many obligate herbivores belong to the Asterinidae, a family that has a well-developed cardiac stomach (Jangoux 1982a). Similarly, some other starfish described here, such as *Oreaster reticulatus* (Oreasteridae), have a relatively larger and expandable cardiac stomach, which has been associated with efficient feeding behaviour on biofilms (Jangoux 1982a,b). For an animal that feeds by digesting the food underneath its stomach, a greater expansion ability of the cardiac stomach would provide a greater feeding surface area. The potential relationship between digestive tract structure and herbivorous starfish feeding behaviour deserves further investigation.

In addition to differences in body profile discussed above (pentagonal vs. stellate) there are other features that separate herbivorous and predatory starfish. Major predatory starfish that are able to manipulate their prey, those in the order Forcipulatida (e.g. *Asterias* and *Coscinasterias*), have maximally developed disc-ending tube feet and complex pedicellariae (Feuda & Smith 2015). The presence of these pedicellariae is an important evolutionary innovation that sets aside these starfish. Forcipulate pedicellariae are intricate jaw-like structures that are used to catch and secure motile prey, including fish. The prey is then delivered to the mouth. In contrast in the order Valvatida, which includes the largely herbivorous asterinids (e.g. *Parvulastra*) as well as the oreasterids (e.g. *Oreaster*) and ophidiasterids (e.g. *Linckia*), pedicellariae are absent or less complex. When present, they are largely used to defend the surface from interfering objects. The tube feet of the valvatids are also less muscular than in the forcipulatids. Pedicellariae are absent in the order Spinulosida (e.g. *Echinaster*) (Feuda & Smith 2015).

While the taxonomic development of disc-ending tube feet and pedicellariae are morphological features along with body profile that can be used to separate some predatory starfish from herbivorous species, there remains considerable overlap in feeding behaviour in the carnivory-herbivory dichotomy, as detailed above. It is likely that other differences between carnivorous and herbivorous starfish will be discerned with investigation of digestive physiology and biology (e.g. enzyme biochemistry and internal digestive structure) and details of the skeleton of the mouth frame (e.g. Gale 2011).

Effects of starfish herbivory on benthic assemblages

The impacts of feeding by herbivorous starfish on benthic communities may be less predictable than by other herbivores because of the mechanism by which starfish feed. Not many studies have investigated the effects of starfish herbivory on benthic communities, but the potential of starfish to modify the structure of algal assemblages has been recognized (Burgett 1982, 1988, Leonard 1994, Fujita 1999, Jackson et al. 2009, Pillay et al. 2010, Dawson & Pillay 2011).

One of the first studies to document that algal distribution could be affected by starfish grazing was done by Burgett (1982), who showed that *Patiriella regularis* could inhibit establishment of some macroalgal species on rocky shores in New Zealand by eating spores and germlings of newly-settled algae. Likewise, *Patiria miniata* was shown to control recruitment of giant kelp (*Macrocystis pyrifera*) along the coast of California by intensively grazing on young sporophytes when they were microscopic (Leonard 1994). Due to the low densities of *Patiria miniata*, however, this affect may be limited to small spatial scales and areas with low kelp recruitment (Leonard 1994). This study also showed that the starfish increased survival of macroscopic juvenile kelp in the surroundings of the grazed kelp. These observations suggest that grazing by starfish may play an important role in the dynamic of benthic communities on kelp forests and rocky shores.

The most studied starfish, in terms of grazing impacts, is *Parvulastra exigua*, which is abundant on intertidal rocky shores of South and South Eastern Australia and South Africa where it forages mostly on biofilms (Branch & Branch 1980, Arrontes & Underwood 1991, Stevenson 1992, Hart et al. 2006, Martinez et al. 2016). Previously it was demonstrated that *P. exigua* decreased the standing stock of algae, but did not suppress growth of macroalgae (Branch & Branch 1980, Arrontes & Underwood 1991). The strength of the grazing impact of *P. exigua*, however, could not be conclusively determined since both studies had problems with experimental artefacts from the caging methods used, in that the starfish would either escape (Branch & Branch 1980) or feed on the biofilm growing on the cages (Arrontes & Underwood 1991). More recently, experiments have demonstrated that *P. exigua* can consume 40 to 70% of the available biofilm on rocks during one period of low tide (Jackson et al. 2009). This calculation was based on natural densities of starfish in the field using measurements of area grazed and time spent per feeding event as well as the frequency of feeding events per hour.

A key question that arises is how grazing pressure by *Parvulastra exigua* is affected by other factors such as competition, predation, tides etc. Alteration of the composition of algal assemblages by herbivorous starfish has been recently demonstrated on intertidal sand flats and rocky shores. Grazing by *P. exigua* increases the richness and diversity of microalgal assemblages on sand flats (Pillay et al. 2010, Dawson & Pillay 2011). This was associated with the increase in extra-polymeric substances (EPS) from *P. exigua*, substances that are an important component for settlement and growth of organisms in biofilms (Wotton 2004). The consumption of dominant competing species in the biofilm matrix by *P. exigua* was another factor that may contribute to the increase in biofilm diversity (Pillay et al. 2010, Dawson & Pillay 2011). Moreover, the disturbance caused by starfish feeding on rocky shores resulted in successional changes in algal composition due to the different ages of starfish feeding marks (Figure 1C,D), which may cause local spatial variation in algal diversity (Jackson et al. 2009). More recently, Martinez (2016) demonstrated that *P. exigua* has a grazing impact within rock pools as strong as that of the key grazing limpet *Cellana tramoserica* on open rocks. This starfish also reduced abundance of primary producers on open rock, but the magnitude of the effect was weaker. The results from this research also showed that *Parvulastra exigua* may promote the development of red microalgae within rock pools and it was suggested that the starfish could increase the productivity of biofilm. Based on the foraging behaviour of *P. exigua* during high tide, the spatial grazing effects by *P. exigua* will depend on the availability of shelters (i.e. refugia at low tide) on rocky shores (Martinez 2016).

Grazing by starfish has also been shown to impact benthic assemblages on seagrass beds. *Oreaster reticulatus* significantly decreases chlorophyll concentrations on the surface of sediments in grazed areas (Scheibling 1980d, 2013). This impact can be intensified and occur over a great spatial extent when individuals aggregate and form dense grazing fronts (Scheibling 1980b,d). As *O. reticulatus* and *Protoreaster nodosus* accumulate sediment under their mouth during feeding and extrude their stomach to ingest associated biofilm and meiofauna, they constantly disturb the surficial sediment (Scheibling 1980b, d, 1982, Scheibling & Metaxas 2008). Thus, it appears that oreasterids in soft sediment systems have an important role in altering the structure of benthic communities as well as in nutrient recycling.

While there is evidence for negative and positive effects of starfish herbivory on benthic assemblages, the interactions between starfish grazing and algae are not well understood. Many starfish forage on calcareous algae and the discoloration of the algae in the grazed area is reported across studies (Barker 1979, Burgett 1982, 1988, Arrontes & Underwood 1991, Fujita 1999). Investigations of the effects of grazing on calcareous algae show that the starfish can damage the top layer of encrusting and turf calcareous algae (Figure 1A,B). It is not known, however, if starfish grazing causes macroalgal mortality (Barker 1979, Burgett 1982, Fujita 1999). It is suggested that grazing on calcareous turf by *Patiriella regularis* may negatively affect the growth of new branches on the apex of the algae because the enzymes secreted by the starfish attack meristematic tissues (Burgett 1982). In addition, following grazing by starfish, epiphytic organisms colonize the damaged areas on turf algae (Burgett 1982). This probably has a negative effect on the photosynthetic activities of the grazed area on the calcareous algae. On the other hand, calcareous algae might be able to recover from the damage if grazing does not persist (Burgett 1982). Fujita (1999) showed that grazing by *Patiria pectinifera* stimulated deep-layer sloughing on encrusting calcareous algae. This process can compensate for grazing damage by regenerating new cover cells. Thus, for calcareous algae grazing by starfish may promote growth, although it can cause some damage to the tissue. Grazing effects on the tissue of non-calcareous algae is not known, although some authors suggest that the grazed area might cause tissue death (Araki 1964, Bak 1981, Levin et al. 1987). Further investigations are required to elucidate the effects of starfish grazing on the tissue of different groups of algae.

The impact of starfish herbivory on algal assemblages indicates that starfish might play an important role in altering the structure of benthic communities. By foraging on biofilm, it is well known that molluscan grazers can control the distribution of algal assemblages as well as of some sessile macrofauna (e.g. barnacles, bivalves) as those grazers ingest not only spores of macroalgae, but also other unicellular microorganisms and newly-settled invertebrates (Hawkins & Hartnoll 1983, Hawkins et al. 1992, Poore et al. 2012). Starfish grazing might have a similar impact, as demonstrated for *Parvulastra exigua* (Martinez 2016). Thus, studies on grazing by herbivorous starfish linked to changes in benthic assemblages are of great importance to elucidate the ecological niche of starfish herbivory in the diverse ecosystems that they inhabit.

Interactions between starfish and herbivores

Competitive interactions can influence spatial distribution and foraging activity of grazers, which may result in spatial and temporal variability in grazing pressure on algae (e.g. Branch 1981, Chapman & Underwood 1992, Johnson et al. 2008, Aguilera & Navarrete 2011). There is, however, a lack of knowledge on the interactions between herbivorous starfish and other herbivorous taxa.

There are only 3 studies to date where hypotheses of competitive interactions have been tested, all investigating the putative competitive interaction between *Parvulastra exigua* and the limpet *Cellana tramoserica* on intertidal rocky shores (Branch & Branch 1980, Arrontes & Underwood 1991, Martinez 2016). The motivation for these studies was that *C. tramoserica* is a strong competitor for food compared to other gastropod grazers and it was expected that the limpet would outcompete the starfish for food resources (Underwood 1978, Underwood & Jernakoff 1981, Steneck & Watling 1982). The results of these early studies, however, were far from conclusive. According to Branch & Branch (1980) *Parvulastra exigua* appears to be a weak competitor when competing with the limpet *Cellana tramoserica*, while the results from Arrontes & Underwood (1991) indicate that competition was not evident. The results from these studies are not only contradictory, but also confounded by experimental artefacts. Branch & Branch (1980) reported issues in caging the starfish because individuals escaped from the cages whereas Arrontes & Underwood (1991) observed starfish feeding on biofilms/microalgae growing on the cages. The recent study of Martinez (2016) eliminated the confounding effects of caging the starfish on her experiment and provided evidence of competition between these grazers. *Parvulastra exigua* was shown to be as strong a competitor as *Cellana tramoserica* with the two species having similar effects on each other. This finding may change our view of the dynamics of rocky shores, since the dominance of limpets in this system is a well-established concept (Branch 1984, Underwood 1992).

Many other benthic grazers, especially of intertidal rocky shores, forage on biofilm or soft foliose and filamentous algae (e.g. Hawkins & Hartnoll 1983, Kennish & Williams 1997, Coleman et al. 2006). The habitat and food resources exploited overlap with the distribution and feeding habits of most obligate herbivorous starfish (see above). This suggests that starfish might compete for food resources with other herbivores, the outcomes of which may affect the structure of benthic communities. The potential competitive interactions between herbivorous starfish and other herbivores are new, open areas of investigation. This information is relevant to understanding the effects of interaction between different groups of herbivores (e.g. starfish, gastropods) on foraging behaviour, growth and distribution of species in benthic communities.

Overview and future directions

There is strong evidence that grazing by starfish plays an important role in modifying the structure of algal/plant assemblages in some systems, e.g. biofilm on rocky and sandy shores (Jackson et al. 2009, Pillay et al. 2010, Dawson & Pillay 2011, Martinez 2016), kelp beds (Leonard 1994), coralline algae on coral reefs (Laxton 1974) and seagrass beds (Scheibling 1980b, d). There are many gaps, however, that need to be investigated to understand the ecological niche of herbivorous starfish in different ecosystems.

First, detailed information on the diet of starfish is required. Most reports of starfish feeding on algae include obligate herbivores as well as omnivores and lack details of prey items. Part of this problem may be the difficulty to actually identify the food item, as most starfish evert their stomach and digest externally. Application of techniques such as stable isotope analysis (e.g. Alfaro 2008), comparison of fatty acid profiles (e.g. Alfaro 2008, Wessels et al. 2012) and even genetic comparisons of stomach contents with the genomes of potential food items (e.g. Pompanon et al. 2012) would be helpful in correctly assigning food types to the diet of these organisms. Also traditional methods such as collecting samples of material under the extruded stomach of starfish might be useful to gather information on starfish diet as long as these data are collected in a systematic way.

Determining the dietary composition of herbivorous starfish species will prompt questions on the factors that might affect their feeding habits. It has been shown that many starfish forage on biofilm, but some species may also forage on macroalgae (e.g. Scheibling 1979, Bak 1981, Stevenson 1992). Investigations of starfish nutritional ecology might be of great importance to elucidate intrinsic characteristics of herbivory in starfish that will give insights into the type of food that different species are likely to exploit. Further studies should focus on the relationships among digestibility of different components of plant/algal resources, morphological and physiological adaptations of starfish species to forage on plant/algae, and herbivore tolerance to chemical defences of primary producers.

It appears that digestive enzymes play an important role in the ability of starfish to forage on algae, such that starfish might forage on algae that are easier to digest e.g. microalgae and soft filamentous / foliose algae (Steneck & Dethier 1994). It has also been shown that starfish can partially digest more robust algae and it is suggested that those might require a longer digestion time (Kristensen 1972, Scheibling 1980a). Thus, feeding on tough macroalgae might be more energetically costly than soft algae or microalgae. It is also possible that starfish respond to algal chemical defences. Digestive enzymes could induce the production of chemical defences in the macroalgae (Amsler 2001), thus starfish would graze less on algae. Alternatively, starfish could be responding negatively to inherently chemically-defended macroalgae. It is well known that chemical defences and palatability of algae are important components of algae-herbivore interactions, which can affect feeding choices and behaviour of herbivores (Hay 1996). There is no information, however, on how algal traits affect herbivory in starfish. Thus, studies should give priority to investigating the ability of different species of starfish to digest different groups of algae and explore the relationship between algal chemical defences and feeding by starfish.

Digestive capabilities, however, would not solely explain why some species feed on algae instead of animal resources because many starfish, including carnivores, possess carbohydrases and proteinases that are potentially capable of breaking down animal or plant/algal material (Araki 1964, Araki & Giese 1970, Kristensen 1972, Scheibling 1980a). Theoretically starfish should be able to exploit any available resource. That said, some species have specialized digestive enzyme systems, as evident for *Acanthaster planci*, which is able to avail of the wax esters produced by its coral prey (Benson et al. 1975, Brahimi-Horn et al. 1989). Detailed comparison of the digestive physiology of herbivorous and predatory starfish may reveal diet-related differences in enzyme biochemistry.

Even though starfish do not have specialized morphological structures related to exploitation of a specific food type, in contrast to other herbivores, there may be some morphological features/ traits associated with body size that influence food selection. Herbivory appears to be influenced by starfish size and shape, since starfish feeding on algae are often of similar size or with the same size of stomach (i.e. small disc sizes). It is also suggested that herbivorous species lack the ability and strength to attack or manipulate more mobile or shelled prey (Birkeland et al. 1971, Sloan 1980). The fact that many starfish that forage on algae might opportunistically eat moribund or dead animals may support this hypothesis. If this is true, herbivory is likely to be more common in starfish than thought and would occur in all small species such as in many asterinid starfish, including at least 25 species of the genus *Aquilonastra* (O'Loughlin & Rowe 2006), species with small oral discs. Juveniles of carnivorous species also appear to be initially herbivorous. Thus, comparative morphological studies between herbivorous, carnivorous and omnivorous starfish linked to feeding habits are an important area for future research.

Factors other than the intrinsic characteristics of starfish must also influence their foraging behaviour. According to the studies on feeding habits of facultative herbivores, foraging on algae seems to be a response to the availability of animal resources and predation risk. It seems that herbivory will occur in the absence or low availability of animal prey or when accessing animal resource incurs a high risk of being preyed on (e.g. Mauzey et al. 1968, Yamaguchi 1973, De'ath & Moran 1998, Scheibling 2013). These arguments would explain the great variability in diet (algal/ animals proportion consumed) across different locations and seasons. Manipulative experiments are needed to test these hypotheses.

In addition to predation risk, competitive interactions can strongly affect foraging behaviour of herbivores and consequently the structure of algal assemblages within benthic communities (Lawrence 1975, Underwood & Denley 1984, Underwood 1992). Because starfish have the potential to forage on any food resource, they may avoid competition by foraging on resources not used by competitors (Arrontes & Underwood 1991). Otherwise, if starfish select a specific food resource, it is likely that competition might take place (Martinez 2016). Due to the paucity of studies, it is not possible to make inferences about competitive interactions between herbivorous starfish and other herbivorous taxa and the consequent outcomes on foraging behaviour and changes in algal assemblages.

In general, the knowledge of herbivory in starfish is currently limited to descriptive investigations of starfish diet and to a few studies of food digestion, changes in diet and grazing effects of starfish on algal assemblages. Future studies on starfish herbivory should investigate feeding habits in more detail and especially expand investigations to areas with a paucity of information, such as foraging behaviour related to algal chemical defences, competitive interactions and predation risk. Discovering the mechanisms that modulate herbivory in starfish is particularly important in consideration of current changes in species ranges and distribution in response to climate change

(García Molinos et al. 2015). There is already evidence that natural invasions of herbivorous species in new habitats in the marine environment are causing strong modifications in habitat structure (Johnson et al. 2011, Vergés et al. 2014a,b, Ling et al. 2015). For example there is evidence that tropical herbivorous fish, which disperse to higher latitudes during summer, are now surviving over winter and their grazing is probably causing changes in algal assemblages (Vergés et al. 2014b). The kelp beds that are usually grazed by sea urchins during warm seasons are not being able to recover during winter, probably because tropical fishes are feeding on young, settled kelp, and therefore kelp beds are being replaced by turf algae assemblages (Vergés et al. 2014b).

Changes in the distribution of herbivorous species are also predicted to affect economic activities, such as fisheries, since changes in the structure of assemblages can cause declines in the stocks of target species (Johnson et al. 2011). Perhaps this is already a concern for abalone fisheries in Eastern Tasmania. Due to the expansion and intense grazing of the sea urchin *Centrostephanus rodgersii*, kelp beds are declining and abalones are losing their habitat as a consequence (Johnson et al. 2011, Ling et al. 2015).

Therefore, there is an urge to investigate the ecology of some groups of herbivores, such as starfish species, that are poorly understood compared to other marine herbivores (Poore et al. 2012). As starfish have no apparent specialized structures adapted to forage on specific food resources, understanding the mechanism that affects their feeding behaviour towards herbivory can be crucial to predict the effects of species expansion on native benthic communities in different ecosystems.

Acknowledgements

We would like to thank Alistair Poore for helping with the organization of the review structure, and Augustine Porter and Rebecca Morris for comments and critique during the production of this review. AM was funded by an Endeavour Award from the Australian Government and a CNPq (Science without Borders scholarship) from the Brazilian Government. Thanks to Phil Mercurio and Marcela Rosa for providing images. This is contribution number 196 of the Sydney Institute of Marine Science.

References

Aguilera, M.A. & Navarrete, S.A. 2011. Distribution and activity patterns in an intertidal grazer assemblage: influence of temporal and spatial organization on interspecific associations. *Marine Ecology Progress Series* **431**, 119–136.

Alexander, S.K. 1986. Diet of the blue crab, *Callinectes sapidus* Rathbun, from nearshore habitats of Galveston Island, Texas. *Texas Journal of Science* **38**, 85–89.

Alfaro, A.C. 2008. Diet of *Littoraria scabra*, while vertically migrating on mangrove trees: gut content, fatty acid, and stable isotope analyses. *Estuarine Coastal and Shelf Science* **79**, 718–726.

Amsler, C.D. 2001. Induced defenses in macroalgae: the herbivore makes a difference. *Journal of Phycology* **37**, 353–356.

Anderson, M.J. 1995. Variations in biofilms colonizing artificial surfaces: seasonal effects and effects of grazers. *Journal of the Marine Biological Association of the United Kingdom* **75**, 705–714.

Araki, G.S. 1964. *On the physiology of feeding and digestion in the sea star, Patiria miniata*. PhD thesis. Stanford University, United States.

Araki, G.S. & Giese, A.C. 1970. Carbohydrases in sea stars. *Physiological Zoology* **43**, 296–305.

Arrontes, J. 1990. Diet, food preference and digestive efficiency in intertidal isopods inhabiting macroalgae. *Journal of Experimental Marine Biology and Ecology* **139**, 231–249.

Arrontes, J. & Underwood, A.J. 1991. Experimental studies on some aspects of the feeding ecology of the intertidal starfish *Patiriella exigua*. *Journal of Experimental Marine Biology and Ecology* **148**, 255–269.

Baird, A.H., Pratchett, M.S., Hoey, A.S., Herdiana, Y. & Campbell, S.J. 2013. *Acanthaster planci* is a major cause of coral mortality in Indonesia. *Coral Reefs* **32**, 803–812.

Bak, H.P. 1981. Feeding habits of the sea star *Asterina pectinifera* (Muller et Troshel) and its grazing effect on the eelgrass *Zostera marina* L. *Publications from the Amakusa Marine Biological Laboratory Kyushu University* **6**, 1–8.

Barker, M.F. 1977. Observations on the settlement of Brachiolaria larvae of *Stichaster australis* (Verrill) and *Coscinasterias calamaria* (Gray) (Echinodermata: Asteroidea) in laboratory and on the shore. *Journal of Experimental Marine Biology and Ecology* **30**, 95–108.

Barker, M.F. 1979. Breeding and recruitment in a population of the New Zealand starfish *Stichaster australis* (Verrill). *Journal of Experimental Marine Biology and Ecology* **41**, 195–211.

Barker, M.F. & Nichols, D. 1983. Reproduction, recruitment and juvenile ecology of the starfish, *Asterias rubens* and *Marthasterias glacialis*. *Journal of the Marine Biological Association of the United Kingdom* **63**, 745–765.

Benedetti-Cecchi, L., Vaselli, S., Maggi, E. & Bertocci, I. 2005. Interactive effects of spatial variance and mean intensity of grazing on algal cover in rock pools. *Ecology* **86**, 2212–2222.

Benson, A.A., Patton, J.S. & Field, C.E. 1975. Wax digestion in a crown-of-thorns starfish. *Comparative Biochemistry and Physiology* **52**, 339–340.

Birkeland, C. 1989. The influence of echinoderms on coral-reef communities. *Echinoderm Studies* **3**, 1–79.

Birkeland, C., Chia, F.-S. & Strathmann, R.R. 1971. Development, substratum selection, delay of metamorphosis and growth in the seastar, *Mediaster aequalis* Stimpson. *Biological Bulletin* **141**, 99–108.

Blake, D.B. 1989. Asteroidea: Functional morphology, classification and phylogeny. *Echinoderm Studies* **3**, 179–223.

Brahimi-Horn, M.C., Guglielmino, M.L., Sparrow, L.G., Logan, R.I. & Moran, P.J. 1989. Lipolytic enzymes of the digestive organs of the crown-of-thorns starfish (*Acanthaster planci*): comparison of the stomach and pyloric caeca. *Comparative Biochemistry and Physiology* **92**, 637–643.

Branch, G.M. 1981. The biology of limpets: physical factors, energy flow, and ecological interactions. *Oceanography and Marine Biology: An Annual Review* **19**, 235–380.

Branch, G.M. 1984. Competition between marine organisms: ecological and evolutionary implications. *Oceanography and Marine Biology: An Annual Review* **22**, 429–593.

Branch, G.M. & Branch, M.L. 1980. Competition between *Cellana tramoserica* (Sowerby) (Gastropoda) and *Patiriella exigua* (Lamarck) (Asteroidea), and their influence on algal standing stocks. *Journal of Experimental Marine Biology and Ecology* **48**, 35–49.

Brun, E. 1972. Food and feeding habits of *Luidia ciliaris* Echinodermata: Asteroidea. *Journal of the Marine Biological Association of the United Kingdom* **52**, 225–236.

Burgett, J.M. 1982. *The feeding ecology of Patiriella regularis (Verrill) in the rocky intertidal*. MSc thesis. University of Auckland, New Zealand.

Burgett, J.M. 1988. Effects of digestive grazing by the sea star *Patirella regularis* on communities of coralline algae. *Pacific Science* **42**, 116.

Chapman, M.G. & Underwood, A.J. 1992. Foraging behaviour of marine benthic grazers. *in* D.M. John et al. (eds). *Plant-Animal Interactions in the Marine Benthos*. United Kingdom: Oxford University, 289–317.

Chen, B.Y. & Chen, C.P. 1992. Reproductive cycle, larval development, juvenile growth and population dynamics of *Patiriella pseudoexigua* (Echinodermata: Asteroidea) in Taiwan. *Marine Biology* **113**, 271–280.

Chesher, R.H. 1969. Destruction of pacific corals by sea star *Acanthaster planci*. *Science* **165**, doi:10.1126/science.1165.3890.1280.

Chia, F.-S. & Amerongen, H. 1975. On the prey-catching pedicellariae of a starfish, *Stylasterias forreri* (de Loriol). *Canadian Journal of Zoology* **53**, 748–755.

Choat, J.H. 1982. Fish feeding and the structure of benthic communities in temperate waters. *Annual Review of Ecology and Systematics* **13**, 423–449.

Coleman, N. 1977. *A Field Guide to Australian Marine Life*. Australia: Rigby Limited.

Coleman, R.A., Underwood, A.J., Benedetti-Cecchi, L., Aberg, P., Arenas, F., Arrontes, J., Castro, J., Hartnoll, R.G., Jenkins, S.R., Paula, J., Della Santina, P. & Hawkins, S.J. 2006. A continental scale evaluation of the role of limpet grazing on rocky shores. *Oecologia* **147**, 556–564.

Crump, R.G. 1969. *Aspects of the biology of some New Zealand echinoderms: feeding, growth and reproduction in the asteroids, Patiriella regularis (Verrill, 1867) and Coscinasterias calamaria (Gray, 1840)*. PhD thesis. University of Otago, New Zealand.

Crump, R.G. & Emson, R.H. 1978. Some aspects of the population dynamics of *Asterina gibbosa* (Asteroidea). *Journal of the Marine Biological Association of the United Kingdom* **58**, 451–466.

Crump, R.G. & Emson, R.H. 1983. The natural history, life history and ecology of the two British species of *Asterina*. *Field Studies* **5**, 867.

Dartnall, A.J. 1971. Australian sea-stars of the genus *Patiriella* (Asteroidea, Asterinidae). *Proceedings of the Linnean Society of New South Wales* **96**, 39–49.

Dartnall, A.J., Byrne, M., Collins J. & Hart, M.W. 2003. A new viviparous species of asterinid (Echinodermata, Asteroidea, Asterinidae) and a new genus to accommodate the species of pan-tropical exiguoid sea stars. *Zootaxa* **359**, 1–14.

Dawson, J. & Pillay, D. 2011. Influence of starfish grazing on lagoonal microalgal communities: non-competitive mechanisms for unimodal effects on diversity. *Marine Ecology Progress Series* **435**, 75–82.

Dayton, P.K., Currie, V., Gerrodette, T., Keller, B.D., Rosenthal, R. & Tresca, D.V. 1984. Patch dynamics and stability of some California kelp communities. *Ecological Monographs* **54**, 254–289.

De'ath, G. & Moran, P.J. 1998. Factors affecting the behaviour of crown-of-thorns starfish (*Acanthaster planci* L.) on the Great Barrier Reef: 1: patterns of activity. *Journal of Experimental Marine Biology and Ecology* **220**, 83–106.

Decho, A.W. 2000. Microbial biofilms in intertidal systems: an overview. *Continental Shelf Research* **20**, 1257–1273.

Dee, L.E., Witman, J.D. & Brandt, M. 2012. Refugia and top-down control of the pencil urchin *Eucidaris galapagensis* in the Galápagos Marine Reserve. *Journal of Experimental Marine Biology and Ecology* **416–417**, 135–143.

Dolecal, R.E. & Long, J.D. 2013. Ephemeral macroalgae display spatial variation in relative palatability. *Journal of Experimental Marine Biology and Ecology* **440**, 233–237.

Duffy, J.E. & Hay, M.E. 1990. Seaweed adaptations to herbivory. *Bioscience* **40**, 368–375.

Duffy, J.E. & Hay, M.E. 1991. Food and shelter as determinants of food choice by an herbivorous marine amphipod. *Ecology* **72**, 1286–1298.

Duyverman, H. 1976. *Factors influencing the local distribution of the sea star Patiriella exigua L.* MSc thesis. Flinders University of South Australia, Australia.

Emson, R.H. & Crump, R.G. 1979. Description of a new species of *Asterina* (Asteroidea), with an account of its ecology. *Journal of the Marine Biological Association of the United Kingdom* **59**, 77–94.

Estes, J.A., Terborgh, J., Brashares, J.S., Power, M.E., Berger, J., Bond, W.J., Carpenter, S.R., Essington, T.E., Holt, R.D., Jackson, J.B.C., Marquis, R.J., Oksanen, L., Oksanen, T., Paine, R.T., Pikitch, E.K., Ripple, W.J., Sandin, S.A., Scheffer, M., Schoener, T.W., Shurin, J.B., Sinclair, A.R.-E., Soule, M.E., Virtanen, R. & Wardle, D.A. 2011. Trophic downgrading of planet earth. *Science* **333**, 301–306.

Eylers, J.P. 1976. Aspects of skeletal mechanics of the starfish *Asterias forbesii*. *Journal of Morphology* **149**, 353–367.

Farias, N.E., Meretta, P.E. & Cledón, M. 2012. Population structure and feeding ecology of the bat star *Asterina stellifera* (Möbius, 1859): omnivory on subtidal rocky bottoms of temperate seas. *Journal of Sea Research* **70**, 14–22.

Feuda, R. & Smith, A.B. 2015. Phylogenetic signal dissection identifies the root of starfishes. *PloS ONE* **10**, e0123331; doi:0123310.0121371/journal.pone.0123331.

Fujita, D. 1999. The sea star *Asterina pectinifera* causes deep-layer sloughing in *Lithophyllum yessoense* (Corallinales, Rhodophyta). *Hydrobiologia* **398–399**, 261–266.

Gale, A.S. 2011. *The Phylogeny of Post-Palaeozoic Asteroidea (Neoasteroidea, Echinodermata)*. Wales, UK: John Wiley and Sons.

García Molinos, J., Halpern, B.S., Schoeman, D.S., Brown, C.J., Kiessling, W., Moore, P.J., Pandolfi, J.M., Poloczanska, E.S., Richardson, A.J. & Burrows, M.T. 2015. Climate velocity and the future global redistribution of marine biodiversity. *Nature Climate Change*, doi:10.1038/nclimate2769.

Gaymer, C.F. & Himmelman, J.H. 2008. A keystone predatory sea star in the intertidal zone is controlled by a higher-order predatory sea star in the subtidal zone. *Marine Ecology Progress Series* **370**, 143–153.

Gerard, V.A. 1976. *Some aspects of material dynamics and energy flow in a kelp forest in Monterey Bay, California*. PhD thesis. University of California, United States.

Grace, R.V. 1967. An underwater survey of two starfish species in the entrance to the Whangateau harbour. *Tane* **13**, 13–19.

Griffin, J.N., Noel, L., Crowe, T.P., Burrows, M.T., Hawkins, S.J., Thompson, R.C. & Jenkins, S.R. 2010. Consumer effects on ecosystem functioning in rock pools: roles of species richness and composition. *Marine Ecology Progress Series* **420**, 45–56.

Hart, M.W., Keever, C.C., Dartnall, A.J. & Byrne, M. 2006. Morphological and genetic variation indicate cryptic species within Lamarck's little sea star, *Parvulastra* (=*Patiriella*) *exigua*. *Biological Bulletin* **210**, 158–167.

Hawkins, S.J. & Hartnoll, R.G. 1983. Grazing of intertidal algae by marine invertebrates. *Oceanography and Marine Biology: An Annual Review* **21**, 195–282.

Hawkins, S.J., Hartnoll, R.G., Kain, J.M. & Norton, T.A. 1992. Plant-animal interactions on hard substrata in the North-East Atlantic, *in* D.M. John et al. (eds). *Plant-Animal Interactions in the Marine Benthos*. United Kingdom: Clarendon Press, 1–32.

Hawkins, S.J., Watson, D.C., Hill, A.S., Harding, S.P., Kyriakides, M.A., Hutchinson, S. & Norton, T.A. 1989. A comparison of feeding mechanisms in microphagous, herbivorous, intertidal, prosobranchs in relation to resource partitioning. *Journal of Molluscan Studies* **55**, 151–165.

Hay, M.E. 1984. Patterns of fish and urchin grazing on Caribbean coral reefs: are previous results typical? *Ecology* **65**, 446–454.

Hay, M.E. 1996. Marine chemical ecology: what's known and what's next? *Journal of Experimental Marine Biology and Ecology* **200**, 103–134.

Hay, M.E. 2009. Marine chemical ecology: chemical signals and cues structure marine populations, communities, and ecosystems. *Annual Review of Marine Science* **1**, 193–212.

Hempson, G.P., Archibald, S., Bond, W.J., Ellis, R.P., Grant, C.C., Kruger, F.J., Kruger, L.M., Moxley, C., Owen-Smith, N., Peel, M.J.S., Smit, I.P.J. & Vickers, K.J. 2015. Ecology of grazing lawns in Africa. *Biological Reviews* **90**, 979–994.

Henderson, J.A. & Lucas, J.S. 1971. Larval development and metamorphosis of *Acanthaster planci* (Asteroidea). *Nature* **232**, 655–657.

Hennebert, E., Haesaerts, D., Dubois, P. & Flammang, P. 2010. Evaluation of the different forces brought into play during tube foot activities in sea stars. *Journal of Experimental Biology* **213**, 1162–1174.

Hennebert, E., Jangoux, M. & Flammang, P. 2013. Functional biology of asteroid tube feet, *in* J.M. Lawrence (ed.). *Starfish: Biology and Ecology of the Asteroidea*. United States: The Johns Hopkins University Press, 24–36.

Himmelman, J.H., Dutil, C. & Gaymer, C.F. 2005. Foraging behavior and activity budgets of sea stars on a subtidal sediment bottom community. *Journal of Experimental Marine Biology and Ecology* **322**, 153–165.

Huntly, N. 1991. Herbivores and the dynamics of communities and ecosystems. *Annual Review of Ecology and Systematics* **22**, 477–503.

Jackson, A.C., Murphy, R.J. & Underwood, A.J. 2009. *Patiriella exigua*: grazing by a starfish in an overgrazed intertidal system. *Marine Ecology Progress Series* **376**, 153–163.

Jangoux, M. 1982a. Digestive systems: Asteroidea, *in* M. Jangoux & J.M. Lawrence (eds). *Echinoderm Nutrition*. Netherlands: A.A. Balkema, 235–272.

Jangoux, M. 1982b. Food and feeding mechanisms: Asteroidea, *in* M. Jangoux & J.M. Lawrence (eds). *Echinoderm Nutrition*. Netherlands: A.A. Balkema, 117–159.

Jenkins, S.R., Moore, P., Burrows, M.T., Garbary, D.J., Hawkins, S.J., Ingólfsson, A., Sebens, K.P., Snelgrove, P.V.R., Wethey, D.S. & Woodin, S.A. 2008. Comparative ecology of North Atlantic shores: do differences in players matter for process? *Ecology* **89**, S3-S23.

Johnson, C.R., Banks, S.C., Barrett, N.S., Cazassus, F., Dunstan, P.K., Edgar, G.J., Frusher, S.D., Gardner, C., Haddon, M., Helidoniotis, F., Hill, K.L., Holbrook, N.J., Hosie, G.W., Last, P.R., Ling, S.D., Melbourne-Thomas, J., Miller, K., Pecl, G.T., Richardson, A.J., Ridgway, K.R., Rintoul, S.R., Ritz, D.A., Ross, D.J., Sanderson, J.C., Shepherd, S.A., Slotwinski, A., Swadling, K.M. & Taw, N. 2011. Climate change cascades: shifts in oceanography, species' ranges and subtidal marine community dynamics in eastern Tasmania. *Journal of Experimental Marine Biology and Ecology* **400**, 17–32.

Johnson, C.R., Sutton, D.C., Olson, R.R. & Giddins, R. 1991. Settlement of crown-of-thorns starfish: role of bacteria on surfaces of coralline algae and a hypothesis for deepwater recruitment. *Marine Ecology Progress Series* **71**, 143–162.

Johnson, M.P., Hanley, M.E., Frost, N.J., Mosley, M.W.J. & Hawkins, S.J. 2008. The persistent spatial patchiness of limpet grazing. *Journal of Experimental Marine Biology and Ecology* **365**, 136–141.

Kamya, P.Z., Byrne, M., Graba-Landry, A. & Dworjanyn, S.A. 2016. Near future ocean acidification enhances the feeding rate and development of the herbivorous juveniles of the Crown of Thorns Starfish, *Acanthaster planci*. *Coral Reefs* DOI 10.1007/s00338–016–1480–6

Kayal, M., Vercelloni, J., de Loma, T.L., Bosserelle, P., Chancerelle, Y., Geoffroy, S., Stievenart, C., Michonneau, F., Penin, L., Planes, S. & Adjeroud, M. 2012. Predator crown-of-thorns starfish (*Acanthaster planci*) outbreak, mass mortality of corals, and cascading effects on reef fish and benthic communities. *PloS ONE* **7**, doi:10.1371/journal.pone.0047363.

Keesing, J.K. 1995. Temporal patterns in the feeding and emergence behaviour of the crown-of-thorns starfish *Acanthaster planci*. *Marine and Freshwater Behaviour and Physiology* **25**, 209–232.

Keesing, J.K. & Halford, A.R. 1992. Field measurement of survival rates of juvenile *Acanthaster planci*: techniques and preliminary results. *Marine Ecology Progress Series* **85**, 107–114.

Keesing, J.K. & Lucas, J.S. 1992. Field measurement of feeding and movement rates of the crown-of-thorns starfish *Acanthaster planci* (L.). *Journal of Experimental Marine Biology and Ecology* **156**, 89–104.

Kennish, R. & Williams, G.A. 1997. Feeding preferences of the herbivorous crab *Grapsus albolineatus:* the differential influence of algal nutrient content and morphology. *Marine Ecology Progress Series* **147**, 87–95.

Kenyon, J.C. & Aeby, G.S. 2009. Localized outbreak and feeding preferences of the crown-of-thorns seastar *Acanthaster planci* (Echinodermata, Asteroidea) on reefs off Oahu, Hawaii. *Bulletin of Marine Science* **84**, 199–209.

Keough, M.J. & Dartnall, A.J. 1978. A new species of viviparous asterinid asteroid from Gyre Peninsula, South Australia. *Records of the South Australian Museum* **17**, 407–416.

Kristensen, J.H. 1972. Carbohydrases of some marine invertebrates with notes on their food and on natural occurrence of carbohydrates studied. *Marine Biology* **14**, 130–142.

Kurihara, T. 1999. Effects of sediment type and food abundance on the vertical distribution of the starfish *Asterina pectinifera*. *Marine Ecology Progress Series* **181**, 269–277.

Laughlin, R.A. 1982. Feeding habits of the blue crab, *Callinectes sapidus* Rathbun, in the Apalachicola Estuary, Florida. *Bulletin of Marine Science* **32**, 807–822.

Lawrence, J.M. 1975. On the relationships between marine plants and sea urchins. *Oceanography and Marine Biology: An Annual Review* **13**, 213–286.

Lawrence, J.M. 2012. Form, function, food and feeding in stellate echinoderms. In *Echinoderm Research 2010: Proceedings of the Seventh European Conference on Echinoderms*, A. Kroh & M. Reich (eds). Auckland: Magnolia Press, 33–42.

Lawrence, J.M. 2013. The asteroid arm, *in* J.M. Lawrence (ed.). *Starfish: Biology and Ecology of Asteroidea*. United States: The Johns Hopkins University Press, 15–23.

Laxton, J.H. 1974. A preliminary study of the biology and ecology of the blue starfish *Linckia laevigata* (L.) on the Australian Great Barrier Reef and an interpretation of its role in the coral reef ecosystem. *Biological Journal of the Linnean Society* **6**, 47–64.

Leonard, G.H. 1994. Effect of the bat star *Asterina miniata* (Brandt) on recruitment of the giant kelp *Macrocystis pyrifera* C. Agardh. *Journal of Experimental Marine Biology and Ecology* **179**, 81–98.

Levin, V.S., Ivin, V.V. & Fadeev, V.I. 1987. Ecology of the starfish *Patiria pectinifera* (Mueller et Troschel) in Possiet Bay, Sea of Japan. *Asian Marine Biology* **4**, 49–60.

Ling, S.D., Scheibling, R.E., Rassweiler, A., Johnson, C.R., Shears, N., Connell, S.D., Salomon, A.K., Norderhaug, K.M., Pérez-Matus, A., Hernández, J.C., Clemente, S., Blamey, L.K., Hereu, B., Ballesteros, E., Sala, E., Garrabou, J., Cebrian, E., Zabala, M., Fujita, D. & Johnson, L.E. 2015. Global regime shift dynamics of catastrophic sea urchin overgrazing. *Philosophical Transactions of the Royal Society of London B: Biological Sciences* **370**, doi:10.1098/rstb.2013.0269.

Littler, M.M. & Arnold, K.E. 1982. Primary productivity of marine macroalgal functional-form groups from Southwestern North America. *Journal of Phycology* **18**, 307–311.

Littler, M.M. & Littler, D.S. 1980. The evolution of thallus form and survival strategies in benthic marine macroalgae: field and laboratory tests of a functional form model. *American Naturalist* **116**, 25–44.

Lubchenco, J. & Cubit, J. 1980. Heteromorphic life histories of certain marine algae as adaptations to variations in herbivory. *Ecology* **61**, 676–687.

Lubchenco, J. & Gaines, S.D. 1981. A unified approach to marine plant-herbivore interactions. I. Populations and communities. *Annual Review of Ecology and Systematics* **12**, 405–437.

Mah, C. & Foltz, D. 2011a. Molecular phylogeny of the Forcipulatacea (Asteroidea: Echinodermata): systematics and biogeography. *Zoological Journal of the Linnean Society* **162**, 646–660.

Mah, C. & Foltz, D. 2011b. Molecular phylogeny of the Valvatacea (Asteroidea: Echinodermata). *Zoological Journal of the Linnean Society* **161**, 769–788.

Mah, C.L. & Blake, D.B. 2012. Global diversity and phylogeny of the Asteroidea (Echinodermata). *PloS ONE* **7**, e35644. doi:35610.31371/journal.pone.0035644.

Martin, R.B. 1970. *Asteroid feeding biology*. MSc thesis. The University of Auckland, New Zealand.

Martin, A., Penchaszadeh, P. & Atienza, D. 2001. Population density and feeding habits of *Oreaster reticulatus* (Linnaeus, 1758) (Echinodermata, Asteroidea) living in seagrass beds off Venezuela. *Boletin Instituto Espanol de Oceanografia* **17**, 203–208.

Martinez, A.S. 2016. *Reevaluating the dynamics of intertidal rocky ecosystems: the foraging and behavioural ecology of an understudied grazer and its effects on benthic assemblages*. PhD thesis. The University of Sydney, Australia.

Martinez, A.S., Byrne, M. & Coleman, R.A. 2016. What and when to eat? Investigating the feeding habits of an intertidal herbivorous starfish. *Marine Biology* **163**, 1–13.

Mauzey, K.P., Birkelan.C & Dayton, P.K. 1968. Feeding behavior of asteroids and escape responses of their prey in Puget Sound region. *Ecology* **49**, 603–619.

Menge, B.A. & Sanford, E. 2013. Ecological role of sea stars from populations to meta-ecosystems, *in* J.M. Lawrence (ed.). *Starfish: Biology and Ecology of Asteroidea*. United States: The Johns Hopkins University Press, 67–80.

Morgan, M.B. & Cowles, D.L. 1997. The effects of temperature on the behaviour and physiology of *Phataria unifascialis* (Gray) (Echinodermata, Asteroidea) implications for the species' distribution in the Gulf of California, Mexico. *Journal of Experimental Marine Biology and Ecology* **208**, 13–27.

Mueller, B., Bos, A.R., Graf, G. & Gumanao, G.S. 2011. Size-specific locomotion rate and movement pattern of four common Indo-Pacific sea stars (Echinodermata; Asteroidea). *Aquatic Biology* **12**, 157–164.

Norton, T.A., Hawkins, S.J., Manley, N.L., Williams, G.A. & Watson, D.C. 1990. Scraping a living: a review of littorinid grazing. *Hydrobiologia* **193**, 117–138.

O'Loughlin, P.M. & Rowe, F.W.E. 2006. A systematic revision of the asterinid genus *Aquilonastra* O'Loughlin, 2004 (Echinodermata: Asteroidea). *Memoirs of Museum Victoria* **63**, 257–287.

Orth, R.J. & Vanmontfrans, J. 1987. Utilization of a seagrass meadow and tidal marsh creek by blue crabs *Callinectes sapidus*. 1. Seasonal and annual variations in abundance with emphasis on post-settlement juveniles. *Marine Ecology Progress Series* **41**, 283–294.

Paine, R.T. 1974. Intertidal community structure: experimental studies on relationship between a dominant competitor and its principal predator. *Oecologia* **15**, 93–120.

Paine, R.T. 1976. Size-limited predation: an observational and experimental approach with *Mytilus-Pisaster* interaction. *Ecology* **57**, 858–873.

Paul, V.J., Cruz-Rivera, E. & Thacker, R.W. 2001. Chemical mediation of macroalgal-herbivore interactions: ecological and evolutionary perspectives. *Marine Chemical Ecology*, 227–265.

Phillips, N.W. 1984. Role of different microbes and substrates as potential suppliers of specific, essential nutrients to marine detritivores. *Bulletin of Marine Science* **35**, 283–298.

Pillay, D., Branch, G.M. & Steyn, A. 2010. Unexpected effects of starfish grazing on sandflat communities following an outbreak. *Marine Ecology Progress Series* **398**, 173–182.

Polanowski, A. 2002. *The feeding behaviour, distribution and population genetics of the endangered sea star Patiriella vivipara*. Hons thesis. University of Tasmania, Australia.

Pompanon, F., Deagle, B.E., Symondson, W.O.C., Brown, D.S., Jarman, S.N. & Taberlet, P. 2012. Who is eating what: diet assessment using next generation sequencing. *Molecular Ecology* **21**, 1931–1950.

Poore, A.B., Gutow, L., F. Pantoja, J., Tala, F., Jofré Madariaga, D. & Thiel, M. 2014. Major consequences of minor damage: impacts of small grazers on fast-growing kelps. *Oecologia* **174**, 789–801.

Poore, A.G.B. 1994. Selective herbivory by amphipods inhabiting the brown alga *Zonaria angustata*. *Marine Ecology Progress Series* **107**, 113–123.

Poore, A.G.B., Campbell, A.H., Coleman, R.A., Edgar, G.J., Jormalainen, V., Reynolds, P.L., Sotka, E.E., Stachowicz, J.J., Taylor, R.B., Vanderklift, M.A. & Emmett Duffy, J. 2012. Global patterns in the impact of marine herbivores on benthic primary producers. *Ecology Letters* **15**, 912–922.

Porter, J.W. 1972. Predation by *Acanthaster* and its effect on coral species diversity. *The American Naturalist* **106**, 487–492.

Pratchett, M.S., Caballes, C.F., Rivera-Posada, J.A. & Sweatman, H.P.A. 2014. Limits to understanding and managing outbreaks of crown-of-thorns starfish (*Acanthaster* spp.). *Oceanography and Marine Biology: An Annual Review* **52**, 133–199.

Pratchett, M.S., Schenk, T.J., Baine, M., Syms, C. & Baird, A.H. 2009. Selective coral mortality associated with outbreaks of *Acanthaster planci* L. in Bootless Bay, Papua New Guinea. *Marine Environmental Research* **67**, 230–236.

Prestedge, G.K. 1998. The distribution and biology of *Patiriella vivipara* (Echinodermata: Asteroidea: Asterinidae) a sea star endemic to southeast Tasmania. *Records of the Australian Museum* **50**, 161–170.

Scheibling, R. & Lauzon-Guay, J.-S. 2007. Feeding aggregations of sea stars (*Asterias* spp. and *Henricia sanguinolenta*) associated with sea urchin (*Strongylocentrotus droebachiensis*) grazing fronts in Nova Scotia. *Marine Biology* **151**, 1175–1183.

Scheibling, R.E. 1979. *The ecology of Oreaster reticulatus (L.) (Echinodermata: Asteroidea) in the Caribbean*. PhD thesis. McGill University, Canada.

Scheibling, R.E. 1980a. Carbohydrases of the pyloric ceca of *Oreaster reticulatus* (L) (Echinodermata: Asteroidea). *Comparative Biochemistry and Physiology B-Biochemistry & Molecular Biology* **67**, 297–300.

Scheibling, R.E. 1980b. Dynamics and feeding activity of high-density aggregations of *Oreaster reticulatus* (Echinodermata: Asteroidea) in a sand patch habitat. *Marine Ecology Progress Series* **2**, 321–327.

Scheibling, R.E. 1980c. Homing movements of *Oreaster reticulatus* (L.) (Echinodermata: Asteroidea) when experimentally translocated from a sand patch habitat. *Marine Behaviour and Physiology* **7**, 213–223.

Scheibling, R.E. 1980d. The microphagous feeding behavior of *Oreaster reticulatus* (Echinodermata: Asteoridea). *Marine Behaviour and Physiology* **7**, 225–231.

Scheibling, R.E. 1981. Optimal foraging movements of *Oreaster reticulatus* (L.) (Echinodermata: Asteroidea). *Journal of Experimental Marine Biology and Ecology* **51**, 173–185.

Scheibling, R.E. 1982. Feeding habits of *Oreaster reticulatus* (Echinodermata: Asteroidea). *Bulletin of Marine Science* **32**, 504–510.

Scheibling, R.E. 2013. *Oreaster reticulatus*, *in* J.M. Lawrence (ed.). *Starfish: Biology and Ecology of Asteroidea*. United States: The Johns Hopkins University Press, 142–152.

Scheibling, R.E. & Metaxas, A. 2008. Abundance, spatial distribution, and size structure of the sea star *Protoreaster nodosus* in Palau, with notes on feeding and reproduction. *Bulletin of Marine Science* **82**, 221–235.

Shepherd, S.A. 1967. A review of the starfish genus *Nectria* (Asteroidea; Goniasteridae). *Records of the South Australian Museum* **15**, 463–482.

Shepherd, S.A. 1968. The shallow water echinoderm fauna of South Australia – Part I: the asteroids. *Records of the South Australian Museum* **15**, 729–756.

Shepherd, S.A. 2013. Echinoderms, *in* S.A. Shepherd & G.J. Edgar (eds). *Ecology of Australian Temperate Reefs: The Unique South*. Australia: CSIRO, 233–258.

Sloan, N.A. 1980. Aspects of the feeding biology of asteroids. *Oceanography and Marine Biology: An Annual Review* **18**, 57–124.

Sloan, N.A. & Robinson, S.M.C. 1983. Winter feeding by asteroids on a subtidal sandbed in British Columbia. *Ophelia* **22**, 125–141.

Steinman, A.D. 1996. Effects of grazers on freshwater benthic algae, *in* R.J. Stevenson et al. (eds). *Algal Ecology: Freshwater Benthic Ecosystem*. United States: Elsevier, 341–374.

Steneck, R.S. & Dethier, M.N. 1994. A functional group approach to the structure of algal-dominated communities. *Oikos* **69**, 476–498.

Steneck, R.S. & Watling, L. 1982. Feeding capabilities and limitation of herbivorous mollusks: a functional group approach. *Marine Biology* **68**, 299–319.

Stevenson, J.P. 1992. A possible modification of the distribution of the intertidal seastar *Patiriella exigua* (Lamarck) (Echinodermata: Asteroidea) by *Patiriella calcar* (Lamarck). *Journal of Experimental Marine Biology and Ecology* **155**, 41–54.

Thomassin, B.A. 1976. Feeding behaviour of the felt-, sponge-, and coral-feeder sea stars, mainly *Culcita schmideliana*. *Helgoländer wissenschaftliche Meeresuntersuchungen* **28**, 51–65.

Town, J.C. 1980. Movement, morphology, reproductive periodicity, and some factors affecting gonad production in the seastar *Astrostole scabra* (Hutton). *Journal of Experimental Marine Biology and Ecology* **44**, 111–132.

Underwood, A.J. 1978. Experimental evaluation of competition between three species of intertidal prosobranch gastropods. *Oecologia* **33**, 185–202.

Underwood, A.J. 1992. Competition and marine-plant interactions, *in* D.M. John & S.J. Hawkins (eds). *Plant-animal Interactions in the Marine Benthos*. United Kingdom: Oxford University, 443–475.

Underwood, A.J. & Denley, E.J. 1984. Paradigms, explanations and generalizations in models for the structure of intertidal communities on rocky shores, *in* D.R. Strong, Jr. et al. (eds). *Ecological Communities: Conceptual Issues and the Evidence*. United States: Blackwell Scientific, 151–180.

Underwood, A.J. & Jernakoff, P. 1981. Effects of interactions between algae and grazing gastropods on the structure of a low-shore intertidal algal community. *Oecologia* **48**, 221–233.

Uthicke, S., Schaffelke, B. & Byrne, M. 2009. A boom and bust phylum? Ecological and evolutionary consequences of large population density variations in echinoderms. *Ecological Monographs* **79**, 3–24.

Vergés, A., Alcoverro, T. & Ballesteros, E. 2009. Role of fish herbivory in structuring the vertical distribution of canopy algae *Cystoseira* spp. in the Mediterranean Sea. *Marine Ecology Progress Series* **375**, 1–11.

Vergés, A., Steinberg, P.D., Hay, M.E., Poore, A.G.B., Campbell, A.H., Ballesteros, E., Heck, K.L., Booth, D.J., Coleman, M.A., Feary, D.A., Figueira, W., Langlois, T., Marzinelli, E.M., Mizerek, T., Mumby, P.J., Nakamura, Y., Roughan, M., van Sebille, E., Gupta, A.S., Smale, D.A., Tomas, F., Wernberg, T. & Wilson, S.K. 2014a. The tropicalization of temperate marine ecosystems: climate-mediated changes in herbivory and community phase shifts. *Proceedings of the Royal Society B: Biological Sciences* **281**, doi:10.1098/rspb.2014.0846.

Vergés, A., Tomas, F., Cebrian, E., Ballesteros, E., Kizilkaya, Z., Dendrinos, P., Karamanlidis, A.A., Spiegel, D. & Sala, E. 2014b. Tropical rabbitfish and the deforestation of a warming temperate sea. *Journal of Ecology* **102**, 1518–1527.

Wessels, H., Karsten, U., Wiencke, C. & Hagen, W. 2012. On the potential of fatty acids as trophic markers in Arctic grazers: feeding experiments with sea urchins and amphipods fed nine diets of macroalgae. *Polar Biology* **35**, 555–565.

Williams, A.H., Coen, L.D. & Stoelting, M.S. 1990. Seasonal abundance, distribution, and habitat selection of juvenile *Callinectes sapidus* (Rathbun) in the northern Gulf of Mexico. *Journal of Experimental Marine Biology and Ecology* **137**, 165–183.

Wilson, C. 2004. *Population ecology and nocturnal foraging patterns of the seastar Paritriella parvivipara (Keough & Dartnall) (Echinodermata: Asteroidea: Asterinidae)*. Hons thesis. University of Adelaide, Australia.

Wolcott, D.L. & O'Connor, N.J. 1992. Herbivory in crabs: adaptations and ecological considerations. *American Zoologist* **32**, 370–381.

Wotton, R.S. 2004. The essential role of exopolymers (EPS) in aquatic systems. *Oceanography and Marine Biology: An Annual Review* **42**, 57–94.

Wulff, J.L. 1995. Sponge-feeding by the Caribbean starfish *Oreaster reticulatus*. *Marine Biology* **123**, 313–325.

Yamaguchi, M. 1973. Early life histories of coral reef asteroids, with special reference to *Acanthaster planci* (L.), *in* O.A. Jones & R. Endean (eds). *Biology and Geology of Coral Reefs*. United States: Academic Press, 369–387.

Yamaguchi, M. 1974. Growth of juvenile *Acanthaster planci* (L.) in the laboratory. *Pacific Science* **28**, 123–138.

Yamaguchi, M. 1975. Coral-reef asteroids of Guam. *Biotropica* **7**, 12–23.

Yamaguchi, M. 1977. Estimating the length of the exponential growth phase: growth increment observations on the coral-reef asteroid *Culcita novaeguineae*. *Marine Biology* **39**, 57–59.

Yamaguchi, M. 1977. Population structure, spawning, and growth of the coral reef asteroid *Linckia laevigata* (Linnaeus). *Pacific Science* **31**, 13–30.

Appendix 1   Current accepted name of species cited in the review and previous species name that may be found in the references cited

Family	Current species name	Previous species name
Asterinidae	*Aquilonastra anomala* (H.L. Clark, 1921)	*Asterina anomala*
	Asterina stellifera (Möbius, 1859)	*Patiria stellifera*
	Cryptasterina pacifica (Hayashi, 1977)	*Patiriella pseudoexigua pacifica*
	Cryptasterina hystera (Dartnall & Byrne, 2003)	*Patiriella pseudoexigua*
	Cryptasterina pentagona (Muller & Troschel, 1842)	*Patiriella pseudoexigua pseudoexigua, Patiriella obscura*
	Meridiastra calcar (Lamarck, 1816)	*Patiriella calcar*
	Meridiastra gunnii (Gray, 1840)	*Patiriella brevispina*
	Parvulastra parvivipara (Keough & Dartnall, 1978)	*Patiriella parvivipara*
	Parvulastra exigua (Lamarck, 1816)	*Patiriella exigua*
	Parvulastra vivipara (Dartnall, 1969)	*Patiriella vivipara*
	Patiria miniata (Brandt, 1835)	*Asterina miniata*
	Patiria pectinifera (Muller & Troschel, 1842)	*Asterina pectinifera*
Ophidiasteridae	*Ophidiaster cribrarius* (Lütken, 1871)	*Ophidiaster robillardi*
	Ophidiaster hemprichi (Müller & Troschel, 1842)	*Ophidiaster squameus*
Oreasteridae	*Pentaceraster cumingi* (Gray, 1840)	*Oreaster occidentalis*

Oceanography and Marine Biology: An Annual Review, 2017, **55**, 35-54
© S. J. Hawkins, D. J. Hughes, I. P. Smith, A. C. Dale, L. B. Firth, and A. J. Evans, Editors
Taylor & Francis

INTERTIDAL BOULDER-FIELDS: A MUCH NEGLECTED, BUT ECOLOGICALLY IMPORTANT, INTERTIDAL HABITAT

M.G. CHAPMAN

School of Life and Environmental Sciences, University of Sydney, NSW 2006, Australia
E-mail: gee.chapman@sydney.edu.au

Intertidal boulder-fields form important habitat because they support a wide range of fauna, many species of which are rare or not found in other habitats. Over many decades, they have been studied to test general ecological theories, such as effects of disturbance, succession and species-area relationships, because of the ease with which the habitats themselves and their biota can be manipulated experimentally. Much less is known about their unique characteristics and biota. Experimental studies are limited to few parts of the world, primarily south-eastern Australia. Intertidal boulder-fields support a diverse biota of often related rare and common species, which typically show extremely aggregated patterns of dispersion, but little consistent association among different species. This review summarizes experimental research in these habitats, which has led to current understanding of factors affecting spatial and temporal patterns of species living in boulder-fields, in addition to the contributions that have been made to broader ecological concepts. Finally, directions for future research are suggested, particularly with respect to conservation and restoration of intertidal boulder-fields in a rapidly changing world.

Introduction

Intertidal boulder-fields are important habitats worldwide. They can occur overlying soft sediments and, in places, grade into cobble shores. Elsewhere they lie adjacent to rocky platforms. Boulder-fields can be extensive (kilometres long), or only occupy tens of metres of shoreline. They can occur on shores that are very exposed or those that are very sheltered, and from highshore levels down to low on the shore where they extend into subtidal areas. They can be natural areas or created by anthropogenic activities, either intentionally to create new habitat (Chapman 2012), or unintentionally through the disposal of waste (Chapman 2006).

The boulders themselves are used as habitats by a mix of widespread, common marine invertebrates and algae and by a suite of specialist, rare species, particularly invertebrates (Chapman 2002a, 2005). Nevertheless, boulder-fields have not been extensively studied by ecologists. A preliminary search of Web of Science 2005–2015 for the numbers of published papers describing research in intertidal habitats gave more than 46,000 records. Of these, fewer than 0.1% described research in boulder-fields (Figure 1) and not all of these were ecological studies. The ecology of this habitat is thus poorly described compared to many other intertidal habitats. Boulder-fields are susceptible to many natural and anthropogenic disturbances, including wave action (Lieberman et al. 1979, Sousa 1979a), burial by sand (McGuinness 1987a,b, Kurihara et al. 2001) and harvesting (Cryer et al. 1987, Addessi 1994). Many of these perturbations are expected to increase with both climatic change and urbanization of the coast.

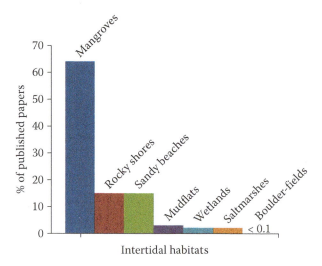

Figure 1   The proportion of more than 46,000 scientific publications between 2005 and 2015 categorized as reporting research in different intertidal habitats.

Boulder-fields contain at least four types of microhabitats, three of which are unique to this habitat. The exposed tops and sides of intertidal boulders have many biota in common with intertidal rock platforms; although the boulders are small isolated patches of habitat, rather than an extensive patch of similar habitat (Figure 2A). The undersurfaces of the boulders, the rocky or sedimentary substratum lying immediately below a boulder and the crevices created between adjacent boulders are, however, microhabitats unique to these habitats, which are composed of movable, isolated units. They tend to support a very different biota to those living on their upper surfaces (Figure 2B). The number of unique microhabitats in boulder-fields has been extended to eight by some authors (Le Hir & Hily 2005), and are distinguished by subtle differences in the degree of shelter offered by the boulders. Many of these additional microhabitats are occupied by very similar biota.

Because boulders are natural patches of isolated habitat, which can be relatively easily manipulated in the field, they have proved ideal units for experimental manipulations to test general ecological ideas. Here, I briefly review the more important early studies in boulder-fields, where boulders were used primarily to investigate general ecological theories, often in conjunction with or parallel to similar studies on rocky shores. Their uniqueness, in being composed of isolated patches of habitat of different size and structure, which could be experimentally manipulated in the field, added considerable insight to current knowledge. I then follow this by describing studies of the general ecology of intertidal boulder-fields themselves, where the primary focus has been on understanding how their ecology differs from that of rocky shores and how we can best ensure the persistence of these habitats in a rapidly changing world.

A brief history of ecological studies in boulder-fields

Using boulder-fields to develop general ecological theories

Many of the early ecological studies in boulder-fields used boulders as natural units of intertidal habitat to test models about general ecological processes, such as the establishment and succession of faunal (Osman 1977) or algal (Sousa 1979b) assemblages. Especially important were investigations into the role of physical disturbance, due to the vulnerability of boulders to movement and overturning, in resetting an assemblage to an earlier successional stage (Sousa 1979a). Sousa (1979b) demonstrated a successional sequence from quick-growing, rapidly-reproducing species,

Figure 2 (A) Diverse assemblages on the tops of boulders resemble those living on rocky shores, whereas those species living under boulders (B) are not generally found together in other intertidal habitats. (C) Many of the species are very overdispersed, being crowded onto relatively few of the available boulders. (Photos: M.G. Chapman)

such as the green alga, *Ulva lactuca*, and barnacles, which were gradually replaced by a range of green and red algae, before the boulders became dominated by the red alga, *Chondracanthus canaliculatus* (=*Gigartina canaliculata*). Despite this general pattern, there was considerable local variation, depending on the season and the size of area (created on the substratum by various forms and intensities of disturbance) available for colonization at different successional stages. The successional stage found on a boulder was related to its size, with early successional species common on small, easily-disturbed boulders and large boulders dominated by *Chondracanthus canaliculatus*. Medium-sized boulders supported a range of algae. Although each small boulder only supported a few species, a set of such boulders could support more species than a similar set of larger stable boulders, because each small boulder was not necessarily occupied by the same species, whereas large boulders tended each to be dominated by the same set of few species (Sousa 1979a; but see Douglas & Lake 1994 for different patterns on boulders in streams). This work was very important in providing experimental evidence for the 'Intermediate Disturbance Hypothesis' (Connell 1978).

A few years later, McGuinness (1984) used boulders of different sizes to investigate hypotheses derived from four models that were currently being used to explain species-area relationships. Using boulders allowed him to examine abundances of species of marine organisms across natural patches of habitat of different sizes. This was in contrast to studies that examined accumulation of species when patches of similar sizes of a shore were summed to create larger areas (Hawkins & Hartnoll

1980). McGuinness (1984) also tested these hypotheses using the mobile and sessile components of assemblages on boulders at different shore levels and in different sites to test for generality of any patterns. Each of the models examined (Random Placement, Intermediate Disturbance, Habitat Diversity, Equilibrium Theory) could explain the number of species on boulders of different sizes and relative to similarly-sized patches of adjacent rock platforms. This, however, depended on the site, shore height and the assemblage examined, thus emphasizing the danger of generalizing ecological concepts from limited data. This contrasts with boulders in rivers, where small boulders were always depauperate, even when their areas were summed to equal that of larger boulders (Douglas & Lake 1994, Downes et al. 1998).

In some terrestrial habitats, boulders have been considered inappropriate to test theories such as island biogeography, because of small-scale disturbances interrupting succession (Kimmerer & Driscoll 2000). Similar patchy disturbances may have contributed to the variable results described by McGuinness (1984), but spatially-variable disturbances are natural features of most habitats. In a smaller study, Londoño-Cruz & Tokeshi (2007) showed a linear relationship between species diversity of molluscs and sizes of boulders, but the abundance-area relationships were dominated by very large abundances on only a few large boulders. It is likely, however, that had this study been replicated under many conditions, as was done by McGuinness (1984), then similar spatial variability in the patterns would have been found.

This research was later extended by McGuinness & Underwood (1986) in their experimental study of the importance of habitat complexity in determining diversity of species on boulders, standardizing habitat by using concrete blocks of different thickness with added pits and/or grooves in place of natural boulders (see also Douglas & Lake 1994, Downes et al. 1998 for similar studies in streams). McGuinness & Underwood (1986) demonstrated considerable spatial variability in patterns of colonization and increased habitat complexity did not consistently lead to increased diversity as predicted by theory. Its importance varied with the types of species examined (algae, sessile fauna, mobile fauna), the height on the shore (which influences the types and intensities of disturbances) and individual sites in complex and unpredictable ways.

Because of the sparse distribution of many taxa living under boulders and the common pattern of extreme overdispersion shown by both rare and common taxa (e.g. chitons, Figure 2C), it can be necessary (depending on local conditions) to sample a large number of boulders to measure temporal and spatial patterns with any degree of precision (Chapman 2002a, 2005). Such data can be used to test more general ideas. Such a dataset of sessile and mobile animals on 100 boulders from each of six locations in the Sydney region of New South Wales (NSW), Australia, allowed a powerful evaluation of the accuracy and precision of six non-parametric methods and five regression models used for deriving species estimates from samples (Chapman & Underwood 2009). This extensive study (which also included samples of species living on rock platforms and small invertebrates colonizing artificial units of habitat), tested how well random samples of different sizes estimated the known number of species from all samples combined. Analyses showed that many of the commonly-used methods estimated the species number very poorly (Table 1). In addition, the accuracy (how close was the predicted total number of species to the true measured number) and the precision (what percentage of the estimated values lay within 10% of the measured value) differed among the different metrics according to site, the sample size and whether one was examining the sessile or mobile component of the assemblage on the boulders (details in Chapman & Underwood 2009). This comparison raises questions about using any single metric to estimate the number of species in an area from a single sample, especially in cases with small replication.

Using the same extensive dataset, Chapman et al. (2009) developed indices for ranking sites for conservation, under conditions where the objective is to choose which sites to conserve among a number of sites for which there are quantitative data on abundances of organisms. Individual species are given scores for each site, according to the proportion of sites that each species is found in, the proportion of patches of habitat (or samples) in which it is found in each site and the proportion

Table 1 Total number of species on 100 boulders in six sites (LR, LB, CB, CH, BH, RB), the mean number of species estimated for 100 boulders from 500 random samples, each of 20 boulders and (in parentheses) the percentage of the 500 estimates that were within 10% of the true value for mobile and sessile species

Site	Total spp.	Non-parametric estimators					Regression models					
		HF	SB	J_2	SJ_1	SJ_2	Chao 2	Power	Exp.	Neg. Exp.	Asymp.	Rational
Mobile species												
LR	53	44 (26)	37 (<1)	51 (43)	35 (<1)	48 (36)	50 (24)	88 (<1)	44 (30)	33 (0)	36 (2)	44 (25)
LB	54	45 (30)	39 (<1)	53 (43)	36 (0)	47 (34)	55 (28)	79 (4)	47 (46)	33 (0)	36 (<1)	43 (19)
CB	48	39 (22)	33 (<1)	45 (44)	32 (<1)	40 (23)	45 (25)	77 (<1)	42 (38)	30 (0)	32 (<1)	38 (17)
CH	47	35 (11)	28 (<1)	42 (29)	28 (<1)	38 (21)	49 (21)	68 (10)	34 (7)	26 (<1)	28 (3)	34 (11)
BH	34	23 (7)	19 (0)	29 (30)	20 (2)	29 (20)	35 (12)	53 (5)	21 (1)	20 (3)	22 (8)	13 (<1)
RB	35	26 (13)	21 (0)	32 (34)	22 (3)	30 (24)	34 (16)	60 <1	25 (5)	22 (4)	23 (8)	15 (2)
Sessile species												
LR	49	46 (62)	41 (17)	49 (49)	38 (2)	46 (55)	48 (39)	75 (0)	51 (71)	36 (0)	37 (<1)	44 (45)
LB	36	32 (45)	27 (6)	36 (45)	25 (<1)	34 (44)	37 (30)	48 (8)	33 (56)	23 (0)	26 (5)	31 (36)
CB	51	49 (70)	44 (30)	52 (49)	41 (4)	49 (63)	49 (44)	81 (0)	56 (57)	38 (0)	41 (4)	48 (69)
CH	52	50 (72)	45 (31)	53 (61)	43 (4)	48 (56)	50 (47)	75 (0)	57 (56)	39 (0)	42 (2)	47 (59)
BH	44	41 (60)	37 (16)	44 (48)	34 (1)	39 (38)	43 #$	62 (<1)	45 (76)	31 (0)	34 (2)	39 (36)
RB	65	60 (64)	55 (14)	63 (65)	52 (2)	57 (31)	61 (43)	94 (0)	70 (62)	48 (0)	51 (0)	57 (31)

Note: Estimates were made from six non-parametric estimates (HF, SB, J_2, SJ_1, SJ_2 and Chao 2) and five regression models (Power, Exponential, Negative exponential, Asymptotic and Rational). Details of the metrics used, sites and methods are in Chapman and Underwood (2009).

of individuals in each site. This information is used to create an index which identifies those species with very limited distributions within and among sites and, when summed across species, identifies those sites containing a large proportion of such species. Although developed for assemblages in boulder-fields, the results are applicable to any conservation issue where sites need to be ranked, using quantitative data from diverse assemblages.

There have thus been many studies in boulder-fields in which the boulders simply represented intertidal habitat, with no questions about their features as boulders (e.g. testing interactions among grazers, barnacles and algae; Van Tamelen 1987). This has contributed to and extended the well-documented early research on rocky shores in the UK (Hawkins & Hartnoll 1982, Hawkins 1983), Australia (Underwood & Jernakoff 1981, Underwood 1984), South Africa (Branch 1975) and in the USA (Dayton 1971, Menge 1976), increasing our understanding of the complexity of interactions among rocky shore assemblages. Green & Crowe (2014) examined the effects of the invasive oyster, *Magallana gigas* (=*Crassostrea gigas*), on intertidal biodiversity, using manipulations of oysters on individual boulders. The questions being addressed, however, regarded the invasive oyster, not the boulder habitat *per se*. Similarly, Altieri et al. (2010) used intertidal cobble beaches that were colonized by cordgrass and mussels to investigate relationships between abiotic stress, species diversity and invasibility.

Understanding the ecology of the biota living under boulders

The upper surfaces of intertidal boulders tend to have much bare space or be dominated by common algae or sessile invertebrates depending on the exposure of the site and the sizes of the boulders. In contrast, the undersurfaces tend to be dominated by encrusting algae and sessile animals, including sponges, bryozoans and tube worms (described by McGuinness 1987a,b, Chapman 2002a, 2005) (Figure 2B). Even in extremely harsh environments, where the tops of boulders are bare, the undersurfaces support a few common epifauna, particularly bryozoans and polychaetes (Barnes et al. 1996, Kuklinski et al. 2006, Waller 2012). Where boulders are colonized by species that provide biogenic habitat (e.g. plants or sessile species, such as mussels), species diversity is increased in such harsh environments, due to amelioration of these environmental stresses (Altieri et al. 2010).

In more benign environments, the undersurfaces of boulders and any rocky substratum on which they rest, support a diverse sessile assemblage of invertebrates and many mobile species, particularly molluscs and echinoderms. Many of these are widespread species, but others appear to be habitat-specialists and can be relatively rare under boulders and even sparser in other habitats (McGuinness 1987a, Chapman 2002a, 2005). Chitons are particularly speciose and abundant, most notably species of the genus *Ischnochiton* (Kangas & Shepherd 1984, Grayson & Chapman 2004). McGuinness (1987a) showed that although the abundances of individual species were sometimes related to size of the boulder, patterns were very variable, with increases in abundance with size on one shore (NSW, Australia), but the opposite pattern found on another nearby shore. This result was, however, confounded by the rock-type differing between the two shores. Most studies of biota living under boulders in NSW have shown no relationship between either abundances of individual species or numbers of species with the size of the boulder (Smith & Otway 1997, Chapman 2002a, 2005, Grayson & Chapman 2004, Palmer 2012).

An obvious characteristic of many species living under boulders is their extreme overdispersion at a number of spatial scales. In any location, many boulders are unoccupied by any particular species and many individuals of that species are crowded on to few boulders (described by Smith & Otway 1997, Chapman 2002a, Grayson & Chapman 2004, Smoothey & Chapman 2007) (Figures 2C and 3). Nevertheless, very few boulders are completely devoid of species (as few as 5%; Chapman 2005) in many locations and the mix of species co-existing on boulders in a location is very variable (shown for a sample of boulders from Cape Banks, NSW, Australia in Figure 4), indicating that many more boulders may be suitable habitat than are actually occupied by any particular species. Up to 90% of the spatial variation at multiple scales, from individual boulders to locations hundreds of kilometres apart, is found at the scale of metres, among individual boulders. Assemblages in adjacent sites in one location can differ more than assemblages that are kilometres apart in different boulder-fields (Chapman 2005). In addition, the species that most contribute to variability among boulders also vary spatially, with different species showing quite different degrees of overdispersion in different places, in quite unpredictable ways.

Palmer (2012) compared the spatial patterns of a brooding chiton and one which was a broadcast fertilizer in a number of boulder-fields in NSW, Australia, expecting that the brooding species would be more aggregated than the broadcast fertilizer. Each species was overdispersed, but the brooder was not more aggregated and did not show greater genetic relationships among animals under individual boulders than did the broadcast fertilizer. In fact, there was a very large amount of genetic variability among individuals for both breeding types, under individual boulders and within patches of a boulder-field. This suggests large amounts of dispersal among boulders by both species, probably as small juveniles or adults, despite their apparent 'preference' for certain boulders.

Patterns of overdispersion are established very early in the development of an assemblage. Chapman (2002b) examined colonization of newly quarried rocks (thus, with no biofilm, which might act as an attractant for some species) placed on a sandy or algal/rocky substratum in replicate areas. Many species that rapidly colonized these boulders—68 taxa within five days—are generally only seen under boulders or in biogenic habitat and not on the substratum among the boulders (e.g. the chitons, *Ischnochiton australis* and *I. smaragdinus*), although other early colonizers were common and widespread species (e.g. the gastropod, *Austrocochlea porcata*). From relatively early in the experiment, however, common and rare species were each aggregated on to some of the new boulders, but all species did not colonize the same subset of boulders. Therefore, nearly all boulders had some species aggregated under them, with different species colonizing a different subset of boulders. It was never clear how or why they located those particular boulders, why they moved on

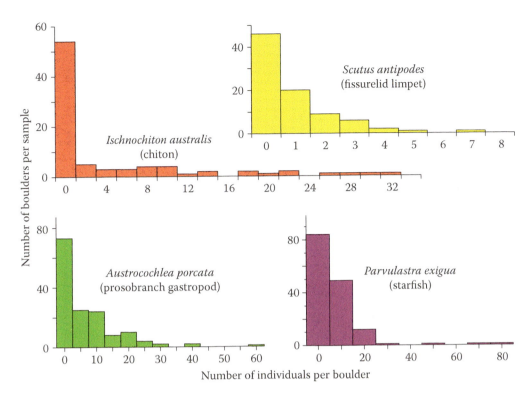

Figure 3 Overdispersion among boulders illustrated for a species of chiton (*Ischnochiton australis*), fissurelid limpet (*Scutus antipodes*), starfish (*Parvulastra exigua*) and prosobranch gastropod (*Austrocochlea porcata*) sampled in NSW, Australia. Most boulders are not occupied by any particular species, with many individuals crowded onto a few boulders, which differ among the different species. (Chapman, unpublished data.)

to the novel habitat, nor the mechanisms by which they dispersed among boulders. Some species (e.g. *Ischnochiton australis* and the gastropod, *Stomatella impertusa*) detach from the surfaces of boulders into the water column when the boulders are disturbed. They are rapidly washed away with the moving water, although *S. impertusa* appears to readily adhere to new substratum that it encounters while carried in the water column (G. Chapman, pers. obs).

Aggregation also occurs at scales larger than among individual boulders. This may occur because different species tend to live at different shore heights (e.g. porcelain crabs, Emparanza 2007; chitons, Palmer 2012), but patchiness is often not related to height, with large and, so far, unexplained variability in assemblages among patches of apparently similar boulders at the same height (Chapman 2005). At a smaller scale, different species distribute themselves non-randomly on individual boulders. For example, in Australia and South Africa, chitons were more common near the edges than the centres of undersurfaces of boulders lying over fine sediments (Liversage et al. 2012), and the distribution of coralline crusts under boulders depends on their shape (Liversage 2016). Experiments in which sediments under the boulders were altered showed increased movement of chitons towards the edges when boulders were placed on fine sediment, perhaps because the centres of boulders tended to bury more deeply in fine than in coarse sediment. Choi & Ginsberg (1983) similarly showed greater densities of coelobites towards the edges of coral rubble rather than towards the middle of the undersurfaces.

Temporal variation of assemblages in boulder-fields is also very difficult to explain, because there are often no clear seasonal or annual patterns of abundance nor recruitment, with both short- and long-term temporal variability in numbers of most species interacting with numerous spatial

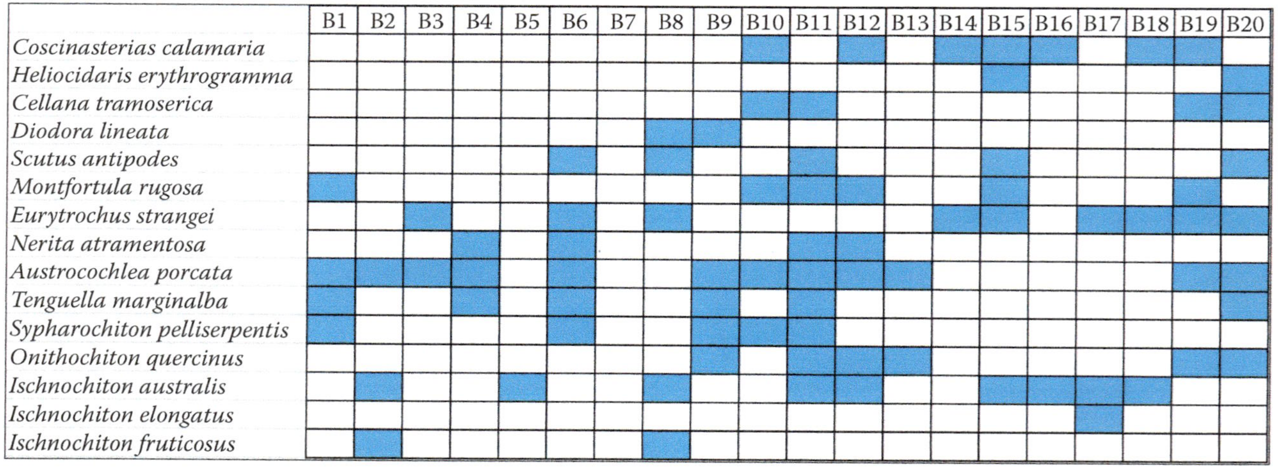

Figure 4 Different combinations of species live on different boulders, with very little pattern of general association among species; data from a subset of species on 20 boulders sampled at Cape Banks, NSW, Australia. (Chapman, unpublished data.)

scales, from metres to kilometres (Chapman 2005). In areas where seasons are more extreme, there may, however, be seasonal patterns in abundance which are more predictable (Gianguzza et al. 2013).

Rarely studied microhabitats in boulder-fields

Some specific microhabitats found in boulder-fields have been little studied. In the spaces between adjacent boulders, crevices can be used as refuges by invertebrates, such as sea urchins (Smoothey & Chapman 2007) or fish, many of which are quite rare (Kovačič et al. 2012). Female blue-ringed octopuses, *Hapalochlaena lunulata*, lay egg masses in the crevices among boulders, which they brood until hatching (G. Chapman, pers. obs.). Similarly, freshwater crayfish, *Cambarus chasmodactylus*, maintain territories in such crevices and under boulders in rivers in Florida, USA (Loughman et al. 2013).

The presence of boulders may also affect species living in soft-sedimentary intertidal habitats, for example in saltmarshes, where patches of boulders protect rare forbs from wave action (Bruno 2000), or in mangroves where boulders provide hard substrata for settlement and grazing (Underwood & Barrett 1990). Where boulders overlay sediment, they lead to altered grain-size and increased amounts of organic matter in the sediment (Cruz-Motta et al. 2003). Either these changes or other factors associated with the presence of boulders can affect the infauna in the sediment directly under the boulders. Again, such influences are spatially variable, although they can be strongly influenced by the state of the tide (Cruz-Motta 2005).

Factors affecting diversity and abundances of biota on boulders

Abiotic factors

Disturbances

Individual boulders are vulnerable to numerous disturbances, both anthropogenic and natural. Small boulders in dynamic environments are overturned by wave action (Osman 1977, Sousa 1979a, McGuinness 1987a,b), or scoured (Littler & Littler 1984) or buried by sediment (McGuinness 1987a). These boulders support few species compared to more stable boulders. Sousa (1979a) showed experimentally, by stabilizing small boulders, that the numbers of species occupying boulders was directly related to the frequency and extent of disturbance. McGuinness (1987b) similarly used experimentally-stabilized or buried boulders of different sizes to illustrate the effects of disturbance on species diversity. In contrast to Sousa (1979a), he showed no effects of the experimental treatments on algal diversity on the tops of boulders, probably because his sites were comparatively undisturbed and/or grazers had strong effects on diversity of algae in some areas. These disturbances did, however, affect some species living under the boulders. In the absence of such disturbances, assemblages under boulders were dominated by ascidians and sponges, which readily overgrow other taxa.

In some areas, boulders are overturned by humans while harvesting (Cryer et al. 1987), which can kill the biota on the upper and lower surfaces (Addessi 1994) and reduces abundances of harvested species, such as urchins and crabs. Harvested species can, but do not always, increase in areas with intermediate amounts of such disturbance (Addessi 1994), although crabs can rapidly move on to boulders from which individuals have been removed by harvesters (Cryer et al. 1987). If boulders are not left overturned, but are replaced in their normal orientation, then as long as disturbances are not too frequent, both mobile and sessile assemblages are likely to be unaltered by the disturbance (Chapman & Underwood 1996). When boulders are overturned on sequential days, however, even if they are replaced each time, abundances of many taxa are likely to decrease as mobile species apparently move away from the disturbed area. This is an important consideration when sampling organisms that live under boulders, because boulders must be overturned before the undersurfaces can be sampled. Nevertheless, this form of disturbance has not received a lot of attention from ecologists working in this habitat (but see Chapman & Underwood 1996).

There are also potential cascading effects of harvesting on other species living under boulders. For example, in NSW, Australia, people frequently harvest sea urchins from under boulders (G. Chapman, pers. obs.). The chiton, *Ischnochiton australis*, is very strongly associated with urchins, being more abundant on boulders with urchins and, on the surface of the boulder, being crowded into the areas under the spines of the urchins (Chapman & Smoothey 2014). Experiments showed that the chitons left boulders from which urchins had been removed at a faster rate than from boulders with urchins, but removal of chitons did not affect numbers of urchins. This suggests that urchins provide important habitat for the chitons, but it is not known what particular resources they provide, considering that the boulder itself appears to protect the chitons from predation or strong water movement. Whatever these resources are, they appear to be removed by removal of urchins.

Features of the boulders or their immediate surroundings

In experiments to try to identify features of the habitat that affect dispersal of animals on to and among boulders, most species that colonized newly quarried boulders aggregated rapidly on some boulders and not others, despite the lack of a sessile assemblage that could influence dispersal (Chapman 2002b). Overdispersion for most species was established very early—within days. These patterns were mainly due to dispersal, not settlement, because most colonizers were not new recruits. Later work compared colonization on to denuded natural boulders or sandstone blocks (which provided reduced complexity of habitat). The blocks had either no or one of two ages of established sessile assemblages (thus providing different levels of biotic complexity). These treatments were either placed adjacent to or away from a natural boulder, with the latter treatments placed in a position from which a boulder was removed, or in a position where there was no natural boulder (potentially influencing proximity to a pool of potential colonizers). Colonization was measured after the first week to nine months after deployment of the treatments. There was rapid colonization of all habitats, with no effects of either features of boulder, nor its position (Chapman 2003a) and assemblages on all habitats converged within six months (development of the mobile assemblage shown as trends in the centroids calculated at each time of sampling for selected treatments in Figure 5). Again, most species were overdispersed, with the detailed patterns showing no similarity among species. Later research showed that neither size of sandstone block, time of deployment, nor length of deployment showed clear effects on animals colonizing the experimental units, although there was some suggestion that the position in the boulder-field may be more important than the habitat itself (Chapman 2007).

The most obvious feature of a boulder that would be expected to affect diversity of species occupying it is its size, but generally, neither abundances of individual taxa nor numbers of species appear related to size of the boulder (Smith & Otway 1997, Chapman 2002a, 2005, Grayson & Chapman 2004), or results are very spatially variable (McGuinness 1984, 1987a,b). In contrast to much of the unexplained variation, experiments have shown that rock-type may be important (McGuinness & Underwood 1986, Green et al. 2012), but it cannot explain patterns of distribution for all species. Liversage & Benkendorff (2013) sampled limestone and basalt boulder-fields and showed species were most abundant in areas with basalt shores and that some species (e.g. *Ischnochiton australis*) were never found on limestone boulders. Patterns of overdispersion for some species also varied with rock-type, but this was not consistent among locations for the different species.

In addition, neither abundances nor diversity may be related to the characteristics of the substratum on which the boulders lie (e.g. the grain-size of the sediments; Smith & Otway 1997), although in a comparative study between Australia and South Africa, Liversage et al. (2012) showed more chitons under boulders lying on fine than on coarse sediment. Changing the sediment showed more immigration to boulders lying on fine sediment, but densities were not controlled in the experimental treatments, so the results were confounded. Chapman (2002b) also showed differences in developing

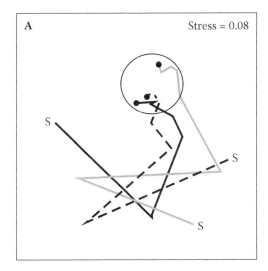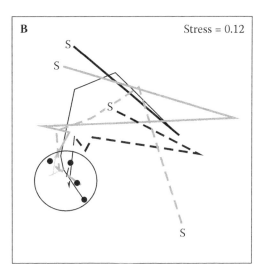

Figure 5 Convergence of the mobile assemblage over nine months from one week after plates and boulders were deployed at the start of the experiment (S). Trends are shown by the centroids calculated from independent samples of each treatment at each time of sampling. (A) Artificial plates with no biofilm, placed at the start of the experiment adjacent to an existing boulder (black dashed line), at least 1 m away from an existing boulder (gray solid line) or in a position where there was originally a natural boulder (black solid line). (B) Treatment placed adjacent to an existing boulder; samples of new plates with no biofilm (gray dashed line), plates with a six-month-old sessile assemblage (black solid line), plates with a nine-month-old sessile assemblage (gray solid line), denuded natural boulder with existing sessile assemblage (black dashed line). (From Chapman 2003a.)

assemblages depending on whether the underlying substratum was sand or algal/rocky, but patterns varied between replicate areas and few general patterns could be identified (Chapman 2003a).

Combinations of abiotic stresses

Cobbles on semi-sheltered beaches in New England, USA, can be colonized by large areas of cord grass and mussels. Experiments have shown that the former provide shade and the latter stabilize the cobbles (Altieri et al. 2010). Both of these mechanisms ameliorate abiotic stresses, increasing the number of species and abundances of the invasive crab, *Hemigrapsus sanguineus*, associated with the cobble beds (Altieri et al. 2010). Similarly, disturbances due to wave action or water movement can vary with the depth of the water or distance off shore, altering the size-range of subtidal boulders that are frequently overturned (Osman 1977). Intertidally, the small-scale configuration of the shoreline and presence of offshore shallow reefs similarly reduce disturbance on intertidal boulders, although these reductions do not occur in very stormy conditions (G. Chapman, pers. obs.). Exposure to waves has been shown to influence disturbance to intertidal coral rubble, directly in relationship to size of the boulders and the height on the shore (Walker et al. 2008). This did not directly relate to the numbers of species under the boulders, which, it was suggested, may also be affected by competition. Experiments are needed, however, to unravel the complexity of interacting abiotic factors, such as height on the shore, wave exposure and size of boulders on the richness of species. Such experiments are still, unfortunately, relatively rare.

Biotic factors

Recruitment

Initial patterns of recruitment can be very important in determining later assemblages on boulders (Osman 1977). Recruitment can be affected by the composition of the rock, as shown by

McGuinness & Underwood (1986), who used a reciprocal transplant of sandstone or shale rocks between two shores to test the hypotheses that differences in the distribution of the alga, *Ulva lactuca*, and spirorbid polychaetes between shores could be explained by recruitment in response to rock-type. As predicted, the alga colonized more onto sandstone than onto shale boulders, with spirorbids showing the opposite pattern. At a larger scale, Green et al. (2012) showed more foliose algae colonizing sandstone boulders and more barnacles colonizing granite boulders in artificial boulder-fields created at the bases of sea walls.

An experimental extension of the research by McGuinness & Underwood (1986), where upper and lower surfaces of the two types of rocks were painted different colours to match different rock-types, showed that spirorbids recruited in response to a dark lower surface, irrespective of the type of rock or the colour of the upper surface (James & Underwood 1994). Colour alone was not, however, adequate to explain all of the variation and on natural boulders, both the upper and lower surfaces tend to be multicoloured, depending on the sessile assemblages that they support. Recruitment in response to rock-type could not explain patterns of distribution in the coexisting common polychaete, *Galeolaria caespitosa* (McGuinness 1988).

Competition

There have been few studies of competition in assemblages on or under boulders, largely because, other than on the tops of stable boulders (Sousa 1979b), there is often considerable free space (McGuinness 1987a,b). Despite available free space, species may, however, be crowded into very small areas of the boulders, as described above for chitons under boulders lying on fine sediments (Liversage et al. 2012). Barnes & Lehane (2001), however, showed strong competitive interactions, including overgrowth, by numerous sessile species of invertebrates under boulders on South Atlantic islands, where the fauna are crowded into the central areas of the undersurfaces. There was, thus, strong competition for space, despite most of the boulders being unoccupied and abundant space being available.

Extreme aggregation of many mobile species among individual boulders (Smith & Otway 1997, Chapman 2002a, 2005) and overdispersion on the undersurfaces of individual boulders themselves (Chapman & Smoothey 2014) suggests that space under boulders may be limiting for attachment and feeding by mobile species, even when many boulders in the vicinity are sparsely or unoccupied. The chiton, *Ischnochiton australis*, can occur in large abundances under some boulders and are often lying in layers on top of each other (Palmer 2012), or crowded under other species, such as urchins (Chapman & Smoothey 2014). This suggests that there could be strong local competition for space or food, but neither intra- nor interspecific competition has been examined in these overdispersed species living under boulders, so this question remains unanswered.

Grazing and predation

Grazers may not be as important as are physical disturbances in controlling cover of algae on boulders, which contrasts to its importance on many rocky shores (Underwood 1980, Hawkins & Hartnoll 1983). Nevertheless, grazing may explain the lack of relationship between algal cover and size of boulders in some lowshore areas, where grazers are more abundant on larger boulders. They may reduce algal cover on these boulders in a similar way that physical disturbance controls cover of algae on smaller or highshore boulders (McGuinness 1987b), thus disrupting the expected species-area relationship.

It is not clear on what and where many of the species that live under boulders feed. Many molluscs, including those living under boulders, are algal grazers, but measures of micro- or macroalgal food under boulders are lacking. Other species, such as *Ischnochiton smaragdinus*, feed on bryozoans and sponges that are common under boulders, and *Loricella angasi* can catch and eat amphipods (Kangas & Shepherd 1984). Some species have been seen to move on to the upper surfaces to feed (e.g. *Ischnochiton australis*; Kangas & Shepherd 1984) and others migrate into the surrounding area at night before returning to boulders during daylight (J. Grayson, pers. comm.). It is assumed, but not known, that these animals are feeding.

There have been few studies of predation in intertidal boulder-fields, other than human harvesting, although Rogers & Elliott (2013) examined predation of starfish by gulls in numerous intertidal habitats in Puget Sound, USA. Small starfish were only found in the boulder-fields, where there was also the largest incidence of predation, with gulls observed pulling small starfish out from under the boulders. Experiments showed that gulls tended to feed only on the small starfish. These may have been eaten in parts of the shore where they are easily visible and were, thus, only found under boulders, or they may not have recruited to those other habitats. Thus, gulls concentrated their predation in the only habitat where young starfish lived.

Post-settlement dispersal

Some experiments on dispersal of organisms among boulders have shown strong behavioural responses for some species. For example, Cryer et al. (1987), in an experimental study of the effects of harvesting, showed that boulders which originally had crabs sheltering under them, were more likely to be recolonized by crabs, even after the crabs had been removed. The colonizing crabs were presumably responding to some features of the boulders (or possibly cues remaining after removal of the crabs), but what these were is unknown. Smoothey & Chapman (2007) showed similar behaviour for sea urchins: more dispersal to boulders under which there were originally urchins, even after the urchins had been removed, than to boulders without urchins at the start of the experiment.

In experiments to try to identify what are the features of boulders or their surroundings that affect dispersal of animals, newly quarried boulders were deployed within boulder-fields (Chapman 2002b). These boulders had no biota or biofilm at the start of the experiment. Nevertheless, they were quickly colonized with most animals arriving as adults or juveniles, rather than settling larvae. They apparently moved under the new boulders from the surrounding habitat (Chapman 2002b, 2003a). An overdispersed pattern of distribution among individual units of habitat was established very quickly—within days or weeks.

Restoration of boulder-fields in response to degradation or loss

Because boulders support a wide diversity of biota and create complex habitat that changes rates of water movement and sediment deposition (Cruz Motta 2005), they have been deployed in streams and rivers as part of restoration programmes, often in conjunction with the addition of vegetation and removal of channelization. The aim has been to increase local biodiversity or improve habitat for fish (Roni et al. 2006, Branco et al. 2013) because boulders are effective at retaining floating vegetation at relatively low rates of discharge (Koljonen et al. 2012), thus increasing the availability of local patches of complex habitat and resources.

Boulders, or artificial surrogates such as concrete blocks, are similarly used to build artificial reefs in subtidal areas to mitigate for loss of habitat, predominantly to provide habitat for exploited or charismatic species, or to increase habitat for fish (reviewed by Baine 2001). Such reefs, however, frequently support unique assemblages and do not form surrogates for any loss of natural habitats. For example, a 5-year-long comparison of a built boulder-reef and natural reef in Florida, USA, showed that the former consistently had larger abundances of fish and supported a different assemblage to that found on the natural reef (Kilfoyle et al. 2013).

Intertidally, there has been little research on the restoration of boulder-fields, with most intertidal restoration efforts focused on mangroves and wetlands. Because boulder-fields are vulnerable to so many disturbances and support such a diverse and specialized fauna, consideration should be given to the possibility of creating novel boulder-fields to replace those that have been lost or are under threat (e.g. due to urban development or potentially through a rise in sea level). The sessile assemblages that naturally inhabit boulders often include many fouling species that readily colonize bare space, but characteristics of the sessile assemblage do not seem important in influencing colonization by the mobile species, many of which are habitat-specialists (Chapman 2002b). When novel habitat, such as sandstone blocks, are placed within boulder-fields, they are rapidly colonized (Chapman 2003a), suggesting that boulder-fields may be easy to duplicate if there is a nearby source of colonists, and boulders or their surrogates can be deployed with minimal disturbances. This view is supported by examining the small artificial boulder-fields that are created from building rubble at the base of many intertidal sea walls. These support many species which do not live on the walls themselves, thus increasing local diversity of species (Chapman 2006). Stony rubble, if left in place, may therefore compensate to some extent for the negative effects on native biodiversity of armoring shorelines by replacing natural habitat with sea walls (Chapman 2003b). This is not to suggest, however, that it is acceptable to simply discard waste materials in the sea and call them artificial reefs or surrogate habitat (Chou 1997).

Research in NSW, Australia, has created new, relatively small boulder-fields from quarried rock in areas near to, but not within, natural boulder-fields (Chapman 2012). Sessile assemblages colonized these patches very slowly and variably and had not reached the equivalence of natural boulders within a year. Nevertheless, both rare and common mobile animals, primarily molluscs and echinoderms which are the main taxa that occupy natural boulders, rapidly colonized these patches. There was no clear successional sequence, with colonization patchy at the scales of individual boulders, among patches of boulders 20 m apart, between sites 500 m apart and between locations 2 km apart, although diversities and abundances of some species matched those of natural boulders within a few months. Both rare and common animals mainly colonized as adults from surrounding areas, despite the fact that there were no visible individuals of many of the rare species in these areas. The size of the newly created patches, composed of 50 or 100 large boulders, showed no effect of patch size on either rates of colonization, or the suite of species that colonized the different patches (Chapman 2013). Most species were randomly distributed between the two patch sizes, some species were more abundant on the larger patches, but other closely-related species showed the opposite pattern. After a few weeks, most species were as abundant in these patches as on natural boulders; colonization was rapid.

Future research directions

Intertidal boulder-fields have many characteristics which make them ideal for experimental studies to further our understanding of ecology. A fertile field for further research is the causes for the extreme patterns of overdispersion that are found in this habitat. Despite considerable research into responses of biota to different features of habitat associated with the boulders themselves, these strong patterns of overdispersion are still little understood. Because the fauna occupy and move among discrete patches of habitat (the boulders), which can be experimentally manipulated with respect to their features (size, complexity, sessile assemblage), their positions relative to each other, and the timing of manipulations relative to timing of settlement, weather, etc., they are ideal habitats in which to test experimentally complex ecological models of dispersal and habitat requirements. In addition, because many of the more common species are also found on continuous rocky shores, they are ideally placed to compare dispersal and recolonization of denuded areas in continuous versus patchy habitats. With increased disturbances on natural rocky shores potentially disrupting and fragmenting natural populations, such information could be invaluable in determining how best to manage such populations.

Despite the well-documented species–area relationship for many habitats, size of boulders appears to have little consistent effect on either abundances or numbers of species found on them. Because this theory is a cornerstone of much ecological theory, this habitat is an ideal arena in which to explore this relationship (or lack thereof) because the habitat itself can be easily experimentally manipulated.

There is still much to learn about the persistence of rare species because their spatial and temporal dynamics, responses to habitat, body-size abundance patterns and many other characteristics of life history are all thought to differ fundamentally from those of common species (Gaston 1994). Because boulder-fields support a number of species which are closely related, which are functionally similar, or which have differing modes of reproduction, they and their habitat can be manipulated in the field with relatively little disturbance (Chapman & Underwood 1996). This makes them ideal for investigating many of the questions raised by Gaston (1994) and others who attempt to understand the dynamics of rare species.

Finally, there is still much to understand in order to manage, restore or recreate these habitats for the conservation of their biota, both with respect to repairing damage to existing boulder-fields, or building new habitat in mitigation for that lost or threatened. In addition, it has been suggested that boulder-fields may be built to compensate for loss of alternative habitat, such as seagrasses through urban development (Iversen & Bannerot 1984). Boulder-fields are being built to protect shorelines (Green et al. 2012) as part of softening armoured shores to create hybrid designs. Yet, to date, there has been little research into the ecological value of such created habitat relative to that lost by urbanization and shoreline development.

Conclusions

Despite their prevalence along some coastlines and the large number of species that live in intertidal boulder-fields, they have been little studied compared to habitats such as mangrove forests or intertidal rock platforms. Yet, because boulder-fields are composed of natural units of habitat (the boulders), which can be transplanted among sites (with or without their very diverse sessile and mobile assemblages), and in addition, experimentally manipulated in many different ways to test a variety of hypotheses, they have been very rich areas of research for tests of such ecological concepts as succession, the role of disturbance, species-area relationships, and the effects of habitat complexity on diversity. As such, research in boulder-fields has added important data to the generality of many ecological theories.

In addition, the undersurfaces of boulders, especially in relatively sheltered areas, are habitat for a wide range of species, particularly of molluscs, many of which are seldom or not found in other habitats. Many of these are rare species, with limited range and/or small abundances. As such, they are vulnerable to many current and predicted disturbances. Most species have extremely strong patterns of habitat association, but in general, the factors causing these overdispersed patterns of abundance, at scales of metres to kilometres, are not known. With few exceptions, animals do not appear to respond to obvious features of the boulders, or to the other biota inhabiting the boulders, but they aggregate on to a limited number of boulders at the stage of initial colonization. To maximize our potential to protect these fauna, it is important to generate better understanding of the responses to habitat of these species, especially the rare species. This requires careful, well thought-out experiments that minimize disturbance to the habitat (Chapman & Underwood 1996) and the rare species (Chapman & Smoothey 2014).

References

Addessi, L. 1994. Human disturbance and long-term changes on a rocky intertidal community. *Ecological Applications* **4**, 786–797.

Altieri, A.H., Van Wesenbeeck, B.K., Bertness, M.D. & Silliman, B.R. 2010. Facilitation cascade drives positive relationship between native biodiversity and invasion success. *Ecology* **91**, 1269–1275.

Baine, M. 2001. Artificial reefs: a review of their design, application, management and performance. *Ocean & Coastal Management* **44**, 241–259.

Barnes, D.K.A. & Lehane, C. 2001. Competition, mortality and diversity in South Atlantic coastal boulder communities. *Polar Biology* **24**, 200–208.

Barnes, D.K.A., Rothery, P. & Clarke, A. 1996. Colonisation and development in encrusting communities from the Antarctic intertidal and sublittoral. *Journal of Experimental Marine Biology and Ecology* **196**, 251–265.

Branch, G.M. 1975. Mechanisms reducing intraspecific competition in *Patella* species: migration, differentiation and territorial behavior. *Journal of Animal Ecology* **44**, 575–600.

Branco, P., Boavida, I., Santos, J.M., Pinheiro, A. & Ferreira, M.T. 2013. Boulders as building blocks: improving habitat and river connectivity for stream fish. *Ecohydrology* **6**, 627–634.

Bruno, J.F. 2000. Facilitation of cobble beach plant communities through habitat modification by *Spartina alterniflora*. *Ecology* **81**, 1179–1192.

Chapman, M.G. 2002a. Patterns of spatial and temporal variation of macrofauna under boulders in a sheltered boulder field. *Austral Ecology* **27**, 211–228.

Chapman, M.G. 2002b. Early colonization of shallow subtidal boulders in two habitats. *Journal of Experimental Marine Biology and Ecology* **275**, 95–116.

Chapman, M.G. 2003a. The use of sandstone blocks to test hypotheses about colonization of intertidal boulders. *Journal of the Marine Biological Association of the United Kingdom* **83**, 415–423.

Chapman, M.G. 2003b. Paucity of mobile species on constructed seawalls: effects of urbanization on biodiversity. *Marine Ecology Progress Series* **264**, 21–29.

Chapman, M.G. 2005. Molluscs and echinoderms under boulders: tests of generality of patterns of occurrence. *Journal of Experimental Marine Biology and Ecology* **325**, 65–83.

Chapman, M.G. 2006. Intertidal seawalls as habitats for molluscs. *Journal of Molluscan Studies* **72**, 247–257.

Chapman, M.G. 2007. Colonization of novel habitat: tests of generality of patterns in a diverse invertebrate assemblage. *Journal of Experimental Marine Biology and Ecology* **348**, 97–110.

Chapman, M.G. 2012. Restoring intertidal boulder-fields as habitat for "specialist" and "generalist" animals. *Restoration Ecology* **20**, 277–285.

Chapman, M.G. 2013. Constructing replacement habitat for specialist and generalist molluscs – the effect of patch size. *Marine Ecology Progress Series* **473**, 201–214.

Chapman, M.G. & Smoothey, A.F. 2014. Sea urchins provide habitat for rare chitons in intertidal boulder-fields. *Journal of Experimental Marine Biology and Ecology* **459**, 31–37.

Chapman, M.G. & Underwood, A.J. 1996. Experiments on effects of sampling on biota under intertidal and shallow subtidal boulders. *Journal of Experimental Marine Biology and Ecology* **207**, 103–126.

Chapman, M.G. & Underwood, A.J. 2009. Evaluating accuracy and precision of species-area relationships for multiple estimators and different marine assemblages. *Ecology* **90**, 754–766.

Chapman, M.G., Underwood, A.J. & Clarke, K.R. 2009. New indices for ranking conservation sites using 'relative endemism'. *Biological Conservation* **142**, 3154–3162.

Choi, D.R & Ginsberg, R.N. 1983. Distribution of coelobites (cavity-dwellers) on coral rubble across a Florida reef tract. *Coral Reefs* **2**, 165–172.

Chou, L.M. 1997. Artificial reefs of Southeast Asia – do they enhance or degrade the marine environment? *Environmental Monitoring and Assessment* **44**, 45–52.

Connell, J.H. 1978. Diversity in tropical rainforests and coral reefs. *Science* **199**, 1302–1310.

Cruz Motta, J.J. 2005. Diel and tidal variations of benthic assemblages in sediments associated with boulder fields. *Marine Ecology Progress Series* **290**, 97–107.

Cruz Motta, J.J., Underwood, A.J., Chapman, M.G. & Rossi, F. 2003. Benthic assemblages in sediments associated with intertidal boulder-fields. *Journal of Experimental Marine Biology and Ecology* **285–286**, 383–401.

Cryer, M., Whittle, G.N. & Williams, R. 1987. The impact of bait collection by anglers on marine intertidal invertebrates. *Biological Conservation* **42**, 83–93.

Dayton, P.K. 1971. Competition, disturbance, and community organization: the provision and subsequent utilization of space in a rocky intertidal community. *Ecological Monographs* **41**, 351–389.

Douglas, M. & Lake, P.S. 1994. Species richness of stream stones: an investigation of the mechanisms generating the species-area relationship. *Oikos* **69**, 387–396.

Downes, B.J., Lake, P.S., Schreiber, E.S.G. & Glaister, A. 1998. Habitat structure and regulation of local species diversity in a stony, upland stream. *Ecological Monographs* **68**, 237–257.

Emparanza, E.J.M. 2007. Patterns of distribution of dominant porcelain crabs (Decapoda: Porcellamidae) under boulders in the intertidal of northern Chile. *Journal of the Marine Biological Association of the United Kingdom* **87**, 523–531.

Gaston, K.J. 1994. *Rarity*. London: Chapman & Hall.

Gianguzza, P., Jensen, K.R., Bonaviri, C., Agnetta, D. & Chemello, R. 2013. Hiding behavior of *Oxynoe olivacea* (Mollusca: Opisthobranchia: Sacoglossa) in the invasive seaweed *Caulerpa taxifolia*. *Italian Journal of Zoology* **3**, 437–442.

Grayson, J.E. & Chapman, M.G. 2004. Patterns of distribution and abundance of chitons of the genus *Ischnochiton* in intertidal boulder fields. *Austral Ecology* **29**, 363–373.

Green, D.S., Chapman, M.G. & Blockley, D.J. 2012. Ecological consequences of the type of rock used in the construction of artificial boulder-fields. *Ecological Engineering* **46**, 1–10.

Green, D.S. & Crowe, T.P. 2014. Physical and biological effects of introduced oysters on biodiversity in an intertidal boulder field. *Marine Ecology Progress Series* **482**, 119–132.

Hawkins, S.J. 1983. Interactions of *Patella* and macroalgae with settling *Semibalanus balanoides* (L.). *Journal of Experimental Marine Biology and Ecology* **71**, 55–72.

Hawkins, S.J. & Hartnoll, R.G. 1980. A study of the small-scale relationship between species number and area on a rocky shore. *Estuarine and Coastal Marine Science* **10**, 201–214.

Hawkins, S.J. & Hartnoll, R.G. 1982. The influence of barnacle cover on the numbers, growth and behavior of *Patella vulgata* on a vertical pier. *Journal of the Marine Biological Association of the United Kingdom* **62**, 855–868.

Hawkins, S.J. & Hartnoll, R.G. 1983. Grazing of intertidal algae by invertebrates. *Oceanography and Marine Biology: An Annual Review* **21**, 195–282.

Iversen, E.S. & Bannerot, S.P. 1984. Artificial reefs under marine docks in southern Florida. *North American Journal of Fisheries Management* **4**, 294–299.

James, R.L. & Underwood, A.J. 1994. Influence of colour of substratum on recruitment of spirorbid tubeworms to different types of intertidal boulders. *Journal of Experimental Marine Biology and Ecology* **181**, 105–115.

Kangas, M. & Shepherd, S.A. 1984. Distribution and feeding of chitons in a boulder habitat at West Island, South Australia. *Journal of the Malacological Society of Australia* **6**, 101–111.

Kilfoyle, A.K., Freeman, J., Jordan, L.K.B., Quinn, T.P. & Spieler, R.E. 2013. Fish assemblages on a mitigation boulder reef and neighbouring hardbottom. *Ocean & Coastal Management* **75**, 53–62.

Kimmerer, R.W. & Driscoll, M.J.L. 2000. Bryophyte species richness on insular boulder habitats: the effect of area, isolation, and microsite diversity. *The Bryologist* **103**, 748–756.

Koljonen, S., Louhi, A., Mäki-Petäys, A., Huusko, A. & Muotka, T. 2012. Quantifying the effects of instream habitat structure and discharge on leaf retention: implications for stream restoration. *Freshwater Science* **31**, 1121–1130.

Kovačič, M., Patzner, R.A. & Schliewen, U. 2012. A first quantitative assessment of the ecology of cryptobenthic fishes in the Mediterranean Sea. *Marine Biology* **159**, 2731–2742.

Kuklinski, P., Barnes, D.K.A. & Taylor, P.D. 2006. Latitudinal patterns of diversity and abundance in North Atlantic intertidal boulder-fields. *Marine Biology* **149**, 1577–1583.

Kurihara, T., Kosuge, T., Kobayashi, M., Katoh, M. & Mito, K.-I. 2001. Spatial and temporal fluctuations in densities of gastropods and bivalves on subtropical cobbled shores. *Bulletin of Marine Science* **68**, 409–426.

Le Hir, M. & Hily, C. 2005. Macrofaunal diversity and habitat structure in intertidal boulder fields. *Biodiversity and Conservation* **14**, 233–250.

Lieberman, M., John, D.M. & Lieberman, D. 1979. Ecology of subtidal algae on seasonally devastated cobble substrates off Ghana. *Ecology* **60**, 1151–1161.

Littler, M.M. & Littler, D.S. 1984. Relationships between macroalgal functional form groups and substrata stability in a subtropical rocky-intertidal system. *Journal of Experimental Marine Biology and Ecology* **74**, 13–34.

Liversage, K. 2016. The influence of boulder shape on the spatial distribution of crustose coralline algae (Corallinales, Rhodophyta). *Marine Ecology* **37**, 459–462.

Liversage, K. & Benkendorff, K. 2013. A preliminary investigation of diversity, abundance and distributional patterns of chitons in intertidal boulder fields of differing rock type in South Australia. *Molluscan Research* **33**, 24–33.

Liversage, K., Cole, V.J., McQuaid, C.D. & Coleman, R.A. 2012. Intercontinental tests of the effects of habitat patch type on the distribution of chitons within and among patches in intertidal boulder field landscapes. *Marine Biology* **159**, 2777–2786.

Londoño-Cruz, E. & Tokeshi, M. 2007. Testing scale variance in species-area and abundance-area relationships in a local assemblage: an example from a subtropical boulder shore. *Population Ecology* **49**, 275–285.

Loughman, Z.J., Skalican, K.T. & Taylor, N.D. 2013. Habitat selection and movement of *Cambarus chasmodactylus* (Decapoda: Cambaridae) assessed via radio telemetry. *Freshwater Science* **32**, 1288–1297.

McGuinness, K.A. 1984. Species-area relations of communities on intertidal boulders: testing the null hypothesis. *Journal of Biogeography* **11**, 439–456.

McGuinness, K.A. 1987a. Disturbance and organisms on boulders. I. Patterns in the environment and the community. *Oecologia* **71**, 409–419.

McGuinness, K.A. 1987b. Disturbance and organisms on boulders. II. Causes in patterns of diversity and abundance. *Oecologia* **71**, 420–430.

McGuinness, K.A. 1988. Explaining patterns in abundances of organisms on boulders: the failure of 'natural' experiments. *Marine Ecology Progress Series* **48**, 199–204.

McGuinness, K.A. & Underwood, A.J. 1986. Habitat structure and the nature of communities on intertidal boulders. *Journal of Experimental Marine Biology and Ecology* **104**, 97–123.

Menge, B.A. 1976. Organization of the New England rocky intertidal community: role of predation, competition and environmental heterogeneity. *Ecological Monographs* **46**, 355–393.

Osman, R.W. 1977. The establishment and development of a marine epifaunal community. *Ecological Monographs* **47**, 37–63.

Palmer, A.N.S. 2012. Spatial and genetic investigation of aggregation in *Ischnochiton* (Polyplacophora; Neoloricata; Ischnochitonina; Ischnochitonida; Ischnochitoninae) species with different larval development. *Austral Ecology* **37**, 110–124.

Rogers, T.L. & Elliott, J.K. 2013. Differences in relative abundance and size structure of the sea stars *Pisaster ochraceus* and *Evasterias trochelii* among habitat types in Puget Sound, Washington, USA. *Marine Biology* **160**, 835–865.

Roni, P., Bennett, T., Morley, S., Pess, G.R., Hanson, K., Van Slyke, D. & Olmstead, P. 2006. Rehabilitation of bedrock stream channels: the effect of boulder weir placement on aquatic habitat and biota. *River Research and Applications* **22**, 967–980.

Smith, K.A. & Otway, N.M. 1997. Spatial and temporal patterns in abundance and the effects of disturbance on under-boulder chitons. *Molluscan Research* **18**, 43–57.

Smoothey, A.F. & Chapman, M.G. 2007. Small-scale variability in the dispersion of the sea urchin *Heliocidaris erythrogramma* among boulders. *Marine Ecology Progress Series* **340**, 89–99.

Sousa, W.P. 1979a. Disturbance in marine intertidal boulder fields: the nonequilibrium maintenance of species diversity. *Ecology* **60**, 1225–1239.

Sousa, W.P. 1979b. Experimental investigations of disturbance and ecological succession in a rocky intertidal algal community. *Ecological Monographs* **49**, 227–254.

Underwood, A.J. 1980. The effects of grazing by gastropods and physical factors on the upper limits of distribution of intertidal macroalgae. *Oecologia* **46**, 201–213.

Underwood, A.J. 1984. Vertical and seasonal patterns in competition for microalgae between intertidal gastropods. *Oecologia* **64**, 211–222.

Underwood, A.J. & Barrett, G. 1990. Experiments on the influence of oysters on the distribution, abundance and sizes of the gastropod *Bembicium auratum* in a mangrove swamp in New South Wales, Australia. *Journal of Experimental Marine Biology and Ecology* **137**, 25–45.

Underwood, A.J. & Jernakoff, P. 1981. Effects of interactions between algae and grazing gastropods on the structure of a lowshore algal community. *Oecologia* **48**, 221–233.

Van Tamelen, P.G. 1987. Early successional mechanisms in the rocky intertidal: the role of direct and indirect interactions. *Journal of Experimental Marine Biology and Ecology* **112**, 39–48.

Walker, S.J., Degnan, B.M., Hooper, J.N.A. & Skilleter, G.A. 2008. Will increased storm disturbance affect the biodiversity of intertidal, nonscleractinian sessile fauna on coral reefs? *Global Change Biology* **14**, 1–16.

Waller, C.L. 2012. Zonation in a cryptic Antarctic intertidal macrofaunal community. *Antarctic Science* **25**, 62–66.

Oceanography and Marine Biology: An Annual Review, 2017, **55**, 55-86
© S. J. Hawkins, D. J. Hughes, I. P. Smith, A. C. Dale, L. B. Firth, and A. J. Evans, Editors
Taylor & Francis

ECOLOGICAL DOMINANCE ALONG ROCKY SHORES, WITH A FOCUS ON INTERTIDAL ASCIDIANS

MARC RIUS[1,2]*, PETER R. TESKE[2], PATRICIO H. MANRÍQUEZ[3],
ROCÍO SUÁREZ-JIMÉNEZ[4], CHRISTOPHER D. MCQUAID[5] & JUAN CARLOS CASTILLA[6]

[1]*Ocean and Earth Science, University of Southampton, National Oceanography Centre,
European Way, Southampton, SO14 3ZH, United Kingdom*
[2]*Centre for Ecological Genomics and Wildlife Conservation, Department of Zoology,
University of Johannesburg, Auckland Park, 2006, South Africa*
[3]*Laboratorio de Ecología y Conducta de la Ontogenia Temprana (LECOT),
Centro de Estudios Avanzados en Zonas Áridas (CEAZA),
Coquimbo, Avenida Ossandón 877, Coquimbo, Chile*
[4]*Department of Botany, University of Otago, Dunedin 9054, New Zealand*
[5]*Coastal Research Group, Department of Zoology and Entomology,
Rhodes University, Grahamstown, South Africa*
[6]*Centro de Conservación Marina, Estación Costera de Investigaciones Marinas (ECIM),
Núcleo de Conservación Marina. Iniciativa Científica Milenio. Facultad de Ciencias Biológicas,
Pontificia Universidad Católica de Chile, Santiago, Chile*
*Corresponding author: Marc Rius
e-mail: M.Rius@soton.ac.uk

The role of dominant species is of central importance in ecology. Such species play a key role in ecosystem structure, stability and function, regulating resource allocation across trophic levels and overall ecosystem productivity. Although ecological interactions between dominant and subordinate species are often considered to influence the latter negatively, the presence of dominant species can also be beneficial. These species commonly act as ecosystem engineers and enhance biodiversity by creating habitat for other species. Along rocky coastlines, dominant species are often sessile suspension-feeding organisms that can monopolize all available substrata. This is particularly noticeable in intertidal and shallow subtidal habitats where the number of species that achieve ecological dominance is limited. Here, we review the ecological and evolutionary mechanisms that facilitate dominance along rocky coastlines. We then focus on a prominent example, the members of the *Pyura stolonifera* species complex (Tunicata), which are an emerging model system for studying ecological dominance. These ascidians achieve the highest biomass levels ever reported in rocky intertidal habitats and, when invasive, can fundamentally transform entire ecosystems. Finally, we discuss conservation implications and conclude with directions for future research.

Introduction

Ecological dominance can be defined as "the exertion of a major controlling influence of one or more species upon all other species by virtue of their number, size, productivity or related activities" (United Nations 1997). Interest in ecological dominance extends across a wide range of fields including, for example, paleontology (Clapham et al. 2006) and anthropology (Flinn et al. 2005).

In ecology, dominance describes the opposite of ecosystem evenness (Hillebrand et al. 2008) and dominant species are generally the most abundant components of natural communities as a result of their competitive superiority. These species have the ability to structure communities in terms of species composition, diversity, biomass, spatial arrangement and occupancy. Additionally, dominant species often function as ecosystem engineers or bioengineer species (Jones et al. 1994, 1997, Nilsson & Wardle 2005) as they provide habitat for, and regulate the distribution and abundance of other species. Ecological dominance could be construed as including keystone predators or habitat-forming seaweeds, but within the context of this chapter we limit it to spatial dominance by sessile or sedentary animals. The persistence of dominant species may depend on certain levels of environmental stochasticity (e.g. disturbance), which maintain stable levels of species diversity and hierarchy (Connell 1978, but see Fox 2013). Consequently, dominant species and disturbance may collectively determine ecosystem stability and levels of functional diversity (Loreau et al. 2001, Smith & Knapp 2003).

In marine benthic ecosystems, dominant species are often sessile suspension-feeders. These organisms are able to gather and incorporate allochthonous pelagic energy into benthic communities with remarkable efficiency (Gili & Coma 1998) and monopolize food and spatial resources (Sarà 1986). This is especially noticeable in rocky intertidal and shallow subtidal ecosystems where dominance is often achieved by one or a few suspension-feeding species (e.g. Dayton 1971, Paine 1971, Sousa 1979, Paine et al. 1985, Underwood et al. 1991, Castilla et al. 2000). Dominant suspension-feeders are ecosystem engineer species (Wright & Jones 2006) and are present in most marine ecosystems around the world (Jones et al. 1994, Crooks 2002, Gutiérrez et al. 2003).

Here, we review the literature to unravel the ecological and evolutionary mechanisms that facilitate dominance. We then focus on a specific group of dominant marine invertebrate species that are of considerable ecological importance in rocky shore communities of the southern hemisphere, the members of the *Pyura stolonifera* species complex. We conclude with the role of ecological dominance in conservation efforts and outline future directions for research on dominant species.

The theory behind ecological dominance

Ecological dominance is strongly linked to competitive ability (e.g. Dayton 1975, Steneck et al. 1991, Baird & Hughes 2000), which is often seen as having negative effects on species richness as competing species work towards eliminating one another. In order to understand how ecological dominance can influence overall community structure, it is important to recognize that competition can take various forms. Theoretically, competition only occurs if specific resources are in limited supply and, in the case of competition for food or space, it can take the form of either exploitation or interference competition (Schoener 1983, Yodzis 1989). Competition for renewable resources such as food often exclusively involves interference, whereas competition for space has two components that operate across different spatial scales and interact with one another: actual competition for space, which operates at small scales through interference, and exploitation competition (Steinwascher 1978) through dispersal, which takes place at larger scales. Dispersal is required to find and to monopolize available space.

In the marine realm, organisms exhibit markedly different scales of propagule dispersal (Kinlan & Gaines 2003), and this has important consequences for the likelihood of coexistence of competing (and potentially dominant) species (Berkley et al. 2010, Aiken & Navarrete 2014). When dispersal is minimal, two species can theoretically coexist, as patches of habitat often operate independently of one another (Leibold et al. 2004, Tilman 1994). If dispersal scales are very large, however, the distinction among patches of habitat is lost and coexistence is less likely. Therefore, it is important to measure and define scales carefully when describing the effects of dispersal and to recognize that ecological processes are affected by dispersal type (Kinlan et al. 2005, Aiken & Navarrete 2014). Nevertheless, there are clearly different implications for space occupiers that are assumed to have the potential to disperse over scales of hundreds of kilometres, such as mussels or barnacles (Tapia & Pineda 2007, Teske et al. 2016) as opposed to those with abbreviated larval development, such as ascidians (Millar 1971, Clarke et al. 1999). Scales of dispersal also have implications for the type of guild responsible for dominating space. Where species are capable of outcompeting others through functional dominance, as is often the case for intertidal communities (e.g. Dayton 1975, Lubchenco & Menge 1978), the guild is composed of dominant space occupiers. In this case, weakening of dominance results in an increase in the number of species that can coexist so that non-selective mortality, common in cases of mass mortality due to wave or heat stress (e.g. Tsuchiya 1983, Erlandsson et al. 2006, Garrabou et al. 2009), has a positive effect on species richness. In contrast, in ecosystems without clear competitive dominance, the system is shaped by dispersal and colonization events and is considered to be founder-controlled. Such systems include coral reef fish communities (e.g. Sale 1977, 1979, Almany et al. 2007) and in this case, mortality will decrease richness (Paine 1966, Sousa 1979, Yodzis 1989).

There is a vast body of both theoretical and empirical literature on resource-mediated interactions in communities dominated by superior competitors, including the implications for the control of species richness, the persistence of subordinate species (Dayton 1975), shifting competitive dominance (Paine 1969, Lubchenco 1978) and ecological consequences of body size (Brown & Maurer 1986). Of course, dominant species interact with other drivers of community structure, including keystone species (Paine 1969, Paine & Suchanek 1983, Menge et al. 1994, Castilla 1999), recruitment limitation (Connolly & Roughgarden 1999) and disturbance (Lubchenco & Menge 1978, Sousa 1979). Importantly, and partly through their effects on other species, dominant species can regulate ecosystem function, trophic complexity and community stability (Paine 1969, Smith & Knapp 2003), which leads to community-level impacts (Harley 2006). More recently, it has been recognized that species that occupy primary space interact not only with other space-occupiers, but also have a key role in enhancing species richness through facilitation. The inclusion of facilitation in ecological models can completely alter predictions of the effects of environmental stress, disturbance or predation on species richness and the probability of success of biological invasions (Bruno et al. 2003). This builds on the recognition of the importance of within-species group effects for space-occupiers (Bertness & Leonard 1997) and their role as ecological engineers (Jones et al. 1994). Dominant species can provide habitat for associated species, offer protection from predation (Stachowicz & Hay 1999, Crain & Bertness 2006), mitigate environmental stress (Rius & McQuaid 2006, 2009) and enhance recruitment success of conspecifics (Erlandsson & McQuaid 2004) as well as other species. For example, successful settlement of mussel recruits can be enhanced by the presence of macroalgae (Bayne 1964, McQuaid & Lindsay 2005).

Resource availability is a critical mechanism modulating ecosystems. It determines community structure, ecological interactions and phenotypic traits (Coley et al. 1985), and it shapes levels of energy transfer across trophic levels that are required to maintain niche differentiation and functional diversity. Along rocky coastlines, the primary limiting factor for benthic organisms is space, which in the intertidal zone can be dominated by both sessile animals and algae. However, the balance between faunal and algal dominance is often mediated by wave action (McQuaid & Branch 1984). At the subtidal fringe, space is mostly dominated by suspension-feeders, with ascidians dominating many temperate coastlines, especially in the Southern Hemisphere (see below).

The monopolization of a specific resource by a single species is generally ascribed to certain attributes (Paine & Suchanek 1983, Guiñez & Castilla 2001). Accordingly, it is expected that competitive dominance will positively correlate with degree of gregariousness and the species' ability to occupy space (Figure 1). Among the species traits that enhance ecological dominance (Figure 2), gregariousness and a sessile or sedentary life strategy are tightly linked (i.e. it is difficult to have one trait without the other), and collectively lead to a specific ecological trade-off. A gregarious

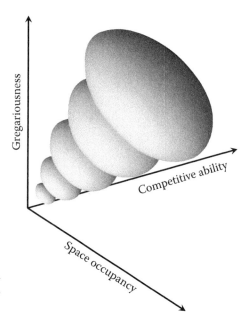

Figure 1 Drivers and consequences of ecological dominance in benthic animals. Nested ellipsoids represent isobars of likelihood of becoming a dominant species.

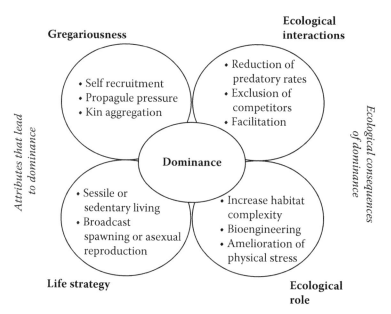

Figure 2 Drivers and consequences of ecological dominance in benthic animals. Shown are main attributes leading to dominance, together with consequences of the presence of dominant species.

species can act as a bioengineer and occupy space while the ability of its predators or competitors to fulfil their roles diminishes. A key aspect for dominance is to maximize the extent and duration of resource monopolization. Species such as ephemeral algae are rarely considered dominant species as they are short-lived and space occupancy is only transient.

The monopolization of resources, particularly space, can be mitigated by compensatory mortality, with dominant species suffering higher rates of mortality through disturbance (e.g. Connell

1978, Paine 1979, Sousa 1979, Erlandsson et al. 2006) or predation (Paine 1976, Symondson et al. 2002). The effects of predation can depend on timing in terms of ecological succession (Vieira et al. 2012) and recruitment rates high enough to swamp predators can ultimately allow monopolization (Navarrete & Berlow 2006). Thus, disturbance and predation generally tend to free resources and ameliorate the subordination of inferior species. This reduces interference competition and allows species to exploit spatial and temporal variability in resources, minimizing dominance by a particular species.

Evolutionary implications of ecological dominance

The formation of aggregations by propagules that disperse freely in a particular environment requires certain behavioural abilities. For example, behaviour-mediated recruitment has been reported in many taxa (Toonen & Pawlik 1994) and is often a response to the presence of conspecific adults (Toonen & Pawlik 1996, Alvarado et al. 2001). However, other external stimuli such as light, biofilm, substratum orientation or flow conditions may be more relevant (Pawlik et al. 1991, Keough & Raimondi 1995, Wieczorek & Todd 1997, Rius et al. 2010a). In addition, evolutionary mechanisms such as Allee effects may be critical for understanding ecological dominance. Allee effects normally appear when there is a decline in population size or density that leads to a loss of overall fitness (Courchamp et al. 1999, Berec et al. 2007). These effects can operate through increased difficulty in finding mates (e.g. Kuussaari et al. 1998) or susceptibility to predators (e.g. Bertness & Grosholz 1985). Additionally, because most broadcast-spawning benthic suspension-feeders are sessile (e.g. ascidians) or near-sessile (e.g. mussels), populations often exhibit Allee effects because the likelihood of gamete encounters resulting in fertilization is significantly lower when adults are present at low densities (Levitan 1991, Babcock & Keesing 1999).

For many organisms, ecological dominance is largely a consequence of the behaviour of their dispersive larvae, which enables the formation of long-lasting aggregations. Aggregated settlement can be facilitated by conspecific cues following successful colonization by a founder (Toonen & Pawlik 1994). Indeed, a possible evolutionary consequence of Allee effects is conspecific attraction. Although the idea of linking Allee effects, recruitment and conspecific attraction was developed in the context of vertebrates, especially colonial- and non-colonial-nesting birds (Reed & Dobson 1993), the insights gained are applicable to species with external fertilization. Stephens & Sutherland (1999) regard conspecific attraction as a "direct product" of Allee effects and Donahue (2006) suggests that conspecific cues and Allee effects jointly lead to conspecific attraction. Another aspect that may be relevant is kin aggregation (Grosberg & Quinn 1986, Veliz et al. 2006), although the reverse situation (kin avoidance) has also been reported (Johnson & Woollacott 2010) and thus requires further investigation.

Propagule attraction to conspecifics can be facilitated by increased habitat complexity. For example, studies have shown increased settlement rates in structurally-complex mussel beds (Alvarado & Castilla 1996, Alvarado 2004), though there is responsiveness to conspecifics in the absence of structural complexity that can change with settler age (von der Meden et al. 2010). Attraction to adult conspecifics can have negative consequences in the case of suspension-feeders that feed indiscriminately, such as adult mussels which are able to consume >70% of potential settlers, including conspecifics, through larviphagy (Lehane & Davenport 2004, Porri et al. 2008, Troost et al. 2008). In addition, self-recruitment (i.e. recruitment of progeny to the parental population or patch) may increase levels of inbreeding (potentially promoting low levels of genetic diversity), which is known to negatively affect population persistence (Keller & Waller 2002). Despite the potential negative effects, settling close to parents seems to have remarkable fitness benefits in some taxa. The positive aspects of attraction of settling larvae to adults are chiefly due to the enhancement of fertilization success via adult aggregation. In the case of broadcast spawners, fertilization success is often correlated with the degree of aggregation (Levitan et al. 1992, Downing et al. 1993). In addition, aggregations can have evolutionary benefits by providing group defence against predators. For example, mussels use byssal threads to trap predatory whelks (Day et al. 1991, Farrell & Crowe 2007) and to mutually protect conspecifics from wave action (van de Koppel et al. 2005). The latter can also involve facilitative interspecific effects between ecologically homologous species (e.g. Rius & McQuaid 2009).

Ecological dominance on rocky shores

Intertidal habitats have long been model systems for the study of ecological dynamics and principles (Paine 1966, 1969, Stephenson & Stephenson 1972, Lubchenco & Gaines 1981, Hawkins & Hartnoll 1983b, Branch 1984, Castilla & Durán 1985, Menge & Sutherland 1987, Menge et al. 1994, Underwood 2000, Navarrete & Castilla 2003). They can support sessile consumers because food can be transported through the aquatic medium itself, and in many systems, suspension-feeders maintain extremely dense populations (Monteiro et al. 2002, Castilla et al. 2004b). These species dominate overall biomass (e.g. McQuaid & Branch 1984, Castilla et al. 2000) and energy flow (Newell et al. 1982) because of their high rates of secondary production (Baird et al. 2004). Dominant rocky shore species filter large volumes of water and suspended particles (mainly originating from primary production), creating a habitat for diverse associated biota. This gives such species a number of critical roles in ecosystem functioning. Firstly, they act as primary consumers, linking primary production and secondary consumers (Gili & Coma 1998), critically contributing to remineralization (Eriksson et al. 2010) and benthic-pelagic coupling. This creates a two-way interaction between the water column and the benthos through both the consumption of suspended particles (McQuaid & Branch 1985, Loo & Rosenberg 1989, 1996) and the benthic recruitment of planktonic larvae (Navarrete et al. 2005). Secondly, dominant rocky shore species can act as autogenic ecological engineers (*sensu* Lawton & Jones 1995), occupying all available primary space (e.g. Castilla et al. 2004a), and in doing so increasing architectural complexity (Hughes & Griffiths 1988, Guiñez & Castilla 1999, 2001), which enhances species richness (Cerda & Castilla 2001, Cole & McQuaid 2010). All these characteristics make this group a unique and important component of benthic communities.

Mass mortalities of dominant suspension-feeding species as a result of extreme environmental stress or disease (e.g. Hanekom et al. 1999), can have important implications for the entire rocky shore ecosystem. One direct consequence is the loss of ecological networks and function. For example, drastic reductions of intertidal suspension-feeding species directly modify the intertidal community structure and zonation, impacting on key ecosystem services (Castilla et al. 2014, Manríquez et al. 2016). Another consequence of major disturbance events is the decrease of structural complexity upon which other species depend. This is particularly critical when the spatially-dominant species lack a hard calcareous skeleton, such as ascidians (Cerrano & Bavestrello 2009). However, in the case of calcareous species such as barnacles, habitat complexity can persist after mortality. Patch fragmentation as a result of disturbance has major effects on dominant suspension-feeding species and associated communities, contributing to a non-random community assembly in intertidal areas.

Ecological dominance along rocky shores is achieved by a small, taxonomically-diverse group of species. Some examples include bivalves, tubeworms, bryozoans and solitary ascidians (Figure 3), all broadcast-spawning organisms with well-studied life histories (Marshall & Keough 2008, Marshall et al. 2012). These taxa have a wide variety of dissimilar characteristics (Figure 3), indicating that ecological dominance is not due to analogous combinations of traits.

The *Pyura stolonifera* species complex: A model system for studying ecological dominance

Members of the *Pyura stolonifera* species complex (*sensu* Rius & Teske 2011, Phylum Chordata, Subphylum Tunicata) (hereafter the *P.s.s.c.*) are amongst the few intertidal solitary ascidian species that form extensive and dense monospecific aggregations, dominating all available substrata. These

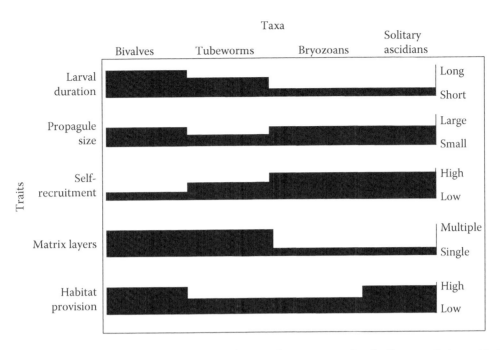

Figure 3 Generalized trait variation of four sessile or sedentary suspension-feeding taxa that are able to dominate rocky shores. The selected traits are all key for understanding ecological dominance and they influence different life-history stages, from early ontogenetic life stages (top) to adulthood (bottom). The right x-axis provides a general idea of rankings and allows overall comparison. Propagule size refers to the size of motile early life-history stages, which in broadcast spawning species is generally correlated with the size of the juvenile and adult stages (Rius et al. 2010b). Self-recruitment refers to recruitment of progeny to the parental population or patch. Matrix layers refer to the ability of dominant species to form multilayer assemblages, which provide secondary habitat for other species.

species are roughly barrel-shaped tunicates that can grow to >30 cm in height (Paine & Suchanek 1983, Fielding et al. 1994, Castilla et al. 2000). They form extensive cemented aggregations or 3-dimensional matrices (Guiñez & Castilla 2001) in the form of collective packed units, although isolated individuals also occur (Castilla & Camaño 2001, Monteiro et al. 2002). Members of the *P.s.s.c.* produce the highest intertidal biomass per unit surface area ever reported in the literature, with dry tissue biomass of >20 kg m^{-2} and densities of up to 1800 individuals m^{-2} (Fielding et al. 1994, Castilla et al. 2000). Such biomass is an order of magnitude higher than the maximum values reported for other suspension-feeding species along rocky shores (e.g. McQuaid & Branch 1985). Ecological theory predicts that the area on either side of the Low Water Springs level is dominated by highly competitive species that are generally free from predators (Hawkins & Hartnoll 1983). Accordingly, intertidal populations of the members of the *P.s.s.c.* achieve the highest densities and biomass in this particular area (Castilla et al. 2000). The members of the *P.s.s.c.* each present unique bioengineer habitat architectures in terms of the number, size and shape of individuals and the arrangements of habitable secondary space. Overall, these ascidians represent good models for the study of ecological dominance in benthic communities (Monteiro et al. 2002, Castilla et al. 2004b, Teske et al. 2011, Manríquez et al. 2016).

Information on the *P.s.s.c.* has been accumulating over the past 130 years or so, from taxonomic (Heller 1878, Van Name 1945, Millar 1955, 1966, Monniot & Bitar 1983, Rius & Teske 2011) and ecological studies (Guiler 1959, Stephenson & Stephenson 1972, Paine & Suchanek 1983, Clarke et al. 1999, Castilla et al. 2000, Monteiro et al. 2002, Castilla et al. 2004a, Castilla et al. 2004b, Knott et al. 2004, Rius et al. 2010a) to recent studies of its physiology (Rius et al. 2014a), genetics

(Castilla et al. 2002, Astorga et al. 2009, Teske et al. 2011, Rius & Teske 2013), invasion biology (Castilla et al. 2004a, Hayward & Morley 2009, Teske et al. 2011, Rius & Teske 2013) and exploitation by humans (Kyle et al. 1997, Castilla et al. 2014, Manríquez et al. 2016). Below, we analyze the biological attributes that have allowed this group to become successful in dominating rocky shores.

Biogeography and evolutionary history of the species complex

Despite the conspicuous nature of the members of the *P.s.*s.c. in both intertidal and subtidal environments, the taxonomy of the group has been fiercely contested until very recently (Kott 2006, Rius & Teske 2011). Many papers referred to all the members of the *P.s.*s.c. as *Pyura stolonifera* (e.g. Kott 1985, Marshall et al. 2000) despite taxonomic (Millar 1962, Monniot & Bitar 1983, Monniot et al. 2001), ecological (Dalby 1997) and genetic (Castilla et al. 2002) evidence pointing to the existence of multiple species. Recent studies employing a combination of morphological and genetic analyses (Rius & Teske 2011, 2013, Teske et al. 2011) have revealed that *Pyura stolonifera* (Heller, 1878) as defined by Kott (2006) is a species complex that in fact represents at least five distinct species. The species presently accepted as valid are the African representatives *P. stolonifera* and *P. herdmani* (Drasche, 1884), and the Australian *P. praeputialis* (Heller, 1878), *P. dalbyi* (Rius & Teske, 2011) and *P. doppelgangera* (Rius & Teske, 2013).

Members of the *P.s.*s.c. are predominantly found along temperate rocky shores of the Southern Hemisphere. In particular, most species are distributed along southern African (Millar 1955, Monniot & Monniot 2001) and Australian (Kott 1985) coasts, but one member of the *P.s.*s.c. (*Pyura herdmani*) is also present in the Northern Hemisphere (Monniot & Bitar 1983, Lafargue & Wahl 1986–1987, Teske et al. 2011). The different species are typically allopatric and some exhibit disjunct distributions, with populations that are separated by large geographic distances (Castilla & Guiñez 2000, Rius & Teske 2013), but there are also instances of sympatric distributions in southern Africa and Australia (Figure 4). Reports from South America (Clarke et al. 1999) and New Zealand (Hayward & Morley 2009) that are corroborated by genetic evidence (Teske et al. 2011), as well as recent sightings in Europe (see further details below), suggest that the species found in these regions originated from elsewhere and were most likely introduced through human activities.

Temperate coastlines characterized by upwelling systems are often preferred habitats for members of the *P.s.*s.c. (Figure 4). Some species, such as *Pyura herdmani* in Africa, are widespread and occur across several biogeographic provinces of differing temperature regimes, providing an interesting system to study population connectivity and physiological tolerance across ecoregions. Another interesting case is *P. dalbyi*, which shows a large distribution gap between the southwestern and southeastern coasts of Australia (Figure 4). However, much of the intermediate region is part of the Great Australian Bight, which is highly inaccessible to study, so this species may be more widespread.

The presence of members of the *P.s.*s.c. on land masses that formed part of the former supercontinent of Gondwanaland (Africa, Australasia and South America) suggests that the present species shared a common ancestor during the Mesozoic. Interestingly, despite treating all as a single species, Kott (1985, 2006) favoured a Gondwanan origin for the group, notwithstanding the fact that tens of millions of years are ample time for speciation to occur. While it is now believed that the populations in South America and New Zealand are the product of recent anthropogenic introductions from Australia (see below), phylogenetic work indicates that there is an ancient split between evolutionary lineages comprising the African species (*Pyura stolonifera* and *P. herdmani*) on one hand, and two of the Australian (*P. praeputialis* and *P. doppelgangera*) species on the other (Teske et al. 2011). This split has not yet been dated, and a shared Gondwanan ancestry of these two lineages thus remains a possibility. However, such a scenario would have involved extinctions on all of the remaining Gondwanan land masses (Madagascar, India, South America, New Zealand and

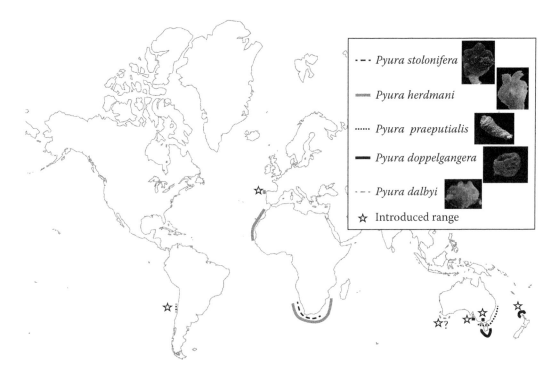

Figure 4 Global distribution of the members of the *Pyura stolonifera* species complex, with the introduced ranges of the different species indicated.

Antarctica), which is a less parsimonious solution than a single long-distance colonization event (e.g. from Africa to Australia via the West Wind Drift). The gametes and larvae of ascidians are unable to disperse over greater distances because of their very short planktonic propagule durations (Millar 1971, Clarke et al. 1999, Rius et al. 2010b), but it is well known that the adults can travel attached to vessels or other floating objects (Lambert 2007, Locke 2009). There is thus little reason to rule out the possibility that the disjunct distribution of the African and Australian lineages was the result of an ancient long-distance colonization event.

Reproductive cycle, early life-history stages, settlement and fertilization

The members of the *P.s.*s.c. have a multiphasic life-cycle (Figure 5) and are broadcast-spawning simultaneous hermaphrodites, releasing male and female gametes mainly during low tides (Marshall 2002, Manríquez & Castilla 2010). After release, currents disperse the gametes, reducing the likelihood of inbreeding. In addition, members of the *P.s.*s.c. have blocks to self-fertilization since eggs of *Pyura praeputialis* fertilized with self-sperm fail to complete development (Manríquez & Castilla 2010). As a result, fertilization and developmental success rely mainly on allogametes encountering one another, which in turn depends on the concentration of allosperm and the viability of eggs (see Marshall 2002, Manríquez & Castilla 2010). Fertilization success in *P. praeputialis* is at its highest with newly-shed sperm and declines as sperm and eggs age (Manríquez & Castilla 2010). Since allogamete limitation as a result of the rapid dispersion of gametes may occur along exposed rocky shores, some members of the *P.s.*s.c. have developed strategies to counteract gamete dilution and mitigate the difficulties of fertilization in such environments. Gamete retention close to the parents following spawning has been observed in *P. praeputialis* both in Australia (Marshall 2002) and Chile (Manríquez & Castilla 2010) and, although this has not yet been described in other members

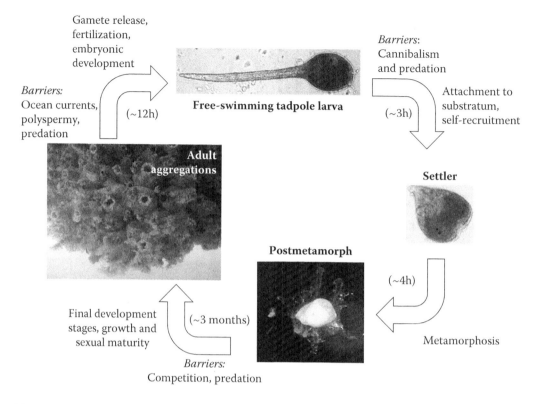

Gamete release, fertilization, embryonic development

Barriers: Cannibalism and predation

Barriers: Ocean currents, polyspermy, predation

(~12h)

Free-swimming tadpole larva

(~3h)

Attachment to substratum, self-recruitment

Adult aggregations

Settler

Postmetamorph

(~4h)

Final development stages, growth and sexual maturity

(~3 months)

Metamorphosis

Barriers: Competition, predation

Figure 5 Life cycle of *Pyura herdmani* depicting different life-history stages and key barriers that may preclude ecological dominance. Early life-history stages were obtained via artificial fertilization in the laboratory (see details in Rius et al. 2010a). All photographs were taken in South Africa.

of this complex, it is highly plausible that it occurs. The gametes of *P. praeputialis* are often shed in a viscous matrix (Marshall 2002, Castilla et al. 2007b, Manríquez & Castilla 2010) and once they come into contact with seawater a biofoam is formed (Castilla et al. 2007b, Manríquez & Castilla 2010). Such biofoam also retains a high concentration of developing embryos and larvae of *P. praeputialis* (Castilla et al. 2007b, Manríquez & Castilla 2010). This suggests that biofoam formation may be an adaptive mechanism that enhances fertilization success and self-recruitment.

After fertilization, embryonic development occurs rapidly (see details in Rius et al. 2010a) and results in tadpole-like larvae that have a very short swimming period (Figure 5). The larvae are lecithotrophic and dependence on limited yolk-reserves limits dispersal range. As a result of the short pelagic duration of gametes, embryos and larvae, settlement will occur in the vicinity of the parental habitat, and colonization of new areas (away from the parental populations) requires the dispersal of adults that have settled on moving objects, such as boats or floating debris (Teske et al. 2015). Since direct observation of pelagic dispersal is challenging, genetic tools are often used to estimate population connectivity (e.g. Teske 2014) and confirm that levels of self-recruitment are high (Teske et al. 2015).

Although experimental trials using *Pyura stolonifera* and *P. herdmani* larvae showed that settlement occurs irrespective of the presence of adult tunic extracts (Rius et al. 2010a), the highest recruitment in the field is consistently reported on the tunics of adults or on substrata in the immediate vicinity (Alvarado et al. 2001, Marshall 2002, Monteiro et al. 2002, Castilla et al. 2014, Manríquez et al. 2016). More studies are however needed to confirm if self-recruitment is consistently present in aggregations of these ascidians. Taken together, gamete release synchrony, as well as strategies to retain early life-history stages, facilitate ecological dominance by members of the *P.s.s.c.* along rocky shores.

Suspension-feeding and diet

Species of the *P.s.s.c.* are, like the great majority of ascidians, ciliary-mucus active sieving suspension-feeders (Bone et al. 2003). Water filtration is extremely efficient, even for particles as small as 2–3 μm, with food items including detrital organic matter, diatoms and other phytoplankton, and suspended bacteria (Millar 1971, Monniot et al. 1991, Bak et al. 1998, Tyree 2001, Lambert 2005). Although no evidence of particle selection has yet been reported (Randløv & Riisgård 1979), studies of the gut of *Pyura stolonifera* suggest that phytoplankton is a much more important energy source than macroalgal detritus (Seiderer & Newell 1988). Other potential food items include developing stages and larvae of other invertebrate species, including self- and allogametes. Gut content analysis of adult individuals of *P. praeputialis* shows the presence of annelid, crustacean and mollusc larvae (Table 1). In addition, faecal pellets of *P. praeputialis* collected in the field contained tadpole larvae of *P. praeputialis*, mytilid larvae and newly hatched veliger larvae of the gastropod *Concholepas concholepas* (Table 2). Regardless of whether or not these items are digested, the available information suggests that members of the *P.s.s.c.* interact with other species inhabiting the same area by reducing food availability and by directly consuming early life-history stages.

The rates of filtration by ascidians generally depend on body size (Monniot et al. 1991) and seawater temperature (Fiala-Médioni 1978, Petersen & Riisgård 1992, Ribes et al. 1998). For *Pyura stolonifera*, filtration rates increase with the size of the branchial sacs (Klumpp 1984) and ciliary bands lining the stigmatal openings (Petersen & Svane 2002). Large individuals of *P. stolonifera* can filter up to 18 litres seawater h^{-1} (Klumpp 1984).

Table 1 Stomach content (mean number of individual food items counted ± SE) of *Pyura praeputialis* collected at two sampling sites in Antofagasta Bay, Chile

	El Way (n = 21)	Las Conchitas (n = 20)
Foraminifera [j, a]	11.67 ± 1.67	3.40 ± 1.02
Tintinnida [j, a]	1.00 ± 10.29	0
Nematoda [j, a]	0.10 ± 0.07)	0.10 ± 0.10
Gastropoda [j]	2.10 ± 0.48	2.05 ± 0.84
Mytilidae [l, j]	1.38 ± 0.30	18.25 ± 10.23
Annelida [l, j]	0.10 ± 0.10	0
Crustacea (unidentified)	0.29 ± 0.20	0.40 ± 0.24
Nauplii	1.86 ± 0.46	0.60 ± 0.60
Cyprids	1.48 ± 0.35	0.25 ± 0.16
Copepoda		
Harpacticoida [j, a]	0.38 ± 0.13	0.40 ± 0.23
Calanoida [j, a]	0.67 ± 0.17	1.60 ± 0.60
Cyclopoida [j, a]	0.33 ± 0.17	0.15 ± 0.08
Bacillariophyceae	3.33 ± 0.87	0.55 ± 0.17
Dinoflagellata		
Protoperidinium sp.	0.81 ± 0.25	0.70 ± 0.47
Ceratium sp.	0.14 ± 0.10	0.20 ± 0.20
Algal detritus	p	p
Silt	p	p
Faecal pellets	p	p

Note: Sampling was conducted between February and May 1998 (El Way, 23°45' S; 70°26' W) and March and May 1998 (Las Conchitas, 23°31' S; 70°32' W). Key: p = present but not quantified, [l] = larvae, [j] = juveniles, [a] = adults.

Table 2 *Pyura praeputialis* consumption of embryos and larvae when biofoam is present and absent

	Pools with biofoam (n = 15)	Pools without biofoam (n = 17)
Developing embryos[a]	0.83 ± 0.21	0
Tadpole larvae[a]	0.41 ± 0.13	0.03 ± 0.03
Mytilidae larvae[b]	0.84 ± 0.22	0.10 ± 0.06
Gastropod larvae[c]	0.15 ± 0.21	0.03 ± 0.21

Note: The abundance (mean ± SE) of developing embryos or larvae per unit length of the faecal pellet is indicated. Faecal pellets were collected from the field in May 2004 at El Way (23° 45' S, 70° 26' W) in Antofagasta Bay, Chile, approximately two hours after a spawning of *Pyura praeputialis*. They were collected from the vicinity of the exhalant siphon of individuals present in small rocky intertidal pools. Faecal pellets were collected from several pools with (n = 50 pellets) and without (n = 50 pellets) surface biofoam.

[a] Embryonic stages (ca. 200 μm, total length) and hatched *Pyura praeputialis* tadpole larvae (120 × 1200 μm; trunk and total length) (see Clarke et al. 1999).

[b] Prodisoconch and disoconch larvae, presumably *Perumytilus purpuratus* with sizes ranging from 100–180 μm (see Ramorino & Campos 1983).

[c] Newly hatched larvae of *Concholepas concholepas* of about 250 μm in size (see Manríquez et al. 2014).

Laboratory studies of *Pyura stolonifera* showed 100% retention efficiency when individuals were offered cells of the alga *Dunaliella primolecta* of sizes ranging from 4 to 6.35 μm (Stuart & Klumpp 1984). The same was found by Klumpp (1984) when food particles from the field were analyzed. These studies suggest that *Pyura stolonifera* is a non-selective suspension-feeder and that this may contribute to their competitive superiority over coexisting species, such as bivalves and sponges (Stuart & Klumpp 1984).

The risk of predation is a major selective pressure driving the evolution of larval settlement strategies in marine invertebrates (Thorson 1950). Young (1988) reported that gregarious species such as *Pyura haustor* rejected their own eggs and larvae as food, which can be seen as an adaptive strategy to avoid cannibalism. The large inhalant siphon of the members of the *P.s.*s.c. (diameter up to 1.5 cm) does not allow discrimination among suspended particles, and high levels of consumption of conspecific offspring have been reported (see Table 2). In addition, rates of cannibalism of larvae are extremely high when mechanisms of gamete and larval retention (e.g. biofoam) are present (Castilla et al. 2007b, Manríquez & Castilla 2010). This suggests that in the absence of biofoam the gametes spawned are not present or are considerably diluted, so there is little or no opportunity for cannibalism.

Community structure and ecological interactions

Members of the *P.s.*s.c. are fierce competitors for space, outcompeting individuals at intra-specific (Dalby 1995, Guiñez & Castilla 2001) and inter-specific (Castilla et al. 2004a, Caro et al. 2011, Manríquez et al. 2016) levels. Experimental studies of intertidal aggregations of *Pyura praeputialis* in Antofagasta Bay, Chile (where this species is invasive), showed that as aggregates reach high population densities, a negative relationship exists between the number of individuals per unit area and mean individual mass (Guiñez & Castilla 2001). As a result, competition in the form of severe crowding affects the morphological characteristics and the energy/tissue allocation of *P. praeputialis* (Guiñez & Castilla 2001). Experiments analyzing competition for space between the introduced *P. praeputialis* and the native mussel *Perumytilus purpuratus* in Antofagasta Bay have shown that the tunicate significantly affects native rocky intertidal biota, as well as several ecological processes.

Figure 6 Dense aggregations of *Pyura praeputilalis* in La Rinconada (23°28'16.41"S, 70°30'47.83"W), Chile. A bar with red and white sections (10 centimetres each) was positioned for scale.

For example, *Pyura praeputialis* has completely modified the intertidal zonation pattern of the bay by monopolizing the low and mid rocky intertidal zones, constraining the distribution of the native mussel to the mid-upper intertidal fringe (Castilla et al. 2004a, Caro et al. 2011, Manríquez et al. 2016). In Antofagasta Bay, Ortiz et al. (2013), analized rocky intertidal communities using Ecopath, Ecosim and Loop Analysis (Levins, 1974) to better understand the properties of keystone species (e.g. biomass, food consumption) and concluded that *P. praeputialis* is not a superior bioengineer compared to the kelp *Lessonia nigrescens*. In fact, the model indicated that these two bioengineer species hosted ecologically-similar species but relied on different ecological processes to carry out their ecosystem role.

Another consequence of dominance by members of the *P.s.s.c.* is the creation of habitat structure that allows a diverse associated community to thrive. In South Africa, 64 intertidal and 61 subtidal taxa of benthic macroinvertebrates, representing 10 phyla, were associated with aggregations of *Pyura stolonifera* (Fielding et al. 1994). Similar research on aggregations of *P. praeputialis* in Antofagasta Bay reported 96 associated benthic invertebrate species (Castilla et al. & Camaño 2001), with polychaetes, decapods and bivalves being the most speciose (see Cerda & Castilla 2001). In addition to epifauna, amphipods, copepods and nemerteans may be present inside the branchial sac of large individuals (e.g. Oldewage 1994, Dalby 1996). Van Driel & Steyl (1978) showed that in Algoa Bay (South Africa), the composition of communities associated with *P. stolonifera* was determined by levels of wave exposure. Similarly, Ramírez & Mena (1984) found differences in the distribution and abundance of macroalgae that grow on top of *Pyura praeputialis* aggregations in Antofagasta Bay across different levels of wave exposure. Although high densities of *Ulva* spp. are frequently observed growing on top of *P. praeputialis* (see Castilla et al. 2014), the presence of algae does not appear to affect the ascidian (Castilla et al. 2004b). The density of *P. praeputialis* individuals also affects the composition of associated communities, with clumped and sparse *P. praeputialis* individuals having different alga- and invertebrate-associated assemblages

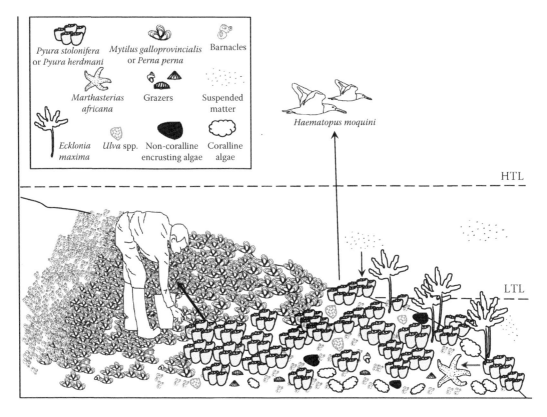

Figure 7 Food web and energy transfer on intertidal rocky shores in South Africa (southwestern coast, native range) where members of the *Pyura stolonifera* species complex can be found. Low tidal level (LTL) and high tidal level (HTL) are indicated. *(Continued)*

(Monteiro et al. 2002). Coralline algae and the limpet *Cellana* sp. are typically found in low-density aggregations of *Pyura praeputialis*, whereas species of whelks are more common in denser aggregations (Monteiro et al. 2002).

Predation upon members of the *P.s.*s.c. has rarely been studied (see Alvarado 2004). In Chile, apart from human harvesting (see below), the main intertidal predators are the sunstar *Heliaster helianthus* (Castilla et al. 2013) and the muricid gastropod *Concholepas concholepas* (Alvarado 2004), which both appear to regulate populations of *Pyura praeputialis* in the lower intertidal zone (Castilla et al. 2004a) (Figure 7). Further, the oystercatcher *Haematopus palliatus pitanay* is also an active predator of *Pyura praeputialis* in Chile (Pacheco & Castilla 2001, Goss-Custard et al. 2006) (Figure 7). In Australia, the triton shell *Cabestana spengleri* and the sooty oystercatcher *Haematopus fuliginosus* have been reported as preying on *Pyura praeputialis* (Schultz 1989, Fairweather 1991, Chafer 1992). Finally, in South Africa, known predators include the oystercatcher *Haematopus moquini* and the seastar *Marthasterias africana* (Wright et al. 2016) (Figure 7), but more work is needed to determine the extent of their predatory role in intertidal communities.

Apart from competition and predation, two additional factors can negatively affect the population dynamics of the *P.s.*s.c.. The first is mass mortality, as reported for *Pyura stolonifera* along the South African coast (Hanekom et al. 1999, Hanekom 2013). Such mortalities were suspected to occur as a result of infection by an unidentified microbe, potentially as an indirect result of abnormally high temperatures (Hanekom 2013). The second factor is patch dynamics that are directly influenced by mechanical forces. Intertidal and subtidal aggregations of *P. praeputalis* are

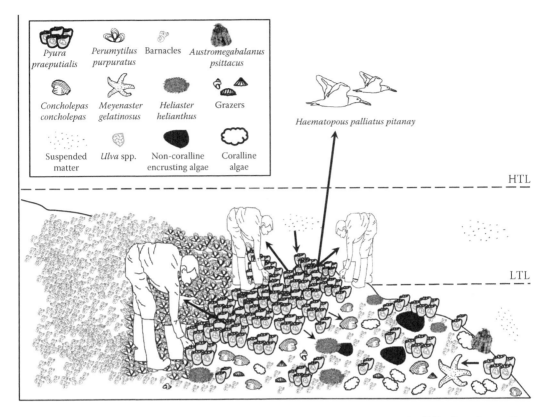

Figure 7 (Continued) Food web and energy transfer on intertidal rocky shores in Chile (Antofagasta Bay, introduced range) where members of the *Pyura stolonifera* species complex can be found. Low tidal level (LTL) and high tidal level (HTL) are indicated.

constantly exposed to removal by wave action. An important patch-filling mechanism is propagule retention, which increases recruitment in the patch-border zone next to adults (Alvarado et al. 2001).

Invasion biology and the Pyura stolonifera *species complex*

Species that are dominant in their native habitat are expected to have a particularly high likelihood of becoming invasive once established elsewhere (Simberloff 2010). Therefore, understanding dominance is particularly important for predicting the changes in ecosystem structure and function caused by biological invasions. Members of the *P.s.s.c.* dominate intertidal and subtidal areas in their native ranges, and when these species are introduced to new areas, they can strongly alter local communities. Introductions of members of the *P.s.s.c.* are being reported with increasing frequency, including the colonization of the northern part of New Zealand's North Island by *Pyura doppelgangera* (Hayward & Morley 2009, Fletcher 2014) and the recent introduction of *P. herdmani* to northwestern Spain (X. Turon, personal communication).

A growing debate exists in the literature around the concept of invasiveness and the impact of invasive species on recipient communities (see Cronon 1983, Katz 1992, Soulé 1995, Jordan 2000, Katz 2000, Cafaro 2001, Castilla & Neill 2009, Simberloff 2012, Simberloff & Vitule 2014). This debate includes conceptual aspects such as understanding what is 'native', 'harmful' or 'wild', and even philosophical aspects; for example whether non-indigenous sentient (vertebrate) or non-sentient species (e.g. invertebrates, plants, fungi) have intrinsic value (see Varner 1990, 1998, Justus et al. 2009). Some non-indigenous species are considered to be innocuous, while others can have dramatic ecological effects (although the latter may have critical economic value for humans, e.g. in terms of food security). Non-indigenous marine organisms can have positive, neutral or negative impacts on humans (Castilla & Neill 2009, Manríquez et al. 2016). For example, the non-indigenous intertidal macroalgae *Porphyra linearis, Pyropia pseudolinearis* (=*Porphyra pseudolinearis*) and *Mastocarpus papillatus* are commercially important resources in Chile, as they are extracted by small-scale (artisanal) fishermen, and as yet no negative ecological impact has been reported. Similarly, both *Pyura praeputialis* and subtidal red algae of the genus *Gracilaria* are introduced species in Chile but have not yet caused adverse economic effects (Castilla et al. 2002, Castilla et al. 2014, Manríquez et al. 2016). In South Africa, the introduced mussel *Mytilus galloprovincialis* has been farmed for decades (Heasman et al. 1998), but even though this species is now of considerable economic importance it dominates extensive stretches of the South African coastline, causing drastic ecological changes to native communities (Robinson et al. 2005). Another example of dominant non-indigenous species causing harm to natural communities is *Pyura praeputialis* in Antofagasta Bay. This species has reduced the abundance of native species of mussels through displacement by competition for space (Castilla 1998, Castilla & Guiñez 2000, Castilla et al. 2004a, Castilla 2008, Caro et al. 2011, Castilla et al. 2014, Manríquez et al. 2016). In turn, however, the presence of this introduced species has resulted in an increase of overall local biodiversity (i.e. an increase in macroinvertebrate richness) via the provision of a new habitat (Castilla et al. 2004b, Castilla et al. 2005).

Once a new geographic area is colonized, establishment success depends on the inability of the resident community to repel newcomers (Rius et al. 2014b). Specifically, certain types of habitats (e.g. marine hard infrastructure, Airoldi et al. 2015) and community attributes (e.g. levels of native species diversity, Crutsinger et al. 2008) may facilitate biological invasions in some regions but not others. This is illustrated by considerable differences in the colonization success of the non-indigenous populations of *Pyura doppelgangera* (Teske et al. 2014). In the North Island of New Zealand this species was introduced less than 20 years ago (Hayward & Morley 2009) and has now spread along 100 km of exposed rocky shores (Fletcher 2014). In contrast, introduced populations of the same species in two regions of the Australian mainland (Adelaide in South Australia, and Corner Inlet in Victoria) have failed to expand their ranges beyond the immediate points of introduction. Taxonomic and genetic evidence suggests that this species has recently been introduced to mainland Australia (Kott 1952, Teske et al. 2014) from northern Tasmania. Both Adelaide and Corner Inlet lack rocky shores and are dominated by sandy shores (Bowman & Harvey 1986), with artificial structures such as jetties and piers representing the only habitat suitable for settlement. Such marine infrastructures are spaced a few kilometres apart and thus large gaps of sandy beach seem to limit connectivity among suitable habitats. Small-scale dispersal has been assessed in the *P. doppelgangera* population of the Adelaide metropolitan area and revealed high levels of self-recruitment, with most larvae settling on the structure occupied by their parents (Teske et al. 2015). This is also consistent with findings suggesting effective gamete retention mechanisms and spawning synchrony of *P. praeputialis* in southeastern Australia (Marshall 2002).

The short larval duration typical of solitary ascidians (Figure 5) considerably reduces colonization success when substrata suitable for settlement are located far from one another, and suggests that *Pyura doppelgangera* will only spread rapidly where habitat is more continuous. In New Zealand, the sea star *Stichaster australis* and a whelk of the genus *Cabestana* prey upon the abundant new food resource provided by *Pyura doppelgangera* (Fletcher 2014). However, such biotic resistance effects are clearly insufficient to counteract the invasion of this ascidian species. Localized removal of *P. doppelgangera* patches have been conducted by local communities, but it is likely to prove too late for the complete eradication of this invasive species.

Cases of naturalization (i.e. species that are able to self-sustain populations but that have failed to spread beyond the immediate point of introduction, Richardson et al. 2000) have also been reported in the *P.s.s.c.* For example, while *Pyura praeputialis* is found along thousands of kilometres of rocky shore in its native habitat in Australia, the Chilean distribution is restricted to a single bay (Clarke et al. 1999, Castilla & Guiñez 2000, Castilla et al. 2002). Other examples of limited distribution include those of *P. dalbyi* in Western Australia (Teske et al. 2011) and *P. doppelgangera* in mainland Australia (Teske 2014, Teske et al. 2015).

Human exploitation and conservation
of the Pyura stolonifera *species complex*

Human activities such as rocky intertidal harvesting and trampling are well-known stressors of several marine taxa including algae (Bally & Griffiths 1989, Castilla & Bustamante 1989, Castilla et al. 2007a) and various invertebrates (Castilla & Durán 1985, Roy et al. 2003, Smith & Murray 2005, Rius et al. 2006). Although human exploitation of intertidal resources targets a wide range of species (Moreno et al. 1984, Lasiak 1991, Keough et al. 1993, Castilla 1999), dominant species (e.g. mussels and tunicates) are often an important proportion of the overall catch (Kyle et al. 1997, Rius & Cabral 2004). This is not surprising as the gregarious nature of these organisms allows maximization of the catch. The selective removal of large adults in limited quantities may allow sustainable exploitation of intertidal resources (as is sometimes seen in subsistence exploitation, Castilla et al. 2014), but human harvesting is often unrestrained and can be a major conservation threat to intertidal communities.

Humans harvest members of the *P.s.s.c.* for subsistence exploitation and/or recreational activities (e.g. bait collection). Such activities have been reported in Australia (Otway 1989, Fairweather 1991, Kingsford et al. 1991, Chapman & Underwood 1994, Monteiro et al. 2002), South Africa (Kyle et al. 1997) and Chile (Castilla et al. 2004a, Castilla et al. 2014, Manríquez et al. 2016). Fairweather (1991) studied the exploitation of *Pyura praeputialis* at seven intertidal sites in New South Wales, Australia, showing that changes in density of *P. praeputialis* were temporally asynchronous among sites but that *P. praeputialis* recovery was consistently slow. The study concluded that the population dynamics of *P. praeputialis* are modulated by human harvesting, episodic storms and recruitment patterns. Monteiro et al. (2002) studied habitat structure in patches of *P. praeputialis* in Sydney, Australia, where fishermen collect this species for bait, and found that changes to the structure of these patches resulted in changes in the composition of associated biota. The authors identified 19 algal and 45 invertebrate species in habitats provided by *P. praeputialis*, and species assemblages differed significantly between sparse and dense patches. In South Africa, *P. herdmani* is heavily exploited by intertidal food-gatherers (Fielding et al. 1994), as well as anglers who collect this species for bait. Along some parts of the South African coast *P. herdmani* is the second most important harvested species (after mussels) (Kyle et al. 1997). In Chile, *P. praeputialis* is considered a delicacy and has a high market value, and studies indicate that the rate of extraction by professional tunicate gatherers can be up to 750 individuals h^{-1} during low tide (Castilla et al. 2014, Manríquez et al. 2016). The harvesting of *P. praeputialis* is in fact so continuous and intense that it impairs recruitment. The shrinking of *P. praeputialis* aggregations in certain sites has allowed mussels to recover intertidal dominance (Castilla et al. 2014, Manríquez et al. 2016).

Reductions of intertidal aggregations of members of the *P.s.s.c.* can significantly affect the associated intertidal community (Fielding et al. 1994, Cerda & Castilla 2001, Monteiro et al. 2002, Castilla et al. 2004b, Manríquez et al. 2016). In Chile, the crevices and gaps between individuals of *Pyura praeputialis* create microhabitats for the settlement of the species *Concholepas concholepas*, a commercially important gastropod (Castilla & Jerez 1986, Castilla et al. 1998, Castilla 1999, Castilla & Defeo 2001, Manríquez et al. 2008, Gelcich et al. 2017), and they are used by females of *C. concholepas* to lay thousands of egg capsules during the reproductive season (authors' unpublished data). In addition, individuals of *C. concholepas* and *Octopus mimus* have traditionally been collected from aggregations of *Pyura praeputialis* during low tide. However, overexploitation of

P. praeputialis during the past decade has resulted in a considerable decrease in the number of these associated species (authors' unpublished data). Similarly, the scarcity of *P. stolonifera* and *P. herdmani* along some sections of the South African coast may be an indication of overharvesting (e.g. Kyle et al. 1997, Majiza & Lasiak 2010). Future repopulation initiatives may be key for restoring ecosystem functioning, as these aggregations play an important role as bioengineers (Castilla et al. 2001, Castilla et al. 2004b). Considering that ecological dominance influences many fundamental aspects of ecosystem health, such as coexistence and metacommunity dynamics (Hillebrand et al. 2008), human activities reducing the dominance of members of the *P.s.*s.c. are likely to result in alterations of biodiversity patterns.

Conclusions and future research directions

Dominant species are superior competitors that often generate exceptional levels of biomass. Although many attributes (e.g. gregariousness, sessile life strategy, broadcast spawning) may be linked to ecological dominance (Figure 2), it cannot be readily explained by any specific combination of traits (Figure 3). The presence of dominant sessile invertebrates generally increases habitat complexity, directly benefiting a wide range of associated biota. Therefore, dominant species are key components for the conservation of biodiversity and ecosystem functioning along rocky shores.

Although knowledge of ecological dominance has been accumulating for decades, more research is needed to understand fully some of the underlying ecological and evolutionary mechanisms. For example, little is known about how kin selection affects gregariousness, and there is limited information on possible links between Allee effects and ecological dominance. To date, there is little empirical evidence for Allee effects in natural populations (Gascoigne & Lipcius 2004), and studies are particularly scarce in the context of dominant species.

This review focused on members of the *P.s.*s.c. as a key example for the study of ecological dominance. Among many consequences that derive from the presence of these dominant species, creation of architectural complexity is one of the most striking, as it influences the hydrodynamics of intertidal zones, ameliorates physical stress and creates habitat for a wide range of associated species. Other aspects remain largely unexplored, however. For example, there is as yet no information on the impact of the *P.s.*s.c. on planktonic communities. The high rates of filtration and particle retention achieved by the members of the *P.s.*s.c. suggest that they may extract massive amounts of suspended particles from seawater (Klumpp 1984, Seiderer & Newell 1988). This, together with their aggregated nature and large adult size, strongly suggest that these species have an important role in ecosystems. High clearance rates that significantly alter seston composition and reduce food availability may directly affect adult survival, growth and reproductive potential of competing or subordinate species, such as mussels and barnacles. Another unexplored consequence of such high filtration capacity is the consumption of heterospecific gametes and larvae. Studies have assessed the role of cannibalism by members of the *P.s.*s.c. and found that it reduces the conspecific larval pool. However, little is known about how such feeding may influence the abundance and distribution of heterospecific gametes and larvae. Species that could be directly displaced include suspension-feeders but also primary producers (e.g. seaweeds). Thus, possible impacts of the presence of these dominant ascidians could go beyond a specific trophic level, and influence entire food chains. A negative association between tunicate abundance and the settlement of mussel larvae has been found (LeBlanc et al. 2007), which suggests tunicate predation on mussel larvae, as well as a reduction of available food particles. This may have been the mechanism through which the invasive population of *Pyura praeputialis* in Antofagasta Bay outcompeted the native mussel species *Perumytilus purpuratus* (Castilla et al. 2004a, Castilla et al. 2014). Finally, the multiphasic life cycle of the members of the *P.s.*s.c. (Figure 5) implies that the size and type of particles consumed by each life-history stage (i.e. postmetamorph, juvenile and adult) vary, so the study of dietary shifts (see Sherrard & LaBarbera 2005) may reveal important insights into this possible form of food competition. Taken together, filtration capacity, when fully investigated, may reveal key aspects facilitating the establishment and maintenance of dense aggregations of these dominant species.

Comprehensive information is available on the intertidal distribution, population structure, biomass, energy/tissue allocation and phenotypic traits of the members of the *P.s.s.c.*, especially from Antofagasta Bay, Chile (e.g. Clarke et al. 1999, Castilla et al. 2000). The restricted distribution of *Pyura praeputialis* in South America (a range of only about 70 km; Castilla et al. 2002) and the high densities attained there (intertidal belts over 10 m wide, with the highest densities towards the centre of the belt) makes Antofagasta Bay a unique location to study ecological aspects of this species. However, our present ecological knowledge is limited to studies of *P. praeputialis* and *P. stolonifera*. Further research is therefore required to study the ecology of the remaining members of the *P.s.s.c.*, as well as to understand the influence of coastal geography and oceanography on ecological dominance. Nearshore larval retention of *P. praeputialis* has been reported in Antofagasta Bay (Castilla & Largier 2002), but similar studies are needed in the native range. Distances between suitable habitats may be too large to be crossed by larval transport, particularly when oceanographic conditions are not favourable, and consequently, self-recruitment seems to be the norm (Teske et al. 2015). Regions where the effects of geography and oceanography could be tested include False Bay in South Africa and the coastline around Adelaide in Australia. Taken together, it remains uncertain how geographic and oceanographic singularities affect ecological dominance.

In order to study ecological dominance, a detailed understanding of the taxonomy and evolutionary history of the studied organism is required. For example, failing to correctly identify cryptic species or hybrids could lead to erroneous interpretation of ecological data. For the members of the *P.s.s.c.*, the combined study of morphological and genetic data has greatly facilitated the resolution of phylogenetic relationships (Teske et al. 2011, Rius & Teske 2013). However, numerous challenges remain. First, it is presently not established whether the species complex is reciprocally monophyletic. Thus, more work is needed to understand whether all its members have arisen from a single ancestor, or whether the inclusion of some species is merely an artefact of similar morphology. For example, the phylogenetic placement of *Pyura dalbyi* is poorly resolved (Teske et al. 2011), and it is possible that this species is more closely related to the morphologically very different *P. spinifera* (Quoy & Gaimard 1834). It is also uncertain whether any of the presently-accepted species comprise additional 'cryptic' species that should be scientifically described. Phylogenetic data based on mitochondrial DNA sequences indicate that *P. herdmani* comprises four reciprocally monophyletic genetic lineages (Teske et al. 2011). One lineage occurs in northwestern Africa, one in subtropical/tropical southern Africa, and two lineages have overlapping ranges in temperate southern Africa. While these lineages may be morphologically difficult to distinguish, different geographical ranges or habitat preferences support the hypothesis that they may be different species (see Rius & Teske 2011). Of the temperate southern African populations, one has a sister-taxon relationship with the northwest African population of *P. herdmani* and has so far been exclusively found on rocky shores, while the other also occurs on sandy sediments. Given that *P. herdmani* can hybridize with *P. stolonifera* (Rius & Teske 2013), it cannot be ruled out that hybridization is also common among the individual southern African lineages of *P. herdmani*, which would considerably complicate attempts at resolving their taxonomy. The existence of hybrids in regions where multiple species coexist (i.e. southern Africa and southeastern Australia) could provide important insights into understanding recent range expansions. Human activities are known to facilitate interbreeding among divergent lineages (Chunco 2014, Vallejo-Marín & Hiscock 2016), which may create hybrids with enhanced ability to colonize new habitats (Ruis & Darling 2014).

Genetic data have been particularly useful in confirming the non-indigenous status of populations of the members of the *P.s.s.c.* Genetic evidence often falls into two categories: 1) lack of genetic differentiation among non-indigenous populations that contrasts with well-defined native population structure, and 2) recent divergence between native and introduced ranges (on the basis of molecular dating) since the start of human-mediated transoceanic transport. In addition, circumstantial evidence for the non-indigenous status of a particular population could include small distribution ranges (e.g. limited to harbours as most likely points of introduction, Carlton & Geller 1993) and settlement on marine infrastructures in regions that lack rocky shores. DNA sequence data have been used to identify the lack of genetic differentiation among introduced populations that are separated by vast distribution gaps. For example, a genetic study of *Pyura praeputialis* samples collected in Chile revealed that these were genetically indistinguishable from eastern Australian populations (Castilla et al. 2002), and all mtDNA haplotypes found in the single Western Australian population of *P. dalbyi* were also found in southeastern Australia (Teske et al. 2011). The use of polymorphic microsatellites can be much more informative than sequence data in revealing the colonization history of non-indigenous populations (e.g. Rius et al. 2012). The high mutation rate of these genetic markers make them suitable for distinguishing ancient natural colonization events from introductions that have occurred since humans started navigating the seas. For example, microsatellite data confirmed that all non-Tasmanian populations of *P. doppelgangera* diverged from closely related northern Tasmanian populations no more than a few hundred years ago (Teske et al. 2014). Genetic data able to provide information on recent changes are thus required for understanding recent colonization events by these dominant species. For example, many uncertainties remain concerning the introduction of *P. praeputialis* from Australia to Chile. Fine-scale and temporal genetic studies have the potential to not only reveal important information on the colonization history of this species, but also to provide key insights into the community effects of this species over time.

Members of the *P.s.*s.c. are often reported as introduced species around the world (Figure 4) but few studies have focused on reconstructing invasion routes or identifying source populations or the presence of recurrent introductions. A particularly interesting example is the recent introduction of *Pyura doppelgangera* in New Zealand (Hayward & Morley 2009), where it has spread across continuous rocky shores, replacing native assemblages. Since limited information is available to date (Rius & Teske 2013), a multilocus genetic or genomic study would help to explain why and how this invasion is particularly successful. New introductions by members of the *P.s.*s.c. provide unplanned replicated experiments to study the consequences of ecological dominance for rocky shore ecosystems.

Acknowledgements

We are grateful to Steve Hawkins, Louise Firth and Hanna Schuster for inviting us to submit this review, as well as an anonymous reviewer for insightful comments that helped improve the final version. We are grateful to John Largier for discussion during the early planning of this review. MR acknowledges ASSEMBLE (an EU FP7 research infrastructure initiative comprising a network of marine research stations) for awarding a grant to conduct research at the 'Estación Costera de Investigaciones Marinas—Las Cruces', Chile. JCC is grateful for support from the project entitled 'Núcleo Milenio en Conservación Marina CCM RC 130004, Iniciativa Científica Milenio, Ministerio de Economía' de Chile and the 'Minera Escondida Ltda' for long-term logistic and financial support. PHM acknowledges the National Fund for Scientific and Technological Development (FONDECYT) of Chile for the financial support that has allowed him to conduct, through several grants, research activities in Antofagasta Bay from year 2000. This work is based upon research supported by the South African Research Chairs Initiative of the Department of Science and Technology and the National Research Foundation.

References

Aiken, C.M. & Navarrete, S.A. 2014. Coexistence of competitors in marine metacommunities: environmental variability, edge effects, and the dispersal niche. *Ecology* **95**, 2289–2302.

Airoldi, L. Turon, X., Perkol-Finkel, S. & Rius, M. 2015. Corridors for aliens but not for natives: effects of marine urban sprawl at a regional scale. *Diversity and Distributions* **21**, 755–768.

Almany, G.R., Berumen, M.L., Thorrold, S.R., Planes, S. & Jones, G.P. 2007. Local replenishment of coral reef fish populations in a marine reserve. *Science* **316**, 742–744.

Alvarado, J.L. 2004. *Patrones de distribución de organismos intermareales en la Bahía de Antofagasta: patrones simples y causas complejas*. PhD thesis, Pontificia Universidad Católica de Chile.

Alvarado, J.L. & Castilla, J.C. 1996. Tridimensional matrices of mussels *Perumytilus purpuratus* on intertidal platforms with varying wave forces in central Chile. *Marine Ecology Progress Series* **133**, 135–141.

Alvarado, J.L., Pinto, R., Marquet, P., Pacheco, C., Guiñez, R. & Castilla, J.C. 2001. Patch recolonization by the tunicate *Pyura praeputialis* in the rocky intertidal of the Bay of Antofagasta, Chile: evidence for self-facilitation mechanisms. *Marine Ecology Progress Series* **224**, 93–101.

Astorga, M.P., Guiñez, R. & Castilla, J.C. 2009. Genetic divergence in the ascidian *Pyura praeputialis* (= *Pyura stolonifera*) (Heller, 1878) from mainland Australia and Tasmania. *Papers and Proceedings of the Royal Society of Tasmania* **143**, 101–104.

Babcock, R. & Keesing, J. 1999. Fertilization biology of the abalone *Haliotis laevigata*: laboratory and field studies. *Canadian Journal of Fisheries and Aquatic Sciences* **56**, 1668–1678.

Baird, A.H. & Hughes, T.P. 2000. Competitive dominance by tabular corals: an experimental analysis of recruitment and survival of understorey assemblages. *Journal of Experimental Marine Biology and Ecology* **251**, 117–132.

Baird, D., Asmus, H. & Asmus, R. 2004. Energy flow of a boreal intertidal ecosystem, the Sylt-Rømø Bight. *Marine Ecology Progress Series* **279**, 45–61.

Bak, R.P.M., Joenje, M., de Jong, I., Lambrechts, D.Y.M. & Nieuwland, G. 1998. Bacterial suspension feeding by coral reef benthic organisms. *Marine Ecology Progress Series* **175**, 285–288.

Bally, R. & Griffiths, C.L. 1989. Effects of human trampling on an exposed rocky shore. *International Journal of Environmental Studies* **34**, 115–125.

Bayne, B.L. 1964. Primary and secondary settlement in *Mytilus edulis* L. (Mollusca). *Journal of Animal Ecology* **33**, 513–523.

Berec, L., Angulo, E. & Courchamp, F. 2007. Multiple Allee effects and population management. *Trends in Ecology & Evolution* **22**, 185–191.

Berkley, H.A., Kendall, B.E., Mitarai, S. & Siegel, D.A. 2010. Turbulent dispersal promotes species coexistence. *Ecology Letters* **13**, 360–371.

Bertness, M.D. & Grosholz, E. 1985. Population dynamics of the ribbed mussel, *Geukensia demissa*: the costs and benefits of an aggregated distribution. *Oecologia* **67**, 192–204.

Bertness, M.D. & Leonard, G.H. 1997. The role of positive interactions in communities: lessons from intertidal habitats. *Ecology* **78**, 1976–1989.

Bone, Q., Carré, C. & Chang, P. 2003. Tunicate feeding filters. *Journal of the Marine Biological Association of the UK* **83**, 907–919.

Bowman, G. & Harvey, N. 1986. Geomorphic evolution of a Holocene beach-ridge complex, LeFevre Peninsula, South Australia. *Journal of Coastal Research* **2**, 345–362.

Branch, G.M. 1984. Competition between marine organisms: ecological and evolutionary implications. *Oceanography and Marine Biology: An Annual Review* **22**, 429–593.

Brown, J.H. & Maurer, B.A. 1986. Body size, ecological dominance and Cope's rule. *Nature* **324**, 248–250.

Bruno, J.F., Stachowicz, J.J. & Bertness, M.D. 2003. Inclusion of facilitation into ecological theory. *Trends in Ecology & Evolution* **18**, 119–125.

Cafaro, P. 2001. For a grounded conception of wilderness and more wilderness on the ground. *Ethics and the Environment* **6**, 1–17.

Carlton, J.T. & Geller, J.B. 1993. Ecological roulette: the global transport of nonindigenous marine organisms. *Science* **261**, 78–82.

Caro, A.U., Guiñez, R., Ortiz, V. & Castilla, J.C. 2011. Competition between a native mussel and a non-indigenous invader for primary space on intertidal rocky shores in Chile. *Marine Ecology Progress Series* **428**, 177–185.

Castilla, J.C. 1998. Las comunidades intermareales de la Bahía San Jorge: estudios de linea base y el programa ambiental de Minera Escondida Ltda. en Punta Coloso. In *Mineria del Cobre, Ecologia y Ambiente Costero*, D. Arcos (ed.). Editorial Anibal Pinto SA, 221–244.

Castilla, J.C. 1999. Coastal marine communities: trends and perspectives from human exclusion experiments. *Trends in Ecology & Evolution* **14**, 280–283.

Castilla, J.C. 2008. Fifty years from the publication of the first two papers on Chilean rocky intertidal assemblages: honoring Professor Eric R. Guiler. *Revista de Biología Marina y Oceanografía* **43**, 457–467.

Castilla, J.C. & Bustamante, R.H. 1989. Human exclusion from rocky intertidal of Las Cruces, central Chile: effects on *Durvillaea antarctica* (Phaeophyta, Durvilleales). *Marine Ecology Progress Series* **50**, 203–214.

Castilla, J.C., & Camaño, A. 2001. El piure de Antofagasta, *Pyura praeputialis* (Heller, 1878): un competidor dominante e ingeniero de ecosistemas. In *Sustentabilidad de la Biodiversidad.* K. Alveal & T. Antezana (eds). Universidad de Concepcíon, Chile, 719–729.

Castilla J.C., Campos, M.A. & Bustamante, R.H. 2007a. Recovery of *Durvillaea antarctica* (Durvilleales) inside and outside Las Cruces Marine Reserve. *Ecological Applications* **17**, 1511–1522.

Castilla, J.C., Collins, A.G., Meyer, C.P., Guiñez, R. & Lindberg, D.R. 2002. Recent introduction of the dominant tunicate, *Pyura praeputialis* (Urochordata, Pyuridae) to Antofagasta, Chile. *Molecular Ecology* **11**, 1579–1584.

Castilla, J.C. & Defeo, O. 2001. Latin-American benthic shellfisheries: emphasis on co-management and experimental practices. *Reviews in Fish Biology and Fisheries* **11**, 1–30.

Castilla, J.C. & Durán, L.R. 1985. Human exclusion from the rocky intertidal zone of central Chile: the effects on *Concholepas concholepas* (Gastropoda). *Oikos* **45**, 391–399.

Castilla, J.C. & Guiñez, R. 2000. Disjoint geographical distribution of intertidal and nearshore benthic invertebrates in the Southern Hemisphere. *Revista Chilena de Historia Natural* **73**, 585–603.

Castilla, J.C., Guiñez, R., Alvarado, J.L., Pacheco, C. & Varas, M. 2000. Distribution, population structure, population biomass and morphological characteristics of the tunicate *Pyura stolonifera* in the Bay of Antofagasta, Chile. *Marine Ecology* **21**, 161–174.

Castilla, J.C., Guiñez, R., Caro, A.U. & Ortiz, V. 2004a. Invasion of a rocky intertidal shore by the tunicate *Pyura praeputialis* in the Bay of Antofagasta, Chile. *Proceedings of the National Academy of Sciences of the USA* **101**, 8517–8524.

Castilla, J.C. & Jerez, G. 1986. Artisanal fishery and the development of a data base for managing the loco (*Concholepas concholepas*) in Chile. *Canadian Journal of Fisheries and Aquatic Sciences* **92**, 133–139.

Castilla, J.C., Lagos, N.A. & Cerda, M. 2004b. Marine ecosystem engineering by the alien ascidian *Pyura praeputialis* on a mid-intertidal rocky shore. *Marine Ecology Progress Series* **268**, 119–130.

Castilla, J.C. & Largier, J. 2002. *The Oceanography and Ecology of the Nearshore and Bays in Chile.* Ediciones Universidad Católica de Chile.

Castilla, J.C., Manríquez, P., Alvarado, J., Rosson, A., Pino, C., Soto, R., Oliva, D. & Defeo, O. 1998. Artisanal caletas as units of production and co-managers of benthic invertebrates in Chile. *Canadian Journal of Fisheries and Aquatic Sciences* **125**, 407–413.

Castilla, J.C., Manríquez, P.H., Delgado, A., Ortiz, V., Jara, M.E. & Varas, M. 2014. Rocky intertidal zonation pattern in Antofagasta, Chile: invasive species and shellfish gathering. *PLoS ONE* **9**, e110301.

Castilla, J.C., Manríquez, P.H., Delgado, A.P., Gargallo, L., Leiva, A. & Radic, D. 2007b. Bio-foam enhances larval retention in a free-spawning marine tunicate. *Proceedings of the National Academy of Sciences of the USA* **104**, 18120–18122.

Castilla, J.C., Navarrete, S.A., Manzur, T. & Barahona, M. 2013. *Heliaster helianthus.* In *Starfish*, J.M. Lawrence (ed.). Baltimore, USA: Johns Hopkins University Press, 153–160.

Castilla, J.C. & Neill, P.E. 2009. Marine bioinvasions in the southeastern Pacific: status, ecology, economic impacts, conservation and management. In *Biological Invasions in Marine Ecosystems,* G. Rilov & J.A. Crooks (eds). Berlin & Heidelberg: Springer-Verlag, 439–457.

Castilla, J.C., Uribe, M., Bahamonde, N., Clarke, M., Desqueyroux-Faúndez, R., Kong, I., Moyano, H., Rozbaczylo, N., Santelices, B., Valdovinos, C. & Zavala, P. 2005. Down under the southeastern Pacific: marine non-indigenous species in Chile. *Biological Invasions* **7**, 213–232.

Cerda, M. & Castilla, J.C. 2001. Diversity and biomass of macro-invertebrates in intertidal matrices of the tunicate *Pyura praeputialis* (Heller, 1878) in the Bay of Antofagasta, Chile. *Revista Chilena de Historia Natural* **74**, 841–853.

Cerrano, C. & Bavestrello, G. 2009. Mass mortalities and extinctions. In *Marine Hard Bottom Communities*, M. Wahl (ed.). Berlin & Heidelberg: Springer-Verlag, 295–307.

Chafer, C.J. 1992. Ascidian predation by the sooty oystercatcher *Haematopus fuliginosus*: further observation. *Stilt* **20**, 20–21.

Chapman, M.G. & Underwood, A.J. 1994. Dispersal of the intertidal snail, *Nodilittorina pyramidalis*, in response to the topographic complexity of the substratum. *Journal of Experimental Marine Biology and Ecology* **179,** 145–169.

Chunco, A.J. 2014. Hybridization in a warmer world. *Ecology and Evolution* **4,** 2019–2031.

Clapham, M.E., Bottjer, D.J., Powers, C.M., Bonuso, N., Fraiser, M.L., Marenco, P.J., Dornbos, S.Q. & Pruss, S.B. 2006. Assessing the ecological dominance of Phanerozoic marine invertebrates. *Palaios* **21,** 431–441.

Clarke, M., Ortiz, V. & Castilla, J.C. 1999. Does early development of the chilean tunicate *Pyura praeputialis* (Heller, 1878) explain the restricted distribution of the species? *Bulletin of Marine Science* **65,** 745–754.

Cole, V.J. & McQuaid, C.D. 2010. Bioengineers and their associated fauna respond differently to the effects of biogeography and upwelling. *Ecology* **91,** 3549–3562.

Coley, P.D., Bryant, J.P. & Chapin, F.S. 1985. Resource availability and plant antiherbivore defense. *Science* **230,** 895–899.

Connell, J.H. 1978. Diversity in tropical rain forests and coral reefs. *Science* **199,** 1302–1310.

Connolly, S.R. & Roughgarden, J. 1999. Theory of marine communities: competition, predation, and recruitment-dependent interaction strength. *Ecological Monographs* **69,** 277–296.

Courchamp, F., Clutton-Brock, T. & Grenfell, B. 1999. Inverse density dependence and the Allee effect. *Trends in Ecology & Evolution* **14,** 405–410.

Crain, C.M. & Bertness, M.D. 2006. Ecosystem engineering across environmental gradients: implications for conservation and management. *BioScience* **56,** 211–218.

Cronon, W. 1983. *Changes in the Land: Indians, Colonists and the Ecology of New England.* New York: Hill and Wang.

Crooks, J.A. 2002. Characterizing ecosystem-level consequences of biological invasions: the role of ecosystem engineers. *Oikos* **97,** 153–166.

Crutsinger, G.M., Souza, L. & Sanders, N.J. 2008. Intraspecific diversity and dominant genotypes resist plant invasions. *Ecology Letters* **11,** 16–23.

Dalby, J.E.J. 1996. Nemertean, copepod, and amphipod symbionts of the dimorphic ascidian *Pyura stolonifera* near Melbourne, Australia: specificities to host morphs, and factors affecting prevalences. *Marine Biology* **126,** 231–243.

Dalby, J.E.J. 1997. Dimorphism in the ascidian *Pyura stolonifera* near Melbourne, Australia, and its evaluation through field transplant experiments. *Marine Ecology* **18,** 253–271.

Day, R.W., Barkai. A. & Wickens, P.A. 1991. Trapping of three drilling whelks by two species of mussel. *Journal of Experimental Marine Biology and Ecology* **149,** 109–122

Dayton, P.K. 1971, Competition, disturbance, and community organization – provision and subsequent utilization of space in a rocky intertidal community. *Ecological Monographs* **41,** 351–389.

Dayton, P.K. 1975. Experimental evaluation of ecological dominance in a rocky intertidal algal community. *Ecological Monographs* **45,** 137–159.

Donahue, M.J. 2006. Allee effects and conspecific cueing jointly lead to conspecific attraction. *Oecologia* **149,** 33–43.

Downing, J.A., Rochon, Y., Pérusse, M. & Harvey, H. 1993. Spatial aggregation, body size, and reproductive success in the freshwater mussel *Elliptio complanata. Journal of the North American Benthological Society* **12,** 148–156.

Eriksson, B.K., van der Heide, T., van de Koppel, J., Piersma, T., van der Veer, H.W. & Olff, H. 2010. Major changes in the ecology of the Wadden Sea: human impacts, ecosystem engineering and sediment dynamics. *Ecosystems* **13,** 752–764.

Erlandsson, J. & McQuaid, C.D. 2004. Spatial structure of recruitment in the mussel *Perna perna* at local scales: effects of adults, algae and recruit size. *Marine Ecology Progress Series* **267,** 173–185.

Erlandsson, J., Pal, P. & McQuaid, C.D. 2006. Re-colonisation rate differs between co-existing indigenous and invasive intertidal mussels following major disturbance. *Marine Ecology Progress Series* **320,** 169–176.

Fairweather, P.G. 1991. A conceptual framework for ecological studies of coastal resources: an example of a tunicate collected for bait on Australian Seashores. *Ocean and Shoreline Management* **15,** 125–142.

Farrell, E.D. & Crowe, T.P. 2007. The use of byssus threads by *Mytilus edulis* as an active defence against *Nucella lapillus. Journal of the Marine Biological Association of the United Kingdom* **87,** 559–564.

Fiala-Médioni, A. 1978. Filter-feeding ethology of benthic invertebrates (ascidians). IV. Pumping rate, filtration rate, filtration efficiency. *Marine Biology* **48**, 243–249.

Fielding, P.J,, Weerts, K.A. & Forbes, A.T. 1994. Macroinvertebrate communities associated with intertidal and subtidal beds of *Pyura stolonifera* (Heller) (Tunicata: Ascidiacea) on the Natal coast. *South African Journal of Zoology* **29**, 46–53.

Fletcher, L. 2014. Background information on the sea squirt *Pyura doppelgangera* to support regional response decisions. Cawthron Institute, report prepared for Marlborough District Council, Nelson, New Zealand.

Flinn, M.V., Geary, D.C. & Ward, C.V. 2005. Ecological dominance, social competition, and coalitionary arms races. *Evolution and Human Behavior* **26**, 10–46.

Fox, J.W. 2013. The intermediate disturbance hypothesis should be abandoned. *Trends in Ecology & Evolution* **28**, 86–92.

Garrabou, J., Coma, R., Bensoussan, N., Bally, M., Chevaldonné, P., Cigliano, M., Diaz, D., Harmelin, J.G., Gambi, M.C., Kersting, D.K., Ledoux, J.B., Lejeusne, C., Linares, C., Marschal, C., Pérez, T., Ribes, M., Romano, J.C., Serrano, E., Teixido, N., Torrents, O., Zabala, M., Zuberer, F. & Cerrano, C. 2009. Mass mortality in Northwestern Mediterranean rocky benthic communities: effects of the 2003 heat wave. *Global Change Biology* **15**, 1090–1103.

Gascoigne, J. & Lipcius, R.N. 2004. Allee effects in marine systems. *Marine Ecology Progress Series* **269**, 49–59.

Gelcich, S., Cinner, J., Donlan, C.J., Tapia-Lewin, S., Godoy, N. & Castilla, J.C. 2017. Fishers' perceptions on the Chilean coastal TURF system after two decades: problems, benefits, and emerging needs. *Bulletin of Marine Science* **93**, 53–67.

Gili, J.-M. & Coma, R. 1998. Benthic suspension feeders: their paramount role in littoral marine food webs. *Trends in Ecology & Evolution* **13**, 316–321.

Goss-Custard, J.D., West, A.D., Yates, M.G., Caldow, R.W., Stillman, R.A., Bardsley, L., Castilla, J., Castro, M., Dierschke, V., Durell, S.E., Eichhorn, G., Ens, B.J., Exo, K.M., Udayangani-Fernando, P.U., Ferns, P.N., Hockey, P.A., Gill, J.A., Johnstone, I., Kalejta-Summers, B., Masero, J.A., Moreira, F., Nagarajan, R.V., Owens, I.P., Pacheco, C., Perez-Hurtado, A., Rogers, D., Scheiffarth, G., Sitters, H., Sutherland, W.J., Triplet, P., Worrall, D.H., Zharikov, Y., Zwarts, L. & Pettifor, R.A. 2006. Intake rates and the functional response in shorebirds (Charadriiformes) eating macro-invertebrates. *Biological Reviews* **81**, 501–529.

Grosberg, R.K. & Quinn, J.F. 1986. The genetic control and consequences of kin recognition by the larvae of a colonial marine invertebrate. *Nature* **322**, 456–459.

Guiler, E.R. 1959. Intertidal belt-forming species on the rocky coasts of northern Chile. *Papers and Proceedings of the Royal Society of Tasmania* **93**, 33–58.

Guiñez, R. & Castilla, J.C. 1999. A tridimensional self-thinning model for multilayered intertidal mussels. *American Naturalist* **154**, 341–357.

Guiñez, R. & Castilla, J.C. 2001. An allometric tridimensional model of self-thinning for a gregarious tunicate. *Ecology* **82**, 2331–2341.

Gutiérrez, J.L., Jones, C.G., Strayer, D.L. & Iribarne, O.O. 2003. Mollusks as ecosystem engineers: the role of shell production in aquatic habitats. *Oikos* **101**, 79–90.

Hanekom, N. 2013. Environmental conditions during mass mortalities of the ascidian *Pyura stolonifera* (Heller) in the Tsitsikamma Marine Protected Area. *African Zoology* **48**, 167–172.

Hanekom, N., Harris, J.M., Branch, G.M. & Allen, J.C. 1999. Mass mortality and recolonization of *Pyura stolonifera* (Heller) on the south coast of South Africa. *South African Journal of Marine Science* **21**, 117–133.

Harley, C.D.G. 2006. Effects of physical ecosystem engineering and herbivory on intertidal community structure. *Marine Ecology Progress Series* **317**, 29–39.

Hawkins, S.J. & Hartnoll, R.G. 1983. Grazing of intertidal algae by marine invertebrates. *Oceanography and Marine Biology: An Annual Review* **21**, 195–282.

Hayward, B.W. & Morley, M.S. 2009. Introduction to New Zealand of two sea squirts (Tunicata, Ascidiacea) and their subsequent dispersal. *Records of the Auckland Museum* **46**, 5–14.

Heasman, K.G., Pitcher, G.C., McQuaid, C.D. & Hecht, T. 1998. Shellfish mariculture in the Benguela system: raft culture of *Mytilus galloprovincialis* and the effect of rope spacing on food extraction, growth rate, production, and condition of mussels. *Journal of Shellfish Research* **17**, 33–39.

Heller, C. 1878. Beiträge zur näheren Kenntniss der Tunicaten. *Sitzungsberichte der Academie der Wissenschaften in Wien* **77**, 83–109.

Hillebrand, H., Bennett, D.M. & Cadotte, M.W. 2008. Consequences of dominance: a review of evenness effects on local and regional ecosystem processes. *Ecology* **89**, 1510–1520.

Hughes, R.N. & Griffiths, C.L. 1988. Self-thinning in barnacles and mussels: the geometry of packing. *American Naturalist* **132**, 484–491.

Johnson, C.H. & Woollacott, R.M. 2010. Larval settlement preference maximizes genetic mixing in an inbreeding population of a simultaneous hermaphrodite (*Bugula stolonifera*, Bryozoa). *Molecular Ecology* **19**, 5511–5520.

Jones, C.G., Lawton, J.H. & Shachak, M. 1994. Organisms as ecosystem engineers. *Oikos* **69**, 373–386.

Jones, C.G., Lawton, J.H. & Shachak, M. 1997. Positive and negative effects of organisms as physical ecosystem engineers. *Ecology* **78**, 1946–1957.

Jordan, W.R. 2000. Restoration, community, and wilderness. In *Restoring Nature. Perspectives from the Social Sciences and Humanities*, P.H. Gobster & R.B. Hull (eds). Washington, DC: Island Press, 21–36.

Justus, J., Colyvan, M., Regan, H. & Maguire, L. 2009. Buying into conservation: intrinsic versus instrumental value. *Trends in Ecology & Evolution* **24**, 187–191.

Katz, E. 1992. The big lie: human reconstuction of nature. *Research in Philosophy and Technology* **12**, 231–241.

Katz, E. 2000. Another look at restoration: technology and artificial nature. In *Restoring Nature. Perspectives from the Social Sciences and Humanities*, P.H. Gobster & R.B. Hull (eds). Washington, DC: Island Press, 37–48.

Keller, L.F. & Waller, D.M. 2002. Inbreeding effects in wild populations *Trends in Ecology & Evolution* **17**, 230–241.

Keough, M.J., Quinn, G.P. & King, A. 1993. Correlations between human collecting and intertidal mollusc populations on rocky shores. *Conservation Biology* **7**, 378–390.

Keough, M.J. & Raimondi, P.T. 1995. Responses of settling invertebrate larvae to bioorganic films: effects of different types of films. *Journal of Experimental Marine Biology and Ecology* **185**, 235–253.

Kingsford, M.J., Underwood, A.J. & Kennelly, S.J. 1991. Humans as predators on rocky reefs in New South Wales, Australia. *Marine Ecology Progress Series* **72**, 1–14.

Kinlan, B.P. & Gaines, S. 2003. Dispersal in marine and terrestrial environments: a community perspective. *Ecology* **84**, 2007–2020.

Kinlan, B.P., Gaines, S.D. & Lester, S.E. 2005. Propagule dispersal and the scales of marine community process. *Diversity and Distributions* **11**, 139–148.

Klumpp, D.W. 1984. Nutritional ecology of the ascidian *Pyura stolonifera*: influence of body size, food quantity and quality on filter-feeding, respiration, assimilation efficiency and energy balance. *Marine Ecology Progress Series* **19**, 269–284.

Knott, N.A., Davis, A.R. & Buttemer, W.A. 2004. Passive flow through an unstalked intertidal ascidian: orientation and morphology enhance suspension feeding in *Pyura stolonifera*. *Biological Bulletin* **207**, 217–224.

Kott, P. 1952. The ascidians of Australia I. *Stolidobranchiata lahille* and *Phlebobranchiata lahille*. *Australian Journal of Marine and Freshwater Research* **3**, 205–333.

Kott, P. 1985. The Australian Ascidiacea, Part 1. Phlebobranchia and Stolidobranchia. *Memoirs of the Queensland Museum* **23**, 1–440.

Kott, P. 2006. Observations on non-didemnid ascidians from Australian waters. *Journal of Natural History* **40**, 169–234.

Kuussaari, M., Saccheri, I., Camara, M. & Hanski, I. 1998. Allee effect and population dynamics in the Glanville fritillary butterfly. *Oikos* **82**, 384–392.

Kyle, R., Pearson, B., Fielding, P.J., Robertson, W.D. & Birnie, S.L. 1997. Subsistence shellfish harvesting in the Maputaland marine reserve in northern Kwazulu-Natal, South Africa: rocky shore organisms. *Biological Conservation* **82**, 183–192.

Lafargue, F. & Wahl, M. (1986–1987) Contribution to the knowledge of littoral ascidians (Ascidiacea, Tunicata) of the Senegalese coast. *Bulletin de l'IFAN* **46**, 385–402.

Lambert, G. 2005. Ecology and natural history of the protochordates. *Canadian Journal of Zoology* **83**, 34–50.

Lambert, G. (2007) Invasive sea squirts: a growing global problem. *Journal of Experimental Marine Biology and Ecology* **342**, 3–4.

Lasiak, T. 1991. The susceptibility and/or resilience of rocky littoral molluscs to stock depletion by the indigenous coastal people of Transkei, Southern Africa. *Biological Conservation* **56**, 245–264.

Lawton, J.H. & Jones, C.G. 1995. Linking species and ecosystems: organisms as ecosystem engineers. In *Linking Species and Ecosystems*, C.G. Jones & J.H. Lawton (eds). Springer, 141–150.

LeBlanc, N., Davidson, J., Tremblay, R., McNiven, M. & Landry, T. 2007. The effect of anti-fouling treatments for the clubbed tunicate on the blue mussel, *Mytilus edulis*. *Aquaculture* **264**, 205–213.

Lehane, C. & Davenport, J. 2004. Ingestion of bivalve larvae by *Mytilus edulis*: experimental and field demonstrations of larviphagy in farmed blue mussels. *Marine Biology* **145**, 101–107.

Leibold, M.A., Holyoak, M., Mouquet, N., Amarasekare, P., Chase, J.M., Hoope, M.F., Holt, R.D., Shurin, J.B., Law, R., Tilman, D., Loreau, M. & Gonzalez, A. 2004. The metacommunity concept: a framework for multi-scale community ecology. *Ecology Letters* **7**, 601–613.

Levins, R. 1974. The qualitative analysis of partially specified systems. *Annals of the New York Academy of Sciences* **231**, 123–138.

Levitan, D.R. 1991. Influence of body size and population density on fertilization success and reproductive output in a free-spawning invertebrate. *Biological Bulletin* **181**, 261–268.

Levitan, D.R., Sewell, M.A. & Chia, F.-S. 1992. How distribution and abundance influence fertilization success in the sea urchin *Strongylocentotus franciscanus*. *Ecology* **73**, 248–254.

Locke, A. 2009. A screening procedure for potential tunicate invaders of Atlantic Canada. *Aquatic Invasions* **4**, 71–79.

Loo, L.-O. & Rosenberg, R. 1989. Bivalve suspension-feeding dynamics and benthic-pelagic coupling in an eutrophicated marine bay. *Journal of Experimental Marine Biology and Ecology* **130**, 253–276.

Loo, L.-O. & Rosenberg, R. 1996. Production and energy budget in marine suspension feeding populations: *Mytilus edulis, Cerastoderma edule, Mya arenaria* and *Amphiura filiformis*. *Journal of Sea Research* **35**, 199–207.

Loreau, M., Naeem, S., Inchausti, P., Bengtsson, J., Grime, J.P., Hector, A., Hooper, D.U., Huston, M.A., Raffaelli, D., Schmid, B., Tilman, D. & Wardle, D.A. 2001. Biodiversity and ecosystem functioning: current knowledge and future challenges. *Science* **294**, 804–808.

Lubchenco, J. 1978. Plant species diversity in a marine intertidal community: importance of herbivore food preference and algal competitive abilities. *American Naturalist* **112**, 23–39.

Lubchenco, J. & Gaines, S.D. 1981. A unified approach to marine plant-herbivore interactions. I. Populations and communities. *Annual Review of Ecology and Systematics* **12**, 405–437.

Lubchenco, J. & Menge, B.A. 1978. Community development and persistence in a low rocky intertidal zone. *Ecological Monographs* **48**, 67–94.

Majiza, V.N. & Lasiak, T.A. 2010. The influence of site, season and day of the week on exploitation of rocky intertidal biota in central Transkei, South Africa. *South African Journal of Marine Science* **24**, 57–64.

Manríquez, P.H. & Castilla, J.C. 2010. Fertilization efficiency and gamete viability in the ascidian *Pyura praeputialis* in Chile. *Marine Ecology Progress Series* **409**, 107–119.

Manríquez, P.H., Castilla, J.C., Ortiz, V. & Jara, M.E. 2016. Empirical evidence for large-scale human impact on intertidal aggregations, larval supply and recruitment of *Pyura praeputialis* around the Bay of Antofagasta, Chile. *Austral Ecology* **41**, 701–714.

Manríquez, P.H., Delgado, A.P., Jara, M.E. & Castilla, J.C. 2008. Field and laboratory pilot rearing experiments with early ontogenic stages of *Concholepas concholepas* (Gastropoda: Muricidae). *Aquaculture* **279**, 99–107.

Manríquez, P.H., Jara, M.E., Mardones, M.L., Torres, R., Lagos, N.A., Lardies, M.A., Vargas, C.A., Duarte, C. & Navarro, J.M. 2014. Effects of ocean acidification on larval development and early post-hatching traits in *Concholepas concholepas* (loco). *Marine Ecology Progress Series* **514**, 87–103.

Marshall, D.J. 2002. In situ measures of spawning synchrony and fertilization success in an intertidal, free-spawning invertebrate. *Marine Ecology Progress Series* **236**, 113–119.

Marshall, D.J. & Keough, M.J. 2008. The evolutionary ecology of offspring size in marine invertebrates. *Advances in Marine Biology* **53**, 1–60.

Marshall, D.J., Krug, P.J., Kupriyanova, E.K., Byrne, M. & Emlet, R.B. 2012. The biogeography of marine invertebrate life histories. *Annual Review of Ecology, Evolution, and Systematics* **43**, 97–114.

Marshall, D.J., Styan, C.A. & Keough, M.J. 2000. Intraspecific co-variation between egg and body size affects fertilisation kinetics of free-spawning marine invertebrates. *Marine Ecology Progress Series* **195**, 305–309.

McQuaid, C.D. & Branch, G.M. 1984. The influence of sea temperature, substratum and wave exposure on rocky intertidal communities: an analysis of faunal and floral biomass. *Marine Ecology Progress Series* **19**, 145-151.

McQuaid, C.D. & Branch, G.M. 1985. Trophic structure of rocky intertidal communities: response to wave action and implications for energy flow. *Marine Ecology Progress Series* **22**, 153-161.

McQuaid, C.D. & Lindsay, J.R. 2005. Interacting effects of wave exposure, tidal height and substratum on spatial variation in densities of mussel *Perna perna* plantigrades. *Marine Ecology Progress Series* **301**, 173–184.

McQuaid, C.D. & Phillips, T.E. 2000. Limited wind-driven dispersal of intertidal mussel larvae: *in situ* evidence from the plankton and the spread of the invasive species *Mytilus galloprovincialis* in South Africa. *Marine Ecology Progress Series* **201**, 211–220.

Menge, B.A., Berlow, E.L., Blanchette, C.A., Navarrete, S.A. & Yamada, S.B. 1994. The keystone species concept – variation in interaction strength in a rocky intertidal habitat. *Ecological Monographs* **64**, 249–286.

Menge, B.A. & Sutherland, J.P. 1987. Community regulation – variation in disturbance, competition, and predation in relation to environmental-stress and recruitment. *American Naturalist* **130**, 730–757.

Millar, R.H. 1955. On a collection of ascidians from South Africa. *Proceedings of the Zoological Society of London* **125**, 169–221.

Millar, R.H. 1962. Further descriptions of South African ascidians. *Annals of the South African Museum* **46**, 113–221.

Millar, R.H. 1966. Port Phillip survey 1957–1963. Ascidiacea. *Memoirs of the National Museum of Victoria* **27**, 357–384.

Millar, R.H. 1971. The biology of ascidians. *Advances in Marine Biology* **9**, 1–100.

Monniot, C. & Bitar, G. 1983. Sur la présence de *Pyura stolonifera* (Tunicata, Ascidiacea) à Ras Achaccar (côte nord atlantique marocaine). Comparaison anatomique distinctive avec *Pyura praeputialis*. *Bulletin de l'Institut Scientifique, Rabat* **7**, 83–91.

Monniot, C., Monniot, F., Griffiths, C.L. & Schleyer, M. 2001. South African ascidians. *Annals of the South African Museum* **108**, 1–141.

Monniot, C., Monniot, F. & Laboute, P. 1991. *Coral Reef Ascidians of New Caledonia*. Paris: Éditions de L'Orstom.

Monniot, F. & Monniot, C. 2001. Ascidians from the tropical western Pacific. *Zoosystema* **23**, 201–383.

Monteiro, S.M., Chapman, M.G. & Underwood, A.J. 2002. Patches of the ascidian *Pyura stolonifera* (Heller, 1878): structure of habitat and associated intertidal assemblages. *Journal of Experimental Marine Biology and Ecology* **270**, 171–189.

Moreno, C.A., Sutherland, J.P. & Jara, H.F. 1984. Man as a predator in the intertidal zone of southern Chile. *Oikos* **42**, 155–160.

Navarrete, S.A. & Berlow, E.L. 2006. Variable interaction strengths stabilize marine community pattern. *Ecology Letters* **9**, 526–536.

Navarrete, S.A. & Castilla, J.C. 2003. Experimental determination of predation intensity in an intertidal predator guild: dominant versus subordinate prey. *Oikos* **100**, 251–262.

Navarrete, S.A., Wieters, E.A., Broitman, B.R. & Castilla, J.C. 2005. Scales of benthic-pelagic coupling and the intensity of species interactions: from recruitment limitation to top-down control. *Proceedings of the National Academy of Sciences of the USA* **102**, 18046–18051.

Newell, R.C., Field, J.G. & Griffiths, C.L. 1982. Energy balance and significance of microorganisms in a kelp bed community. *Marine Ecology Progress Series* **8**, 103–113.

Nilsson, M.-C. & Wardle, D.A. 2005. Understory vegetation as a forest ecosystem driver: evidence from the northern Swedish boreal forest. *Frontiers in Ecology and the Environment* **3**, 421–428.

Oldewage, W.H. 1994. Description of *Doropygus pyurus* n.sp. (Copepoda, Notodelphyidae) from *Pyura stolonifera* (Echinodermata, Ascidiacea) in South Africa. *South African Journal of Zoology* **29**, 212–216.

Ortiz, M., Campos, L., Berrios, F., Rodriguez, F., Hermosillo, B. & González, J. 2013. Network properties and keystoneness assessment in different intertidal communities dominated by two ecosystem engineer species (SE Pacific coast): a comparative analysis. *Ecological Modelling* **250**, 307–318.

Otway, N.M. 1989. *The effects of grazing by chitons on mid and low shore intertidal communities*. PhD thesis, University of Sydney, Australia.

Pacheco, C.J. & Castilla, J.C. 2001. Foraging behavior of the American oystercatcher *Haematopus pallia-tus pitanay* (Murphy 1925) on the intertidal ascidian *Pyura praeputialis* (Heller 1878) in the Bay of Antofagasta, Chile. *Journal of Ethology* **19**, 23–26.

Paine, R.T. 1966. Food web complexity and species diversity. *American Naturalist* **100**, 65–75.

Paine, R.T. 1969. A note on trophic complexity and community stability. *American Naturalist* **103**, 91–93.

Paine, R.T. 1971. Short-term experimental investigation of resource partitioning in a New Zealand rocky intertidal habitat. *Ecology* **52**, 1096–1106.

Paine, R.T. 1976. Size-limited predation: an observational and experimental approach with the *Mytilus-Pisaster* interaction. *Ecology* **57**, 858–873.

Paine, R.T. 1979. Disaster, catastrophe, and local persistence of the sea palm *Postelsia palmaeformis*. *Science* **205**, 685–687.

Paine, R.T., Castilla, J.C. & Cancino, J. 1985. Perturbation and recovery patterns of starfish-dominated inter-tidal assemblages in Chile, New Zealand, and Washington State. *American Naturalist* **125**, 679–691.

Paine, R.T. & Suchanek, T.H. 1983. Convergence of ecological processes between independently evolved competitive dominants: a tunicate-mussel comparison. *Evolution* **37**, 821–831.

Pawlik, J.R., Butman, C.A. & Starczak, V.R. 1991. Hydrodynamic facilitation of gregarious settlement of a reef-building tube worm. *Science* **251**, 421–423.

Petersen, J.K. & Riisgård, H.U. 1992. Filtration capacity of the ascidian *Ciona intestinalis* and its grazing impact in a shallow fjord. *Marine Ecology Progress Series* **88**, 9–17.

Petersen, J.K. & Svane, I. 2002. Filtration rate in seven Scandinavian ascidians: implications of the morphol-ogy of the gill sac. *Marine Biology* **140**, 397–402.

Porri, F., Jordaan, T. & McQuaid, C.D. 2008. Does cannibalism of larvae by adults affect settlement and con-nectivity of mussel populations? *Estuarine, Coastal and Shelf Science* **79**, 687–693.

Quoy, J. & Gaimard, A. 1834. Animaux Mollusques. In *Voyages de Decouvertes de l'Astrolabe Executes par Orde du Roi, Pendant les Annees 1826–29, sous le Commandement de M.J. Dumont d'Urville, Zoologie*, Vol 3, 559–625.

Ramírez, M.E. & Mena, O. 1984. Distribucion, abundancia y estructuras de las comunidades de algas que crecen sobre *Pyura praeputialis* Heller, 1878 en Caleta Coloso (Antofagasta). *Boletin Museo Nacional Historia Nacional Chile* **40**, 7–21.

Ramorino, L. & Campos, B. 1983. Larvas y postlarvas de Mytilidae de Chile (Mollusca: Bivalvia). *Revista de Biología Marina* **19**, 143–192.

Randløv, A. & Riisgård, H.U. 1979. Efficiency of particle retention and filtration rate in four species of ascid-ians. *Marine Ecology Progress Series* **1**, 55–59.

Reed, J.M. & Dobson, A.P. 1993. Behavioural constraints and conservation biology: conspecific attraction and recruitment. *Trends in Ecology & Evolution* **8**, 253–256.

Ribes, M., Coma, R. & Gili, J.M. 1998. Seasonal variation of in situ feeding rates by the temperate ascidian *Halocynthia papillosa*. *Marine Ecology Progress Series* **175**, 201–213.

Richardson, D.M., Pyšek, P., Rejmánek, M., Barbour, M.G., Panetta, F.D. & West, C.J. 2000. Naturalization and invasion of alien plants: concepts and definitions. *Diversity and Distributions* **6**, 93–107.

Rius, M., Branch, G.M., Griffiths, C.L. & Turon, X. 2010a. Larval settlement behaviour in six gregarious ascidians in relation to adult distribution. *Marine Ecology Progress Series* **418**, 151–163.

Rius, M. & Cabral, H.N. 2004. Human harvesting of *Mytilus galloprovincialis* Lamarck, 1819, in the central coast of Portugal. *Scientia Marina* **68**, 545–551.

Rius, M., Clusella-Trullas, S., McQuaid, C.D., Navarro, R.A., Griffiths, C.L., Matthee, C.A., von der Heyden, S. & Turon, X. 2014a. Range expansions across ecoregions: interactions of climate change, physiology and genetic diversity. *Global Ecology and Biogeography* **23**, 76–88.

Rius, M. & Darling, J.A. 2014. How important is intraspecific genetic admixture to the success of colonising populations? *Trends in Ecology & Evolution* **29**, 233–242.

Rius, M., Kaehler, S. & McQuaid, C.D. 2006. The relationship between human exploitation pressure and condition of mussel populations along the south coast of South Africa. *South African Journal of Science* **102**, 130–136.

Rius, M. & McQuaid, C.D. 2006. Wave action and competitive interaction between the invasive mussel *Mytilus galloprovincialis* and the indigenous *Perna perna* in South Africa. *Marine Biology* **150**, 69–78.

Rius, M. & McQuaid, C.D. 2009. Facilitation and competition between invasive and indigenous mussels over a gradient of physical stress. *Basic and Applied Ecology* **10**, 607–613.

Rius, M., Potter, E.E., Aguirre, J.D. & Stachowicz, J.J. 2014b. Mechanisms of biotic resistance across complex life cycles. *Journal of Animal Ecology* **83**, 296–305.

Rius, M. & Teske, P.R. 2011. A revision of the *Pyura stolonifera* species complex (Tunicata, Ascidiacea), with a description of a new species from Australia. *Zootaxa* **2754**, 27–40.

Rius, M. & Teske, P.R. 2013. Cryptic diversity in coastal Australasia: a morphological and mitonuclear genetic analysis of habitat-forming sibling species. *Zoological Journal of the Linnean Society* **168**, 597–611.

Rius, M., Turon, X., Dias, G.M. & Marshall, D.J. 2010b. Propagule size effects across multiple life-history stages in a marine invertebrate. *Functional Ecology* **24**, 685–693.

Rius, M., Turon, X., Ordóñez, V. & Pascual, M. 2012. Tracking invasion histories in the sea: facing complex scenarios using multilocus data. *PLoS ONE* **7**, e35815.

Robinson, T.B., Griffiths, C.L., McQuaid, C.D. & Rius, M. 2005. Marine alien species of South Africa – status and impacts. *African Journal of Marine Science* **27**, 297–306.

Roy, K., Collins, A.G., Becker, B.J., Begovic, E. & Engle, J.M. 2003. Anthropogenic impacts and historical decline in body size of rocky intertidal gastropods in southern California. *Ecology Letters* **6**, 205–211.

Sale, P.F. 1977. Maintenance of high diversity in coral reef fish communities. *American Naturalist* **111**, 337–359.

Sale, P.F. 1979. Recruitment, loss and coexistence in a guild of territorial coral reef fishes. *Oecologia* **42**, 159–177.

Sarà, M 1986. Sessile macrofauna and marine ecosystem. *Bolletino di Zoologia* **53**, 329–337.

Schoener, T.W. 1983. Field experiments on interspecific competition. *American Naturalist* **122**, 240–285.

Schultz, M. 1989. Sooty oystercatcher feeding on washed-up cunjevoi. *Stilt* **14**, 66–67.

Seiderer, L.J. & Newell, R.C. 1988. Exploitation of phytoplankton as a food resource by the kelp bed ascidian *Pyura stolonifera*. *Marine Ecology Progress Series* **50**, 107–115.

Sherrard, K.M. & LaBarbera, M. 2005. Form and function in juvenile ascidians II. Ontogenetic scaling of volumetric flow rates. *Marine Ecology Progress Series* **287**, 139–148.

Simberloff, D. 2010. Invasive species. In *Conservation Biology for All*, N.S. Sodhi & P.R. Ehrlich (eds). Oxford: Oxford University Press, 131–152.

Simberloff, D. 2012. Nature, natives, nativism, and management: Worldviews underlying controversies in invasion biology. *Environmental Ethics* **34**, 5–25.

Simberloff, D. & Vitule, J.R.S. 2014. A call for an end to calls for the end of invasion biology. *Oikos* **123**, 408–413.

Smith, J.R. & Murray, S.N. 2005. The effects of experimental bait collection and trampling on a *Mytilus californianus* mussel bed in southern California. *Marine Biology* **147**, 699–706.

Smith, M.D. & Knapp, A.K. 2003. Dominant species maintain ecosystem function with non-random species loss. *Ecology Letters* **6**, 509–517.

Soulé, M. 1995. The social siege of nature. In *Reinventing Nature? Responses to Postmodern Deconstruction*. M. Soulé & G. Lease (eds). Washington DC, Island Press, 137–170.

Sousa, W.P. 1979. Experimental investigations of disturbance and ecological succession in a rocky intertidal algal community. *Ecological Monographs* **49**, 227–254.

Stachowicz, J.J. & Hay, M.E. 1999. Reducing predation through chemically mediated camouflage: indirect effects of plant defenses on herbivores. *Ecology* **80**, 495–509.

Steinwascher, K. 1978. Interference and exploitation competition among tadpoles of *Rana utricularia*. *Ecology* **59**, 1039–1046.

Steneck, R.S., Hacker, S.D. & Dethier, M.N. 1991. Mechanisms of competitive dominance between crustose coralline algae: an herbivore-mediated competitive reversal. *Ecology* **72**, 938–950.

Stephens, P.A. & Sutherland, W.J. 1999. Consequences of the Allee effect for behaviour, ecology and conservation. *Trends in Ecology & Evolution* **14**, 401–405.

Stephenson, T.A. & Stephenson, A. 1972. *Life Between Tidemarks on Rocky Shores*. San Francisco: W.H. Freeman and Company.

Stuart, V. & Klumpp, D.W. 1984. Evidence for food-resource partitioning by kelp-bed filter feeders. *Marine Ecology Progress Series* **16**, 27–37.

Symondson, W.O.C., Sunderland, K.D. & Greenstone, M.H. 2002. Can generalist predators be effective biocontrol agents? *Annual Review of Entomology* **47**, 561–594.

Tapia, F.J. & Pineda, J. 2007. Stage-specific distribution of barnacle larvae in nearshore waters: potential for limited dispersal and high mortality rates. *Marine Ecology Progress Series* **342**, 177–190.

Teske, P.R. 2014. Connectivity in solitary ascidians: is a 24-h propagule duration sufficient to maintain large-scale genetic homogeneity? *Marine Biology* **161**, 2681–2687.

Teske, P.R., Rius, M., McQuaid, C.D., Styan, C.A., Piggott, M.P., Benhissoune, S., Fuentes-Grünewald, C., Walls, K., Page, M., Attard, C.R.M., Cooke, G.M., McClusky, C.F., Banks, S.C., Barker, N.P. & Beheregaray, L.B. 2011. "Nested" cryptic diversity in a widespread marine ecosystem engineer: a challenge for detecting biological invasions. *BMC Evolutionary Biology* **11**, 176.

Teske, P.R., Sandoval-Castillo, J., Sasaki, M. & Beheregaray, L.B. 2015. Invasion success of a habitat-forming marine invertebrate is limited by lower-than-expected dispersal ability. *Marine Ecology Progress Series* **536**, 221–227.

Teske, P.R., Sandoval-Castillo, J., van Sebille, E., Waters, J. & Beheregaray, L.B. 2016. Oceanography promotes self-recruitment in a planktonic larval disperser. *Scientific Reports* **6**, 34205.

Teske, P.R., Sandoval-Castillo, J., Waters, J.M. & Beheregaray, L.B. 2014. Can novel genetic analyses help to identify low-dispersal marine invasive species? *Ecology and Evolution* **4**, 2848–2866.

Thorson, G. 1950. Reproductive and larval ecology of marine bottom invertebrates. *Biological Reviews* **25**, 1–45.

Tilman, D. 1994. Competition and biodiversity in spatially structured habitats. *Ecology* **75**, 2–16.

Toonen, R.J. & Pawlik, J.R. 1994. Foundations of gregariousness. *Nature* **370**, 511–512.

Toonen, R.J. & Pawlik, J.R. 1996. Settlement of the tube worm *Hydroides dianthus* (Polychaeta: Serpulidae): cues for gregarious settlement. *Marine Biology* **126**, 725–733.

Troost, K., Kamermans, P. & Wolff, W.J. 2008. Larviphagy in native bivalves and an introduced oyster. *Journal of Sea Research* **60**, 157–163.

Tsuchiya, M. 1983. Mass mortality in a population of the mussel *Mytilus edulis* L. caused by high temperature on rocky shores. *Journal of Experimental Marine Biology and Ecology* **66**, 101–111.

Tyree, S. 2001. *The Ascidians (Sea Squirts). Their Biology, Physiology and Natural Filtration Integration.* Rancho Cucamonga: DE Publishing.

Underwood, A.J. 2000. Experimental ecology of rocky intertidal habitats: what are we learning? *Journal of Experimental Marine Biology and Ecology* **250**, 51–76.

Underwood, A.J., Kingsford, M.J. & Andrew, N.L. 1991. Patterns in shallow subtidal marine assemblages along the coast of New South Wales. *Australian Journal of Ecology* **16**, 231–249.

United Nations 1997. *Glossary of Environment Statistics.* New York.

Vallejo-Marín, M. & Hiscock, S.J. 2016. Hybridization and hybrid speciation under global change. *New Phytologist* **211**, 1170–1187.

van de Koppel, J., Rietkerk, M., Dankers, N. & Herman, P.M.J. 2005. Scale-dependent feedback and regular spatial patterns in young mussel beds. *American Naturalist* **165**, E66-E77.

Van Driel, C.D. & Steyl, C.D. 1978. A quantitative similarity analysis of fauna associated with growth forms of red bait (*Pyura stolonifera*) in Algoa Bay. In *Proceedings of the First Interdisciplinary Conference on Marine and Freshwater Research in Southern Africa, 5–10 July 1976, University of Port Elizabeth, Summerstrand Campus, Port Elizabeth, South Africa.* South African National Committee for Oceanographic Research.

Van Name, W.G. 1945. The North and South American ascidians. *Bulletin of the American Museum of Natural History* **84**, 1–476.

Varner, G. 1990. Biological functions and biological interests. *Southern Journal of Philosophy* **17**, 251–270.

Varner, G. 1998. *In Nature's Interest? Interest, Animal Rights and Environmental Ethics.* Oxford: Oxford Univeristy Press.

Veliz, D., Duchesne, P., Bourget, E. & Bernatchez, L. 2006. Genetic evidence for kin aggregation in the intertidal acorn barnacle (*Semibalanus balanoides*). *Molecular Ecology* **15**, 4193–4202.

Vieira, E.A., Duarte, L.F.L. & Dias, G.M. 2012. How the timing of predation affects composition and diversity of species in a marine sessile community? *Journal of Experimental Marine Biology and Ecology* **412**, 126–133.

von der Meden, C.E.O., Porri, F., McQuaid, C.D., Faulkner, K. & Robey, J. 2010. Fine-scale ontogenetic shifts in settlement behaviour of mussels: changing responses to biofilm and conspecific settler presence in *Mytilus galloprovincialis* and *Perna perna*. *Marine Ecology Progress Series* **411**, 161–171.

Wieczorek, S.K. & Todd, C.D. 1997. Inhibition and facilitation of bryozoan and ascidian settlement by natural multi-species biofilms: effects of film age and the roles of active and passive larval attachment. *Marine Biology* **128**, 463–473.

Wright, A., Pérez-Portela, R. & Griffiths, C.L. 2016. Determining the correct identity of South African *Marthasterias* (Echinodermata: Asteroidea). *African Journal of Marine Science* **38,** 443–455.

Wright, J.P. & Jones, C.G. 2006. The concept of organisms as ecosystem engineers ten years on: Progress, limitations, and challenges. *BioScience* **56,** 203–209.

Yodzis, P. 1989. *Introduction to Theoretical Ecology.* New York: Harper and Row.

Young, C.M. 1988. Ascidian cannibalism correlates with larval behavior and adult distribution. *Journal of Experimental Marine Biology and Ecology* **117,** 9–26.

Oceanography and Marine Biology: An Annual Review, 2017, **55**, 87-388
© S. J. Hawkins, D. J. Hughes, I. P. Smith, A. C. Dale, L. B. Firth, and A. J. Evans, Editors
Taylor & Francis

GIANT CLAMS (BIVALVIA: CARDIIDAE: TRIDACNINAE): A COMPREHENSIVE UPDATE OF SPECIES AND THEIR DISTRIBUTION, CURRENT THREATS AND CONSERVATION STATUS

MEI LIN NEO[1,11]*, COLETTE C.C. WABNITZ[2,3], RICHARD D. BRALEY[4],
GERALD A. HESLINGA[5], CÉCILE FAUVELOT[6], SIMON VAN WYNSBERGE[7],
SERGE ANDRÉFOUËT[6], CHARLES WATERS[8], AILEEN SHAU-HWAI TAN[9],
EDGARDO D. GOMEZ[10], MARK J. COSTELLO[8] & PETER A. TODD[11]*

*[1]St. John's Island National Marine Laboratory, c/o Tropical Marine Science Institute,
National University of Singapore, 18 Kent Ridge Road, Singapore 119227, Singapore*
[2]The Pacific Community (SPC), BPD5, 98800 Noumea, New Caledonia
*[3]Changing Ocean Research Unit, Institute for the Oceans and Fisheries,
The University of British Columbia, AERL, 2202 Main Mall, Vancouver, BC, Canada*
[4]Aquasearch, 6–10 Elena Street, Nelly Bay, Magnetic Island, Queensland 4819, Australia
[5]Indo-Pacific Sea Farms, P.O. Box 1206, Kailua-Kona, HI 96745, Hawaii, USA
*[6]UMR ENTROPIE Institut de Recherche pour le développement, Université de La Réunion,
CNRS; Centre IRD de Noumea, BPA5, 98848 Noumea Cedex, New Caledonia*
*[7]UMR ENTROPIE Institut de Recherche pour le développement,
Université de La Réunion, CNRS; Centre IRD de Tahiti,
BP529, 98713 Papeete, Tahiti, French Polynesia*
[8]Institute of Marine Science, University of Auckland, P. Bag 92019, Auckland 1142, New Zealand
[9]School of Biological Sciences, Universiti Sains Malaysia, Penang 11800, Malaysia
*[10]Marine Science Institute, University of the Philippines, Diliman,
Velasquez Street, Quezon City 1101, Philippines*
*[11]Experimental Marine Ecology Laboratory, Department of Biological Sciences,
National University of Singapore, 14 Science Drive 4, Singapore 117557, Singapore*
Corresponding authors:
Mei Lin Neo
e-mail: tmsnml@nus.edu.sg
Peter A. Todd
e-mail: dbspat@nus.edu.sg

Giant clams, the largest living bivalves, play important ecological roles in coral reef ecosystems and provide a source of nutrition and income for coastal communities; however, all species are under threat and intervention is required. Here, we re-examine and update their taxonomy, distribution, abundance and conservation status as a contribution to the protection, rebuilding and management of declining populations. Since the first comprehensive review of the Tridacnidae by Rosewater (1965), the taxonomy and phylogeny of giant clams have evolved, with three new species descriptions and rediscoveries since 1982 represented by *Tridacna squamosina* (formerly known as *T. costata*), *T. noae* and *T. lorenzi*. Giant clams are distributed along shallow coasts and coral reefs from South

Africa to the Pitcairn Islands (32°E to 128°W), and from southern Japan to Western Australia (24°N to 15°S). Geographic distribution of the 12 currently recognized species is not even across the 66 localities we review here. *Tridacna maxima* and *T. squamosa* are the most widespread, followed by the intermediate-range species, *T. gigas*, *T. derasa*, *T. noae*, *T. crocea* and *Hippopus hippopus*, and the restricted-range species, *Tridacna lorenzi*, *T. mbalavuana*, *T. squamosina*, *T. rosewateri* and *Hippopus porcellanus*. The larger species, *Tridacna gigas* and *T. derasa* are the most endangered, with >50% of wild populations either locally extinct or severely depleted. The smaller and boring species, such as *T. maxima* and *T. crocea*, remain relatively abundant despite ongoing fishing activities. Population density also varies across localities. Areas with the lowest densities generally correspond with evidence of high historical exploitation intensity, while areas with the highest densities tend to be within marine reserves, remote from human populations or have low historical fishing pressures. Exploitation continues to be the main threat and conservation challenge for giant clams. Harvesting for subsistence use or local sale remains an important artisanal fishery in many localities; however, increased commercial demand as well as advances in fishing, transport and storage practices, are in large part responsible for the ongoing loss of wild populations. Habitat loss and a suite of other anthropogenic stressors, including climate change, are potentially accelerating stock depletions. Despite these challenges, global efforts to protect giant clams have gained momentum. CITES Appendix II listings and IUCN conservation categories have raised awareness of the threats to giant clams and have contributed to stemming their decline. The continued development of mariculture techniques may also help improve stock numbers and lend populations additional resilience. However, more effective implementation of conservation measures and enforcement of national and international regulations are needed. It is clear that active management is necessary to prevent the extinction of giant clam species as they continue to face threats associated with human behaviours.

Introduction

Giant clams ('tridacnines', of the subfamily Tridacninae) are the largest and most conspicuous sessile molluscs on coral reefs, where their presence can be traced back to possibly the Upper Cretaceous (Keen 1969), and from the late Eocene and Oligocene (Oppenheim 1901, Cox 1941, Harzhauser et al. 2008). These highly specialized bivalves have the ability to both filter feed and photosynthesize via symbionts (zooxanthellae, *Symbiodinium* spp.) living within their mantle tissues (Yonge 1936, 1982, Fankboner 1971, Fitt 1988). All species of giant clams are considerably larger than most other bivalves, from the smallest species, *Tridacna crocea*, that measures up to 15 cm, to the largest, *T. gigas*, that can grow to over 1 m long and weigh over 300 kg (Rosewater 1965). Tridacnines are effective ecosystem engineers that play numerous ecological roles on coral reefs (Neo et al. 2015a). For example, the high tissue biomass of giant clams makes them attractive to a wide range of predators (Perron et al. 1985, Alcazar 1986, Cumming 1988, Heslinga et al. 1990, Govan 1992), while opportunistic feeders exploit their expelled zooxanthellae, gametes and faeces (Ricard & Salvat 1977, Maboloc & Mingoa-Licuanan 2011). Tridacnine shells provide extensive surfaces for epibiont colonization (Vicentuan-Cabaitan et al. 2014), and their large mantle cavities host a diversity of reef fish, as well as commensal and parasitic organisms (Rosewater 1965, Bruce 2000). Collectively, giant clams can increase topographic relief of coral reefs (Cabaitan et al. 2008), act as reservoirs of zooxanthellae (DeBoer et al. 2012), and potentially counteract eutrophication via water filtering (Klumpp & Griffiths 1994). Finally, dense populations of tridacnines produce large quantities of calcium carbonate shell material that may eventually become incorporated into the reef framework (Gilbert et al. 2006a). Given the wide range of ecological contributions giant clams make to coral reefs, they are unique among reef organisms and their conservation yields benefits beyond the preservation of a single taxon.

Giant clams have been utilized by humans for millennia. Human artefacts (at least 2500 years old) made from their shells, such as adzes and engraved shell discs, have featured strongly in numerous excavation finds in the Middle East, Italy and Japan (Reese 1988, Asato 1991, Reese & Sease 1993). In modern times, tridacnine shells have been used to make terrazzo/terasa tiles (Brown & Muskanofola 1985, Juinio et al. 1989), domestic tools (Hviding 1993, Richards & Roga 2004), beads and other craft ware (Lai 2015, Gomez 2015a). Tridacnines are also commercially valuable in the aquarium trade (Brown & Muskanofola 1985, Teitelbaum & Friedman 2008) and the flesh is a popular food (Hviding 1993). During the past few decades, the increase in demand for their adductor muscles as an ingredient in Asian gastronomy, and their shells for carving and for the preparation of seed used in the freshwater pearl-farming industry have made giant clams highly valuable (Dawson & Philipson 1989, Shang et al. 1991, Heslinga 1995, Kinch & Teitelbaum 2010, Hambrey Consulting 2013, Larson 2016). This has resulted in a period of intensive exploitation by locals and illegal harvesting by foreign fishers, and has been responsible for rapid stock reductions across the Indo-Pacific (Bryan & McConnell 1976, Pearson 1977, Gomez 2015a, Larson 2016). Increased fishing pressure can result in tridacnine densities below levels required for successful reproduction and recruitment (Lucas 1988, Munro 1992), thereby impeding natural recovery of stocks and the possible collapse of entire populations (Neo et al. 2013a).

Early concerns over the heavy exploitation of giant clams and their threatened status throughout the Indo-Pacific fuelled scientific interest, particularly in the development of mariculture techniques to assist in their conservation (Jameson 1976, Yamaguchi 1977, Beckvar 1981, Heslinga et al. 1984, 1990, Crawford et al. 1987, Heslinga & Fitt 1987, Braley et al. 1988), symbiosis as a biological phenomenon (Fitt & Trench 1981, Trench et al. 1981, Norton et al. 1992, Maruyama & Heslinga 1997), physiology (Yonge 1936, Morton 1978) and biochemistry (Baldo & Uhlenbruck 1975, Reid et al. 1984). Yamaguchi (1977) was the first to mention the lack of conservation measures to curb extensive exploitation of giant clams. The International Union for Conservation of Nature (IUCN) first engaged with this issue in 'The IUCN Invertebrate Red Data Book' (Wells et al. 1983), which highlighted the various human pressures on tridacnine populations, and how each species was threatened worldwide. The IUCN Red List of Threatened Species then re-assessed nine species in 1996 and listed them as either 'Least Concern' or 'Vulnerable'. The IUCN status of tridacnine species, however, is in need of updating. The first giant clams to be listed in Appendix II of the Convention on International Trade in Endangered Species of Wild Fauna and Flora (CITES) were *Tridacna derasa* and *T. gigas* in 1983. The other species, *Hippopus hippopus*, *H. porcellanus*, *Tridacna squamosa*, *T. maxima* and *T. crocea*, were listed in 1985—regulating international trade in any of their parts (shells, tissues, alive or dead). In 1988, CITES re-examined whether trade levels could pose problems for wild populations (Wells 1997). Key literature reviews on giant clams reiterated their threatened status, and highlighted the role that mariculture could play in sustainable exploitation and restocking (Munro & Heslinga 1983, Heslinga & Fitt 1987, Munro 1989, Lucas 1994, 1997, Braley 1996, Bell et al. 2005). Based on results from earlier hatchery programmes in the Pacific Islands (Heslinga et al. 1990), Australia (Braley 1992) and the Philippines (Calumpong & Solis-Duran 1993), these studies emphasized domestication as an aid to giant clam conservation.

Despite the efforts to promote the sustainable exploitation and conservation of giant clams outlined above, Lucas (2014, p. R184) highlighted that "giant clams species are extinct or in danger of extinction in many parts of their distributions". Othman et al. (2010) published the most recent review on the status of giant clams worldwide but, while cited widely, it requires significant updates. Moreover, there remains a paucity of published data on tridacnines from lesser-known regions such as East Asia, the Indian Ocean and East Africa. Here we synthesize the recent taxonomy of giant clams and their global distribution, collate the information available on their exploitation and the laws that protect them, review the impacts that harvesting rates may have on wild populations, and summarize the outcomes of past and ongoing mariculture programmes. We also re-examine the current conservation approaches for all tridacnine species and identify key knowledge gaps for future research.

Taxonomy

Giant clams are morphologically derived cardiids (true cockles) which have evolved an obligate symbiotic association with photosynthetic dinoflagellate algae (Schneider 1998, Morton 2000). The current, and most widely accepted, scientific classification of giant clams is: Order Venerida Gray, 1854, Family Cardiidae Lamarck, 1809, Subfamily Tridacninae Lamarck, 1819, and two genera: *Hippopus* Lamarck, 1799 and *Tridacna* Bruguière, 1797 (Rosewater 1965, 1982, Schneider 1998, Schneider & Ó Foighil 1999). Giant clams, however, were formerly regarded as a distinct family, Tridacnidae Lamarck, 1819, within the Order Venerida. Lamarck (1809) was the first to recognize a close relationship between cardiids and giant clams. Yonge (1936) and Stasek (1962), using anatomical characters, similarly proposed that the ancestry of *Tridacna*, was close to that of *Cerastoderma* Poli, 1795, which is the least derived of the Lymnocardiinae Stoliczka, 1870. The results of successive cladistic analyses of shell, anatomical, sperm ultrastructural, and molecular characters have revealed that giant clams indeed form a monophyletic group within the Cardiidae (Schneider 1992, 1998, Braley & Healy 1998, Maruyama et al. 1998, Schneider & Ó Foighil 1999, Keys & Healy 2000, Herrera et al. 2015). Tree topologies by Schneider (1992, 1998) also suggested sister taxa relationships between the azooxanthellate Lymnocardiinae (*Cerastoderma*) and the zooxanthellate Tridacninae (*Hippopus* and *Tridacna*) and Fragiinae Stewart, 1930 (*Fragum* Röding, 1798), although Herrera et al. (2015) cast some doubts over this possibility as only a single representative and a single genetic marker (18S rRNA) were used for the analysis. In general, evidence over the last two decades supports earlier proposals that giant clams should be considered a subfamily (Tridacninae) of the Cardiidae, but the sister taxa relationships within cardiids still need to be resolved. It must be noted that others have argued to maintain Tridacnidae as a full family, based mainly on its highly distinct morphology (Huber 2010, Huber & Eschner 2011, Penny & Willan 2014).

The number of described tridacnine species continues to expand with some new additions since Rosewater's (1965) seminal paper listing *Hippopus hippopus*, *Tridacna gigas*, *T. derasa*, *T. squamosa*, *T. maxima* and *T. crocea*. In 1982, a new *Hippopus* species, *H. porcellanus*, was described from the Sulu Archipelago, Philippines (Rosewater 1982) and in 1991, a new *Tridacna* species, *T. rosewateri* was described from the Saya de Malha Bank, Indian Ocean (Sirenko & Scarlato 1991). Lucas et al. (1990, 1991) also discovered and described a new species '*Tridacna tevoroa*' in 1991, apparently unaware of an earlier description of the same species as *Tridacna mbalavuana*. *T. mbalavuana* was first described from fossils on Viti Levu, Fiji (Ladd 1934), and was already commonly known to the locals as 'tevoro', the devil clam. After closer examination of their morphological characters the two species are now considered synonymous, with *T. tevoroa* the junior synonym of *T. mbalavuana* (Newman & Gomez 2000). In the late 2000s, Richter et al. (2008) discovered a new Red Sea species '*Tridacna costata*'. A subsequent morphological comparison of *T. squamosina* of Sturany (1899) and *T. costata* of Richter et al. (2008) suggest, however, that the two species are identical (Huber & Eschner 2011). Hence, *T. squamosina* is now recognized as the lectotype and *T. costata* as a junior synonym.

Finally, the recent use of molecular tools has led to the rediscovery of a cryptic species: *Tridacna noae* (Su et al. 2014, Borsa et al. 2015a). *Tridacna noae* was previously relegated as one of the many variants of *T. maxima* (McLean 1947, Rosewater 1965) owing to morphological similarity. However, McLean (1947) pointed out that *T. noae* had well-spaced scutes on the upper (i.e. ventral) shell compared to the close-set scutes of *T. maxima*. Moreover, in living specimens *T. noae* can also generally be distinguished from *T. maxima* through the presence of discrete teardrop-shaped markings on the mantle, typically bounded by white margins (Wabnitz & Fauvelot 2014). Furthermore, genetic analyses showed that *T. noae* and *T. maxima* are distinct (Su et al. 2014). Another newly described species, '*Tridacna ningaloo*' from Western Australia (Penny & Willan 2014), is similar in appearance to *T. maxima* and *T. noae*, and Borsa et al. (2015a) established that *T. noae* and *T. ningaloo* have no apparent genetic or morphological differences (except, possibly, in mantle patterns). Hence,

T. ningaloo should be regarded as a junior synonym of *T. noae*. Lastly, the most recent species to be described, based purely on morphology, is *T. lorenzi*. *Tridacna lorenzi* is so far recorded only from the outlying territories of Mauritius (Monsecour 2016). It is morphologically similar to *T. maxima* and *T. rosewateri*, but can still be distinguished from both species by its triangular primary ribs and more globose shell (Monsecour 2016). However, considering the high variation typically observed in tridacnine shell morphology, future studies should include genetic comparisons when delimiting Tridacninae species.

Both genera, *Hippopus* and *Tridacna*, were thought to have evolved independently from a now-extinct *Byssocardium*-like ancestor in the early Miocene. *Hippopus* is considered more primitive as it has retained more *Byssocardium*-like ancestral characters than *Tridacna* (Stasek 1962, Schneider 1998). *Hippopus* and *Tridacna* are reciprocally monophyletic sister taxa (Benzie & Williams 1998, Herrera et al. 2015). *Tridacna* is subdivided into three subgenera: *Tridacna* (comprising *T. gigas*), *Persikima* Iredale, 1937 (comprising *Tridacna derasa* and *T. mbalavuana*), and *Chametrachea* Herrmannsen, 1846 (comprising *Tridacna squamosa*, *T. maxima*, *T. crocea*, *T. squamosina* and *T. noae*) (Rosewater 1965, 1982, Lucas et al. 1991, Benzie & Williams 1998, Schneider & Ó Foighil 1999, Nuryanto et al. 2007, Richter et al. 2008, Lizano & Santos 2014, Su et al. 2014, Borsa et al. 2015b). While the phylogenetic relationships among the subgenera remain equivocal, most tree topologies suggest that *T. gigas* is an intermediate between *Chametrachea* and *Persikima* on the basis of morphological characters and genetic markers (Benzie & Williams 1998, Herrera et al. 2015). In addition, the relationship within *Chametrachea* for *Tridacna squamosa*, *T. maxima* and *T. crocea* has been inconsistent across studies using different genetic markers (Benzie & Williams 1998, Maruyama et al. 1998, Schneider & Ó Foighil 1999, Nuryanto et al. 2007, Herrera et al. 2015, see Table 1 for details). However, the latest molecular analysis (using 16S gene sequences), including all five known species from the subgenus *Chametrachea*, place *Tridacna squamosa* and *T. crocea* as sister taxa with a high degree of statistical confidence (Huelsken et al. 2013, DeBoer et al. 2014, Lizano & Santos 2014, Su et al. 2014, Borsa et al. 2015b). These ongoing updates and debates illustrate the need for more robust datasets and analyses (Herrera et al. 2015).

Table 1 Chronology of giant clam taxonomic changes

Year	Description	Character traits	Taxonomic level	Reference
1809	Recognized a close relationship between cardiids and giant clams	Morphology	Familial	Lamarck (1809)
1921	Classified giant clams as family Tridacnidae	Morphology	Familial	Hedley (1921)
1936	Proposed that the ancestry of *Tridacna* was close to that of *Cerastoderma* (family Cardiidae)	Morphology	Familial	Yonge (1936)
1947	Classified giant clams as family Tridacnidae	Morphology	Familial	McLean (1947)
1962	Proposed that the ancestry of *Tridacna* was close to that of *Cerastoderma* (family Cardiidae)	Morphology	Familial	Stasek (1962)
1965	Classified giant clams as family Tridacnidae	Morphology	Familial	Rosewater (1965)
1969	Proposed superfamily Tridacnoidea	Morphology	Familial	Keen (1969)
1982	New species described, *Hippopus porcellanus*	Morphology	Species	Rosewater (1982)
1991	New species described, *Tridacna tevoroa*	Morphology	Species	Lucas et al. (1991)
1991	New species described, *Tridacna rosewateri*	Morphology	Species	Sirenko & Scarlato (1991)
1992	Giant clams formed a monophyletic group within family Cardiidae	Morphology	Familial	Schneider (1992)

Continued

Table 1 (Continued)   Chronology of giant clam taxonomic changes

Year	Description	Character traits	Taxonomic level	Reference
1998	Giant clams formed a monophyletic group within family Cardiidae	Morphology	Familial	Schneider (1998)
1998	Proposed relationship within subgenus *Chametrachea*: (*Tridacna squamosa* (*T. crocea* + *T. maxima*)), (*T. maxima* (*T. crocea* + *T. squamosa*)), (*T. crocea* (*T. squamosa* + *T. maxima*))	Genetic markers (18S)	Genus	Maruyama et al. (1998)
1998	Proposed relationship within subgenus *Chametrachea*: (*Tridacna squamosa* (*T. crocea* + *T. maxima*))	Allozyme variations	Genus	Benzie & Williams (1998)
1999	Proposed relationship within subgenus *Chametrachea*: (*Tridacna maxima* (*T. crocea* + *T. squamosa*))	Genetic markers (partial 16S)	Genus	Schneider & Ó Foighil (1999)
2000	Giant clams formed a monophyletic group within family Cardiidae	Sperm ultrastructure	Familial	Keys & Healy (2000)
2000	Proposed that *Tridacna rosewateri* belong to subgenus *Chametrachea*	Morphology	Genus	Newman & Gomez (2000)
	Tridacna tevoroa a junior synonym of *T. mbalavuana*	Morphology	Species	
2007	Discovered a '*Tridacna maxima*' lookalike in Japan waters but did not identify species	Morphology	Species	Kubo & Iwai (2007)
2007	Proposed relationship within subgenus *Chametrachea*: (*Tridacna maxima* (*T. crocea* + *T. squamosa*))	Genetic markers (CO1)	Genus	Nuryanto et al. (2007)
2008	New species described, *Tridacna costata*	Morphology, Genetic markers (16S)	Species	Richter et al. (2008)
2011	*Tridacna costata* a junior synonym of *T. squamosina*	Morphology	Species	Huber & Eschner (2011)
2014	Rediscovered species, *Tridacna noae*	Morphology, Genetic markers (CO1, 16S, 18S)	Species	Su et al. (2014)
2014	Proposed that *Tridacna noae* and *T. squamosina* belong to subgenus *Chametrachea*	Genetic markers (CO1, 16S)	Species	Lizano & Santos (2014)
2014	New species described, *Tridacna ningaloo*	Morphology, Genetic markers (CO1, 16S)	Species	Penny & Willan (2014)
2015	*Tridacna ningaloo* a junior synonym of *T. noae*	Genetic markers (CO1)	Species	Borsa et al. (2015a)
2015	Giant clams formed a monophyletic group within family Cardiidae	Genetic markers (H3, 16S, 28S)	Familial	Herrera et al. (2015)
	Proposed relationship within subgenus *Chametrachea*: (*Tridacna maxima* (*T. crocea* + *T. squamosa*))		Genus	
2016	New species described, *Tridacna lorenzi*	Morphology	Species	Monsecour (2016)

Distribution of giant clam species

Since Rosewater's (1965) paper, only a few publications have attempted to consolidate global distribution data for giant clams. Early surveys by Dawson (1986) and Munro (1989) list the presence or absence of tridacnine species in 18 and 32 countries, respectively (Table 2), while others provide broad geographic descriptions for individual species (e.g. Wells 1996, Lucas 1997). Othman et al. (2010) compiled the geographic ranges and densities for ten species in 15 countries, but did not discuss the status of tridacnines in certain ranges (i.e. Red Sea, East Africa and the Indian Ocean). Van Wynsberge et al. (2016) extensively reviewed the status of *Tridacna maxima* using 59 studies that reported density estimates for 172 sites across 26 countries in the Indo-Pacific and Red Sea. The present study has identified 66 localities (defined as either countries or regions) globally where giant clams are present or have been present (Table 2, see Supplementary Tables A1 & A2). Tridacnines generally inhabit shallow coastal waters and coral reefs from South Africa to the Pitcairn Islands (32°E to 128°W), and from southern Japan to Western Australia (24°N to 15°S). The extent of the geographic range differs among the 12 known species, with the highest diversity (nine species) within the Coral Triangle (Figure 1). The most widespread species, *T. maxima* and *T. squamosa*, can be found in almost all of the 66 localities reviewed. These are followed by the species with an intermediate geographic range: *T. gigas*, *T. derasa*, *T. noae*, *T. crocea* and *Hippopus hippopus*, while the rare species *Tridacna lorenzi*, *T. mbalavuana*, *T. squamosina*, *T. rosewateri* and *Hippopus porcellanus* are each recorded from only one or a few locations.

In most surveyed areas, the density of tridacnine species typically ranges from 10^{-4} to 10^{-5} individuals per metre squared (m^{-2}), equivalent to $1–10$ ha^{-1}, with occasional exceptions of >10 m^{-2} (see Supplementary Table A3). Such exceptions include atolls of the Eastern Tuamotu in French Polynesia that are characterized by natural densities of *Tridacna maxima* of up to 500 m^{-2} in the early 2000s (Andréfouët et al. 2005, Gilbert et al. 2006b). Reef Check surveys often report densities of 10^{-3} m^{-2} to 1 m^{-2} ($10–10,000$ ha^{-1}) (Reef Check Foundation 2016, see Supplementary Table A4), but these surveys group all *Tridacna* species together. In general, areas with the lowest densities correspond with evidence of high historical exploitation intensity, whereas areas with the highest densities tend to correspond to marine reserves, remoteness from human populations, or low historical fishing pressures (Table 3, see Supplementary Tables A3 and A4).

The following sections examine the 12 known giant clam species and their characteristics, with a summary of their individual geographic distribution, exploitation and conservation status. Table 4 presents species status, exploitation and conservation efforts (if any) by locality.

Table 2 A comparison of survey information on the global status of giant clam stocks provided by the current and past reviews that have considered all species

Study	Species list	Number of localities examined	Density data?
Dawson (1986)	Hh, Hp, Tg, Td, Ts, Tm (6)	18	×
Munro (1989)	Hh, Hp, Tg, Td, Ts, Tm (6)	32	×
Wells (1996)—IUCN	Hh, Hp, Tg, Td, Tmb, Ts, Tr, Tm, Tc (9)	46	×
Othman et al. (2010)	Hh, Hp, Tg, Td, Tmb, Ts, Tsi, Tr, Tm, Tc (10)	15	✓
Present study	Hh, Hp, Tg, Td, Tmb, Ts, Tsi, Tm, Tno, Tr, Tlz, Tc (12)	66	✓

Note: Abbreviations for species: Tg—*Tridacna gigas*, Td—*T. derasa*, Tmb—*T. mbalavuana* (previously *T. tevoroa*), Ts—*T. squamosa*, Tsi—*T. squamosina* (previously *T. costata*), Tr—*T. rosewateri*, Tlz—*T. lorenzi*, Tm—*T. maxima*, Tno—*T. noae*, Tc—*T. crocea*, Hh—*Hippopus hippopus*, Hp—*H. porcellanus*. A specific review on *Tridacna maxima* is provided by Van Wynsberge et al. (2016).

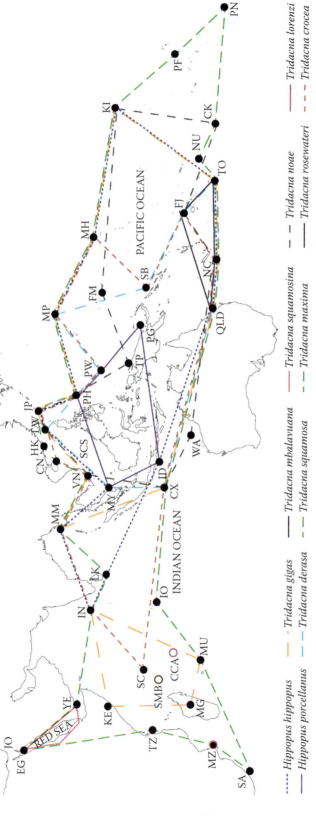

Figure 1 The natural geographic distribution of giant clam (tridacnine) species. Abbreviations for localities: EG—Egypt, JO—Jordan, YE—Yemen, KE—Kenya, TZ—Tanzania, MZ—Mozambique, SA—South Africa, MG—Madagascar, MU—Mauritius, CCA—Cargados Carajos Archipelago, SMB—Saya de Malha Bank, SC—Seychelles, IO—British Indian Ocean Territory, IN—India, LK—Sri Lanka, CX—Christmas Island, MM—Myanmar (Burma), VN—Viet Nam, MY—Malaysia, ID—Indonesia, CN—China, HK—Hong Kong, TW—Taiwan, JP—Japan, PH—Philippines, PW—Palau, TP—East Timor, PG—Papua New Guinea, MP—Northern Mariana Islands, FM—Federated States of Micronesia, MH—Marshall Islands, SB—Solomon Islands, KI—Republic of Kiribati, PF—French Polynesia, PN—Pitcairn Islands, CK—Cook Islands, NU—Niue, TO—Tonga, FJ—Fiji, NC—New Caledonia, QLD—Queensland, Australia, WA—Western Australia, Australia. Abbreviation for sea: SCS—South China Sea.

Table 3 An overview of global records of population density, presenting the highest and lowest densities recorded for all 12 tridacnine species

Species	Record	Locality	Year	Density (ha^{-1})	Reference
Hippopus hippopus	Lowest	Tarawa Atoll, Central Gilbert Islands Group, Republic of Kiribati	1985	0.2	Munro (1988)
	Highest	Helen Reef, Western Caroline Islands, Palau	1976	40.7	Hirschberger (1980)
Hippopus porcellanus	Lowest	Engineer and Conflict Group Islands, Papua New Guinea	1996	0.3	Kinch (2001)
	Highest	Tubbataha Reefs, Cagayancillo, Philippines	2008	97.6	Dolorosa & Jontila (2012)
Tridacna gigas	Lowest	Tarawa Atoll, Central Gilbert Islands Group, Republic of Kiribati	1985	0.2	Munro (1988)
	Highest	Michaelmas Reef, Great Barrier Reef, Australia	1978	431.9	Pearson & Munro (1991)
Tridacna derasa	Lowest	Milne Bay Province, Papua New Guinea	2001	0.3	Kinch (2002)
		North Eastern Lagoon (Poeubo to Hienghène), New Caledonia	2004	0.3	McKenna et al. (2008)
	Highest	Meara Island, Palawan, Philippines	2004	250	Gonzales et al. (2014)
Tridacna mbalavuana	Data Deficient				
Tridacna squamosa	Lowest	Helen Reef, Western Caroline Islands, Palau	1972	0.2	Hester & Jones (1974)
	Highest	Chiriyatapu, Andaman and Nicobar Island (S), India	?	10,000	Ramadoss (1983)
Tridacna squamosina	Lowest	Fayrouza, Nuweiba, Egypt	?	2.9	Richter et al. (2008)
	Highest	Marsa Abu Kalawa, Egypt	?	62.2	Richter et al. (2008)
Tridacna rosewateri	Data Deficient				
Tridacna maxima	Lowest	Pari Island, Indonesia	2003	0.3	Eliata et al. (2003)
	Highest	Tatakoto Atoll, Eastern Tuamotu Archipelago, French Polynesia	2004	5.44×10^6	Gilbert et al. (2005)
Tridacna noae	Lowest	Kavieng lagoonal system, New Ireland Province, Papua New Guinea	2015	27.3	Militz et al. (2015)
	Highest	Mandu Mandu, Ningaloo Marine Park, WA	2014	2,800	Johnson et al. (2016)
Tridacna lorenzi	Data Deficient				
Tridacna crocea	Lowest	Mare, New Caledonia	2010	0.2	Dumas et al. (2011)
	Highest	Cau Island, Con Dao Archipelago, Viet Nam	2011	250,000	Selin & Latypov (2011)

Note: Densities originally published as number of individuals per metre squared have been converted into number of individuals per hectare (ha^{-1}). For more information, please see Supplementary Table A3.

Table 4 Giant clam species presence, abundance and status across their geographic ranges

Region/country	Species Presence and Abundance												Status of giant clams
	Tg	Td	Tmb	Ts	Tsi	Tm	Tno	Tr	Tlz	Tc	Hh	Hp	
Red Sea (22°N 38°E)													
Djibouti				+		++							Two widespread species, but surveys indicate generally small stock sizes. Commercial fisheries are limited, but subsistence fisheries are locally important. Major threats to reefs in Djibouti are coastal development, tourism and sewage discharges. Two marine protected areas (Moucha and Maskali) prohibit the collection of corals and molluscs (with the exception of artisanal fishing of edible species).
Egypt				+	+	++							Tm is most common in shallow waters, while Ts, Tsi inhabit deeper waters. Surveys noted major declines in giant clam populations between 1997 and 2002, attributed to increased sediment load from major construction work. Locals harvest the meat as fish bait while the shells are sold as ornaments. Live specimens are exported for local aquarium markets. Recent surveys indicated patchy distribution with localized declines. Near shore populations are exposed to human impacts such as pollution and tourism.
Eritrea	DD			DD[1]	DD[1]								Reef Check data only listed *Tridacna* spp. No documented data to assess status of giant clams.
Israel				DD[1]	DD[1]	DD[1]							No formal published data on giant clams in Israel, but diver, E. Pszczol (pers. comm.), noted three species (Tm, Ts and possibly Tsi). Coral reefs of Eilat are highly impacted by human pressures, causing damage to the reefs since the 1980s. While not specific to giant clams, pollution most likely caused considerable harm and high larval mortalities in marine invertebrates.

Continued

Table 4 (Continued) Giant clam species presence, abundance and status across their geographic ranges

Region/country	Species Presence and Abundance												Status of giant clams
	Tg	Td	Tmb	Ts	Tsi	Tm	Tno	Tr	Tlz	Tc	Hh	Hp	
Jordan				+	+	+							Surveys suggest that all species are endangered, as they are rare along the Jordanian coast of the Gulf of Aqaba. The scarcity of clams is probably attributable to habitat loss, overfishing, and souvenir collecting. Other major threats include tourism, industry and construction along the coastline.
Saudi Arabia				+	DD[2]	++							Three species are present. A Tsi specimen was collected from Farasan Islands, Tiger Head Island, on 10 March 2013 (G. Paulay, pers. comm.). Tm is more abundant than Ts, but both species are subjected to heavy exploitation. Often collected for food and decorative purposes. Despite this, there are no reports of population decline yet.
Somalia				+		+							Populations are sparse. Locally collected for food by fishermen in coral reef areas. Human disruption and impacts are minimal. Due to the country's political instability, national conservation legislation is non-existent.
Sudan				++		++							*Tridacna* spp. are not common along the coastal and inshore reefs of Sudanese seas, except those found within Sanganeb Marine National Park, where they are very abundant and may represent an unexploited population. No information on clam fishing within Sudan.
Yemen				++	DD[3]	++							Three species recorded, with reported declines in clam abundance due to habitat loss and overfishing. Furthermore, coral reefs in Yemen are generally affected by coastal development such as dredging and land filling. Clam abundances are relatively higher in un-fished and protected areas, such as Socotra Archipelago.

Continued

Table 4 (Continued) Giant clam species presence, abundance and status across their geographic ranges

Region/country	Species Presence and Abundance												Status of giant clams
	Tg	Td	Tmb	Ts	Tsi	Tm	Tno	Tr	Tlz	Tc	Hh	Hp	
South-East Africa (7°N 21°E)													
Cargados Carajos Archipelago				+		+			++				Three species reported. Ts is rare in the archipelago: Tm is uncommon and typically embedded in corals. Tlz said to be locally common and often encountered in shallow waters. Local fishermen harvest giant clams for food and later used their shells as ornaments.
Comoros				++[4]		++[4]							Reef Check data only listed *Tridacna* spp., with two confirmed species (E. de Troyer, pers. comm.). Surveys suggest an abundance of clams, but the reefs also face high fishing pressures (e.g. blast fishing).
Kenya				+		+							Only two species are now observed, although fossilized Tg is omnipresent in the Pleistocene fossil reef complex of the Kenyan coast. In the 1970s, the over-collection of shells on the Kenyan coast denuded reefs, which included giant clam Tm. Both extant species are of interest to local fisheries and are generally harvested by hand.
Madagascar	DD[5]			++		++							Giant clams occurred widely but in small populations. Surveys indicate that offshore reefs (e.g. Nosy Hao, Nosy Fasy) support higher densities of giant clams. Ts is commercially fished, and considered a high-value food.
Mauritius	DD[6]			+		+							Three species recorded. Giant clams remain a major part of the artisanal fishery, where shells are used as birdbaths and holy fonts, and adductor muscles as food. Overfishing of Tm in lagoons has contributed to their low numbers.
Mayotte				DD[7]		DD[7]							Reef Check data only listed *Tridacna* spp., with two confirmed species. Ts is considered rare (S. Andréfouët, pers. obs.). Giant clams are not eaten by locals.
Mozambique				+	DD[8]	+							Two recorded species in the literature, but recent photographic evidence suggests the presence of Tsi in Mozambique waters. Subsistence harvesting reported for Ts.

Continued

Table 4 (Continued) Giant clam species presence, abundance and status across their geographic ranges

Region/country	Species Presence and Abundance												Status of giant clams
	Tg	Td	Tmb	Ts	Tsi	Tm	Tno	Tr	Tlz	Tc	Hh	Hp	
La Réunion				DD[9]		+[10]							Reef Check data only listed *Tridacna* spp., with two confirmed species. No status information.
Saya de Malha Banks								DD					Only one species recorded. Tr was found in a community of *Madrepora* corals and densely covered seagrass. Species record remains ambiguous with no recent living individuals.
Seychelles				+		. +				+[11]			Early surveys in the 1960s indicated three species but generally not abundant. Exploitation of reef species is not a major problem, as locals prefer oceanic pelagic fishes. However, global change such as the bleaching event in 1998 devastated masses of corals, with slow recovery of cover. Efforts to restock clams began in 1980s, with a recent successful transplantation of 30 Tm onto the reefs of Praslin.
South Africa				DD		DD							Reef Check data only listed *Tridacna* spp., with two confirmed species. No status information.
Tanzania				+		++							Two species can still be found in Tanzania. Tm was mentioned as a traditional sea product harvested by local fishing communities. The lucrative shell curio business mainly drives the harvesting pressure on giant clams. Ts shells are frequently sold as curios (collection and trade), and the species may be locally depleted. Fossilized giant clam shell middens are common on Chumbe Island Coral Park (CHICOP), which has been a private nature reserve since 1991. Tm, Ts are found on intertidal areas of CHICOP.
Indian Ocean (20°S 80°E)													
Christmas Island	EX	+		+		++	DD[12]			+			Possibly six species, but reefs naturally have small stock sizes, perhaps due to the lack of suitable habitats (i.e. lagoons). Tg was last recorded in 1932 with no recent sightings. No records of subsistence fishing in appreciable quantities by the local population.

Continued

Table 4 (Continued) Giant clam species presence, abundance and status across their geographic ranges

Region/country	Species Presence and Abundance												Status of giant clams
	Tg	Td	Tmb	Ts	Tsi	Tm	Tno	Tr	Tlz	Tc	Hh	Hp	
Cocos (Keeling) Islands	EX	EX		EX		+++				+			Possibly five species, but presence of Tc and Ts not verified in the later surveys. A culturally important species, Cocos-Malay fishers harvest *Tridacna* spp. for subsistence consumption. Artisanal overfishing appears to be directly responsible for the severe depletion of stocks. Only two Tg were found in 2001 and one Td was found in 2011. Recent surveys in 2014 conclusively identified only Tm, with no sightings of Td, Tg. Higher densities of Tm tend to be found in slightly deeper and less accessible reefs in the lagoon, and around ecotourism hotspots. Recreational harvest of giant clams is currently unregulated.
Chagos				DD		DD							One survey mentioned the presence of two species, with no further information on status. A reef relatively remote from large landmasses and human disturbances.
India	+			+		++				++	+		Five recorded species, but recent presence of Hh, Tg are unconfirmed. Tm is considerably widespread, but Ts appears to be uncommon. Three species (Hh, Tm, Ts) are included in Schedule 1 of Wildlife Protection Act of India (1972). No mention of Tc in Protection Act. Populations are not subjected to extensive commercial exploitation, with occasional subsistence consumption. Populations may be susceptible to local environmental variability.
Maldives				+		++							Only two species found in Maldives, traditionally not fished by locals. A commercial clam fishery started in 1990. The major target species is Ts, while Tm is occasionally taken. Concerns of unsustainable fishing arose when Ts stocks became depleted on numerous atolls. A recent survey in 2009 at Baa atoll suggested otherwise, where both Tm, Ts were widespread and more abundant at depths below 5 m.
Sri Lanka				DD		DD							Two species noted by Munro (1989), but no further status information.

Continued

Table 4 (Continued) Giant clam species presence, abundance and status across their geographic ranges

Region/country	Species Presence and Abundance												Status of giant clams
	Tg	Td	Tmb	Ts	Tsi	Tm	Tno	Tr	Tlz	Tc	Hh	Hp	
East Asia (35°N 136°E)													
China	EX			+		+				+			Three confirmed species within Chinese waters, and possibly Tg. Clam stocks were considered plentiful in the late 1950s, but sharply declined by the 1970s—possibly due to overfishing. By the late 1990s, Tg was no longer observed. Tg sought after for its adductor muscles and shells. First report of successful mariculture of Ts by the South China Sea Institute of Oceanology (SCSIO) in 2016. Imported and popular in the local aquarium trade: Tc, Tm, Tno, Ts.
Hong Kong						DD							Only one species has been definitively recorded, but no recent sightings (Morton & Morton 1983). A market survey in 1980s indicated no known market for giant clam meat and shells. Tm possibly locally extinct. Imported and popular in the local aquarium trade: Tc, Tm, Tno, Ts (M.L. Neo, pers. obs.).
Japan	+			+		++	+			++	+		All species definitively recorded in Japan, although there are no recent records of Hh, Tg. Clams were harvested to supply the demands of domestic market (meat and shells), with a preference for Tc, followed by Hh, Ts. Numbers have declined severely due to overfishing, and regulations are at hand to prevent further decline. Only protected within Okinawa Prefecture. Mariculture of Tc for release into Ryukyu Archipelago has been carried out. Imported and popular in the local aquarium trade: Tc, Tm, Tno, Ts.

Continued

Table 4 (Continued)   Giant clam species presence, abundance and status across their geographic ranges

Region/country	Species Presence and Abundance												Status of giant clams
	Tg	Td	Tmb	Ts	Tsi	Tm	Tno	Tr	Tlz	Tc	Hh	Hp	
Taiwan	EX	EX		+		++	++			+	EX		Hh, Td, Tg have not been recorded over the last three decades and may be locally extinct. Other species are moderately common, occurring in densities 1–5 ind. per 100 m² (Y. Su, pers. comm.). Since the early 1970s, Taiwan has had a well-established market for giant clam adductor muscle, but sources are not local. Taiwanese clam fishing vessels illegally harvest clams, activities which threatened the natural populations in the tropical Pacific (e.g. Australia, Palau, Solomon Islands). Taiwanese government now rejects all requests for clam fishing activities. Locally, reduction in population is attributed to overharvesting for shells by tourist divers and locals. Taitung and Penghu counties have banned the harvesting of their surrounding waters and listed giant clams as protected species. There is ongoing development of conservation plans for replenishing clam stocks. Small-scale mariculture of Tm, Tno has been carried out.
South China Sea (12°N 113°E)													
South China Sea (SCS)	+			++		++	+			++	+		Published surveys of various SCS islands noted the presence of seven species. Harvesting of clams remains common, mainly by fishers from surrounding countries with territorial claims, such as China, Philippines, and Viet Nam. Due to overharvesting, Tg is likely locally extinct within the Spratly and Paracel Islands, and Scarborough Shoal. Illegal vessels have been caught off SCS carrying masses of Tg shells, presumably to be sold in the ornament trade. In recent years, the increasing demand for giant clam shells (particularly Tg) as handicraft decoration in China has led to the rapid extraction and depletion of both live and dead Tg shells within SCS. Island groups such as Swallow Reef (Layang Layang) and Pratas Islands (Dongsha Atoll) are 'claimed' by Malaysia and Taiwan, respectively, and these islands are 'protected' by the military of these countries.

Continued

Table 4 (Continued) Giant clam species presence, abundance and status across their geographic ranges

Region/ country	Species Presence and Abundance												Status of giant clams
	Tg	Td	Tmb	Ts	Tsi	Tm	Tno	Tr	Tlz	Tc	Hh	Hp	
South Asia (12°N 105°E)													
Brunei				DD[13]									Reef Check data only listed *Tridacna* spp., with one confirmed species. No status information.
Cambodia	DD			++[14,15]						++[15]			Earlier surveys reported the presence of Tg, Tc, Ts. However, the records for Tg cannot be verified. Reported subsistence consumption by locals, and overfishing for the trade has depleted stocks (e.g. Koh Rong). Following trade restrictions of wild caught clams for the aquarium trade in Viet Nam, wild live clam exports from Cambodia surged. It is possible that some giant clams from Viet Nam were rerouted for export through Cambodia. Alternatively, Cambodian fishermen may have seized the opportunity and increased extraction activity in Cambodian waters. However, exports have essentially ceased from 2013 onwards.
East Timor	DD[16]						DD[16]			DD[16]			Reef Check data listed only *Tridacna* spp., with three confirmed species (N. Hobgood, pers. comm.). No status information.
Indonesia	+	+		++		+++	+			+++	+	+	Hh, Hp, Td, Tg are presently extremely rare, while Tc, Tm, Tno, Ts can still be found in relatively healthy numbers. All eight species remain heavily exploited for their meat (domestic consumption) and shells, and some for live aquarium trade. The Indonesian government has declared giant clams as protected species. Since the 1990s, the Indonesian Institute of Science (LIPI) has been culturing giant clams (Hh, Td, Ts) for restocking reefs.

Continued

Table 4 (Continued) Giant clam species presence, abundance and status across their geographic ranges

Region/ country	Species Presence and Abundance												Status of giant clams
	Tg	Td	Tmb	Ts	Tsi	Tm	Tno	Tr	Tlz	Tc	Hh	Hp	
Malaysia	+	+		++		+++				+++	+	+	Tg is now only found in Sabah (east Malaysia). Hp and Td are restricted to Sabah and also in Pulau Bidong (east coast of Peninsular Malaysia). Hh is also rare and only reported in Johor Islands. Tc, Tm, Ts are still widespread. Populations are in a state of decline due to the combined effects of pollution, environmental degradation and harvesting for meat and shells. All species are protected under Malaysian Department of Fisheries. Universiti Sains Malaysia (USM) successfully spawned Hh and Ts onsite in 1997. The giant clams produced were restocked in Johor Islands located on the west coast of Peninsular Malaysia. The Marine Ecology Research Centre (MERC) at the Gayana Eco-Resort is the first to successfully produce and restock all seven species of giant clams found in Malaysian waters. Hatchery-produced Tg from the Philippines have been restocked in Johor Islands in 2012, and these Tg have now reached maturity for potential breeding.
Myanmar (Burma)	DD			DD		DD					DD		Four species reported by Munro (1989), which mentioned the presence of relict Tg populations. No further status information.

Continued

104

Table 4 (Continued)   Giant clam species presence, abundance and status across their geographic ranges

Region/country	Species Presence and Abundance												Status of giant clams
	Tg	Td	Tmb	Ts	Tsi	Tm	Tno	Tr	Tlz	Tc	Hh	Hp	
Philippines	+	+		++		+++	+			+++	+	+	Eight species can still be found in the Philippines. Native Tg populations are restricted to the Tubbataha reefs in extremely low abundances. While subsistence harvesting was widespread, commercial exploitation decimated populations of Hh, Hp, Tg, Ts (mainly for international shell trade). In 1996, exports of all species from Philippines were banned. The Bolinao Marine Laboratory pioneered the country's first giant clam mariculture for all native species in the late 1980s, with the aim to restock cultured clams onto denuded reefs. The programme has successfully reintroduced ~40,000 cultured Tg of Australia and Solomon Islands origins. Recent surveys showed Tg recruitment on nearby restocked reefs (E.D. Gomez, pers. obs.). All species are protected within the Philippines.
Singapore	EX			++		+				++	EX		Hh and Tg are locally extinct, while Tc, Tm, Ts occur in low abundances. Exploited since the mid 19th century, particularly for the curio trade. Subsequently, coastal development projects led to habitat degradation and pollution, which further impacted the already low stocks. Funded by the National Parks Board Singapore, the National University of Singapore (NUS) recently established a hatchery for culturing and restocking clams onto local reefs, with a focus on rearing Ts. There are no specific laws protecting giant clams within Singapore.

Continued

Table 4 (Continued) Giant clam species presence, abundance and status across their geographic ranges

| Region/country | Species Presence and Abundance | | | | | | | | | | | | Status of giant clams |
	Tg	Td	Tmb	Ts	Tsi	Tm	Tno	Tr	Tlz	Tc	Hh	Hp	
Thailand	EX			++		++				++			Tg has not been observed alive within Thai waters for at least a century, but their shells were found at Surin Islands and Racha Yai. Ts is rare, Tc and Tm can still be found in relatively good numbers. Consumption of clam meat is limited to locals living along the Thai coast. Mainly harvested for its shells (especially Ts) for ornamental trade, while the adductor muscles are exported. The demand for clam shells led to overexploitation of stocks. Since 1992, all species are protected by law, which has been enforced through CITES. Successful breeding of Ts at Prachuap Khiri Khan Coastal Aquaculture Development Center, with ongoing programmes to replenish depleted stocks off Thai waters.
Viet Nam	DD[17]			++		+				++			Though not formerly recorded, a pair of Tg shells was observed at Ha Long Bay (M.L. Neo, pers. obs.). Three other species are widespread across all reefs, but occur in low to moderate abundances. Long-term surveys noted a significant decrease in clam densities between 1998 and 2007, probably due to overfishing. Up to around 2012, Viet Nam was the most important exporter of live wild-caught clams for the aquarium trade with exports peaking in 2008 (85.561 specimens). The decline in exports from Viet Nam in recent years is related to concerns and regulations about sourcing wild specimens. Since then the government has introduced a quota system. The decline in exports from Viet Nam has been partly compensated by substantially increased exports from Cambodia. It is possible that there has been some re-routing of giant clams through Cambodia, where restrictions may be less tightly implemented. However, exports from Cambodia declined abruptly from 2013 onwards. Côn Dao Archipelago was declared a national park reserve in 1993, to protect the country's marine biodiversity. However, illegal harvesting of clams for sale on the black market remains a problem.

Continued

Table 4 (Continued)   Giant clam species presence, abundance and status across their geographic ranges

Region/ country	Species Presence and Abundance														Status of giant clams
	Tg	Td	Tmb	Ts	Tsi	Tm	Tno	Tr	Tlz	Tc	Hh	Hp			
Australia (25°S 135°E)															
Australia	++	++	DD[18]	++		+++	+			+++	++				All eight species of giant clams found within Australia are protected. Carried out mariculture of Hh, Tg, Td in late 1980s. Despite the early complete protection afforded in Australian waters, extensive illegal harvesting by foreign vessels occurred in the 1970s to 1980s. Today, populations of giant clams in Australia can be considered healthy with some almost pristine examples, but poaching is still prevalent off the Great Barrier Reef. Important exporter of clams for the aquarium market (particularly through a farm at Cocos Keeling), especially in the mid 2000s. Exported important numbers of shells in 2001 and 2007. Sales to the domestic market are prohibited.

Continued

Table 4 (Continued) Giant clam species presence, abundance and status across their geographic ranges

Region/country	Species Presence and Abundance												Status of giant clams
	Tg	Td	Tmb	Ts	Tsi	Tm	Tno	Tr	Tlz	Tc	Hh	Hp	
Pacific Ocean (0°S 160°W)													
Melanesia													
Fiji	REIN[19]	++	+	++		++	+			+[20]	REIN		Tmb is locally endemic. Hh, Tg are thought to be locally extinct, possibly due to previous overexploitation of stocks. While giant clams are still common, they are much less abundant than in the past. Tg specimens reintroduced in 1986, 1987 and 1990 from Australia, and Tg was translocated from Fiji to Samoa in 1999. Hh broodstock was imported from Palau in 1985 and Australia in 1992 to the Makogai hatchery; small village farms were also established in the 1990s. A significant food source for the locals, smaller species: Tm, Ts are still harvested for local subsistence. Ts specimens translocated to Samoa in 1992, 1993 and 1998. Td was not favourably harvested due to perception of toughness of meat and its coarse flavour. Td translocated to Fiji from Palau in 1985 and from Fiji to Samoa in 1992, 1993, 1998 and 1999. Village marine tenure rights regulate clam harvesting to some extent. Fiji bans commercial harvest and export, except domestic harvest of no more than three shells weighing no more than 3 kg per person. Cultured Hh, Tc, Tm, Ts, Td, Tg exported for the aquarium trade until 2002; although CITES records do indicate trade in large numbers of wild specimens until that time as well. Until 2003, Fiji also exported a number of shells of above listed species from both cultured and wild sources. Fiji is a party to CITES.

Continued

Table 4 (Continued)   Giant clam species presence, abundance and status across their geographic ranges

Region/country	Species Presence and Abundance												Status of giant clams
	Tg	Td	Tmb	Ts	Tsi	Tm	Tno	Tr	Tlz	Tc	Hh	Hp	
New Caledonia	EX	++	+[21]	++		+++	+			+++	++		Tg only found as fossils. Tno only found in Loyalty Islands and north-eastern coast of New Caledonia. Hh, Td, Ts preferentially harvested for local consumption, with commercial market for meat only. Common practice to build giant clam 'gardens' for local consumption (mostly Hh and Ts): clams are collected at low tide on fringing reefs and aggregated in front of collectors' properties. Shells are by-products for domestic markets. Populations of larger species showing signs of declines. Some regulations exist in various provinces to control harvest. In the Northern Province: bag limits of five giant clams per vessel per trip for professional fishers, and two for others. In the Southern Province: a maximum bag limit of 40 kg.
Papua New Guinea (PNG)	+	+		+		++	++			++	+	+	Eight species can still be found in PNG, where Tc, Tm, Tno are most common. Previous surveys recorded sparse distributions at most sites, with occasional isolated patches of high population densities. Local extinctions at sites and general low stocks can be attributed to unsustainable practices from commercial harvesting, poaching, and long-standing exploitation observed from archaeological records through to colonial times. No monitoring of populations is taking place, and there are no restrictions regarding fishing seasons, fishing gear and size limits; but PNG now forbids the harvesting of giant clams at night using dive torches. Several other management plans have been proposed but are not yet suitably executed. A commercial fishery for giant clams previously operated in the Milne Bay Province until it closed in 2000. A ban on exports was implemented that same year and appears to have been successful in stemming trade in Tm, Td. Papua New Guinea is a party to CITES.

Continued

109

Table 4 (Continued) Giant clam species presence, abundance and status across their geographic ranges

Region/ country	Species Presence and Abundance														Status of giant clams
	Tg	Td	Tmb	Ts	Tsi	Tm	Tno	Tr	Tlz	Tc	Hh	Hp			
Solomon Islands	+	+		+++		+++	++			+++	++				One of the few island states in the region with relatively good stocks of Tc, Tm, Ts. Overall, however, recent surveys indicate lower densities than previously reported for these species. Td has limited distribution and recent surveys found depleted populations. Tg was formerly widespread and abundant but is now considered depleted. Harvesting for export, with large-scale commercial harvesting, took place in the 1970s to 1980s, and subsistence use was considered a major cause of population declines. Large populations of clams can be found within the only marine protected area: Arnavon Marine Conservation Area. In areas of high population density, there is high fishing pressure on larger species, such as Tg. Poaching off remote reefs was not uncommon in the 1960s to 1980s (Taiwanese vessels), which exacerbated stock depletion. Current legislation is no commercial-scale harvesting and exporting overseas (except for aquaculture species); however official records show trade in high quantities of some wild-sourced live specimens and shells. Td, Tc accounted for the majority of trade. The Solomon Islands is regarded as one of the pioneering countries in the development of clam mariculture: in the 1980s ICLARM (now World Fish Centre) established a hatchery at Aruligo near Honiara and started participatory grow-out trials in villages throughout the islands. Production initially targeted the meat market with a shift to culture clams mainly for the aquarium trade (especially Td). Hatchery production stopped in early 2010s with exports declining abruptly as a consequence, subsequently leading to livelihood loss. The Solomon Islands are a party to CITES.

Continued

110

Table 4 (Continued) Giant clam species presence, abundance and status across their geographic ranges

Region/country	Species Presence and Abundance												Status of giant clams
	Tg	Td	Tmb	Ts	Tsi	Tm	Tno	Tr	Tlz	Tc	Hh	Hp	
Vanuatu (and New Hebrides)	REIN	+		+		++	+			+	++		A definitive survey in 1988 indicated the rarity of Tc, Td, Tg, Ts, while Hh, Tm were relatively common and abundant on most reefs. Td is believed to be locally extinct. It was always very rare in Vanuatu, and a number of individuals were translocated in 1998. Tg was reintroduced in 1998 and 2006. On most islands, giant clams, a prized subsistence food, especially Hh, Tm, are collected for household consumption. Only a small proportion of harvest is for sale in domestic markets. The Ministerial Order of 2000 enacted regulations to protect wild stocks of Tc and limit harvest of other clam species, but enforcement has not been effective. Vanuatu is signatory to CITES and implements its obligations through the International Trade (Flora and Fauna) Act No. 56, 1989, and several other pieces of legislation. The country also has a National Marine Aquarium Trade Management Plan and an Aquaculture Development Plan. The introduction of community-based coastal resource management (CBCRM) measures was relatively successful at a number of sites: giant clam farming in Aneityum, monitoring of Tg in Tassiriki and Sunae, and ocean nursery for Tm in Sunae. Significant numbers of live Tc, Tm, Ts were traded for the aquarium market between the late 1990s to 2007. In 2007, the Department of Fisheries imposed a ban on the harvest and export of wild giant clams (export of cultured specimens is allowed). From 2008 onwards cultured individuals of Tm, Ts were used for restocking of natural areas and for live exports. Between 2008 and 2011, Vanuatu was one of the most important sources of giant clam for the aquarium trade. Production has declined since.

Continued

111

Table 4 (Continued) Giant clam species presence, abundance and status across their geographic ranges

Region/ country	Species Presence and Abundance												Status of giant clams
	Tg	Td	Tmb	Ts	Tsi	Tm	Tno	Tr	Tlz	Tc	Hh	Hp	
Micronesia													
Federated States of Micronesia (FSM) – Kosrae, Pohnpei, Chuuk, Yap States	REIN	INT		++		++	++				++		Hh, Tm, Ts can still be found in the wild, while Tg only in very low numbers. Tn particularly abundant on Yap. Td was introduced from Palau, but wild stocks have only established in Yap. An important traditional resource throughout FSM: primarily collected as a food source and shells for curios. Previous commercial exploitation of wild stocks was mainly for adductor muscles sold to Southeast Asian markets. As a result, wild stock numbers have declined. Currently, seed clams from Palau are used in restocking and reintroduction programmes. In FSM, there is now a ban on commercial harvest and export. FSM has been a significant exporter of live giant clams (Td, Tm, Tc) for many years, contributing around 10% of global supply; though production has been erratic with more recent declines. There are two main production facilities, one in Kosrae and one in Pohnpei. There was, briefly, a third one in Yap from 2013 to 2014. CITES data suggests that most of the clams that are now exported have been farmed or ranched, although significant numbers still appear to be sourced from the wild.
Guam	EX	INT		+		++					EX		Tm is relatively common. Ts is rare. Hh, Tg were reintroduced respectively from Palau in 1982, but may be locally extinct. Td was introduced from Palau in 1984 and 1989. Clams are highly valued as a local delicacy, particularly for their adductor muscles. Harvesting regulations apply, and collection is only permitted for local consumption. The law now prohibits commercial harvest and export; harvesting for subsistence is limited to no more than three clams per person per day.

Continued

Table 4 (Continued)   Giant clam species presence, abundance and status across their geographic ranges

Region/ country	Species Presence and Abundance															Status of giant clams
	Tg	Td	Tmb	Ts	Tsi	Tm	Tno	Tr	Tlz	Tc	Hh	Hp				
Republic of Kiribati	+			++		++	+				+					Tg almost locally extinct in some areas while the remaining species are rare. Tm is still intensively harvested, possibly liable to overexploitation (despite healthy populations). Clam gardens were previously common in Kiribati seascape, but locals are less inclined to invest time in keeping clams. A traditionally important food and shell resource, subsistence fishing alone places a heavy pressure on clam stocks, particularly around South Tarawa. Two local companies are involved in the marketing of giant clams for local consumption. Local laws (e.g. Abemama) prohibit removal of clams by visitors. There is also a ban on commercial harvest and export (except for aquaculture species). One low investment enterprise cultures clams primarily supplying the aquarium trade. Exports began in 2002 and are mainly destined for Europe. Production has been limited in recent years.

Continued

113

Table 4 (Continued) Giant clam species presence, abundance and status across their geographic ranges

Region/ country	Species Presence and Abundance												Status of giant clams
	Tg	Td	Tmb	Ts	Tsi	Tm	Tno	Tr	Tlz	Tc	Hh	Hp	
Marshall Islands	+	INT		+		+					+		All species are widely harvested for subsistence use. Species are generally rare, especially near human population centres, except in the Outer Islands where stocks remain relatively healthy. Tg populations were severely reduced by illegal fishing, but one atoll (Ailinginae atoll) may still boast a healthy Tg population. Pristine populations (e.g. Ailinginae and Rongelap atolls), however remain vulnerable to illegal fishing. Td introduced from Palau in 1985 and 1990 as an aquaculture species. Marshall Islands has a longstanding history of aquaculture production with notable technical support from the US and Japan. Numerous giant clam hatcheries are successfully in operation on Majuro, Likiep, Mili and Arno atolls; production (Tg, Td, Ts, Tm) for restocking purposes and mainly for the aquarium trade through engagement with local community farmers. Over the last decade, Marshall Islands has contributed between 4% and 16% of global supply and has been the largest supplier of cultured giant clams to the global aquarium market. While, production has been erratic, there have been recent efforts to consolidate activities and maintain steadier supply and ensure the diversity of clam products. The government has developed a number of initiatives and regulations to control resource use, enforce policies and ensure protection, including an Aquarium Trade Management Plan.
Nauru						+	+						Previous surveys confirmed the presence of Tm only. However, recent surveys indicate that the specimens found are in fact Tno (D. Thoma, pers. comm.); Tm may therefore be extinct. Populations appear to have disappeared during 1980s, due to overfishing (for subsistence use). Marine areas have little to no protection and implementation of relevant legislation has been slow.

Continued

114

Table 4 (Continued) Giant clam species presence, abundance and status across their geographic ranges

Region/ country	Species Presence and Abundance												Status of giant clams
	Tg	Td	Tmb	Ts	Tsi	Tm	Tno	Tr	Tlz	Tc	Hh	Hp	
Northern Mariana Islands	REIN	INT		+		+				+	REIN		Six species were recorded on IUCN, but Tc, Td unconfirmed in published records. Td was introduced from Palau (1986, 1991); may now be locally extinct. Hh and Tg reintroduced from Palau (1986, 1991). Heavy exploitation resulted in local extinction of Hh, Tg. No commercial fishery, but subsistence harvesting of clams through gleaning. Existing Coral Reef Ecosystem Fisheries Management Plan to help manage the harvest of all reef organisms within Federal economic zone.
Palau	+	+		+		++				+	+	+	Published giant clam surveys for Palau are quite old, with no recent updates. Hh, Td, Tg were highly sought for their shells, while Tm, Ts are in demand for their meat. Tc was rarely utilized for either purpose. Population numbers of larger species (except Tc, Tm) have declined since 1972, mainly due to illegal foreign fishers. Established in the 1970s, the Micronesian Mariculture Demonstration Center (MMDC), later renamed as Palau Mariculture Demonstration Center (PMDC) in 2005, became one of the first institutions to succeed in mass production of giant clams. Cultured clams have been translocated as broodstock to many other countries; helped with natural stock enhancement; and exported for the meat and aquarium trade. Production and exports have been very erratic. All giant clams are protected within Palau, with a complete ban on commercial harvesting. The Marine Protection Act 1994 and its regulations prohibit the exports of wild clams. However, no management is in place to regulate wild harvests outside conservation areas. Palau is a party to CITES and has developed specific laws to address its obligations.

Continued

Table 4 (Continued) Giant clam species presence, abundance and status across their geographic ranges

Region/country	Species Presence and Abundance												Status of giant clams
	Tg	Td	Tmb	Ts	Tsi	Tm	Tno	Tr	Tlz	Tc	Hh	Hp	
United States Minor Outlying Islands						++[22]							Only one species recorded within the marine reserves—Palmyra atoll and Kingman Reef. Reefs are relatively remote with little human disturbance.
Polynesia													
American Samoa	INT	INT		++		+++					REIN		All species are heavily overfished for subsistence use, which has led to local extinction of Hh while Tm, Ts present only in low densities. Hh, has been reintroduced, and Tg and Td have been introduced. Rose Atoll National Wildlife Refuge holds one of the highest densities of Tm in the region. Overfishing and poaching by local and foreign fishers. Harvest regulations imposed in 2009 enforce harvest size limits of 180 mm shell length for all clam species.
Cook Islands	INT	INT		+		++	+[23]				INT		Tm is most common, Ts is rare on reefs. Subsequent to when the giant clam restoration project began in 1991, Td, Tg were given to the Cook Islands Ministry of Marine Resources as a gesture to promote both mariculture and tourism. Hh and Tg introduced from Australia (1991); Td introduced from Palau (1986). A culturally significant food item, Tm is often harvested for subsistence consumption. Previous overharvesting in Aitutaki greatly depleted stocks. Despite all efforts such as reserves, aquaculture and hatchery operations, Tm populations are not recovering in Aitutaki. Clam fishing is banned in Manihiki (except for special occasions, such as independence day, according to quota and size limits based on stock monitoring) and Tongareva (now Penrhyn). A local hatchery on Aitutaki provides clams for restocking purposes and small-scale exports of giant clams for the aquarium trade.

Continued

Table 4 (Continued) Giant clam species presence, abundance and status across their geographic ranges

Region/ country	Species Presence and Abundance														Status of giant clams
	Tg	Td	Tmb	Ts	Tsi	Tm	Tno	Tr	Tlz	Tc	Hh	Hp			
French Polynesia				+		+++									High densities of Tm were reported for some atolls, but decreased during the last decade. Ts is rarer in French Polynesia and only found on outer reef slopes in Tuamotu-Gambier and Austral Archipelago, but not in any of the Society Islands. Tm is a traditional delicacy, and commercial exploitation increased during the past decades to supply the demand for Tahiti (main island of French Polynesia). International exports of wild clams are under CITES control, and allowed for the aquarium trade to some extent. Other threats for Tm include susceptibility to climate stress in enclosed lagoons of Tuamotu Archipelago. A harvest minimum size limit of 120 mm shell length has been implemented for Tm throughout French Polynesia. Large clams (30–45cm) are still collected as prized gifts to officials and families, especially in Tuamotu and Gambier. Current statutes refer only to Tm, and therefore protection measures may not apply to Ts. When Andréfouët et al. (2014) was published, a new text mentioning a harvest maximum legal size for all clams was discussed to protect large Ts, but this was not implemented. Spat collecting has been developed and legally authorized for two atolls of Tuamotu Archipelago (Tatakoto and Reao), and local management measures (No-Take Areas, quotas and restocking) are also implemented in these two atolls. The contribution of spat-collected cultured clams has significantly increased in the last couple of years. Regulations for giant clam farming (spat collection, grow-out, transport and reseeding) were implemented in 2008; they are strictly adhered to and operate within a traceability framework. From a CITES perspective spat-collected cultured clams are considered wild—they should probably be labelled 'ranched'. In 2014, French Polynesia was the largest exporter of clams for the aquarium market.

Continued

Table 4 (Continued) Giant clam species presence, abundance and status across their geographic ranges

Region/ country	Species Presence and Abundance												Status of giant clams
	Tg	Td	Tmb	Ts	Tsi	Tm	Tno	Tr	Tlz	Tc	Hh	Hp	
Pitcairn Islands				++		++							A survey in 1987 described Ts to be fairly common on Ducie Reef while Tm was reportedly intensively harvested near inhabited areas compared to pristine areas (e.g. Oeno Lagoon). Surveys also found an abundance of dead Tm shells embedded in the rocks, suggesting that they may have been more common in the past. Henderson Islanders had previously used giant clam shells to make tools or oven stones. Though present populations appear to be under low threat, large specimens are less frequently seen, while young specimens are occasionally seen.
Niue				+		++							Ts has been absent in surveys since 1998. Rather than for subsistence, clam meat is viewed as luxury food by Niueans. Clam stocks have dramatically reduced since the 1990s, with overharvesting the probable cause of decline. Pristine Tm populations in Niue do exist (e.g. Beveridge reef). Some harvesting bans have been instituted amongst villages (e.g. one-year ban to allow stock recovery from cyclone damage). The Niue Domestic Fisheries Regulations of 1996 also limits the harvest size (180 mm shell length) and catch of clams (a bag limit of ten clams per person per day for subsistence use).

Continued

Table 4 (Continued) Giant clam species presence, abundance and status across their geographic ranges

Region/country	Species Presence and Abundance														Status of giant clams
	Tg	Td	Tmb	Ts	Tsi	Tm	Tno	Tr	Tlz	Tc	Hh	Hp			
Samoa	INT	INT		++		++	++				REIN				Hh seems to be locally extinct. Tm, Ts mainly harvested for domestic consumption with giant clams considered a local delicacy. Shellfish data showed a long-term decline in both species. While Ts harvest has been small, the resource became functionally extinct in 2000. Collection of Tm continues. Overfishing is a major problem. Broodstock for Hh, Td, Tg, Tm, Ts has been translocated at various times since 1988 from various Pacific Island countries or territories including Fiji, Tonga, Palau, and American Samoa. Over the past 15 years, Samoa Department of Marine and Wildlife Resources (DMWR) has also successfully introduced cultured Td, Tg, and reintroduced Hh. Local mariculture has mainly provided for family needs rather than commercial business. For subsistence use, there are harvest size limits of 180 mm shell length for Tm and 160 mm shell length for Ts. Samoa is a recent party to CITES.
Tokelau				+		+++									Tm is still relatively abundant in most atolls, but Ts is very scarce as it is preferentially fished. Ts was translocated from Tokelau to Samoa in 1989. Tm is an important food item in Tokelau. Traditionally, clams are substitute seafood when locals are unable to fish in rough seas. While Tm has been relatively well managed for local use, Tm at Atafu need further management attention. However, the largest threat is harvesting for export to Western Samoa. Further reduction of clam numbers is intensified by the use of modern fishing methods. No laws to regulate traditional clam fishing in Tokelau; but community-based fisheries management plans exist.

Continued

Table 4 (Continued) Giant clam species presence, abundance and status across their geographic ranges

Region/ country	Species Presence and Abundance															Status of giant clams
	Tg	Td	Tmb	Ts	Tsi	Tm	Tno	Tr	Tlz	Tc	Hh	Hp				
Tonga	REIN	+	+	++		++				INT	REIN					Hh, Tg are locally extinct since mid 1970s, while Td is severely exploited. Tmb (endemic species) is typically a rare species. Reintroduced Hh, Tg in 1989–1991, with translocation of Tc from Vanuatu also taking place in 2006, as part of stock enhancement and aquaculture programmes. Tongans highly favour giant clam meat, with clams harvested on both subsistence and commercial basis. Larger species (Td) are commercially more valuable. Modern fishing techniques (hookah gear) have also accelerated fishing efforts. Today, Tm, Ts are most commonly traded. Tonga has cultured giant clams since the late 1990s with cultured individuals supporting local stock enhancement and supplying the aquarium trade market. However, hatchery production has been erratic and exports have significantly declined since the mid 2000s. Community-led initiatives to establish 'clam circles' have helped to promote the restoration of depleted stocks, but efforts have ceased. Tonga also imposed minimum harvest size limits for various species: 260 mm for Td, 155 mm for Tm, and 180 mm for Ts. A provision under the Fisheries Management Regulation 2008 prohibits the selling of giant clams on the local market without its shell to facilitate enforcement of size limits.

Continued

Table 4 (Continued) Giant clam species presence, abundance and status across their geographic ranges

Region/country	Species Presence and Abundance												Status of giant clams
	Tg	Td	Tmb	Ts	Tsi	Tm	Tno	Tr	Tlz	Tc	Hh	Hp	
Tuvalu	EX	INT		++		+++					DD		Early surveys found Tg shells but no recent live specimens. Some question whether Tg ever occurred naturally. Densities of Tm are high, while those of Ts are moderate (Siaosi e al. 2012). Hh was noted by Munro (1989) but not in other literature. Occasionally, clam meat harvested for local consumption. Surveys in 2010 indicated no living clams in Nanumea, Langi, Apinelu, and Naseli, except in Funafuti lagoon. In 1988, 1000 Td were introduced for restocking purposes, but due to exploitation only eight individuals remain in 2011. No regulations exist to protect the remaining clam stocks, though the creation of reserves was advised.
Wallis and Futuna Islands				+++		+++	+						No formal scientific survey conducted to determine population size of any species, but locals suggest that the stocks of clams around the coast of Wallis may be abundant. During low tides, women frequently glean for clam meat, while young men dive for them. However, clam meat does not constitute a significant dietary component, hence large populations of clams are virtually untouched (e.g. southwest of Wallis Island). The reefs however are presently threatened by anthropogenic impacts, especially dynamite fishing.

Continued

Table 4 (Continued)   Giant clam species presence, abundance and status across their geographic ranges

Region/ country	Species Presence and Abundance													Status of giant clams
	Tg	Td	Tmb	Ts	Tsi	Tm	Tno	Tr	Tlz	Tc	Hh	Hp		

Note: Abbreviations for species: Tg—*Tridacna gigas*, Td—*T. derasa*, Tmb—*T. mbalavuana* (formerly *T. tevoroa*), Ts—*T. squamosa*, Tsi—*T. squamosina* (formerly *T. costata*), Tm—*T. maxima*, Tno—*T. noae*, Tr—*T. rosewateri*, Tlz—*T. lorenzi*, Tc—*T. crocea*, Hh—*Hippopus hippopus*, Hp—*H. porcellanus*. Species abundance: +++, Abundant (0.01–1 m^{-2}); ++, Frequent (10^{-3}–10^{-4} m^{-2}); +, Rare (<10^{-5} m^{-2}); EX, locally extinct; INT, introduced species; REIN, reintroduced species; DD, data deficient.

[1] E. Pszczol, pers. comm. (Tm, Ts, Tsi).

[2] G. Paulay, pers. comm. (Tsi).

[3] Huber & Eschner (2011) mentioned that the largest *Tridacna squamosina* specimen examined originated in the southern Red Sea at Kamaran Island, off Yemen.

[4] E. de Troyer, pers. comm. (Tm, Ts).

[5] Hopkins (2009) mentioned Tg but cannot be verified.

[6] Michel et al. (1985) mentioned a 92 cm specimen, and a possible species match is Tg.

[7] S. Andréfouët, pers. obs. (Tm, Ts).

[8] N. Helgason, pers. comm. (Tsi).

[9] C. Peneau, pers. comm. (Ts).

[10] H. Magalon, pers. comm. (Tm).

[11] Only recorded for Coëtivy Island.

[12] Neo & Low (2017) reported five unique individuals sighted in 2010 and 2011.

[13] S. Ng, Oceanic Quest Company, pers. comm.

[14] J. Wong, pers. comm. (Ts).

[15] J.M. Savage, pers. comm. (Ts, Tc).

[16] N. Hobgood, pers. comm. (Tc, Tg, Tno).

[17] M.L. Neo, pers. obs. (shell specimen displayed at Ha Long Bay).

[18] A.M. Ayling, pers. comm. (Tmb).

[19] A Tg was photographed in 2007 (see Supplementary Table A2).

[20] Now very rare, only in Lakeba Island.

[21] Recently seen in Loyalty Islands (Bouchet et al. 2001) and on the north eastern outer reef of New Caledonia (Tiavouane & Fauvelot 2016).

[22] A. Pollock, pers. comm. (Tm).

[23] R. Mayston, pers. comm. and C.C.C. Wabnitz, pers. obs. (Tno).

Species characteristics, distribution, and status

Hippopus hippopus *(Linnaeus, 1758)*

Hippopus hippopus (Figure 2A) has several common names, such as the horse's hoof clam and strawberry clam. Individuals have been reported to grow up to 40 cm (Poutiers 1998), yet an individual within a marine protected area of the north-eastern lagoon in New Caledonia measured 47 cm (C. Fauvelot, pers. obs.) and another one at the Bolinao Marine Laboratory, Philippines, reached 50 cm (Mingoa-Licuanan & Gomez 2007). Unlike the *Tridacna* species, the *Hippopus* species lack hyaline organs (small pinhole eyes) in their mantles, which also do not extend over their shell margins, and they have a narrow byssal orifice with tight-fitting teeth (Rosewater 1965). The thick shells of *H. hippopus* have strong radial ribbing and display reddish blotches in irregular bands. Their mantles usually exhibit green, yellow-brown or grey mottled patterns, and their incurrent siphon bears no guard tentacles. Byssal attachment is present in young individuals, but older ones mostly lie unattached on the substratum (Rosewater 1965). *Hippopus hippopus* often inhabits shallow, nearshore patches of reef, sandy areas and seagrass beds that can be exposed during low tides. It is occasionally found as deep as 10 m (S. Andréfouët, pers. obs.). This species is common throughout the Indo-Pacific, except for the Red Sea and Western Indian Ocean (Figure 1). It has been recorded in at least 25 localities, but at ten of these *H. hippopus* has been reported to be locally extinct (Table 4). *Hippopus hippopus* is a popular species for local harvesting and consumption (Hviding 1993), as it is traditionally favoured as a delicacy, considered as 'high status food' for use on special occasions, or as a reserve food when times are difficult. The nearshore habitats where *H. hippopus* is found are accessible and the species is free-living (i.e. unattached to the substratum), making it an easy target for reef gleaners (Hviding 1993). Consequently, populations are widely depleted. It is currently listed as a species of 'Lower Risk/Conservation Dependent' under the IUCN Red List of threatened species. *Hippopus hippopus* has been cultured in Palau, Australia (Orpheus Island Research Station, north Queensland), Malaysia and the Philippines for purposes of translocation to other areas (e.g. from Palau to American Samoa, Yap, the Cook Islands, Samoa and Tonga) or restocking (Table 4). Maricultured *H. hippopus* specimens in Palau exhibited exceptional hardiness and a short generation time (three years), earning this species the distinction of being the most 'farmer-friendly' of the giant clams (Heslinga 2012, 2013).

Hippopus porcellanus *Rosewater, 1982*

Before its formal description, *Hippopus porcellanus* (Figure 2B), also referred to as the China clam, was already common in the shell trade (Rosewater 1982). Maximum shell length is typically ~40 cm, with the largest specimen recorded at 41.1 cm (Hutsell et al. 1997). Unlike the elaborate shells of *H. hippopus*, *H. porcellanus* has a smoother and thinner shell (Rosewater 1982). This species may be easily mistaken for *Tridacna derasa* due to its similar shell shape and texture, but the mantles of *Hippopus porcellanus* are generally grey or brown, lack hyaline organs, and the incurrent siphon has prominent guard tentacles (Rosewater 1982). As with *H. hippopus*, the mantle does not extend beyond the shell margins, and there is a narrow byssal orifice. *Hippopus porcellanus* is usually found free-living on intertidal reef flats (Pasaribu 1988), and on the shallow reefs along the edges of lagoons (Dolorosa et al. 2014). This species has only been recorded from the Sulu Archipelago and Palawan (Philippines), Sabah (Malaysia), Sulawesi and Raja Ampat (Indonesia), Palau, and Milne Bay Province (Papua New Guinea) (Table 4, Figure 1). Heavy exploitation, from both subsistence and commercial fishing, has decimated populations of *H. porcellanus*, leading to extirpations (Calumpong & Cadiz 1993, Dolorosa et al. 2014). Like *H. hippopus*, it is classified by IUCN as of 'Lower Risk/Conservation Dependent'. The few surveys conducted to date suggest that *H. porcellanus* is rare. Some of the healthiest populations are located within southeast Sulawesi (Indonesia) and the Tubbataha Reef Natural Park (Philippines). At the latter site, 100 individuals of

Figure 2 Giant clam species: (A) *Hippopus hippopus*, (B) *H. porcellanus*, (C) *Tridacna gigas*, (D) *T. derasa*, (E) *T. mbalavuana*, (F) *T. squamosa*, (G) *T. squamosina*, (H) *T. maxima*, (I) *T. noae*, (J) *T. crocea*.

various sizes (shell length = 8.2–31.3 cm) were found tagged and being monitored (Dolorosa et al. 2014). There are few published data on the reproduction of *H. porcellanus* (Alcazar et al. 1987, Calumpong et al. 1993), but ~2000 maricultured F1 *H. porcellanus* individuals were successfully raised to sexual maturity at Palau's Micronesian Mariculture Demonstration Center (MMDC) facility in the mid-1990s (G.A. Heslinga & T.C. Watson, pers. comm.). At present, the Marine Ecology Research Centre in Malaysia produces *H. porcellanus* in limited numbers (E.D. Gomez, pers. obs.).

Tridacna gigas *(Linnaeus, 1758)*

Tridacna gigas (Figure 2C) is the only truly gigantic giant clam species: the largest individual reported was 137 cm long (Rosewater 1965), while the heaviest known specimen (106 cm shell length) weighed approximately 500 kg (Lucas 1994). The species is easily identified by its size and distinctive elongate and triangular projections on the upper shell margins. Mantle colours are mostly dull brown and olive green, and the mantle edge bears numerous iridescent blue-green circles. Unlike the other *Tridacna* species, the incurrent siphon of *T. gigas* bears no tentacles. *Tridacna gigas* typically lives in coral reefs with good light penetration, and is usually free-living on either sand or hard reef substrata (Rosewater 1965). It occurs naturally from Myanmar (Burma) to the Republic of Kiribati (but not the Cook Islands), and the Ryukyus (southern Japan) to Queensland (Australia) (Figure 1). Anecdotal accounts suggest that the historical species range possibly extended to southeast Africa (Kenya: Accordi et al. 2010), Madagascar (Hopkins 2009) and Mauritius (Michel et al. 1985). A living *T. gigas* individual was observed on the fringing reefs of Tonumea Island, an uninhabited island in the southern Haápai group of Tonga in December 1973 (R.D. Braley, pers. obs.). Records have recently been discovered for Singapore, although no living individuals have been encountered in recent memory (Neo & Todd 2012a, 2013). Currently, there are at least 31 localities with natural wild populations of *T. gigas*, but at 26 of them this species is severely depleted, locally extinct or data deficient (Table 4). Globally, the IUCN classifies the conservation status of *T. gigas* as 'Vulnerable'. The Great Barrier Reef (GBR) in Australia is the most extensive area within the natural distribution of *T. gigas* that still supports relatively undisturbed populations (Braley 1984, 1986, 1987a,b, Table 5) and exhibits evidence of natural recruitment (Braley 1988, Braley & Muir

Table 5 A 25-year population data set for pristine populations of *Tridacna gigas* and *T. derasa* from five sites in the far northern Great Barrier Reef, Australia

Species	Site number	Survey area (hectares)	Clam abundance 1982–1985	Clam abundance 2007–2009	Percentage change
Tridacna gigas	1	0.550	136	158	+16.0%
	2	0.730	79	61	−22.7%
	3	0.561	61	28	−54.0%
	4	0.022	9	5	−44.0%
	5	0.120	89	71	−15.7%
Tridacna derasa	1	0.550	29	26	−10.0%
	2	0.730	22	26	+18.8%
	3	0.561	30	17	−43.3%
	4	0.022	6	1	−83.0%
	5	0.120	8	2	−62.5%

Note: Survey sites: 1—Watson's Bay, Lizard Island, 2—Palfrey-South Channel, Lizard Island, 3—West bommie of Rachel Carson Reef (formerly Northern Escape Reef), 4—Small east bommie of Rachel Carson Reef, 5—Southern end of Michaelmas Cay. (R.D. Braley, unpublished data)

1995). Samples of these populations have been monitored over a 25-year period and continue to be monitored today (Table 5). Generally, however, populations of *T. gigas* are dwindling. Extensive surveys in the Pacific Islands indicate that sometimes the presence of this species is limited to one individual (C.C.C. Wabnitz, pers. obs.). Populations typically face high levels of exploitation pressure and habitat deterioration (Gomez 2015a, Larson 2016). *Tridacna gigas* remains a valuable coastal resource for both domestic and commercial markets, as it is highly favoured for its meat as food and large shells for the ornament trade. To assist its conservation, *T. gigas* has been extensively cultivated and reintroduced (albeit in some areas, sometimes limited to a couple of individuals) to Peninsular Malaysia, Sabah, Philippines, Fiji, Northern Mariana Islands, Vanuatu and Tonga, as well as introduced to American Samoa, the Cook Islands, Hawaii (USA) and Samoa (Table 4). The oldest known maricultured *T. gigas* individual is 34 years old and was produced at Palau's MMDC in 1982. It is now on display at the Waikiki Aquarium in Honolulu (Carlson 2012, Heslinga 2013). Unfortunately, there is little information available regarding the outcomes of restocking in these areas (with a notable exception of the Philippines; Gomez & Mingoa-Licuanan 2006, Cabaitan & Conaco 2017).

Tridacna derasa *(Röding, 1798)*

The second largest species, *Tridacna derasa* (Figure 2D), grows up to 60 cm in shell length. It is known as the smooth giant clam because its valves have almost no ribbing (Lucas 1988). *Tridacna derasa* has brilliant mantle colours, displaying shades of blue and green with striped patterns. Its incurrent siphon bears relatively inconspicuous guard tentacles (Lucas et al. 1991). Mostly free-living as adults, this species can be found on reef flats, fore reefs, barrier reefs and in atoll lagoons (S. Andréfouët, pers. obs.) down to depths of 20 m. *Tridacna derasa* occurs from the Cocos (Keeling) Islands to Tonga, and from China to Queensland (Australia) (Figure 1). Of the 16 localities in which the presence of *T. derasa* has been recorded, in 12 of them wild populations are either severely exploited or locally extinct (Table 4). As with *T. gigas*, populations of *T. derasa* on the GBR are virtually undisturbed, and surveys of 57 reefs determined an average density of 4.4 ha^{-1}, with the highest density being 30 ha^{-1} (Braley 1986, Table 5). Similar to *T. gigas*, large *T. derasa* individuals are also highly valued for their meat and shells as food and curios, respectively. *Tridacna derasa* is classified as 'Vulnerable' by the IUCN. *Tridacna derasa* was one of the first giant clam species to be commercially bred, partly owing to its fast growth and durability (Hart et al. 1998) making it better suited for meat production (Heslinga et al. 1984, Leung et al. 1994). Mariculture of this species has been highly successful (e.g. Palau, Marshall Islands, Federated States of Micronesia, the Cook Islands and Solomon Islands). Spats tend to be produced for local enhancement, occasionally for translocation programmes to other countries, for sale 'live' in the aquarium trade and, in Palau, sometimes either as food for local restaurants or export to Japan for sale as sashimi (Table 4). For subsistence and conservation purposes, *T. derasa* has been introduced to island states in Micronesia and Polynesia, and reintroduced to Palau, Indonesia, Malaysia and the Philippines.

Tridacna mbalavuana *Ladd, 1934*

Previously described as *Tridacna tevoroa*, the devil clam has been hypothesized to be a transitional species between *Hippopus* and *Tridacna* due to overlapping characters (Schneider & Ó Foighil 1999). The species has *Hippopus*-like features, such as the absence of a byssal gape, no extension of the mantle over the shells, and the absence of hyaline organs (Lucas et al. 1991). *Tridacna mbalavuana* (Figure 2E) also resembles *T. derasa* in appearance, but is distinguished by its rugose mantle surface, prominent guard tentacles on the incurrent siphon, thinner shell valves, and coloured patches on the shell ribbing. Individuals can normally grow up to ~50 cm, with the largest specimen recorded at 56 cm long (Lucas et al. 1991). *Tridacna mbalavuana* inhabits relatively deep waters

(>20 m) compared to other tridacnines, and is apparently intolerant of conditions in shallow water (Lucas et al. 1991). Previously restricted to Fiji and Tonga, this species has been sighted in the Loyalty Islands, New Caledonia (Bouchet et al. 2001), the main island of New Caledonia (Tiavouane & Fauvelot 2016), and Australia (A.M. Ayling, pers. comm., Newman & Gomez 2000) (Figure 1). *Tridacna mbalavuana* is generally rare throughout its known range: Ledua et al. (1993) reported few live specimens (abundance, N = 20, 1989 to 1991) in the eastern Lau group of Fiji, and a slightly higher abundance in Tonga (N = 50, 1989 to 1992) (see Supplementary Table A3). In Haápai, Tonga, individuals were seen on live coral habitat at >30 m depth in clear water, whilst in the eastern Lau group of Fiji, individuals were never found on live coral habitat, but instead next to rocks on steep slopes (Ledua et al. 1993). Recently, only two living individuals have been reported from New Caledonia, despite exhaustive searches (Tiavouane & Fauvelot 2016). In Fiji, some *T. mbalavuana* have been 'accidentally' collected along with *T. derasa* for commercial exports of its meat (Lewis & Ledua 1988, Lewis et al. 1988). In Tonga, *T. mbalavuana* has been harvested for domestic markets either using SCUBA or traditional Pacific Islands fishing methods (Ledua et al. 1993). Even though their preference for deeper water habitats may have offered some protection from harvesting (Lewis & Ledua 1988, Lucas et al. 1991), the development of SCUBA and hookah gear has facilitated access to previously inaccessible *T. mbalavuana* stocks. The species is classified as 'Vulnerable' by the IUCN. There is little information regarding the mariculture of *T. mbalavuana*, but there was a successful spawning in December 1991 at the Tonga Fisheries Hatchery (Ledua et al. 1993).

Tridacna squamosa *Lamarck, 1819*

Tridacna squamosa (Figure 2F) is commonly known as the fluted giant clam. The valves have well-defined ribs and folds (the ribs also possess distinct protrusions called scutes). This species typically attains shell lengths of ~40 cm, but Hutsell et al. (1997) recorded an individual with a shell length of 42.9 cm. The mantle of *T. squamosa* usually exhibits mottled patterns in combinations of yellow, orange, blue, green and brown, and the incurrent siphon bears distinct tentacles. The valves are often coloured (yellow and orange-pink), which makes the species highly valued in the shell trade (Lucas 1988). Juvenile *T. squamosa* are typically byssally attached to coral rubble, while adults may be byssally attached or free-living. *Tridacna squamosa* inhabits a wide depth range, from reef flats to reef slopes down to 42 m (Jantzen et al. 2008), and is usually found in sheltered sites (e.g. wedged between corals) (Rosewater 1965). Globally, *T. squamosa* is the second most common tridacnine species, present from the Red Sea and eastern Africa in the west to the Pitcairn Islands, southern Japan and Queensland (Australia) in the east (Figure 1). New records for the central Pacific (Australes, Tuamotu and Gambier Archipelagos) have been added recently, although some gaps persist (such as Society Islands, French Polynesia) (Gilbert et al. 2007, Andréfouët et al. 2014). Despite ongoing exploitation, population numbers remain relatively stable across its range, with the exception of Cocos (Keeling) Islands and the Northern Mariana Islands where the species is locally extinct. *Tridacna squamosa* is classified by the IUCN as of 'Lower Risk/Conservation Dependent'. It is mainly harvested for subsistence use in local island communities and has been reported to be preferred in the shell trade due to its attractive colours, appearance and size. This species has been successfully cultured, mainly for restocking purposes in Indonesia, Malaysia, Philippines, Singapore, Thailand, Fiji, Palau, Federated States of Micronesia, Marshall Islands, Tonga, Vanuatu and Solomon Islands (Table 4), but there have been no reports of the outcomes of these endeavours. Individuals were also translocated from Palau to Guam and Tokelau, and Fiji to Samoa to help with local restocking initiatives (Kinch & Teitelbaum 2010). Juveniles from culture efforts in Australia, Palau, Federated States of Micronesia, Marshall Islands, Tonga, Vanuatu and Solomon Islands are (or have been) exported for the aquarium trade. As part of its larger research programme, the Darwin Aquaculture Centre (Northern Territory, Australia) also cultures *T. squamosa* to encourage farming as an economic opportunity for indigenous communities (Darwin Aquaculture Centre, pers. comm.).

Tridacna squamosina *Sturany, 1899*

Tridacna squamosina (Figure 2G) was originally collected during the 'Pola' expedition to the Red Sea in the 1890s (Huber & Eschner 2011). Sturany (1899) first published the results of this expedition, which noted the presence of three *Tridacna* species in the Red Sea: *T. maxima*, *T. squamosa* and a new species: *T. elongata* var. *squamosina*. The species was later rediscovered when living individuals were found in the Red Sea in the late 2000s (Richter et al. 2008), the largest recorded being 32 cm long. The species bears a strong resemblance to *T. squamosa*, but can be distinguished by its asymmetrical shells, crowded scutes, wider byssal orifice, and deep triangular radial folds (Roa-Quiaoit 2005). *Tridacna squamosina* strictly inhabits shallow reef areas and seagrass beds (~5 m depth), and is usually weakly byssally attached to the substratum (Roa-Quiaoit 2005). Presently only known from the Red Sea (i.e. Egypt, Israel, Jordan, Saudi Arabia and Yemen), recent anecdotal sightings of *T. squamosina* in Mozambique suggest that the species may also occur in the Indian Ocean (Table 4, Figure 1). Survey data suggest that live *T. squamosina* are generally rare. For example, only 13 individuals were identified during extensive surveys along the Jordanian Red Sea coastline (Richter et al. 2008). The current low numbers are postulated to be a result of overharvesting in the Red Sea, where it formed an important diet component of early coastal gatherers (>125,000 years ago) (Richter et al. 2008). As *Tridacna* exploitation remains prevalent in the Red Sea, *T. squamosina* is highly vulnerable to extinction. Mariculture of this species may have been carried out in Jordan (Roa-Quiaoit 2005), but the small number of individuals available for broodstock would make any mariculture effort a significant challenge.

Tridacna maxima *(Röding, 1798)*

The small giant clam, *Tridacna maxima* (Figure 2H), usually grows up to ~35 cm, with the largest individual collected (from Fanning Island, Republic of Kiribati) measuring 41.7 cm (Stasek 1965). *Tridacna maxima* is one of the three boring (sometimes referred to as 'burrowing') *Tridacna* species; juveniles are usually fully embedded in the reef substratum, but older individuals eventually outgrow the bored concavity and become partially embedded only. In areas characterized by high densities, such as the enclosed lagoons of French Polynesia, some individuals can be found on sand (Van Wynsberge et al. 2016). A persistent characteristic among the boring tridacnines is the tendency to byssally attach to the inside of the borehole. *Tridacna maxima* is also identified by its close-set scutes on the upper valves, the neat rows of tightly spaced hyaline organs along its mantle margin, and its brilliantly coloured and mottled mantle (usually blue, green and brown). It typically dwells in shallow areas of reefs and lagoons, rarely beyond a depth of 10 m (the deepest record is 21.2 m at the Dongsha atoll, South China Sea; M.L. Neo, pers. obs.). With a similar geographic range to *T. squamosa*, *T. maxima* is also a cosmopolitan species, but with more variable population densities across its range compared to *T. squamosa* (Van Wynsberge et al. 2016). Although *T. maxima* is harvested frequently for either subsistence or commercial purposes, it is still relatively common and hence classified by the IUCN as of 'Lower Risk/Conservation Dependent'. With rapid declines in the populations of larger tridacnine species, *T. maxima* is increasingly being extracted for local consumption and is likely to become more of a target for fisheries in the future (Van Wynsberge et al. 2016). Due to its attractively coloured mantle patterns it is, together with *T. crocea*, the most sought-after species for the aquarium trade. With a current ban on exports of wild-caught individuals for most countries within its range, the majority of individuals that enter the aquarium trade are cultured. While the species has been bred mainly for the aquarium trade (Wabnitz et al. 2003), wherever aquaculture and mariculture efforts exist (or were active), e.g. the Cook Islands (Waters et al. 2013), French Polynesia, Federated States of Micronesia, Samoa, Republic of Kiribati, Marshall Islands, Solomon Islands, Fiji, Vanuatu, Tonga, Palau and Taiwan (L.-L. Liu, pers. comm.), they have also contributed to reef restocking efforts (Table 4).

Tridacna noae *(Röding, 1798)*

The largest individual of *Tridacna noae* (Figure 2I) reported to date, from Kosrae, Micronesia, was 28 cm long (Borsa et al. 2015b). *Tridacna noae* cannot be readily identified by its shell traits, but it exhibits a highly distinctive mantle ornamentation including discrete teardrop patches typically bounded by white margins, sparsely distributed hyaline organs along the mantle margin, and the presence of papillae (Penny & Willan 2014, Su et al. 2014, Borsa et al. 2015a). Nevertheless, the mantle patterns of *T. noae* vary greatly in appearance among individuals (Borsa et al. 2015b). Because of its generally highly distinct and beautiful mantle, *T. noae*, long identified by aquarists as 'teardrop *maxima*', is highly desired for the aquarium trade (Wabnitz & Fauvelot 2014). The habitats of *T. noae* are generally similar to those of *T. maxima*, occupying depths of 1–15 m (Borsa et al. 2015b, Militz et al. 2015). Also a boring species, individuals are often found partially embedded within reef substrata. The known geographic distribution of *T. noae* extends from the Ryukyus (southern Japan), Taiwan, Southeast Asia, Western Australia and the Pacific Islands as far east as Christmas Island (Borsa et al. 2015b, Neo & Low 2017, Figure 1). As a newly resurrected species, data on the habitat and distribution of *T. noae* are scarce, but inferred to be similar to *T. maxima* due to morphological similarities and habitat preferences. A survey by Militz et al. (2015) determined that almost 42% of the specimens recorded as *T. maxima* within the Kavieng Lagoon system, Papua New Guinea, could now be classified as *T. noae*. Also, re-surveys of the Ningaloo Reef Marine Park revealed the presence of *T. noae* only, with no signs of *T. maxima* (Johnson et al. 2016); findings that challenge an earlier survey reporting the presence of (only) *T. maxima* (Black et al. 2011). Snorkel surveys on the reefs in Yap (Federated States of Micronesia), also identified high abundances of *T. noae*, which would have previously been recorded as *T. maxima* (C.C.C. Wabnitz, pers. obs.). Moreover, in Nauru, the only species found on the reefs during dedicated reef invertebrate surveys was recently re-identified as *T. noae* and not *T. maxima* (D. Thoma, pers. comm.). This inadvertent confusion of the two species highlights two problems: 1) the historical and current densities of *T. maxima* are likely to be overestimates in several locations, and 2) commercial exploitation that does not differentiate between the two species could interfere with local extinction risk calculations (Borsa et al. 2015b, Militz et al. 2015, Johnson et al. 2016). There have been a number of *ex situ* attempts to breed *T. noae* in Taiwan for restocking purposes (Su 2013) and some culture trials for mariculture grow-out and subsequent sale for the aquarium trade in the Federated States of Micronesia (C.C.C. Wabnitz, pers. obs.). Embryology, larval development and feeding ecology of *T. noae* in Papua New Guinea have recently been described (Southgate et al. 2016, 2017), while successful hatchery production has been reported in Fiji (P. Southgate, pers. comm.).

Tridacna rosewateri *Sirenko & Scarlato, 1991*

The first and only specimens of *Tridacna rosewateri* were collected from the Saya de Malha Bank (currently administered by Mauritius), Indian Ocean, during a 1984 expedition (Sirenko & Scarlato 1991). Nine individuals were collected measuring 6.7–19.1 cm shell length. The shell morphology of *T. rosewateri* shares features with both *T. maxima* (i.e. large byssal orifice) and *T. squamosa* (i.e. large scutes), but differs from those species in having thinner shell walls, deep triangular valve margin folds, and larger dense scutes on primary radial folds (Sirenko & Scarlato 1991, Monsecour 2016). Little is known about its habitat, but the *T. rosewateri* individuals were found among corals (*Madrepora* sp.) and dense beds of the seagrass *Thalassodendron ciliatum* (Sirenko & Scarlato 1991). *Tridacna rosewateri* is currently classified as 'Vulnerable' by the IUCN. The absence of living individuals makes the validity of *T. rosewateri* as a tridacnine species ambiguous. Benzie & Williams (1998) criticized the poor description of the species and proposed that it is a junior synonym of *T. squamosa*, while Newman & Gomez (2000) and Monsecour (2016) have argued that they could readily distinguish its shells from *T. squamosa* and concluded that it might be a distinct species endemic to Saya de Malha Banks.

Tridacna lorenzi *Monsecour, 2016*

Tridacna lorenzi is the newest species added to the list of Tridacninae. The species was described from the Cargados Carajos Archipelago (St. Brandon), Mascarene Plateau in the outlying territories of Mauritius (Monsecour 2016). A medium-sized species, ten of the largest type specimens measured between 11.3 and 26.0 cm shell length. Monsecour (2016) notes that both *T. maxima* and *T. rosewateri* are likely the closest congeners to *T. lorenzi* on the basis of some overlapping morphological characters. Similar to *T. maxima*, *T. lorenzi* has asymmetric shells, a large byssal orifice, and close-set scutes, but differs in the narrow interstices between primary ribs, its triangular valve margins, and the dull-coloured mantle that does not extend beyond the shell margins (Monsecour 2016). Commercially, this species has previously been misidentified as *T. rosewateri*, since the valve margins of both primary ribs and rib interstices are triangular in both species (Monsecour 2016). *Tridacna lorenzi* can, however, be distinguished from *T. rosewateri* by its more asymmetric, more globose, heavier shell valves, and closer-set scutes. The *T. lorenzi* individuals described by Monsecour (2016) were mostly collected from shallow waters in turbid lagoons of no more than 1 m depth, free-living on sand and among loose rubble. Distribution data are limited, although Monsecour (2016) suggested that *T. lorenzi* was locally common and encountered more often than the rarer *T. squamosa* and uncommon *T. maxima*. Local fishermen reportedly eat the species, and use their shells as saucers or ashtrays. A molecular analysis of *T. lorenzi* to determine its relationship with congeners has yet to be conducted.

Tridacna crocea *Lamarck, 1819*

Of all the tridacnine species, *Tridacna crocea* (Figure 2J) is the smallest with a maximum size of ~15 cm (Rosewater 1965). Commonly known as the 'burrowing' or 'boring' giant clam, *T. crocea* is a rock borer that embeds its entire body into the substratum, leaving only the mantle exposed (Yonge 1936). It appears to be well adapted to low salinity levels, often found in areas that experience freshwater runoff (Hart et al. 1998). As with *T. maxima* and *T. noae*, this species byssally attaches to its bored concavity. *Tridacna crocea* is usually identified by its boring habit, but it also develops well-spaced scutes that become eroded over time within the borehole. The mantles are brightly coloured, exhibiting various shades of blue, green, purple, white and brown (Todd et al. 2009). *Tridacna crocea* mostly inhabits reef flats in shallow waters of depths no more than 10 m (Hamner & Jones 1976, Hamner 1978). The species has a wide distribution (24 localities), ranging from Australia to Japan, east to Palau, and from Vanuatu to the Andaman Islands (Figure 1). It is possibly extinct in Guam and the Northern Mariana Islands (Wells 1997). In most areas, *T. crocea* is still considered reasonably abundant, probably due to its small size and the difficulty of extracting it from reef substrata. Even though *T. crocea* is one of the most easily accessible tridacnine species, exploitation is limited to domestic consumption. It is a popular delicacy in Okinawa, Japan (Okada 1997). The species was considered widespread in the Solomon Islands (Wells 1997) and was preferentially harvested as a source of food (Hviding 1993). More recent surveys indicate that it is much less common than it used to be in the Solomon Islands (Ramohia 2006). It is classified as of 'Lower Risk/Least Concern' by the IUCN. Mariculture of *T. crocea* is well established in Okinawa, Japan, where the spats are distributed to local fishermen for culture and release (Okada 1997). There have also been *ex situ* attempts to culture *T. crocea* in Brazil (Mies et al. 2012). Due to its bright colours, it is highly prized in the aquarium trade (Wabnitz et al. 2003), and mariculture efforts in Palau, the Marshall Islands and the Federated States of Micronesia, for example, have had some success breeding it (Heslinga 1995, 2013). However, because of its comparatively slow growth and poor early survival rates, it is often regarded as less suitable (not cost-effective) for aquaculture or mariculture operations, in spite of its desirability within the aquarium trade.

Contemporary threats and challenges

Throughout their geographic range, representatives of the Tridacninae remain an important and valuable coastal resource to both local fishing communities and commercial markets. The relative abundance, shallow distribution, conspicuous appearance, and sessile nature of giant clams make them easy to harvest with simple fishing gear. During reef gleaning and free-diving (Hviding 1993, Sant 1995), individuals are usually collected opportunistically in areas of low densities, but they can be the main target of fishing trips in areas where densities are high. Their flesh is excised from the shells with knives, wooden sticks or metal stakes (Kinch & Teitelbaum 2010). SCUBA and improvised diving apparatus such as hookah gear (a simple surface air-feed) are used to reach individuals in deeper waters (Hviding 1993, Ledua et al. 1993, Kinch & Teitelbaum 2010). Almost all species of the Tridacninae have been exploited for meat as food, fish bait or animal feed, their shells sold to the curio trade, and exported live for the aquarium trade (Heslinga 1995, Sant 1995, Kinch & Teitelbaum 2010, Neo & Loh 2014).

Prior to the 1980s, commercial exports of tridacnine adductor muscles to Asian markets and illegal poaching by long-range foreign vessels were responsible for the severe stock reductions occurring in the Indo-Pacific (Pearson 1977, Dawson & Philipson 1989, Shang et al. 1991, Sant 1995). Even though commercial exploitation of wild stocks is now banned in most countries, either poorly regulated or unregulated subsistence harvesting can still threaten remaining stocks (Tan & Zulfigar 2003). Large-scale poaching also poses a major and persistent threat for wild populations. Coastal resource authorities from various countries (Australia, Cambodia, Malaysia and the Philippines) have reported an increase in the number of fishing boats harvesting giant clams illegally within the last five years (Krell et al. 2011, Lee 2014, Colbeck 2015, Gomez 2015b). The scale of this harvest is substantial, with almost 20 tonnes of shells reportedly removed from protected areas (Lee 2014). One of the largest *Tridacna* shell markets today is China. Many of the local fishermen from Tanmen, Hainan, have converted from traditional fishing to the more lucrative tridacnine fishing as their main livelihood (Zhang 2014). Shells of giant clams may have become a substitute for ivory, the import of which is now regulated strictly (Gomez 2015a,b, Cavell 2016, Larson 2016). As the shell craft industry flourishes in Tanmen, large quantities of fossilized giant clam shells have been extracted from the sea beds of Scarborough Shoal, the Spratlys and Paracel Islands (South China Sea) to support the handicraft industry (Zhang 2014, Gomez 2015a,b, Larson 2016). Large shells are carved into sculptures, with medium-sized shells processed into beads for jewellery. It is also thought that giant clam shells are increasingly being used to manufacture nuclei for the Chinese freshwater pearl industry (X. Fan, pers. comm.). Even though recent sources suggest that the local Chinese government has banned the harvesting of dead shells (Master 2016), the intense extraction has devastated large tracts of coral reefs within the South China Sea.

The habitats of tridacnines are also threatened as corals reefs throughout the Indo-Pacific become degraded (Huang 2012, Neo & Todd 2012b). The pressure of anthropogenic activities threatens the health of reef environments and hence the survival and growth of the tridacnines that live in them. In a global meta-analysis for *Tridacna maxima*, Van Wynsberge et al. (2016) highlighted that, except for areas with very low human population density (<20 inhabitants ha^{-1} of reef), giant clam densities tended to decrease as human presence increased. Giant clam densities were also strongly dependent on the type of reef (atoll, island, continent)—which is an important natural co-factor. In the northern Red Sea (Egypt), Mekawy & Madkour (2012) showed that the abundance of *T. maxima* was higher at sites further away from anthropogenic sources and proposed that the main stressors were tourism, SCUBA diving, water pollution and contaminants, and the drilling for and production of oil. The survival, growth and photosynthetic performance of giant clams is significantly reduced when exposed to high copper concentration (tested at 50 µg L^{-1}) (Elfwing et al. 2001) and reduced salinities (Eckman et al. 2014). Coastal urbanization also has negative effects on giant clam populations. For example, in Singapore, many of the reefs where giant clams were previously found have been buried as a result of large-scale land reclamation projects (Guest et al. 2008, Neo & Todd 2012a,b). The impacts of sedimentation on tridacnines are not yet well understood, but, in addition to affecting photosynthetic performance, sediment stress has been hypothesized to divert energy away from maximizing photosynthesis (e.g. by transporting inorganic ions to the zooxanthellae) to supporting behavioural responses and increased respiratory demands (Elfwing et al. 2001). A preliminary study by Ang (2014) revealed that juvenile *T. squamosa* was more susceptible to chronic sedimentation than to acute deposition events.

Climate warming may lead to undesirable effects on giant clams, where extremes in either temperature or ultraviolet irradiation can lead to poor growth, bleaching (the expulsion of photosynthetic symbionts), and increased mortality (Buck et al. 2002, Andréfouët et al. 2013, Junchompoo et al. 2013), particularly near the equator (Chaudhary et al. 2016). The few studies relevant to the impacts of climate change on tridacnines have focused on the effects of thermal stress and bleaching responses (Norton et al. 1995, Blidberg et al. 2000, Buck et al. 2002, Leggat et al. 2003), which have been shown to affect their growth negatively. Warming oceans can also lead to bleaching of both juveniles and broodstock individuals, resulting in the loss of productivity or lower survival of 'grow-out' stocks (Wilkinson & Buddemeier 1994, Gomez & Mingoa-Licuanan 1998). In the 2016 global mass coral bleaching event, bleaching incidences among giant clams varied across geographic sites: *Tridacna maxima* did not bleach in Mauritius (R. Bhagooli, pers. comm.), but those in Singapore (M.L. Neo, pers. obs.), Guam (A. Miller, pers. comm.) and East Tuamotu (S. Andréfouët, pers. obs.) were bleached severely. Interestingly, surveys of giant clam populations at Lizard Island, Australia, showed that the 2016 mass coral bleaching event and cyclones during the previous three years resulted in a much lower mortality rate for *T. gigas* compared to either *T. derasa* or *T. squamosa*, suggesting that *T. gigas* may be best able to survive after major perturbations in the GBR (A.D. Lewis, pers. comm.).

The detrimental effects of ocean acidification have also been demonstrated in juvenile giant clams, with experimental evidence showing that they exhibit negative shell growth (dissolution) (Waters 2008) and lower survival rates (Watson et al. 2012) in acidic conditions (~600–1000 μatm [60.8–101.3 Pa] pCO_2). Studies testing the combined effects of increasing temperature and pCO_2 (based on climate projections for the end of this century) for 60 days showed that the shells of juvenile *Tridacna squamosa* were significantly altered with a decrease in calcium and magnesium ions (Armstrong et al. 2014), and lower survival rates (Watson et al. 2012, Watson 2015). Less is known about the effects of climate change stressors on early life-history stages, with only one study conducted to date. Neo et al. (2013b) tested the combined effects of temperature and salinity on *T. squamosa* fertilization and embryo development, and showed that salinity (27 psu and 32 psu) had no significant effect on survival but mortality increased at the higher of the two temperatures tested (22.5°C and 29.5°C). Climate change could also place additional economic and developmental pressures on giant clam mariculture operations. Increased temperatures in hatcheries can cause problems of algal overgrowth (M.L. Neo, pers. obs.), poor shell precipitation (Schwartzmann et al. 2011), and possibly premature spawning patterns, which are all undesired outcomes for spawning and rearing of juveniles.

Impacts due to the threats outlined above lead to the lowering of tridacnine population densities across their ranges in the wild, which has serious repercussions for their ability to reproduce successfully (Munro 1992). Fertilization success depends on the synchronized spawning of conspecifics (Lucas 1988, Gilbert et al. 2006a), as the trigger for sperm release is dependent on the chemical cues found on the eggs (Munro et al. 1983). Upon detection of the inducer, other neighbouring clams may also release eggs, thus encouraging progressive downstream fertilization. The tendency for tridacnines to aggregate has been attributed to their need to be close to each other to reproduce (Braley 1984, Huang et al. 2007, Soo & Todd 2012, 2014). Giant clam populations are therefore highly sensitive to stock depletion, where sparse spawning adult populations can lead to lowered (or zero) fertilization rates and consequently reduced or absent recruitment rates (Munro 1992, Tan & Zulfigar 1999, Neo et al. 2013a). To compound matters, as stocks become more scarce, harvesting size tends to decrease, meaning that individuals may be harvested even before reaching reproductive viability, thereby further affecting the availability of mates and limiting fertilization rates (i.e. component Allee effects) (Stephens et al. 1999). This could lead to the functional extinction and eventual collapse of the entire population (Frank & Brickman 2000, Petersen & Levitan 2001). Wild stocks may recover via the dispersal of planktonic larvae from other reefs brought in by prevailing currents (Benzie & Williams 1992a,b, Tan & Zulfigar 1999, 2001, Neo et al. 2013a). Such recovery, however, may take decades if coral reefs are isolated (due to the short [9-day] pelagic larval duration of giant clams), and/or currents are unfavourable (Yamaguchi 1977). Even in closed lagoons (with high retention rate) and with large stocks the recovery to initial population levels may still take decades. This is the case for Tatakoto Atoll, renowned for supporting the highest clam densities on record (Supplementary Table A3), but now depleted severely after a mass mortality event (Andréfouët et al. 2013). It may be many decades before densities such as those observed in 2004 (Gilbert et al. 2006a, Van Wynsberge 2016) will be witnessed again.

Cryptic species also present another challenge for the management and conservation of remaining tridacnine populations. When cryptic species become confused with contemporary common species, there are implications for commercial giant clam fisheries and their regulation due to the potential for misidentification (e.g. Rosewater 1982, Borsa et al. 2015b, Militz et al. 2015, Monsecour 2016). Additionally, the lack of knowledge regarding these species makes it difficult to implement appropriate conservation measures (Militz et al. 2015, Johnson et al. 2016). Previous systematic research on tridacnines relied heavily on morphological and behavioural characterization (e.g. Rosewater 1982, Lucas et al. 1991). These diagnostic characters can, however, be misleading in that giant clams generally are morphologically plastic and functionally similar (Benzie & Williams 1998, Neo & Todd 2011). During the last decade, the global use of genetic tools and breakthroughs in sequencing have led to the discovery of an increasing number of cryptic lineages (Pfenninger & Schwenk 2007) hidden behind one species name (morphologically close, but genetically divergent). Yet, the conversion of genetically unique lineages into robust and formally named taxonomic entities remains challenging. Considering the recognized variability in tridacnine morphology, they are good candidates for crypticity. In 2008, phylogenetics helped to identify a cryptic Red Sea species: first described as a new species, *Tridacna costata* (Richter et al. 2008) and later synonymized as *T. squamosina* (Huber & Eschner 2011). Subsequently, there has been the rediscovery of *T. noae* using various genetic markers (Su et al. 2014), and *T. noae* has turned out to be a widespread cryptic species in the Indo-Pacific (Borsa et al. 2015a,b, Militz et al. 2015, Johnson et al. 2016). Given the ambiguity of morphological characters among cryptic individuals, the growing body of molecular evidence can help reveal deep lineages across taxa and lead to the (re)discovery of species (Wilson & Kirkendale 2016).

Conservation and management

Legislation and regulations

Convention on International Trade in Endangered Species of Wild Fauna and Flora (CITES)

The Convention on International Trade in Endangered Species of Wild Fauna and Flora (CITES) is recognized internationally as the governing body that oversees the trade exports and imports of selected endangered species. Giant clams are currently listed on Appendix II of CITES, which comprises species that are not necessarily now threatened with extinction, but that may become so unless trade is closely monitored. *Tridacna gigas* and *T. derasa* were first listed in 1983, and the other members of the family Tridacnidae (now subfamily Tridacninae) were listed in 1985 on the basis of so-called 'look alike species', i.e. species whose specimens in trade look like those of species listed for conservation reasons (Wells 1997). CITES states that the international trade in giant clams (whole or any part of the animal) is permitted only if the relevant export/import certifications are issued. The effectiveness of enforcing CITES is, however, largely dependent on whether the countries involved in the trade are signatories to the Treaty, or if a non-signatory is trading with the signatories (Wells 1997). In the past, countries such as Taiwan and the Maldives were involved heavily in the giant clam trade but were not CITES Parties, which impeded the implementation of CITES legislation (Wells 1997). Even in instances where exporting countries are CITES Parties, the trade data provided may be unreliable. In a number of examples capacity within relevant offices has been reduced, at times resulting in omissions, erroneous data entry (e.g. wrong source code, and submission of number of permits issued instead of actual numbers traded), and failure to submit or significant delays in providing trade statistics to the Secretariat (UNEP-WCMC 2011, C.C.C. Wabnitz, pers. obs.). Various workshops and other initiatives have been conducted to strengthen CITES capacity for countries in Oceania, including non-parties to the Convention (Table 4). Another concern, however, is that the scope of the CITES Treaty does not include localized collection and trade of giant clams within countries (which can be substantial), regardless of their status as a party to the convention. Relatedly, these countries may allow a quota of wild tridacnines to be collected and sold for the aquarium trade, but suppliers will usually collect only specimens with the highest value colours. This can result in genes for colour being reduced or lost from wild populations. Although not well understood, it is likely that mantle colours and their varieties (colour polymorphism) are ecologically important in natural reef settings (Todd et al. 2009).

International Union for Conservation of Nature (IUCN) Red List categories of threat

Nine of the 12 species of Tridacninae are on the IUCN Red List of Threatened Species (Neo & Todd 2013). *Tridacna gigas*, *T. derasa* and *T. rosewateri* are listed as 'Vulnerable', due to the rate of decline of remaining wild stocks. *Tridacna mbalavuana* is also listed as 'Vulnerable' on the basis of its small and declining area of occupancy, although it has been suggested that it should be categorized as 'Endangered' (Wells 1997). *Hippopus hippopus*, *H. porcellanus*, *Tridacna maxima* and *T. squamosa* are listed as 'Lower Risk/Conservation Dependent' due to the decline and disappearance of many populations. *Tridacna crocea* was initially excluded in the earlier Red Lists due to insufficient data (Wells 1997), but was reinstated in 1996 and listed as 'Lower Risk/Least Concern' (Molluscs Specialist Group 1996). The IUCN Red List of Threatened Species draws attention to species at risk of extinction and promotes their conservation (Collar 1996), and is frequently used to guide the management of resources (Rodrigues et al. 2006). It is, however, important to point out that 1) the global IUCN classifications for tridacnines are outdated as they were last reviewed by Wells (1996); 2) the reported status may not accurately reflect the situation within individual countries, e.g. Neo & Todd (2013) for Singapore; and 3) recent species, i.e. *T. lorenzi*, *T. noae* and *T. squamosina*, are not yet on the IUCN Red List as their ecology, habitat occupancy and density have not been assessed. Given the decline in tridacnine stocks and their habitat, it is important to produce a definitive update of IUCN classifications for all 12 species, including promoting the use of localized or regional classifications to better represent situations 'on the ground' that are of greater value when planning conservation strategies (Neo & Todd 2013).

Local mitigation measures

Regional efforts to initiate cooperation and collaboration among nations towards the management of sustainable giant clam fisheries have been few (e.g. Kinch & Teitelbaum 2010), but much has been done locally to reduce exploitation. The conservation efforts implemented throughout the Indo-Pacific are listed in Table 4. The localities of Red Sea, Southeast Africa and the Indian Ocean generally lack specific laws to regulate recreational fishing of giant clams. In East Asia, restoration of impacted populations has begun, but mariculture there (except in Japan) is still in its infancy. In the South China Sea, tridacnines are, unfortunately, within disputed territorial waters, which makes agreeing and coordinating ocean governance among the numerous neighbours a substantial challenge. There have been a number of restocking efforts using mariculture in Southeast Asia, but the success of programmes has been variable at each locality (Indonesia, Malaysia, the Philippines, Singapore and Thailand). The management of tridacnine populations is most advanced in Australia and the Pacific Island nations. For example, some coastal communities in the South Pacific have put in place stricter measures to alleviate tridacnine fishing pressures (Table 4), including banning commercial fishing (Fiji, Papua New Guinea, Solomon Islands, Vanuatu, Federated States of Micronesia, Guam, Republic of Kiribati and Palau), setting minimum size limits for subsistence harvesting (French Polynesia, Niue, Samoa and Tonga), imposing harvesting quotas or bag limits (New Caledonia, American Samoa and the Cook Islands), restricting fishing to free diving only and banning the use of mechanical fishing equipment (Chambers 2008, Kinch & Teitelbaum 2010, Andréfouët et al. 2013). Outcomes of these measures vary among the South Pacific nations as they depend on the degree of exploitation (i.e. a highly exploited population will take a longer recovery time), local enforcement measures and capacity, as well as community willingness to adopt these practices (Munro 1989, Lucas 1997). For instance, some Tongan communities set up giant clam 'circles' (i.e. aggregating adult clams into rings) to facilitate reproduction among individuals, and were able to repopulate nearby reefs with juveniles within ten months (Chesher 1993). Unfortunately, efforts do not appear to have been maintained and stocks in Tonga are severely depleted (C.C.C. Wabnitz, pers. obs.)—it is hoped that the regulation of selling giant clams in their shells to enforce size limits, which is widely respected, will help resolve this issue. In general, surveys throughout the region continue to indicate that populations are under severe stress (K. Pakoa, pers. comm.). Australia, India, China, Mauritius, Taiwan, and Japan have their own national protection acts that include giant clams (Table 4). Within Southeast Asia, it is generally recognized that tridacnines need protection, but only the Philippines, Malaysia and Thailand have national legislation regulating their exploitation (Knight et al. 2010, Gomez 2015a). Illegal fishing by coastal communities, however, remains prevalent in many of these countries, probably because of the traditional importance of giant clams as a coastal resource coupled with the lack of manpower and funding to support long-term monitoring, surveillance and law enforcement.

Mariculture for restocking

Giant clam breeding was pioneered in the 1970s at the University of Guam Marine Laboratory and the Micronesian Mariculture Demonstration Centre (MMDC) in Palau. It was further complemented by the work of John Lucas in Australia supported by the Australian Centre for International Agricultural Research in the 1980s, and consolidated by the work of ICLARM (now WorldFish) in the Solomon Islands in the late 1980s and early 1990s and, subsequently, supported the extensive research and technical training throughout the Pacific and Southeast Asia (e.g. Heslinga et al. 1984, Heslinga & Fitt 1987, Heslinga 1991, Copland & Lucas 1988, Braley 1992, Calumpong 1992, Norton & Jones 1992, Tisdell 1992, Fitt 1993). Mariculture is being adopted increasingly for mass production of individuals for the aquarium trade (Heslinga et al. 1990, O'Callaghan 1995, Bell et al. 1997, Heslinga 2013) as well as the restocking of rare species (Neo et al. 2009, 2011, Neo & Todd 2012a, Heslinga 2013) or extirpated populations (Braley & Muir 1995, Gomez & Mingoa-Licuanan 2006). Tridacnine mariculture has no apparent deleterious environmental effects (Lucas 1997), but there remains the possibility of inadvertently introducing exotic parasites, diseases and other biota (Newman & Gomez 2000), especially if broodstocks are imported without appropriate quarantines. Combined with local community farm grow-out operations, such mariculture activities can provide sustainable livelihood opportunities in localities where there are few alternatives (e.g. remote atolls in French Polynesia, remote locations in the Solomon Islands, and outlying islands in the Marshall Islands), as long as projects are conceived and run as sustainable and cost-effective enterprises or projects. In many cases, however, poor survival, limited production, and hatchery expenses result in cost-ineffective production and eventual termination of activities. Nevertheless, as of 2016, there were at least 34 functioning giant clam hatcheries in 25 countries, and hundreds of ocean nurseries and reserves (G.A. Heslinga, pers. obs.).

While most giant clam hatcheries operate on some commercial (or semi-commercial) basis, some, generally with the support of foreign aid or other forms of subsidies, also function as a means to support conservation and facilitate sustainable harvesting (Lucas 1997, Heslinga 2012, see Table 4). In general, the success of these initiatives is neither well studied nor well documented (Teitelbaum & Friedman 2008). Restocking programmes often do not have a set of protocols for fisheries officers and managers to follow, nor do they tend to be accompanied by regular monitoring to ascertain the success of such efforts over time (C.C.C. Wabnitz, pers. obs.). The survivorship of restocked clams varies widely within and among localities, with the main causes of mortality being predators, storms, poaching, and the lack of continuous husbandry (Lucas 1997, Southward et al. 2005, Heslinga 2013). In addition, hatchery-produced juveniles may be less genetically variable, which could increase vulnerability to parasites and diseases (Benzie & Williams 1996). High mortality rates, coupled with the high costs and intensive labour of rearing giant clams to reach 'escape size' (typically ~25 mm long, at which point they are less vulnerable to predators), may explain the waning enthusiasm and funding for restocking in some areas, notably Queensland (Australia) and the Solomon Islands, since the late 1980s (Bell 1999, Southward et al. 2005).

Restocking giant clams requires long-term commitment and monitoring, with examples of this mainly occurring in Palau, the Philippines and Japan, where mariculture, domestication and restocking have maintained momentum for over 20 years (Murakoshi 1986, Bell 1999, Gomez & Mingoa-Licuanan 2006, Heslinga 2013). There are also many examples of maricultured giant clams being shipped around the Indo-Pacific region as juveniles in the 1980s and 1990s, matured in ocean nurseries in the destination countries, and then used as breeding stock in local hatcheries. Firm evidence that restocked clams have produced local juvenile recruitment is either absent of or poorly documented, probably owing in part to the remoteness of the areas under study, and the difficulty and expense of conducting authoritative surveys. Exceptions to this may be found in Yap (Federated States of Micronesia) and the Philippines, where *Tridacna derasa* and *T. gigas*, respectively, were restocked (Table 6) and where new recruits have been reported (Cabitan & Conaco 2017). This is encouraging, as restocking without the creation of new generations will provide few long-term conservation benefits. How to ensure that restocked populations successfully reproduce and recruit is a major challenge for giant clam restoration efforts globally.

Recent conservation approaches

Biophysical modelling for conservation

At national and local (archipelago, island, reef) scales, giant clam conservation management has focused on fishing regulations and restocking (see previous sections). Assessing the effectiveness of such conservation efforts for a particular location requires an understanding, and ideally modelling, of processes and factors that influence their distribution and abundance. These include aspects of the species' biology, population dynamics (e.g. size-structure, density, recruitment, mortality), life-history traits (e.g. growth-fertility, reproduction and spawning occurrences) (Apte & Dutta 2010, Black et al. 2011, Yau et al. 2014, Dolorosa et al. 2014, Neo et al. 2013b, 2015b, Menoud et al. 2016, Van Wynsberge et al. 2017), and larval flux (Neo et al. 2013a). Human uses and impacts are also important factors to consider (Van Wynsberge et al. 2015, 2016). Recently, mass mortality in semi-enclosed atolls due to unusual physical oceanographic conditions has been identified as a key driver of population dynamics (Andréfouët et al. 2013) and climate change is likely to make these events more frequent (Andréfouët et al. 2015). These examples highlight the importance of monitoring physical conditions and their integration into models (Neo et al. 2015b, Van Wynsberge et al. 2017).

Table 6 An overview of reports of local recruitment after restocking efforts

Locality	Restocked species	Restocking period	Number of restocked individuals (life stage; size range)	Recruitment monitoring period	Remarks
Yap, Federated States of Micronesia	*Tridacna derasa*	1984–1991	1984–1989: 8000 (8–11 cm) 1988–1989: 3500 (6–8 cm) April 1991: 2000 (5–6 cm) Nov. 1991: 2000 (10 cm)	1991–2014 (ongoing)	*Tridacna derasa* juveniles were found by local fishermen and international experts from the Secretariat of the Pacific Community (SPC) in the early 1990s (J.O. Fagolimul & P. Dor, pers. comm.) after an extensive reintroduction program initiated in the mid-1980s undertaken with clams cultured at Palau's MMDC (Price & Fagolimul 1988, Heslinga 1991, 1993a,b, 2013, Lindsay 1995, Teitelbaum & Friedman 2008). In 2013–2014, some of these Yapese *T. derasa* recruits reached full maturity and were used with replicated success as breeding stock in a local hatchery managed by Mr. Philip Dor (P. Dor, pers. comm.). Mr. Dor has successfully produced commercial numbers (hundreds of thousands) of macroscopic *T. derasa* juveniles in the Yap hatchery, as verified by international experts.
Philippines	*Tridacna gigas*	1990s to present-day	~45,000 (Sub-adults; >20 cm)	2007–2015 (ongoing)	For >20 years, the Marine Science Institute, University of the Philippines, has been culturing giant clam species for restoration of depleted populations in the Philippines. Several species were initially restocked, but later efforts focused on *Tridacna gigas* (Gomez & Mingoa-Licuanan 2006). Recruits of *T. gigas* were first observed in the vicinity of Bolinao, Pangasinan, where the broodstock are placed (Cabitan & Conaco 2017). Subsequently, occasional reports have been received from at least two other localities where restocking was carried out.

Finally, but this has never been attempted, an ecosystem-based characterization including spatio-temporal variation in predation, competition, and food availability, is also likely to influence the accuracy of models simulating the effectiveness of conservation measures.

A pilot fishery-oriented modelling study on what could be the effects of management measures such as no-take areas, rotational closures, fishing quotas, and maximum or minimum catch sizes, on giant clam populations was undertaken by Van Wynsberge et al. (2013) for two islands of the Austral Archipelago in French Polynesia (Tubuai and Raivavae). This was the first spatially explicit model of giant clam population dynamics, based on maps of densities and habitat-specific age structure of populations. It was calibrated according to stock data quantified a few years apart, and parameterized and validated using limited local life-history traits and population dynamics data. More recently, the initial model was improved by including spatial patterns of fishing, mass mortality occurrences, size-structure per habitats, and refined population dynamics parameters following two years of surveys during which physical conditions were monitored (Van Wynsberge 2016). This more realistic model has been used to test the effects of conservation measures on *Tridacna maxima* populations. While such modelling opens new pathways for conservation and research, it requires intensive fieldwork for calibration/validation and substantial computing resources.

The models described above cannot be implemented easily and duplication at new sites needs caution, but staged efforts and priorities can be recommended. An important aspect is spatial variability. Different locations along either a reef or lagoon, for example, can display different tridacnine densities as a result of the combination of a number of biophysical processes, such as those associated with coastal hydrodynamics, climate change and pollution (Zuschin & Piller 1997, Green & Craig 1999, Andréfouët et al. 2005, Neo & Todd 2012b, Ullmann 2013). It is, therefore, desirable to first map the continuum of giant clam density across a reef system together with the clam size-structure (Andréfouët et al. 2005, 2009, Gilbert et al. 2006b). Ideally, the spatial characterization of density and size-structure should be used to determine where to monitor population dynamics and life traits and, if there is ongoing human exploitation, focal sites should be selected to represent both exploited and refuge areas.

Information about larval dispersal is another critical input for conservation modelling. The priority level for such work is dependent on the degree of closure and isolation of the studied reef, or sets of reefs. In Singapore, for instance, there is a continuum of reefs along the continent and island shores organized in a dense matrix, and understanding larval dispersal of *Tridacna squamosa* among reefs and (meta-)populations is necessary for the sound management of this species (Neo et al. 2013a). Conversely, the populations of *T. maxima* in the east Tuamotu archipelago of French Polynesia presents an opposing scenario, where remote and hydrodynamically closed atoll lagoons are more self-recruiting with limited flux from outside compared to open lagoons. While fluxes between atolls may be important for genetics, they are negligible in term of demography and fishery management (Van Wynsberge et al. 2016).

Biophysical modelling for conservation of giant clams is a new, complex and exciting task; however, it requires diverse spatial and temporal information that is difficult and costly to acquire. Nevertheless, population dynamics modelling and connectivity modelling are needed to create a holistic dynamic framework that can be applied to multiple locations, as well as to foster ambitious informative multidisciplinary studies to enhance knowledge for giant clam conservation.

Genetic information and evolutionary relationships for conservation planning

As molecular genetics techniques become more efficient and cost-effective, it is increasingly common for conservation managers to use genetic data in prioritizing species conservation (e.g. Huang 2012, Neo & Todd 2012b, Beger et al. 2014, von der Heyden et al. 2014). Fundamentally, genetic data offer insights into genetic diversity, population connectivity, and the evolutionary history of species (Beger et al. 2014). Such information provides the opportunity to investigate cryptic species diversity (discussed earlier in 'Contemporary threats and challenges'), spatial ecological interactions (Selkoe et al. 2008), as well as the evolutionary potential of species (Peijnenburg & Goetze 2013). The genetic structure of giant clam populations has been of interest since the 1990s, mainly to differentiate populations (e.g. Benzie & Williams 1992a,b, 1995, Macaranas et al. 1992), although none of these previous studies mentioned the incorporation of genetic information for spatial conservation prioritization. Subsequent giant clam population genetic studies have provided opportunities to develop phylogenetically-informed management strategies (e.g. DeBoer et al. 2008, Kochzius & Nuryanto 2008, Neo & Todd 2012b).

Another genetic-based conservation approach is the consideration of evolutionary relationships within a clade of target species (Faith 1992, 2007), especially for species that may be at risk of extinction and thus lead to loss of phylogenetic diversity (Huang & Roy 2013, Curnick et al. 2015). One such platform is the EDGE (Evolutionarily Distinct and Globally Endangered) of Existence programme that converts IUCN threat categories to probabilities of extinction for phylogenetic conservation prioritization (Redding & Mooers 2006, Mooers et al. 2008). The current programme has applied these metrics to major taxonomic groups such as mammals (Isaac et al. 2007, Safi et al. 2013) and amphibians (Isaac et al. 2012, Safi et al. 2013), but not to invertebrate taxa, with the exception of the Scleractinia (Huang 2012, Huang & Roy 2013). Given that wild tridacnines today are facing an array of threats, the use of phylogenetic diversity and evolutionary distinctiveness could help to hasten the evaluation of species' extinction risk.

Beyond phylogeny, in principle, larval dispersal and population genetic information can contribute to the design of more effective reserve networks by ensuring that all identified (meta-)populations are represented within them and by protecting source areas (Fogarty & Botsford 2007). All published studies thus far have used water circulation models and simulation of passive drifters to predict and explain (or not) the spatial patterns in genetic or demographic observations. In Indonesia, DeBoer et al. (2008) found poor agreement between larval dispersal distances of *Tridacna crocea* inferred from passive larval dispersal modelling and from genetic data. Van Wynsberge (2016) showed that biophysical models are in better agreement with *T. maxima* genetic observations in New Caledonia if habitat distribution and population densities are taken into account. Reaching an agreement between models and empirical *in situ* data is also likely largely dependent on enhanced realistic biophysical model forcing, with the necessary future inclusion of larval behaviour, settlement processes, fine-scale coastal hydrodynamics, habitat distribution, and so on (Dumas et al. 2014, Neo et al. 2013a, 2015b, Soo & Todd 2014, Van Wynsberge 2016). All these represent significant long-term challenges.

The future of giant clams?

This review synthesizes the current state of knowledge on giant clam taxonomy, distribution and abundance, exploitation and other threats, and conservation issues. In general, there exists a global consensus that tridacnines in many localities are endangered, especially the larger species, *Tridacna gigas* and *T. derasa*, where >50% of naturally occurring populations are severely depleted, locally extinct, or data deficient. The combination of increased commercial demand (including large-scale illegal fisheries) coupled with advances in fishing techniques, transport and storage have had significant negative impacts. Overharvesting for human use (consumption and materials) is probably the greatest driver of decline. Climate change, pollution, habitat loss and coastal development are additional factors that can deleteriously influence the survival of remaining stocks. As a result of lowered densities, populations are potentially experiencing component Allee effects (i.e. low-density constraints on fertilization efficacy), thus impairing their capacity to reproduce successfully in the wild (Neo et al. 2013a). Furthermore, the genetic diversity of populations may already have been reduced irretrievably in many areas. CITES listings and the IUCN Red List of Threatened categories have helped to raise awareness of the threats giant clams face, regulate trade and mitigate the decline of remnant populations. Local measures such as the enforcement of laws to regulate (or ban) both subsistence and commercial fishing (i.e. South Pacific), as well as mariculture and restocking to help maintain population numbers (i.e. Southeast Asia, Australia and the Pacific) have had some success. There is, however, a lack of standard protocols and regular monitoring to ascertain success of these mitigation measures on a local scale. Decades of giant clam research have also contributed to our understanding of their systematics, biology, physiology and ecological significance, which has helped to reinforce the case for protecting these charismatic molluscs (Neo et al. 2015a).

Even though substantial effort and resources have been injected into giant clam conservation since the 1970s, positive results are limited. Successes are generally due to the availability of large sums of financial aid to support the continuity of programmes, strong governance to implement fishery policies, as well as the involvement of local communities to take ownership of their coastal resources and help manage them. Updated data and new conservation approaches such as biophysical modelling and molecular genetic tools will be needed to help resolve fundamental issues such as larval dispersal and connectivity, fishery projections, cryptic species and population genetics. Mariculture also has a complementary role in the conservation of giant clams, as it is capable of producing large numbers of individuals to assist the restoration of depleted populations, and it may relieve some fishing pressures. Collectively, these approaches should help to prevent local extinctions of larger species (e.g. *Tridacna gigas* and *T. derasa*) and avoid the population collapse of smaller ones (e.g. *T. maxima*). Towards these important goals, the following fundamental ecological questions need to be resolved:

- What is the minimum number and density of giant clams (i.e. minimum viable population) needed to ensure that a population remains reproductive and yield genetically diverse progeny in the wild? Sexually mature individuals are becoming rare, and are therefore a limiting factor in reproductive success. These data are also key for restocking endeavours.
- Where and how should aggregations of restocked individuals be spatially arranged on reefs to optimize both fertilization success, survival and dispersal of larvae? Giant clams are broadcast spawners and aggregation is necessary to promote both spawning and fertilization success. However, data such as the minimum distances required between spawning individuals remain limited.
- What is the genetic connectivity, and larval dispersal extent, of wild giant clam populations locally, regionally and globally? An understanding of how populations are related promotes appropriate boundary management among populations. Such data can also contribute towards the maintenance of genetic diversity within regions, and will be especially useful for informing translocation and restocking endeavours.
- What are the phylogenetic relationships among giant clam species? This information is fundamental to the correct identification of species and subsequent planning of species-specific policies.
- How might giant clams (both in the wild and mariculture) acclimatize/adapt to anthropogenic threats, such as warming oceans and ocean acidification? There has been some progress on this front, mostly via manipulative experiments, but impacts on wild stocks and mariculture production are poorly understood.

These questions highlight the paucity of essential ecological data available to resource managers trying to improve the success of restocking giant clams, as well as conservation planners designing legislation to ensure sustainable exploitation. In addition to science-based conservation and management, it is critical to engage all stakeholders and increase conservation literacy through education, outreach and capacity building. Emphasizing the ecological benefits of giant clams and the consequences of overexploitation can help bring about changes in attitude and lead to improved fishing practices. Enforcement of existing regulations and the implementation of locally-appropriate new legislation is also crucial if populations are to be protected.

Acknowledgements

This review would not have been possible if not for the following who provided so much invaluable information: Steven Ng, Oceanic Quest Company (Brunei), Jessica Savage (Cambodia), Coral Cay Conservation (Cambodia), Yuehuan Zhang, Ziniu Yu and Xubing Fan (China), Allison Miller (Guam), Sundy Ramah and Ranjeet Bhagooli (Mauritius), D. Thoma (Nauru), Keryea Soong, Li-Lian Liu and Hei-Nin Kwong (Taiwan), Jeffrey Low, Jim Wong, Hiu Fung Wong, Denise Cheong and Youna Lyons (Singapore), and Reef Check Foundation (Worldwide). For the image of *Tridacna squamosina*, we thank Gustav Paulay, Michael Berumen and KAUST (Red Sea), and the research cruise was supported by a KAUST Collaborative Research Grant (URF/1/1389–01–01). Research on giant clam population and mariculture in French Polynesia has been supported by the Direction des Ressources Marines et Minières, under the leadership of Georges Remoissenet. Author C.C.C. Wabnitz would like to acknowledge the financial support from Australia (DFAT) to SPC's FAME division as well as the information provided by her colleagues from the SPC network, particularly John Hambrey, Antoine Teitelbaum, Georges Remoissenet, Ian Bertram and Richard Story. Author A.S.-H. Tan would like to acknowledge the Marine Ecology Research Centre, Sabah for the information provided. Author M.L. Neo would like to acknowledge the National Research Foundation Singapore for supporting her research endeavours at the St. John's Island National Marine Laboratory. This work was partially supported by the National Parks Board's Coastal & Marine Environment grant number R-154–000–568–490, and the L'Oréal-UNESCO For Women in Science National Fellowship 2015.

References

Accordi, G., Brilli, M., Carbone, F. & Voltaggio, M. 2010. The raised coral reef complex of the Kenyan coast: *Tridacna gigas* U-series dates and geological implications. *Journal of African Earth Sciences* **58**, 97–114.

Alcazar, S.N. 1986. Observations on predators of giant clams (Bivalvia: Family Tridacnidae). *Silliman Journal* **33**, 54–57.

Alcazar, S.N., Solis, E.P. & Alcala, A.C. 1987. Serotonin-induced spawning and larval rearing of the china clam *Hippopus porcellanus* Rosewater (Bivalvia, Tridacnidae). *Aquaculture* **66**, 359–368.

Andréfouët, S., Dutheil, C., Menkes, C.E., Bador, M. & Lengaigne, M. 2015. Mass mortality events in atoll lagoons: environmental control and increased future vulnerability. *Global Change Biology* **21**, 195–205.

Andréfouët, S., Friedman, K., Gilbert, A. & Remoissenet, G. 2009. A comparison of two surveys of invertebrates at Pacific Ocean Islands: the giant clam at Raivavae Island, Australes Archipelago, French Polynesia. *ICES Journal of Marine Science* **66**, 1825–1836.

Andréfouët, S., Gilbert, A., Yan, L., Remoissenet, G., Payri, C. & Chancerelle, Y. 2005. The remarkable population size of the endangered clam *Tridacna maxima* assessed in Fangatau atoll (Eastern Tuamotu, French Polynesia) using *in situ* and remote sensing data. *ICES Journal of Marine Science* **62**, 1037–1048.

Andréfouët, S., Van Wynsberge, S., Fauvelot, C., Bruckner, A.W. & Remoissenet, G. 2014. Significance of new records of *Tridacna squamosa* Lamarck, 1819, in the Tuamotu and Gambier Archipelagos (French Polynesia). *Molluscan Research* **34**, 277–284.

Andréfouët, S., Van Wynsberge, S., Gaertner-Mazouni, N., Menkes, C., Gilbert, A. & Remoissenet, G. 2013. Climate variability and massive mortalities challenge giant clam conservation and management efforts in French Polynesia atolls. *Biological Conservation* **160**, 190–199.

Ang, C.F.A. 2014. *Responses of juvenile fluted giant clams (*Tridacna squamosa*) to experimentally enhanced sedimentation.* Honours Thesis, National University of Singapore, Singapore.

Apte, D. & Dutta, S. 2010. Ecological determinants and stochastic fluctuations of *Tridacna maxima* survival rate in Lakshadweep Archipelago. *Systematics and Biodiversity* **8**, 461–469.

Armstrong, E.J., Watson, S.-A., Calosi, P., Munday, P. & Stillman, J. 2014. Increased temperature and lowered pH altered shell mineralogy of the scaled giant clam (*Tridacna squamosa*). *Integrative and Comparative Biology (Society for Integrative and Comparative Biology – Meeting Abstract)* **54**, E238.

Asato, S. 1991. The distribution of *Tridacna* shell adzes in the Southern Ryukyu Islands. In *Indo-Pacific Prehistory 1990*, P. Bellwood (ed.). Indo-Pacific Prehistory Association Bulletin **10**, 1991, 282–291.

Baldo, B.A. & Uhlenbruck, G. 1975. Quantitative precipitin studies on the specificity of an extract from *Tridacna maxima* (Röding). *Carbohydrate Research* **40**, 143–151.

Beckvar, N. 1981. Cultivation, spawning and growth of the giant clams *Tridacna gigas*, *Tridacna derasa* and *Tridacna squamosa* in Palau, Caroline Islands. *Aquaculture* **24**, 21–30.

Beger, M., Selkoe, K.A., Treml, E., Barber, P.H., von der Heyden, S., Crandall, E.D., Toonen, R.J. & Riginos, C. 2014. Evolving coral reef conservation with genetic information. *Bulletin of Marine Science* **90**, 159–185.

Bell, J.D. 1999. Restocking of giant clams: progress, problems and potential. In *Stock Enhancement and Sea Ranching. First International Symposium on Stock Enhancement and Sea Ranching, Bergen, Norway, 8–11 September 1997*, B.R. Howell et al. (eds). Bergen, Norway. Oxford: Blackwell Science, 437–452.

Bell, J.D., Lane, I., Gervis, M., Soule, S. & Tafea, H. 1997. Village-based farming of the giant clam, *Tridacna gigas* (L.), for the aquarium market: initial trials in Solomon Islands. *Aquaculture Research* **28**, 121–128.

Bell, J.D., Rothlisberg, P.C., Munro, J.L., Loneragan, N., Nash, W., Ward, R. & Andrew, N. 2005. Restocking and stock enhancement of marine invertebrate fisheries. *Advances in Marine Biology* **49**, 1–370.

Benzie, J.A.H. & Williams, S.T. 1992a. Genetic structure of giant clam (*Tridacna maxima*) populations from reefs in the Western Coral Sea. *Coral Reefs* **11**, 135–141.

Benzie, J.A.H. & Williams, S.T. 1992b. No genetic differentiation of giant clam (*Tridacna gigas*) populations in the Great Barrier Reef, Australia. *Marine Biology* **113**, 373–377.

Benzie, J.A.H. & Williams, S.T. 1995. Gene flow among giant clam (*Tridacna gigas*) populations in Pacific does not parallel ocean circulation. *Marine Biology* **123**, 781–787.

Benzie, J.A.H. & Williams, S.T. 1996. Limitations in the genetic variation of hatchery produced batches of giant clam, *Tridacna gigas*. *Aquaculture* **139**, 225–241.

Benzie, J.A.H. & Williams, S.T. 1998. Phylogenetic relationships among giant clam species (Mollusca: Tridacnidae) determined by protein electrophoresis. *Marine Biology* **132**, 123–133.

Black, R., Johnson, M.S., Prince, J., Brearley, A. & Bond, T. 2011. Evidence of large, local variations in recruitment and mortality in the small giant clam, *Tridacna maxima*, at Ningaloo Marine Park, Western Australia. *Marine and Freshwater Research* **62**, 1318–1326.

Blidberg, E., Elfwing, T., Plantman, P. & Tedengren, M. 2000. Water temperature influences on physiological behaviour in three species of giant clams (Tridacnidae). *Proceedings 9th International Coral Reef Symposium, Bali, Indonesia, 23–27 October 2000*, M.K. Moosa et al. (eds). Jakarta: Indonesian Institute of Sciences, Jakarta: Ministry of Environment, Honolulu, Hawaii: International Society for Reef Studies, 561–565.

Borsa, P., Fauvelot, C., Andréfouët, S., Chai, T.-T., Kubo, H. & Liu, L.-L. 2015a. On the validity of Noah's giant clam *Tridacna noae* (Röding, 1798) and its synonymy with Ningaloo giant clam *Tridacna ningaloo* Penny and Willan, 2014. *Raffles Bulletin of Zoology* **63**, 484–489.

Borsa, P., Fauvelot, C., Tiavouane, J., Grulois, D., Wabnitz, C., Abdon Naguit, M.R. & Andréfouët, S. 2015b. Distribution of Noah's giant clam, *Tridacna noae*. *Marine Biodiversity* **45**, 339–344.

Bouchet, P., Heros, V., Le Goff, A., Lozouet, P. & Maestrati, P. 2001. Atelier biodiversité Lifou 2000: grottes et récifs coralliens. Rapports de Missions, Science de la Mer, Biologie Marine, No. 26. Institut de recherche pour le développement, Noumea, New Caledonia. Online. http://horizon.documentation.ird.fr/exl-doc/pleins_textes/divers15-06/010031564.pdf (accessed 19 December 2016).

Braley, R.D. 1984. Reproduction in the giant clams *Tridacna gigas* and *T. derasa in situ* on the North-Central Great Barrier Reef, Australia, and Papua New Guinea. *Coral Reefs* **3**, 221–227.

Braley, R.D. 1986. *Reproduction and recruitment of giant clams and some aspects of their larval and juvenile biology*. PhD Thesis, University of New South Wales, Australia.

Braley, R.D. 1987a. Distribution and abundance of the giant clams *Tridacna gigas* and *T. derasa* on the Great Barrier Reef. *Micronesica* **20**, 215–223.

Braley, R.D. 1987b. Spatial distribution and population parameters of *Tridacna gigas* and *T. derasa*. *Micronesica* **20**, 225–246.

Braley, R.D. 1988. Recruitment of the giant clams *Tridacna gigas* and *T. derasa* at four sites on the Great Barrier Reef. In *Giant Clams in Asia and the Pacific*, J.W. Copland & J.S. Lucas (eds). ACIAR Monograph No. 9. Canberra: Australian Centre for International Agricultural Research, 73–77.

Braley, R.D. 1992. *The Giant Clam: Hatchery and Nursery Culture Manual.* ACIAR Monograph No. 15. Canberra: Australian Centre for International Agricultural Research.

Braley, R.D. 1996. The importance of aquaculture and establishment of reserves for the restocking of giant clams on over-harvested reefs in the Indo-Pacific region. In *The Role of Aquaculture in World Fisheries*, T.G. Heggberget (ed.). Proceedings of the World Fisheries Congress, Theme 6. New Delhi: Oxford & IBH Publishing, 136–147.

Braley, R.D. & Healy, J.M. 1998. Superfamily Tridacnoidea. In *Mollusca: The Southern Synthesis, Fauna of Australia. Vol. 5*, P.L. Beesley et al. (eds). Melbourne: CSIRO Publishing, Part A, 332–336.

Braley, R.D. & Muir, F. 1995. The case history of a large natural cohort of the giant clam *Tridacna gigas* (Fam. Tridacnidae) and the implications for restocking depauperate reefs with maricultured giant clams. *Asian Fisheries Science* **8**, 229–237.

Braley, R.D., Nash, W.J., Lucas, J.S. & Crawford, C.M. 1988. Comparison of different hatchery and nursery culture methods for the giant clam *Tridacna gigas*. In *Giant Clams in Asia and the Pacific*, J.W. Copland & J.S. Lucas (eds). ACIAR Monograph No. 9. Canberra: Australian Centre for International Agricultural Research, 110–114.

Brown, J.H. & Muskanofola, M.R. 1985. An investigation of stocks of giant clams (family Tridacnidae) in Java and of their utilization and potential. *Aquaculture and Fisheries Management* **1**, 25–39.

Bruce, A.J. 2000. Biological observations on the commensal shrimp *Paranchistus armatus* (H. Milne Edwards) (Crustacea: Decapoda: Pontoniinae). *Beagle (Records of the Museum and Art Galleries Northern Territory)* **16**, 91–96.

Bryan, P.G. & McConnell, D.B. 1976. Status of giant clam stocks (Tridacnidae) on Helen Reef, Palau, Western Caroline Islands, April 1975. *Marine Fisheries Review* **38**, 15–18.

Buck, B.H., Rosenthal, H. & Saint-Paul, U. 2002. Effect of increased irradiance and thermal stress on the symbiosis of *Symbiodinium microadriaticum* and *Tridacna gigas*. *Aquatic Living Resources* **15**, 107–117.

Cabaitan, P.C. & Conaco, C. 2017. Bringing back the giants: juvenile *Tridacna gigas* from natural spawning of restocked giant clams. *Coral Reefs*, doi:10.1007/s00338-017-1558-9

Cabaitan, P.C., Gomez, E.D. & Aliño, P.M. 2008. Effects of coral transplantation and giant clam restocking on the structure of fish communities on degraded patch reefs. *Journal of Experimental Marine Biology and Ecology* **357**, 85–98.

Calumpong, H.P. 1992. *The Giant Clam: An Ocean Culture Manual.* ACIAR Monograph No. 16. Canberra: Australian Centre for International Agricultural Research.

Calumpong, H.P., Ablan, M.C., Macaranas, J., Solis-Duran, E., Alcazar, S. & Abdon-Naguit, R. 1993. Biochemical evidence of self-fertilisation in *Hippopus* species. In *The Biology and Mariculture of Giant Clams: A Workshop Held in Conjunction with the 7th International Coral Reef Symposium 21–26 June 1992, Guam, USA*, W.K. Fitt (ed.). ACIAR Proceedings No. 47. Canberra: Australian Centre for International Agricultural Research, 99–110.

Calumpong, H.P. & Cadiz, P. 1993. Observations on the distribution of giant clams in protected areas. *Silliman Journal* **36**, 107–113.

Calumpong, H.P. & Solis-Duran, E. 1993. Constraints in restocking Philippine reefs with giant clams. In *The Biology and Mariculture of Giant Clams: A Workshop Held in Conjunction with the 7th International Coral Reef Symposium 21–26 June 1992, Guam, USA*, W.K. Fitt (ed.). ACIAR Proceedings No. 47. Canberra: Australian Centre for International Agricultural Research, 94–98.

Carlson, B.C. 2012. Waikiki Aquarium's giant clams mark 30-year anniversary. *CORAL Magazine* **9**, 54–60.

Cavell, N. 2016. *Blame an ivory ban for China's vanishing giant clams.* WIRED, 10 February 2016. Online. http://www.wired.com/2016/02/blame-an-ivory-ban-for-chinas-vanishing-giant-clams/ (accessed 29 February 2016).

Chambers, C.N.L. 2008. *Pasua* and the politics of environmental management, Tongareva, Cook Islands. *Scottish Geographical Journal* **124**, 192–197.

Chaudhary, C., Saeedi, H. & Costello, M.J. 2016. Bimodality of latitudinal gradients in marine species richness. *Trends in Ecology & Evolution* **31**, 670–676.

Chesher, R.H. 1993. Giant clam sanctuaries in the Kingdom of Tonga. Marine Studies of the University of the South Pacific Technical Report Series 95/2. Suva, Fiji: University of the South Pacific. Online. http://www.tellusconsultants.com/chesher-1993-Giant%20Clam%20Sanctuaries%20in%20the%20Kingdom%20of%20Tonga.pdf (accessed 19 December 2016).

Colbeck, R. 2015. Clams to help slam trade shut. Canberra: Australian Fisheries Management Authority. Online. http://www.afma.gov.au/clams-help-slam-trade-shut/ (accessed 26 February 2016).

Collar, N.J. 1996. The reasons for Red Data Books. *Oryx* **30**, 121–130.

Copland, J.W. & Lucas, J.S. (eds) 1988. *Giant Clams in Asia and the Pacific*. ACIAR Monograph No. 9. Canberra: Australian Centre for International Agricultural Research.

Cox, L.R. 1941. Lamellibranchs from the White Limestone of Jamaica. *Proceedings of the Malacological Society, London* **24**, 135–144.

Crawford, C.M., Lucas, J.S. & Munro, J.L. 1987. The mariculture of giant clams. *Interdisciplinary Science Reviews* **12**, 333–340.

Cumming, R.L. 1988. Pyramidellid parasites in giant clam mariculture systems. In *Giant Clams in Asia and the Pacific*, J.W. Copland & J.S. Lucas (eds). ACIAR Monograph No. 9. Canberra: Australian Centre for International Agricultural Research, 231–236.

Curnick, D.J., Head, C.E.I., Huang, D., Crabbe, M.J.C., Gollock, M., Hoeksema, B.W., Johnson, K.G., Jones, R., Koldewey, H.J., Obura, D.O., Rosen, B.R., Smith, D.J., Taylor, M.L., Turner, J.R., Wren, S. & Redding, D.W. 2015. Setting evolutionary-based conservation priorities for a phylogenetically data-poor taxonomic group (Scleractinia). *Animal Conservation* **18**, 303–312.

Dawson, B. 1986. Report on a study of the market for giant clam products in Taiwan, Japan, Hong Kong and Singapore. FFA Report No. 86/37, Honiara, Solomon Islands: Pacific Islands Forum Fisheries Agency. Online. www.spc.int/DigitalLibrary/Doc/FAME/FFA/Reports/FFA_1986_037.pdf (accessed 13 April 2017).

Dawson, R.F. & Philipson, P.W. 1989. The market for giant clam in Japan, Taiwan, Hong Kong and Singapore. In *The Marketing of Marine Products from the South Pacific*, P.W. Philipson (ed.). Suva, Fiji: Institute of Pacific Studies of the University of the South Pacific, 90–123.

DeBoer, T.S., Abdon-Naguit, M.R., Erdmann, M.V., Ablan-Lagman, M.C.A., Ambariyanto, Carpenter, K.E., Toha, A.H.A. & Barber, P.H. 2014. Concordance between phylogeographic and biogeographic boundaries in the Coral Triangle: conservation implications based on comparative analyses of multiple giant clam species. *Bulletin of Marine Science* **90**, 277–300.

DeBoer, T.S., Baker, A.C., Erdmann, M.V., Ambariyanto, Jones, P.R. & Barber, P.H. 2012. Patterns of *Symbiodinium* distribution in three giant clam species across the biodiverse Bird's Head region of Indonesia. *Marine Ecology Progress Series* **444**, 117–132.

DeBoer, T.S., Subia, M.D., Ambariyanto, Erdmann, M.V., Kovitvongsa, K. & Barber, P.H. 2008. Phylogeography and limited genetic connectivity in the endangered boring giant clam across the Coral Triangle. *Conservation Biology* **22**, 1255–1266.

Dolorosa, R.G., Conales, S.F. & Bundal, N.A. 2014. Shell dimension–live weight relationships, growth and survival of *Hippopus porcellanus* in Tubbataha Reefs Natural Park, Philippines. *Atoll Research Bulletin* **604**, 1–9.

Dolorosa, R.G. & Jontila, J.B.S. 2012. Notes on common macrobenthic reef invertebrates of Tubbataha Reefs Natural Park, Philippines. *Science Diliman* **24**, 1–11.

Dumas, P., Fauvelot, C., Andréfouët, S. & Gilbert, A. 2011. Les beniters en Nouvelle-Caledonie: Statut des populations, impacts de l'exploitation & connectivitié. Rapport final d'opération, Programme ZONECO, Avril 2011. Noumea, New Caledonia: Zonéco. Online. http://www.zoneco.nc/documents/les-beniters-de-nouvelle-caledonie-statut-des-populations-impact-de-lexploitation-et (accessed 19 December 2016).

Dumas, P., Tiavouane, J., Senia, J., Willam, A., Dick, L. & Fauvelot, C. 2014. Evidence of early chemotaxis contributing to active habitat selection by the sessile giant clam *Tridacna maxima*. *Journal of Experimental Marine Biology and Ecology* **452**, 63–69.

Eckman, W., Vicentuan-Cabaitan, K. & Todd, P.A. 2014. Observations on the hyposalinity tolerance of fluted giant clam (*Tridacna squamosa*, Lamarck 1819) larvae. *Nature in Singapore* **7**, 111–116.

Elfving, T., Plantman, P., Tedengren, M. & Wijnbladh, E. 2001. Responses to temperature, heavy metal and sediment stress by the giant clam *Tridacna squamosa*. *Marine and Freshwater Behaviour and Physiology* **34**, 239–248.

Eliata, A., Zahida, F., Wibowo, N.J. & Panggabean, L.M.G. 2003. Abundance of giant clam in coral reef ecosystem at Pari Island: A population comparison of 2003's to 1984's data. *Biota* **8**, 149–152.

Faith, D.P. 1992. Conservation evaluation and phylogenetic diversity. *Biological Conservation* **61**, 1–10.

Faith, D.P. 2007. Threatened species and the potential loss of phylogenetic diversity: Conservation scenarios based on estimated extinction probabilities and phylogenetic risk analysis. *Conservation Biology* **22**, 1461–1470.

Fankboner, P.V. 1971. Intracellular digestion of symbiotic zooxanthellae by host amoebocytes in giant clams (Bivalvia: Tridacnidae), with a note on the nutritional role of the hypertrophied siphonal epidermis. *Biological Bulletin* **141**, 222–234.

Fitt, W.K. 1988. Role of zooxanthellae in the mariculture of giant clams. In *Giant Clams in Asia and the Pacific*, J.W. Copland & J.S. Lucas (eds). ACIAR Monograph No. 9. Canberra: Australian Centre for International Agricultural Research, 166–169.

Fitt, W.K. (ed.) 1993. *The Biology and Mariculture of Giant Clams: A Workshop Held in Conjunction with the 7th International Coral Reef Symposium 21–26 June 1992, Guam, USA*. ACIAR Proceedings No. 47. Canberra: Australian Centre for International Agricultural Research.

Fitt, W.K. & Trench, R.K. 1981. Spawning, development, and acquisition of zooxanthellae by *Tridacna squamosa* (Mollusca, Bivalvia). *Biological Bulletin* **161**, 213–235.

Fogarty, M.J. & Botsford, L.W. 2007. Population connectivity and spatial management of marine fisheries. *Oceanography* **20**, 112–123.

Frank, K.T. & Brickman, D. 2000. Allee effects and compensatory population dynamics within a stock complex. *Canadian Journal of Fisheries and Aquatic Sciences* **57**, 513–517.

Gilbert, A., Andréfouët, S., Yan, L. & Remoissenet, G. 2006b. The giant clam *Tridacna maxima* communities of three French Polynesia islands: comparison of their population sizes and structures at early stages of their exploitation. *ICES Journal of Marine Science* **63**, 1573–1589.

Gilbert, A., Planes, S., Andréfouët, S., Friedman, K. & Remoissenet, G. 2007. First observation of the giant clam *Tridacna squamosa* in French Polynesia: a species range extension. *Coral Reefs* **26**, 229 only.

Gilbert, A., Remoissenet, G., Yan, L. & Andréfouët, S. 2006a. Special traits and promises of the giant clam (*Tridacna maxima*) in French Polynesia. *SPC Fisheries Newsletter* **118**, 44–52.

Gilbert, A., Yan, L., Remoissenet, G., Andréfouët, S., Payri, C. & Chancerelle, Y. 2005. Extraordinarily high giant clam density under protection in Tatakoto Atoll (eastern Tuamotu Archipelago, French Polynesia). *Coral Reefs* **24**, 495 only.

Gomez, E.D. 2015a. Rehabilitation of biological resources: coral reefs and giant clam populations need to be enhanced for a sustainable marginal sea in the western Pacific. *Journal of International Wildlife Law and Policy* **18**, 120–127.

Gomez, E. 2015b. Destroyed reefs, vanishing giant clams. Inquirer.net, 3 May 2015. Online. http://opinion.inquirer.net/84595/destroyed-reefs-vanishing-giant-clams (accessed 26 February 2016).

Gomez, E.D. & Mingoa-Licuanan, S.S. 1998. Mortalities of giant clams associated with unusually high temperatures and coral bleaching. *Reef Encounter* **24**, 23 only.

Gomez, E.D. & Mingoa-Licuanan, S.S. 2006. Achievements and lessons learned in restocking giant clams in Philippines. *Fisheries Research* **80**, 46–52.

Gonzales, B.J., Becira, J.G., Galon, W.M. & Gonzales, M.M.G. 2014. Protected versus unprotected area with reference to fishes, corals, marine invertebrates, and CPUE in Honda Bay, Palawan. *The Palawan Scientist* **6**, 42–59.

Govan, H. 1992. Predators and predator control. In *The Giant Clam: An Ocean Culture Manual*, H.P. Calumpong (ed.). ACIAR Monograph No. 16. Canberra: Australian Centre for International Agricultural Research, 41–49.

Green, A. & Craig, P. 1999. Population size and structure of giant clams at Rose Atoll, an important refuge in the Samoan Archipelago. *Coral Reefs* **18**, 205–211.

Guest, J.R., Todd, P.A., Goh, E., Sivalonganathan, B.S. & Reddy, K.P. 2008. Can giant clam (*Tridacna squamosa*) populations be restored on Singapore's heavily impacted coral reefs? *Aquatic Conservation: Marine and Freshwater Ecosystems* **18**, 570–579.

Hambrey Consulting 2013. Market study on exporting cultured giant clams from French Polynesia. Synthesis report. Report commissioned by Agence Francaise de Developpement in partnership with the Secretariat of the Pacific Community and the French Polynesian Ministry of Marine Resources. Strathpeffer, Scotland, UK: Hambrey Consulting.

Hamner, W.M. 1978. Intraspecific competition in *Tridacna crocea*, a burrowing bivalve. *Oecologia* **34**, 267–281.

Hamner, W.M. & Jones, M.S. 1976. Distribution, burrowing, and growth rates of the clam *Tridacna crocea* on interior reef flats. *Oecologia* **24**, 207–227.

Hart, A.M., Bell, J.D. & Foyle, T.P. 1998. Growth and survival of the giant clams, *Tridacna derasa*, *T. maxima* and *T. crocea*, at village farms in the Solomon Islands. *Aquaculture* **165**, 203–220.

Harzhauser, M., Mandic, O., Piller, W.E., Reuter, M. & Kroh, A. 2008. Tracing back the origin of the Indo-Pacific mollusc fauna: basal Tridacninae from the Oligocene and Miocene of the Sultanate of Oman. *Palaeontology* **51**, 199–213.

Hedley, C. 1921. A revision of the Australian *Tridacna*. *Records of the Australian Museum* **13**, 163–172.

Herrera, N.D., ter Poorten, J.J., Bieler, R., Mikkelsen, P.M., Strong, E.E., Jablonski, D. & Steppan, S.J. 2015. Molecular phylogenetics and historical biogeography amid shifting continents in the cockles and giant clams (Bivalvia: Cardiidae). *Molecular Phylogenetics and Evolution* **93**, 94–106.

Heslinga, G.A. 1991. History and current status of the MMDC giant clam project. A Special Report Prepared for: The House of Delegates, Third Olbill Era Kelulau (Palau National Congress). February 10, 1991. Koror, Palau: Micronesian Mariculture Demonstration Center.

Heslinga, G.A. 1993a. Regional yield trials for commercially valuable giant clam species, Phase I. *Tridacna gigas* and *Tridacna derasa*. Final Project Report U.S. National Marine Fisheries Service (NOAA/NMFS NA16DO335–01). Koror, Palau: Micronesian Mariculture Demonstration Center.

Heslinga, G.A. 1993b. Regional yield trials for commercially valuable giant clam species, Phase II. *Hippopus hippopus* and *Tridacna derasa*. U.S. National Marine Fisheries Service (NOAA/NMFS NA16FDO335–02). Koror, Palau: Micronesian Mariculture Demonstration Center.

Heslinga, G.A. 1995. Clams to cash: how to make and sell giant clam shell products. Publication No. 125, Waimanalo, Hawaii: Center for Tropical and Subtropical Aquaculture (Hawaii). Online. http://www.ctsa.org/files/publications/CTSA_125631672862855825081.pdf (accessed 19 December 2016).

Heslinga, G.A. 2012. The origin and future of farming giant clams. *CORAL Magazine* **9**, 38–52.

Heslinga, G.A. 2013. *Saving Giants (eBook): Cultivation and Conservation of Tridacnid Clams*. Kailua-Kona, Hawaii: Indo-Pacific Sea Farms. http://www.blurb.com/ebooks/374835-saving-giants (accessed 20 July 2016).

Heslinga, G.A. & Fitt, W.K. 1987. The domestication of reef-dwelling clams. *BioScience* **37**, 332–339.

Heslinga, G.A., Perron, F.E. & Orak, O. 1984. Mass culture of giant clams (f. Tridacnidae) in Palau. *Aquaculture* **39**, 197–215.

Heslinga, G.A., Watson, T.C. & Isamu, T. 1990. *Giant Clam Farming*. Honolulu, Hawaii: Pacific Fisheries Development Foundation (NMFS/NOAA).

Hester, F.J. & Jones, E.C. 1974. A survey of giant clams, Tridacnidae, on Helen Reef, a Western Pacific atoll. *Marine Fisheries Review* **36**, 17–22.

Hirschberger, W. 1980. Tridacnid clam stocks on Helen Reef, Palau, Western Caroline Islands. *Marine Fisheries Review* **42**, 8–15.

Hopkins, A. 2009. *Marine invertebrates as indicators of reef health: a study of the reefs in the region of Andavadoaka, South West Madagascar*. MSc Dissertation, Imperial College London, UK.

Huang, D. 2012. Threatened reef corals of the world. *PLoS ONE* **7**, e34459. doi:10.1371/journal.pone.0034459

Huang, D. & Roy, K. 2013. Anthropogenic extinction threats and future loss of evolutionary history in reef corals. *Ecology and Evolution* **3**, 1184–1193.

Huang, D., Todd, P.A. & Guest, J.R. 2007. Movement and aggregation in the fluted giant clam (*Tridacna squamosa* L.). *Journal of Experimental Marine Biology and Ecology* **342**, 269–281.

Huber, M. 2010. *Compendium of Bivalves. A Full-Color Guide to 3,300 of the World's Marine Bivalves. A Status on Bivalvia after 250 Years of Research*. Hackenheim: Conchbooks.

Huber, M. & Eschner, A. 2011. *Tridacna (Chametrachea) costata* Roa-Quiaoit, Kochzius, Jantzen, Al-Zibdah and Richter from the Red Sea, a junior synonym of *Tridacna squamosina* Sturany, 1899 (Bivalvia, Tridacnidae). *Annalen des Naturhistorischen Museums in Wien B* **112**, 153–162.

Huelsken, T., Keyse, J., Liggins, L., Penny, S., Treml, E.A. & Riginos, C. 2013. A novel widespread cryptic species and phylogeographic patterns within several giant clam species (Cardiidae: *Tridacna*) from the Indo-Pacific Ocean. *PLoS ONE* **8**, e80858. doi:10.1371/journal.pone.0080858

Hutsell, K.C., Hutsell, L.L. & Pisor, D.L. 1997. *Registry of World Record Size Shell*. San Diego, California: Snail's Pace Productions.

Hviding, E. 1993. The rural context of giant clam mariculture in Solomon Islands: an anthropological study. ICLARM Technical Report 39. Manila, Philippines: International Center for Living Aquatic Resources Management. Online. http://pdf.usaid.gov/pdf_docs/Pnabq793.pdf (accessed 19 December 2016).

Isaac, N.J.B., Redding, D.W., Meredith, H.M. & Safi, K. 2012. Phylogenetically-informed priorities for amphibian conservation. *PLoS ONE* **7**, e43912. doi:10.1371/journal.pone.0043912

Isaac, N.J.B., Turvey, S.T., Collen, B., Waterman, C. & Baillie, J.E.M. 2007. Mammals on the EDGE: Conservation priorities based on threat and phylogeny. *PLoS ONE* **2**, e296. doi:10.1371/journal.pone.0000296

Jameson, S.C. 1976. Early life of the giant clam *Tridacna crocea* Lamarck, *Tridacna maxima* (Röding), and *Hippopus hippopus* (Linnaeus). *Pacific Science* **30**, 219–233.

Jantzen, C., Wild, C., El-Zibdah, M., Roa-Quiaoit, H.A., Haacke, C. & Richter, C. 2008. Photosynthetic performance of giant clams, *Tridacna maxima* and *T. squamosa*, Red Sea. *Marine Biology* **155**, 211–221.

Johnson, M.S., Prince, J., Brearley, A., Rosser, N.L. & Black, R. 2016. Is *Tridacna maxima* (Bivalvia: Tridacnidae) at Ningaloo Reef, Western Australia? *Molluscan Research* **36**, 264–270.

Juinio, M.A.R., Meñez, L.A.B., Villanoy, C.L. & Gomez, E.D. 1989. Status of giant clam resources of the Philippines. *Journal of Molluscan Studies* **55**, 431–440.

Junchompoo, C., Sinrapasan, N., Penpian, C. & Patsorn, P. 2013. Changing seawater temperature effects on giant clams bleaching, Mannai Island, Rayong Province, Thailand. *KURENAI* **2013–03**, Proceedings of the Design Symposium on Conservation of Ecosystem (2013) (The 12th SEASTAR2000 workshop), Bangkok, Thailand, 71–76. doi:10.13140/2.1.1906.5600

Keen, A.M. 1969. Superfamily Tridacnacea Lamarck, 1819. In *Treatise on Invertebrate Paleontology, Part N, Mollusca 6, Bivalvia, Vol. 2*, R.C. Moore (ed.). Boulder, Colorado: Geological Society of America and Lawrence, Kansas: University of Kansas, N594–N595.

Keys, J.L. & Healy, J.M. 2000. Relevance of sperm ultrastructure to the classification of giant clams (Mollusca, Cardioidea, Cardiidae, Tridacninae). In *The Evolutionary Biology of the Bivalvia*, E.M. Harper et al. (eds). Geological Society Special Publication No. 177. London: Geological Society, 191–205.

Kinch, J. 2001. Clam harvesting, the Convention on the International Trade in Endangered Species (CITES) and conservation in Milne Bay Province, Papua New Guinea. *SPC Fisheries Newsletter* **99**, 24–36.

Kinch, J. 2002. Giant clams: their status and trade in Milne Bay Province, Papua New Guinea. *TRAFFIC Bulletin* **19**, 1–9.

Kinch, J. & Teitelbaum, A. 2010. *Proceedings of the Regional Workshop on the Management of Sustainable Fisheries for Giant Clams (Tridacnidae) and CITES Capacity Building (4–7 August 2009, Nadi, Fiji)*. Noumea, New Caledonia: Secretariat of the Pacific Community.

Klumpp, D.W. & Griffiths, C.L. 1994. Contributions of phototrophic and heterotrophic nutrition to the metabolic and growth requirements of four species of giant clam (Tridacnidae). *Marine Progress Ecology Series* **115**, 103–115.

Knight, R., Watson, K., Dill, J., Moore, P. & Miller, K. 2010. *A Toolkit for Protecting the Environment and Natural Resources in Kuraburi*. Bangkok, Thailand: IUCN Thailand Programme and IUCN Regional Environmental Law Programme, Asia.

Kochzius, M. & Nuryanto, A. 2008. Strong genetic population structure in the boring giant clam, *Tridacna crocea*, across the Indo-Malay Archipelago: implications related to evolutionary processes and connectivity. *Molecular Ecology* **17**, 3775–3787.

Krell, B., Skopal, M. & Ferber, P. 2011. Koh Rong Samloem and Koh Kon Marine Environmental Assessment: Report on Marine Resources and Habitats. Koh Rong Samloen, Mittapheap District, Cambodia: Marine Conservation Cambodia. Online. http://www.marineconservationcambodia.org/marine-reef-research/file/16-koh-rong-samloem-koh-rong-marine-assessment-2011-for-fiacd (accessed 19 December 2016).

Kubo, H. & Iwai, K. 2007. On two sympatric species within *Tridacna "maxima"*. *Annual Report Okinawa Fisheries Oceanography Research Centre* **68**, 205–210.

Ladd, H.S. 1934. Geology of Viti Levu, Fiji. *Bulletin of the Bernice P. Bishop Museum* **119**, 1–263.

Lai, L.T.-A. 2015. A large baroque *Tridacna gigas* (giant clam) pearl. *Gems and Gemology, The Quarterly Journal of the Gemological Institute of America* **L**, 247 only.

Lamarck, J.B. De 1809. *Philosophie Zoologique*. Volumes 1 and 2. Paris: Dentu.

Larson, C. 2016. Shell trade pushes giant clams to the brink. *Science* **351**, 323–324.

Ledua, E., Manu, N. & Braley, R.D. 1993. Distribution, habitat and culture of the recently described giant clam *Tridacna tevoroa* in Fiji and Tonga. In *The Biology and Mariculture of Giant Clams: A Workshop Held in Conjunction with the 7th International Coral Reef Symposium 21–26 June 1992, Guam, USA*, W.K. Fitt (ed.). ACIAR Proceedings No. 47. Canberra: Australian Centre for International Agricultural Research, 147–153.

Lee, S. 2014. Twenty tonnes of giant clams seized from Vietnamese fishermen. The Star Online, 14 April 2014. Online. http://www.thestar.com.my/news/nation/2014/04/14/crime-cops-clam/ (accessed 26 February 2016).

Leggat, W., Buck, B.H., Grice, A. & Yellowlees, D. 2003. The impact of bleaching on the metabolic contribution of dinoflagellate symbionts to their giant clam host. *Plant, Cell and Environment* **26**, 1951–1961.

Leung, P.S., Shang, Y.C., Wanitprapha, K. & Tian, X. 1994. *Production economics of giant clam (*Tridacna *species) culture systems in the U.S.-affiliated Pacific Islands*. Publication No. 114, Waimanalo, Hawaii: Center for Tropical and Subtropical Aquaculture. Online. http://www.ctsa.org/files/publications/CTSA_114631681200866376521.pdf (accessed 19 December 2016).

Lewis, A.D. & Ledua, E. 1988. A possible new species of *Tridacna* (Tridacnidae: Mollusca) from Fiji. In *Giant Clams in Asia and the Pacific*, J.W. Copland & J.S. Lucas (eds). Canberra: Australian Centre for International Agricultural Research, 82–84.

Lewis, A.D., Adams, T.J.H. & Ledua, E. 1988. Fiji's giant clam stocks – a review of their distribution, abundance, exploitation and management. In *Giant Clams in Asia and the Pacific*, J.W. Copland & J.S. Lucas (eds). Canberra: Australian Centre for International Agricultural Research, 66–72.

Lindsay, S. 1995. *Giant Clams Reseeding Programs: Do They Work and Do They Use Limited Resources Wisely? Using Yap State, Federated States of Micronesia as a Model*. Joint FFA/SPC Workshop on the Management of South Pacific Inshore Fisheries, Noumea, New Caledonia, 26 June–7 July 1995. Noumea, New Caledonia: South Pacific Commission.

Lizano, A.M.D. & Santos, M.D. 2014. Updates on the status of giant clams *Tridacna* spp. and *Hippopus hippopus* in the Philippines using mitochondrial CO1 and 16S rRNA genes. *Phillipine Science Letters* **7**, 187–200.

Lucas, J.S. 1988. Giant clams: description, distribution and life history. In *Giant Clams in Asia and the Pacific*, J.W. Copland & J.S. Lucas (eds). Canberra: Australian Centre for International Agricultural Research, 21–32.

Lucas, J.S. 1994. The biology, exploitation, and mariculture of giant clams (Tridacnidae). *Reviews in Fisheries Science* **2**, 181–223.

Lucas, J.S. 1997. Giant clams: mariculture for sustainable exploitation. In *Conservation and the Use of Wildlife Resources*, M. Bolton (ed.). London: Chapman and Hall, 77–95.

Lucas, J.S. 2014. Giant clams. *Current Biology* **24**, R183–R184.

Lucas, J.S., Ledua, E. & Braley, R.D. 1990. *A new species of giant clam (Tridacnidae) from Fiji and Tonga*. ACIAR Working Paper No. 23. Canberra: Australian Centre for International Agricultural Research.

Lucas, J.S., Ledua, E. & Braley, R.D. 1991. *Tridacna tevoroa* Lucas, Ledua and Braley: A recently described species of giant clam (Bivalvia: Tridacnidae) from Fiji and Tonga. *Nautilus* **105**, 92–103.

Maboloc, E.A. & Mingoa-Licuanan, S.S. 2011. Feeding aggregation of *Spratelloides delicatulus* on giant clams' gametes. *Coral Reefs* **30**, 167 only.

Macaranas, J.M., Ablan, C.A., Pante, M.J.R., Benzie, J.A.H. & Williams, S.T. 1992. Genetic structure of giant clam (*Tridacna derasa*) populations from reefs in the Indo-Pacific. *Marine Biology* **113**, 231–238.

Maruyama, T. & Heslinga, G.A. 1997. Fecal discharge of zooxanthellae in the giant clam *Tridacna derasa*, with reference to their *in situ* growth rate. *Marine Biology* **127**, 473–477.

Maruyama, T., Ishikura, M., Yamazaki, S. & Kanai, S. 1998. Molecular phylogeny of zooxanthellate bivalves. *Biological Bulletin* **195**, 70–77.

Master, F. 2016. South China Sea reefs "decimated" as giant clams harvested in bulk. Thomson Reuters, 27 June 2016. http://www.reuters.com/article/us-china-clams-idUSKCN0ZD30F (accessed 12 July 2016).

McKenna S.A., Baillon N., Blaffart H. & Abrusci G. 2008. Une évaluation rapide de la biodiversité marine des récifs coralliens du Mont Panié, Province Nord, Nouvelle Calédonie. Bulletin PER d'évaluation biologique N°42. Arlington, Virginia: Conservation International.

McLean, R.A. 1947. A revision of the Pelecypod family Tridacnidae. *Notulae Naturae of The Academy of Natural Sciences of Philadelphia* **195**, 1–7.

Mekawy, M.S. & Madkour, H.A. 2012. Studies on the Indo-Pacific Tridacnidae (*Tridacna maxima*) from the Northern Red Sea, Egypt. *International Journal of Geosciences* **3**, 1089–1095.

Menoud, M., Van Wynsberge, S., Le Moullac, G., Levy, P., Andréfouët, S., Remoissenet, G. & Gaertner-Mazouni, N. 2016. Identifying robust proxies of gonad maturation for the protandrous hermaphrodite *Tridacna maxima* (Röding, 1798, Bivalvia) from individual to population scale. *Journal of Shellfish Research* **35**, 51–61.

Michel, C., Coowar, M. & Takoor, S. 1985. *Marine Molluscs of Mauritius*. Gland, Switzerland: WWF and IUCN.

Mies, M., Braga, F., Scozzafave, M.S., de Lemos D.E.L. & Sumida, P.Y.G. 2012. Early development, survival and growth rates of the giant clam *Tridacna crocea* (Bivalvia: Tridacnidae). *Brazilian Journal of Oceanography* **60**, 127–133.

Militz, T.A., Kinch, J. & Southgate, P.C. 2015. Population demographics of *Tridacna noae* (Röding, 1798) in New Ireland, Papua New Guinea. *Journal of Shellfish Research* **34**, 329–335.

Mingoa-Licuanan, S.S. & Gomez, E.D. 2007. *Giant Clam Hatchery, Ocean Nursery and Stock Enhancement*. Iloilo, Philippines: Aquaculture Department, Southeast Asian Fisheries Development Center.

Molluscs Specialist Group, 1996. *Tridacna crocea*. In *IUCN Red List of Threatened Species. Version 2012.2*. Cambridge, UK: IUCN Global Species Programme Red List Unit. Online. http://www.iucnredlist.org (accessed 14 July 2016).

Monsecour, K. 2016. A new species of giant clam (Bivalvia: Cardiidae) from the Western Indian Ocean. *Conchylia* **46**, 69–77.

Mooers, A.Ø., Faith, D.P. & Maddison, W.P. 2008. Converting endangered species categories to probabilities of extinction for phylogenetic conservation prioritisation. *PLoS ONE* **3**, e3700. doi:10.1371/journal.pone.0003700

Morton, B. 1978. The diurnal rhythm and the processes of feeding and digestion in *Tridacna crocea* (Bivalvia : Tridacnidae). *Journal of Zoology, London* **185**, 371–388.

Morton, B. 2000. The biology and functional morphology of *Fragum erugatum* (Bivalvia: Cardiidae) from Shark Bay, Western Australia: the significance of its relationship with entrained zooxanthellae. *Journal of Zoology, London* **251**, 39–52.

Morton, B. & Morton, J. 1983. *The Seashore Ecology of Hong Kong*. Hong Kong: Hong Kong University Press.

Munro, J.L. 1988. *Status of Giant Clam Stocks in the Central Gilbert Islands Group, Republic of Kiribati*. Workshop on Pacific Inshore Fishery Resources, Noumea, New Caledonia, 14–25 March 1988. SPC/Inshore Fish Res/BP54. Noumea, New Caledonia: South Pacific Commission.

Munro, J.L. 1989. Fisheries for giant clams (Tridacnidae: Bivalvia) and prospects for stock enhancement. In *Marine Invertebrate Fisheries: Their Assessment and Management*, J.F. Caddy (ed.). New York: Wiley, 541–558.

Munro, J.L. 1992. *Chapter 13 – Giant clams*. FFA Report 92/75. Honiara, Solomon Islands: Pacific Islands Forum Fisheries Agency. Online. http://www.spc.int/DigitalLibrary/Doc/FAME/Reports/Munro_93_GiantClams.pdf (accessed 19 December 2016).

Munro, J.L. & Heslinga, G.A. 1983. Prospects for the commercial cultivation of giant clams (Bivalvia: Tridacnidae). *Proceedings of the Annual Gulf Caribbean Fisheries Institute* **35**, 122–134.

Munro, P.E., Beard, J.H. & Lacanienta, E. 1983. Investigations on the substance which causes sperm release in Tridacnid clams. *Comparative Biochemistry and Physiology* **74C**, 219–223.

Murakoshi, M. 1986. Farming of the boring giant clam, *Tridacna crocea* Lamarck. *Galaxea* **5**, 239–254.

Neo, M.L., Eckman, W., Vicentuan-Cabaitan, K., Teo, S.L.-M. & Todd, P.A. 2015a. The ecological significance of giant clams in coral reef ecosystems. *Biological Conservation* **181**, 111–123.

Neo, M.l.., Erftermeijer, P.L.A., van Beek, J.K.L., van Maren, D.S., Teo, S.L.-M. & Todd, P.A. 2013a. Recruitment constraints in Singapore's fluted giant clam (*Tridacna squamosa*) population – A dispersal model approach. *PLoS ONE* **8**, e58819. doi:10.1371/journal.pone.0058819

Neo, M.L. & Loh, K.S. 2014. Giant clam shells 'graveyard' at Semakau Landfill. *Singapore Biodiversity Records* **2014**, 248–249.

Neo, M.L. & Low, J.K.Y. 2017. First observations of *Tridacna noae* (Röding, 1798) (Bivalvia: Heterodonta: Cardiidae) in Christmas Island (Indian Ocean). *Marine Biodiversity*, doi:10.1007/s12526-017-0678-3

Neo, M.L. & Todd, P.A. 2011. Predator-induced changes in fluted giant clam (*Tridacna squamosa*) shell morphology. *Journal of Experimental Marine Biology and Ecology* **397**, 21–26.

Neo, M.L. & Todd, P.A. 2012a. Giant clams (Mollusca: Bivalvia: Tridacninae) in Singapore: history, research and conservation. *Raffles Bulletin of Zoology* **25**, 67–78.

Neo, M.L. & Todd, P.A. 2012b. Population density and genetic structure of the giant clams *Tridacna crocea* and *T. squamosa* on Singapore's reefs. *Aquatic Biology* **14**, 265–275.

Neo, M.L. & Todd, P.A. 2013. Conservation status reassessment of giant clams (Mollusca: Bivalvia: Tridacninae) in Singapore. *Nature in Singapore* **6**, 125–133.

Neo, M.L., Todd, P.A., Chou, L.M. & Teo, S.L.-M. 2011. Spawning induction and larval development in the fluted giant clam, *Tridacna squamosa* (Bivalvia: Tridacnidae). *Nature in Singapore* **4**, 157–161.

Neo, M.L., Todd, P.A., Teo, S.L.-M. & Chou, L.M. 2009. Can artificial substrates enriched with crustose coralline algae enhance larval settlement and recruitment in the fluted giant clam (*Tridacna squamosa*)? *Hydrobiologia* **625**, 83–90.

Neo, M.L., Todd, P.A., Teo, S.L.-M. & Chou, L.M. 2013b. The effects of diet, temperature and salinity on survival of larvae of the fluted giant clam, *Tridacna squamosa*. *Journal of Conchology* **41**, 369–376.

Neo, M.L., Vicentuan, K., Teo, S.L.M., Erftemeijer, P.L.A. & Todd, P.A. 2015b. Larval ecology of the fluted giant clam, *Tridacna squamosa*, and its potential effects on dispersal models. *Journal of Experimental Marine Biology and Ecology* **469**, 76–82.

Newman, W.A. & Gomez, E.D. 2000. On the status of giant clams, relics of Tethys (Mollusca: Bivalvia: Tridacninae). In *Proceedings of the 9th International Coral Reef Symposium, Bali, Indonesia, 23–27 October 2000, Vol. 2*, M.K. Moosa et al. (eds). Jakarta: Indonesian Institute of Sciences, Jakarta: Ministry of Environment, Honolulu, Hawaii: International Society for Reef Studies, 927–936.

Norton, J.H. & Jones, G.W. 1992. *The Giant Clam: An Anatomical and Histological Atlas*. ACIAR Monograph. Canberra: Australian Centre for International Agricultural Research.

Norton, J.H., Prior, H.C., Baillie, B. & Yellowlees, D. 1995. Atrophy of the zooxanthellal tubular system in bleached giant clams *Tridacna gigas*. *Journal of Invertebrate Pathology* **66**, 307–310.

Norton, J.H., Shepherd, M.A., Long, H.M. & Fitt, W.K. 1992. The zooxanthellal tubular system in the giant clam. *Biological Bulletin* **183**, 503–506.

Nuryanto, A., Duryadi, D., Soedharma, D. & Blohm, D. 2007. Molecular phylogeny of giant clams based on mitochondrial DNA cytochrome C oxidase I gene. *HAYATI Journal of Biosciences* **14**, 162–166.

O'Callaghan, M. 1995. Village-farmed giant clams – From South Pacific Ocean to you, sustainably. *Freshwater and Marine Aquarium* **18**, 8–10.

Okada, H. 1997. *Market Survey of Aquarium Giant Clams in Japan*. South Pacific Aquaculture Development Project (Phase II). FAO Fisheries and Aquaculture Department Field Document No. 8. Rome: Food and Agriculture Organization of the United Nations. Online. http://www.fao.org/docrep/005/ac892e/AC892E00.htm (accessed 19 December 2016).

Oppenheim, P. 1901. Die Priabonaschichten und ihre Fauna im Zusammenhange mit gleichalterigen und analogen Ablagerungen. *Palaeontographica* **47**, 1–348. (In German)

Othman, A.S., Goh, G.H.S. & Todd, P.A. 2010. The distribution and status of giant clams (family Tridacnidae) – a short review. *Raffles Bulletin of Zoology* **58**, 103–111.

Pasaribu, B.P. 1988. Status of giant clams in Indonesia. In *Giant Clams in Asia and the Pacific*, J.W. Copland & J.S. Lucas (eds). Canberra: Australian Centre for International Agricultural Research, 44–46.

Pearson, R.G. 1977. Impact of foreign vessels poaching giant clams. *Australian Fisheries* **36**, 8–11, 23.

Pearson, R.G. & Munro, J.L. 1991. Growth, mortality and recruitment rates of giant clams, *Tridacna gigas* and *T. derasa*, at Michaelmas Reef, central Great Barrier Reef, Australia. *Australian Journal of Marine and Freshwater Research* **42**, 241–262.

Peijnenburg, K.T.C.A. & Goetze, E. 2013. High evolutionary potential of marine zooplankton. *Ecology and Evolution* **3**, 2765–2781.

Penny, S.S. & Willan, R.C. 2014. Description of a new species of giant clam (Bivalvia: Tridacnidae) from Ningaloo Reef, Western Australia. *Molluscan Research* **34**, 201–211.

Perron, F.E., Heslinga, G.A. & Fagolimul, J. 1985. The gastropod *Cymatium muricinum*, a predator on juvenile tridacnid clams. *Aquaculture* **48**, 211–221.

Petersen, C.W. & Levitan, D.R. 2001. The Allee effect: a barrier to recovery by exploited species. In *Conservation of Exploited Species*, J.D. Reynolds et al. (eds). Conservation Biology Series 6. Cambridge: Cambridge University Press, 281–300.

Pfenninger, M. & Schwenk, K. 2007. Cryptic animal species are homogeneously distributed among taxa and biogeographical regions. *BMC Evolutionary Biology* **7**, 121 only.

Poutiers, J.M. 1998. Bivalves. Acephala, Lamellibranchia, Pelecypoda. In *FAO Species Identification Guide for Fishery Purposes. The Living Marine Resources of the Western Central Pacific. Volume 1. Seaweeds, Corals, Bivalves, and Gastropods*, K.E. Carpenter & V.H. Niem (eds). Rome: Food and Agriculture Organization of the United Nations, 123–362.

Price, C.M. & Fagolimul, J.O. 1988. Reintroduction of giant clams to Yap State, Federated States of Micronesia. In *Giant Clams in Asia and the Pacific*, J.W. Copland & J.S. Lucas (eds). Canberra: Australian Centre for International Agricultural Research, 41–43.

Ramadoss, K. 1983. Giant clam (*Tridacna*) resources. *CMFRI Bulletin* **34**, 79–81.

Ramohia, P. 2006. Fisheries resources: commercially important macroinvertebrates. In *Solomon Islands Marine Assessment: Technical Report on Survey Conducted May 13 to June 17, 2004*, A. Green et al. (eds). Arlington, Virginia: The Nature Conservancy, 330–400.

Redding, D.W. & Mooers, A.Ø. 2006. Incorporating evolutionary measures into conservation prioritisation. *Conservation Biology* **20**, 1670–1678.

Reef Check Foundation 2016. Global reef tracker. Marina del Rey, California: Reef Check Foundation. Online. http://data.reefcheck.us/ (accessed 28 December 2016).

Reese, D.S. 1988. A new engraved *Tridacna* shell from Kish. *Journal of Near Eastern Studies* **47**, 35–41.

Reese, D.S. & Sease, C. 1993. Some previously unpublished engraved *Tridacna* shells. *Journal of Near Eastern Studies* **52**, 109–128.

Reid, R.G.B., Fankboner, P.V. & Brand, D.G. 1984. Studies on the physiology of the giant clam *Tridacna gigas* Linné – I. Feeding and digestion. *Comparative Biochemistry and Physiology* **78A**, 95–101.

Ricard, M. & Salvat, B. 1977. Faeces of *Tridacna maxima* (Mollusca: Bivalvia), composition and coral reef importance. In *Proceedings of the Third International Coral Reef Symposium, Volume 1: Biology*, D.L. Taylor (ed.). Miami, Florida: Rosenstiel School of Marine and Atmospheric Science, 495–501.

Richards, R. & Roga, K. 2004. Barava: land title deeds in fossil shell from the western Solomon Islands. *Tuhinga* **15**, 17–26.

Richter, C., Roa-Quiaoit, H., Jantzen, C., Al-Zibdah, M. & Kochzius, M. 2008. Collapse of a new living species of giant clam in the Red Sea. *Current Biology* **18**, 1349–1354.

Roa-Quiaoit, H.A.F. 2005. *The ecology and culture of giant clams (Tridacnidae) in the Jordanian sector of Gulf of Aqaba, Red Sea*. PhD Dissertation, University of Bremen, Germany.

Rodrigues, A.S.L., Pilgrim, J.D., Lamoreux, J.F., Hoffmann, M. & Brooks, T.M. 2006. The value of the IUCN Red List for conservation. *Trends in Ecology & Evolution* **21**, 71–76.

Rosewater, J. 1965. The family Tridacnidae in the Indo-Pacific. *Indo-Pacific Mollusca* **1**, 347–396.

Rosewater, J. 1982. A new species of *Hippopus* (Bivalvia: Tridacnidae). *The Nautilus* **96**, 3–6.

Safi, K., Armour-Marshall, K., Baillie, J.E.M. & Issac, N.J.B. 2013. Global patterns of evolutionary distinct and globally endangered amphibians and mammals. *PLoS ONE* **8**, e63582. doi:10.1371/journal.pone.0063582

Sant, G. 1995. *Marine Invertebrates of the South Pacific: An Examination of the Trade*. Cambridge, UK: TRAFFIC International. Online. https://portals.iucn.org/library/sites/library/files/documents/Traf-024.pdf (accessed 19 December 2016).

Schneider, J.A. 1992. Preliminary cladistic analysis of the bivalve family Cardiidae. *American Malacological Bulletin* **9**, 145–155.

Schneider, J.A. 1998. Phylogeny of the Cardiidae (Bivalvia): Phylogenetic relationships and morphological evolution within the subfamilies Clinocardiidae, Lymnocardiidae, Fraginae and Tridacninae. *Malacologia* **40**, 321–373.

Schneider, J.A. & Ó Foighil, D. 1999. Phylogeny of giant clams (Cardiidae: Tridacninae) based on partial mitochondrial 16S rDNA gene sequences. *Molecular Phylogenetics and Evolution* **13**, 59–66.

Schwartzmann, C., Durrieu, G., Sow, M., Ciret, P., Lazareth, C.E. & Massabuau, J.-C. 2011. *In situ* giant clam growth rate behaviour in relation to temperature: a one-year coupled study of high-frequency noninvasive valvometry and sclerochronology. *Limnology and Oceanography* **56**, 1940–1951.

Selin, N.I. & Latypov, Y.Y. 2011. The size and age structure of *Tridacna crocea* Lamarck, 1819 (Bivalvia: Tridacnidae) in the coastal area of islands of the Cön Dao Archipelago in the South China Sea. *Russian Journal of Marine Biology* **37**, 376–383.

Selkoe, K.A., Henzler, C.M. & Gaines, M.D. 2008. Seascape genetics and the spatial ecology of marine populations. *Fish and Fisheries* **9**, 363–377.

Shang, Y.C., Tisdell, C. & Leung, P.S. 1991. *Report on a Market Survey of Giant Clam Products in Selected Countries*. Publication No. 107. Waimanalo, Hawaii: Center for Tropical and Subtropical Aquaculture.

Siaosi, F., Sapatu, M., Lalavanua, W., Pakoa, K., Yeeting, B., Magron, F., Moore, B., Bertram, I. & Chapman, L. 2012. *Climate Change Baseline Assessment – Funafuti Atoll, Tuvalu. July–August 2011.* Coastal Fisheries Science and Management Section, Secretariat of the Pacific Community, December 2012. Online. http://www.spc.int/DigitalLibrary/Doc/FAME/Reports/Siaosi_12_Tuvalu_Climate_Change_Baseline_Monitoring_Report.pdf (accessed 15 March 2017).

Sirenko, B.I. & Scarlato, O.A. 1991. *Tridacna rosewateri* sp. n. A new species of giant clam from Indian Ocean. *La Conchiglia* **22**, 4–9.

Soo, P. & Todd, P.A. 2012. Nocturnal movement and possible geotaxis in the fluted giant clam (*Tridacna squamosa*). *Contributions to Marine Science* **2012**, 159–162.

Soo, P. & Todd, P.A. 2014. The behaviour of giant clams (Bivalvia: Cardiidae: Tridacninae). *Marine Biology* **161**, 2699–2717.

Southgate, P.C., Braley, R.D. & Militz, T.A. 2016. Embryonic and larval development of the giant clam *Tridacna noae* (Röding, 1798) (Cardiidae: Tridacninae). *Journal of Shellfish Research* **35**, 777–783.

Southgate, P.C., Braley, R.D. & Militz, T.A. 2017. Ingestion and digestion of micro-algae concentrates by veliger larvae of the giant clam, *Tridacna noae*. *Aquaculture*, doi:10.1016/j.aquaculture.2017.02.032

Southward, A.J., Young, C.M. & Fuiman, L.A. 2005. Restocking initiatives. *Advances in Marine Biology* **49**, 9–41.

Stasek, C.R. 1962. The form, growth, and evolution of the Tridacnidae (giant clams). *Archives de Zoologie Expérimentale et Générale* **101**, 1–40.

Stasek, C.R. 1965. Behavioural adaptation of the giant clam *Tridacna maxima* to the presence of grazing fishes. *The Veliger* **8**, 29–35.

Stephens, P.A., Sutherland, W.J. & Freckleton, R.P. 1999. What is the Allee Effect? *Oikos* **87**, 185–190.

Sturany, R. 1899. Expedition S.M. Schiff "Pola" in das Rothe Meer. Nördliche und südliche Hälfte. 1895/96 und 1897/98. Zoologische Ergebnisse XIV Lamellibranchiaten des Rothen Meeres. Berichte der Commission für oceanographische Forschungen. *Sonder druck aus: Denkschriften der mathematisch-naturwissenschaftlichen Classe der Kaiserli chen Akademie der Wissenschaften, Wien* **69**, 255–295.

Su, P.-W. 2013. *The reproductive comparison of giant clams* Tridacna noae *and* Tridacna maxima. MSc Thesis, National Sun Yat-Sen University, Taiwan.

Su, Y., Hung, J.-H., Kubo, H. & Liu, L.-L. 2014. *Tridacna noae* (Röding, 1798) – a valid giant clam species separated from *T. maxima* (Röding, 1798) by morphological and genetic data. *Raffles Bulletin of Zoology* **62**, 124–135.

Tan, S.H. & Zulfigar, Y. 1999. Factors affecting the interchange of *Tridacna squamosa* larvae and gamete material between Pulau Tioman and Johore Islands in the South China Sea. *Proceedings the Tenth Joint Seminar on Marine and Fisheries Sciences, Melaka, Malaysia, 1–3 December 1999.* Tokyo: Japan Society for the Promotion of Science and Kuala Lumpur: Vice-Chancellors' Council of National Universities in Malaysia, 288–304.

Tan, S.H. & Zulfigar, Y. 2001. Factors affecting the dispersal of *Tridacna squamosa* larvae and gamete material in the Tioman Archipelago, The South China Sea. *Phuket Marine Biological Center Special Publication* **25**, 349–356.

Tan, S.H. & Zulfigar, Y. 2003. Status of giant clam in Malaysia. *SPC Trochus Information Bulletin* **10**, 9–10.

Teitelbaum, A. & Friedman, K. 2008. Successes and failures in reintroducting giant clams in the Indo-Pacific region. *SPC Trochus Information Bulletin* **14**, 19–26.

Tiavouane, J. & Fauvelot, C. 2016. First record of the Devil Clam, *Tridacna mbalavuana* Ladd 1934, in New Caledonia. *Marine Biodiversity*, doi:10.1007/s12526-016-0506-1

Tisdell, C. (ed.) 1992. *Giant Clams in the Sustainable Development of the South Pacific: Socioeconomic Issues in Mariculture and Conservation.* ACIAR Monograph No. 18. Canberra: Australian Centre for International Agricultural Research.

Todd, P.A., Lee, J.H. & Chou, L.M. 2009. Polymorphism and crypsis in the boring giant clam (*Tridacna crocea*): potential strategies against visual predators. *Hydrobiologia* **635**, 37–43.

Trench, R.K., Wethey, D.S. & Porter, J.W. 1981. Observations on the symbiosis with zooxanthellae among the Tridacnidae (Mollusca, Bivalvia). *Biological Bulletin* **161**, 180–198.

Ullmann, J. 2013. Population status of giant clams (Mollusca: Tridacnidae) in the northern Red Sea, Egypt. *Zoology in the Middle East* **59**, 253–260.

UNEP-WCMC 2011. *Review of Oceanian Species/Country Combinations Subject to Long-Standing Import Suspensions.* Cambridge: UNEP World Conservation Monitoring Centre. Online. http://ec.europa.eu/environment/cites/pdf/reports/Review_Oceanian_species.pdf (accessed 19 December 2016).

Van Wynsberge, S. 2016. *Approche comparée, intégrée et spatialisée pour la gestion d'une ressource emblématique exploitée en Polynésie française et en Nouvelle-Calédonie: le cas du bénitier (*Tridacna maxima*).* PhD Thesis, Université de la Polynésie française, IRD Centre de Nouméa, Papeete, Nouméa, New Caledonia.

Van Wynsberge, S., Andréfouët, S., Gaertner-Mazouni, N. & Remoissenet, G. 2015. Conservation and resource management in small tropical islands: trade-offs between planning unit size, data redundancy and data loss. *Ocean & Coastal Management* **116**, 37–43.

Van Wynsberge, S., Andréfouët, S., Gaertner-Mazouni, N., Wabnitz, C.C.C., Gilbert, A., Remoissenet, G., Payri, C. & Fauvelot, C. 2016. Drivers of density for the exploited giant clam *Tridacna maxima*: a meta-analysis. *Fish and Fisheries* **17**, 567–584.

Van Wynsberge, S., Andréfouët, S., Gaertner-Mazouni, N., Wabnitz, C.C.C., Menoud, M., Le Moullac, G., Levy, P., Gilbert, A. & Remoissenet, G. 2017. Growth, survival and reproduction of the giant clam *Tridacna maxima* (Röding 1798, Bivalvia) in two contrasting lagoons in French Polynesia. *PLoS ONE* **12**, e0170565. doi:101317/journal.pone.0170565

Van Wynsberge, S., Andréfouët, S., Gilbert, A., Stein, A. & Remoissenet, G. 2013. Best management strategies for sustainable giant clam fishery in French Polynesia Islands: Answers from a spatial modeling approach. *PLoS ONE* **8**, e64641. doi:10.1371/journal.pone.0064641

Vicentuan-Cabaitan, K., Neo, M.L., Eckman, W., Teo, S.L.-M. & Todd, P.A. 2014. Giant clam shells host a multitude of epibionts. *Bulletin of Marine Science* **90**, 795–796.

von der Heyden, S., Beger, M., Toonen, R.J., van Herwerden, L., Juinio-Meñez, M.A., Ravago-Gotanco, R., Fauvelot, C. & Bernardi, G. 2014. The application of genetics to marine management and conservation: examples from the Indo-Pacific. *Bulletin of Marine Science* **90**, 123–158.

Wabnitz, C.C.C. & Fauvelot, C. 2014. *Tridacna noae* is back. *SPC Fisheries Newsletter* **145**, 30 only.

Wabnitz, C., Taylor, M., Green, E. & Razak, T. 2003. *From Ocean to Aquarium: The Global Trade in Marine Ornamental Species.* Cambridge, UK: UNEP World Conservation Monitoring Centre. Online. http://wedocs.unep.org//handle/20.500.11822/8341 (accessed 19 December 2016).

Waters, C.G. 2008. *Biological responses of juvenile* Tridacna maxima *(Mollusca:Bivalvia) to increased* pCO_2 *and ocean acidification.* MSc Thesis, The Evergreen State College, Olympia, Washington, USA.

Waters, C.G., Story, R. & Costello, M.J. 2013. A methodology for recruiting a giant clam, *Tridacna maxima*, directly to natural substrata: a first step in reversing functional extinctions? *Biological Conservation* **160**, 19–24.

Watson, S.-A. 2015. Giant clams and rising CO_2: light may ameliorate effects of ocean acidification on a solar-powered animal. *PLoS ONE* **10**, e0128405. doi:10.1371/journal.pone.0128405

Watson, S.-A., Southgate, P.C., Miller, G.M., Moorhead, J.A. & Knauer, J. 2012. Ocean acidification and warming reduce juvenile survival of the fluted giant clam, *Tridacna squamosa*. *Molluscan Research* **32**, 177–180.

Wells, S. 1996. *The IUCN Red List of Threatened Species 1996.* Cambridge, UK: IUCN Global Species Programme Red List Unit. Online. http://www.iucnredlist.org/ (accessed 28 July 2016).

Wells, S. 1997. *Giant Clams: Status, Trade and Mariculture, and the Roles of CITES Management.* Gland, Switzerland and Cambridge, UK: IUCN. Online. https://portals.iucn.org/library/sites/library/files/documents/1997-076.pdf (accessed 19 December 2016).

Wells, S.M., Pyle, R.M. & Collins, N.M. 1983. *The IUCN Invertebrate Red Data Book.* Gland, Switzerland and Cambridge, UK: International Union for the Conservation of Nature and Natural Resources.

Wilkinson, C.R. & Buddemeier, R.W. 1994. *Global Climate Change and Coral Reefs: Implications for People and Reefs. Report of the UN EP-IOC-ASPEI-IUCN Global Task Team on the Implications of Climate Change on Coral Reefs.* Gland, Switzerland: International Union for Conservation of Nature and Natural Resources.

Wilson, N.G. & Kirkendale, L.A. 2016. Putting the 'Indo' back into the Indo-Pacific: resolving marine phylogeographic gaps. *Invertebrate Systematics* **30**, 86–94.

Yamaguchi, M. 1977. Conservation and cultivation of giant clams in the tropical Pacific. *Biological Conservation* **11**, 13–20.

Yau, A.J., Lenihan, H.S. & Kendall, B.E. 2014. Fishery management priorities vary with self-recruitment in sedentary marine populations. *Ecological Applications* **24**, 1490–1504.

Yonge, C.M. 1936. Mode of life, feeding, digestion and symbiosis with zooxanthellae in the Tridacnidae. *Great Barrier Reef Expedition* **1928–29**, 283–321.

Yonge, C.M. 1982. Functional morphology and evolution in the Tridacnidae (Mollusca: Bivalvia: Cardiacea). *Records of the Australian Museum* **33**, 735–777.

Zhang, H. 2014. Chinese fishermen in troubled waters. *The Diplomat* October 23, 2014. Online. http://thediplomat.com/2014/10/chinese-fishermen-in-troubled-waters/ (accessed 26 February 2016).

Zuschin, M. & Piller, W.E. 1997. Bivalve distribution on coral carpets in the Northern Bay of Safaga (Red Sea, Egypt) and its relation to environmental parameters. *Facies* **37**, 183–194.

Appendix A: Supplementary materials

Table A1 List of localities with giant clams (in alphabetical order)

Locality	Locality	Citations	Was data useful for review?	Was data extracted?
American Samoa	Rose Atoll	Radtke (1985)	Y (survey data)	Y (Table A3)
	—	Bell (1993)	Y	
	—	Nagaoka (1993)	Y	
	Rose Atoll	Green & Craig (1999)	Y (survey data)	Y (Table A3)
	—	Green (2002)	Y	
	—	Kelty & Kuartei (2004)	Y	
	—	Craig (2009)	Y	
	—	Reef Check (1997, 2003)	Y (survey data)	Y (Table A4)
Australia	One Tree Island, Capricorn Group, QL	McMichael (1974)	Y (survey data)	Y (Table A3)
	Orpheus Island, Palm Island Group, QL	Hamner & Jones (1976)	Y (survey data)	Y (Table A3)
	Great Barrier Reef (North & South)	Braley (1987a, b)	Y (survey data)	Y (Table A3)
	Lizard Island, Great Barrier Reef	Alder & Braley (1989)	Y	
	Michaelmas Reef, Great Barrier Reef	Pearson & Munro (1991)	Y (survey data)	Y (Table A3)
	Lizard Island, Great Barrier Reef	Braley & Muir (1995)	Y (insufficient data)	
	Montebello Islands, Western Australia	Wells et al. (2000)	Y	
	Mermaid Reef, Cartier Reef, and Ashmore Reef	Rees et al. (2003)	Y (survey data)	Y (Table A3)
	Heron Island, southern Great Barrier Reef	Strotz et al. (2010)	Y	
	Ningaloo Marine Park, WA	Black et al. (2011)	Y (survey data)	Y (Table A3)
	Solitary Islands Marine Park, NSW	Smith (2011)	Y	
	Ningaloo Marine Park, WA	Penny & Willan (2014)	Y	
	Western Australia	Borsa et al. (2015)	Y (DNA)	
	Ningaloo Marine Park, WA	Johnson et al. (2016)	Y (survey data)	Y (Table A3)
	—	Reef Check (1997–2014)	Y (survey data)	Y (Table A4)
British Indian Ocean Territory	Chagos Archipelago	Sheppard (1984)	Y	
	Chagos Archipelago (Salomon and Peros Banhos atolls)	Chagos Conservation Trust (FaceBook) (2014)	Y	
Brunei	—	Reef Check		
Cambodia	—	Vibol (N.D.)	Y (survey data)	Y (Table A3)
	Koh Rong	Chou (2000)	Y (exploitation)	
	—	Chou et al. (2002)	Y (survey data)	Y (Table A3)
	—	Kim et al. (2004)	Y (survey data)	Y (Table A3)
	—	Van Bochove et al. (2011)	Y (survey data)	Y (Table A3)

Continued

Table A1 (Continued)   List of localities with giant clams (in alphabetical order)

Locality	Locality	Citations	Was data useful for review?	Was data extracted?
Cambodia (*Continued*)	Song Saa Private Island, Koh Rong Archipelago	Savage et al. (2013)	Y (survey data)	Y (Table A3)
	Koh Rong and Koh Koun, Koh Rong Archipelago	Thorne et al. (2015)	Y (survey data)	Y (Table A3)
	—	Reef Check (1998, 2001, 2003, 2009–2010)	Y (survey data)	Y (Table A4)
Cargados Carajos Archipelago	—	Monsecour (2016)	Y (exploitation)	Y
China	Hainan Islands	Hutchings & Wu (1987)	Y	
	Hainan Islands	Fiege et al. (1994)	Y	
	—	Qi (2004)	Y	
	Sanya waters	Tadashi et al. (2008)	Y	
	—	Liu (2013)	Y	
	—	Reef Check (2000, 2002)	Y (survey data)	Y (Table A4)
Christmas Island	Flying Fish Cove	Andrews et al. (1900)	Y	
	—	Tomlin (1934)	Y	
	—	Wells & Slack-Smith (2000)	Y	
	—	Gilligan et al. (2008)	Y (survey data)	Y (Table A3)
	—	Hourston (2010)	Y (survey data)	Y (Table A3)
	—	Huber (2010)	Y	
	—	Tan & Low (2014)	Y	
	—	Reef Check (2003–2007)	Y (survey data)	Y (Table A4)
Cocos (Keeling) Islands	—	Gibson-Hill (1946)	Y	
	—	Abbott (1950)	Y	
	—	Maes (1967)	Y	
	—	Wells (1994)	Y	
	—	Hender et al. (2001)	Y (survey data)	Y (Table A3)
	—	Australian Government (2005)	Y (survey data)	Y (Table A3)
	—	Hourston (2010)	Y (survey data)	Y (Table A3)
	—	Huber (2010)	Y	
	—	Bellchambers & Evans (2013)	Y (survey data)	Y (Table A3)
	—	Tan & Low (2014)	Y	
	—	Evans et al. (2016)	Y (survey data)	Y (Table A3)
	—	Reef Check (1997–1999, 2001–2005, 2007–2008)	Y (survey data)	Y (Table A4)
Comoros	Nioumachouoi site; Ouenefou reef	Bigot et al. (2000)	Y	
Cook Islands	—	Paulay (1987)	Y	
	Aitutaki Lagoon, Manihiki Lagoon, Suwarrow Lagoon, and Penrhyn Lagoon	Sims & Howard (1988)	Y (survey data)	Y (Table A3)
	—	Tisdell & Wittenberg (1992)	Y	

Continued

Table A1 (Continued)   List of localities with giant clams (in alphabetical order)

Locality	Locality	Citations	Was data useful for review?	Was data extracted?
Cook Islands	Tongareva Lagoon	Chambers (2007)	Y (survey data)	Y (Table A3)
(Continued)	Tongareva Lagoon	Chambers (2008)	Y (insufficient data)	
	—	Reef Check (2005)	Y	Y (Table A4)
Djibouti	—	Pilcher & Djama (2000)	Y (survey data)	Y (Table A3)
	—	PERSGA (2010)	Y (survey data)	Y (Table A3)
East Timor	Dili	Flickr	Y	
	—	Reef Check (2004, 2008)	Y (survey data)	Y (Table A4)
Egypt	Northern Bay of Safaga, Red Sea	Zuschin & Pillar (1997)	Y (survey data)	Y (Table A3)
	Northern Red Sea	Kilada et al. (1998)	Y (survey data)	Y (Table A3)
	Northern Red Sea	Ullmann (2013)	Y (survey data)	Y (Table A3)
	Egyptian Red Sea	Mekawy (2014)	Y	
	Red Sea area	Richter et al. (2008)	Y (survey data)	Y (Table A3)
	Red Sea area	Huber & Eschner (2011)	Y	
	Northern Red Sea	Mekawy & Madkour (2012)	Y	
	—	Reef Check (1997, 2000–2015)	Y (survey data)	Y (Table A4)
Eritrea	—	Reef Check (2000)	Y (survey data)	Y (Table A4)
Federated States of Micronesia	Yap State	Price & Fagolimul (1988)	Y	
	—	Smith (1992)	Y	
	Kosrae, part of the Caroline Islands	Borsa et al. (2015)	Y	
	—	Reef Check (2000–2008)	Y (survey data)	Y (Table A4)
Fiji	—	Lewis et al. (1988)	Y	
	Eastern islands (Lau)	Lewis & Ledua (1988)	Y	
	Eastern islands (Lau)	Lucas et al. (1991)	Y	
	Eastern islands (Lau)	Vuki et al. (1992)	Y	
	—	Tacconi & Tisdell (1992) Chapter 13	Y	
	—	Tisdell & Wittenberg (1992)	Y	
	Eastern islands (Lau)	Ledua et al. (1993)	Y (survey data)	Y (Table A3)
	Southwest Viti Levu Island	Seeto et al. (2012)	Y (exploitation)	
	Viti-Levu	Borsa et al. (2015)	Y	
	—	Reef Check (1997, 1999–2011)	Y (survey data)	Y (Table A4)
French Polynesia	Takapoto Atoll	Jaubert (1977)	Y	
	Takapoto Atoll	Richard (1977)	Y (survey data)	Y (Table A3)
	Bora Bora Lagoon	Planes et al. (1993)	Y	
	Moorea, Takapoto, and Anaa	Laurent (2001)	Y (survey data)	Y (Table A3)
	Tatakoto Atoll, Eastern Tuamotu	Gilbert et al. (2005)	Y (survey data)	Y (Table A3)

Continued

Table A1 (Continued)   List of localities with giant clams (in alphabetical order)

Locality	Locality	Citations	Was data useful for review?	Was data extracted?
French Polynesia *(Continued)*	Fangatau Atoll, Eastern Tuamotu	Andréfouët et al. (2005)	Y (survey data)	Y (Table A3)
	Tubuai, Austral Islands	Larrue (2006)	Y	
	Reao, Pukarua, and Raivavae	Gilbert et al. (2006a)	Y (survey data)	Y (Table A3)
	Fangatau Atoll, Tatakoto Atoll, and Tubuai	Gilbert et al. (2006b)	Y (survey data)	Y (Table A3)
	Tubuai, Austral Islands	Newman & Gomez (2007)	Y	
	—	Gilbert et al. (2007)	Y	
	Raivavae Island	Andréfouët et al. (2009)	Y (survey data)	Y (Table A3)
	Tatakoto Atoll	Andréfouët et al. (2013)	Y (survey data)	Y (Table A3)
	—	Van Wynsberge et al. (2013)	Y (survey data)	Y (Table A3)
	Tuamotu and Gambier Archipelago	Andréfouët et al. (2014)	Y	
	—	Reef Check (1999–2014)	Y (survey data)	Y (Table A4)
Guam	—	Stojkovich (1977)	Y	
	—	Munro (1989)	Y	
	—	Hensley & Sherwood (1993)	Y	
	—	Anonymous (1994)	Y	
	—	Paulay (2003)	Y	
	—	Reef Check (1998–1999, 2001, 2004)	Y (survey data)	Y (Table A4)
Hong Kong	Mirs Bay	Morton & Morton (1983)	Y	
	—	Reef Check (2003, 2006, 2011)	Y (survey data)	Y (Table A4)
India	Andaman and Nicobar Islands	Rosewater (1965)	Y	
	Kavaratti Atoll	Namboodiri & Sivadas (1979)	Y	
	Andaman and Nicobar Islands	Ramadoss (1983)	Y (survey data)	Y (Table A3)
	Lakshadweep	George et al. (1986)	Y	
	Lakshadweep	Apte & Dutta (2010)	Y	
	Lakshadweep	Apte et al. (2010)	Y (survey data)	Y (Table A3)
	Lakshadweep	Bijukumar et al. (2015)	Y (legislation)	
	—	Reef Check (1998)	Y (survey data)	Y (Table A4)
Indonesia	Karimun Java	Brown & Muskanofola (1985)	Y (survey data)	Y (Table A3)
	—	Pasaribu (1988)	Y	
	Karimunjawa Islands	Pringgenies et al. (1995)	Y	
	Gulf of Tomini, Sulawesi	Wells (2001)	Y	
	Rajah Ampat Islands, Papua Province	Wells (2002)	Y	
	Pari Island	Eliata et al. (2003)	Y (survey data)	Y (Table A3)

Continued

Table A1 (Continued) List of localities with giant clams (in alphabetical order)

Locality	Locality	Citations	Was data useful for review?	Was data extracted?
Indonesia (Continued)	Anambas and Natuna Islands	Tan & Kastoro (2004)	Y	
	Pari Island	Panggabean (2007)	Y	
	Seribu Islands and Manado waters	Yusuf et al. (2009)	Y (survey data)	Y (Table A3)
	Kei Kecil, Southeast Maluku	Kusnadi et al. (2008)	Y	
	Kei Kecil, Southeast Maluku	Hernawan (2010)	Y (survey data)	Y (Table A3)
	Savu Sea, East Nusa Tenggara Province	Naguit et al. (2012)	Y (survey data)	Y (Table A3)
	Bunaken, Manado and Alor Archipelago, Savu Sea and Doi Island	Borsa et al. (2015)	Y	
	—	Reef Check (1997–2014)	Y (survey data)	Y (Table A4)
Israel	Eilat (southernmost Israel)	Flickr	Y	
	—	Reef Check (1997–1998, 2001)	Y (survey data)	Y (Table A4)
Japan	—	Hirase (1954)	Y	
	Okinawa	Kanno et al. (1976)	Y	
	Okinawa	Okada (1997)	Y	
	Ogasawara National Park	Fujiwara et al. (2000)	Y	
	Okinawa and Ishigaki Islands	Kubo & Iwai (2007)	Y	
	—	Reef Check (1997–2012, 2014)	Y (survey data)	Y (Table A4)
Jordan	Northern Gulf of Aqaba	Roa-Quaoit (2005)	Y (survey data)	Y (Table A3)
	Jordanian coast of Gulf of Aqaba	Al-Horani et al. (2006)	Y (survey data)	Cannot be easily retrieved
	Red Sea area	Richter et al. (2008)	Y (survey data)	Y (Table A3)
	Red Sea area	Huber & Eschner (2011)	Y	
	—	Reef Check (2007)	Y (survey data)	Y (Table A4)
Kenya	—	Evans et al. (1977)	Y (exploitation)	
	Kenyan coastline	Accordi et al. (2010)	Y	
	—	Anam & Mostarda (2012)	Y	
	—	Reef Check (2003–2004)	Y (survey data)	Y (Table A4)
La Réunion	—	Flickr	Y	
	—	Reef Check (2003–2013)	Y (survey data)	Y (Table A4)
Madagascar	Northwest Madagascar	Wells (2003)	Y	
	Andavadoaka region	Harding et al. (2006)	Y (survey data)	Y (Table A3)
	Andavadoaka region	Nadon et al. (2007)	Y (survey data)	Y (Table A3)
	Northern Madagascar	Harding & Randriamanantsoa (2008)	Y (survey data)	Y (Table A3)
	Southwest Madagascar	Barnes & Rawlinson (2009)	Y	

Continued

Table A1 (Continued)   List of localities with giant clams (in alphabetical order)

Locality	Locality	Citations	Was data useful for review?	Was data extracted?
Madagascar (Continued)	Andavadoaka region	Hopkins (2009)	Y (survey data)	Y (Table A3)
	—	Reef Check (2001, 2003–2005, 2007, 2009–2011)	Y (survey data)	Y (Table A4)
Malaysia	Pulau Redang	Mohamed-Pauzi et al. (1994)	Y	
	Pulau Tioman	Tan et al. (1998)	Y (survey data)	Y (Table A3)
	Johore Islands	Zulfigar & Tan (2000)	Y	
	Johore Islands	Tan & Zulfigar (2001)	Y	
	—	Tan & Zulfigar (2003)	Y	
	Tun Sakaran Marine Park, East Sabah	Montagne et al. (2013)	Y (survey data)	Y (Table A3)
	—	Reef Check (1997–2000, 2003–2012, 2014)	Y (survey data)	Y (Table A4)
Maldives	—	Basker (1991)	Y (survey data)	Y
	Baa Atoll	Andréfouët et al. (2012)	Y	
	—	Reef Check (1997, 2001, 2005–2014)	Y (survey data)	Y (Table A4)
Marshall Islands	Rongelap Island	Pinca & Beger (2002)	Y (survey data)	Y (Table A3)
	Mili Atoll, Rongelap Atoll	Beger & Pinca (2003)	Y (survey data)	Cannot be easily retrieved
	—	Beger et al. (2008) [http://www.nras-conservation.org/publications.html]	Y	
	—	Reef Check (2002)	Y (survey data)	Y (Table A4)
Mauritius	—	Michel et al. (1985)	Y	
	Rodrigues Island	Oliver et al. (2004)	Y	
	—	Reef Check (1999–2003)	Y (survey data)	Y (Table A4)
Mayotte	Mayotte	Jana Around the World (2010)	Y	
	—	Reef Check (2003–2007, 2009–2010, 2014)	Y (survey data)	Y (Table A4)
Mozambique	Quirimba Archipelago	Barnes et al. (1998)	Y	
	—	ReefBuilders.com (2015)	Y	
	—	Reef Check (1997, 2000–2002)	Y (survey data)	Y (Table A4)
Myanmar	—	Wells (1997)	Y	
	—	Reef Check (2001, 2003–2005, 2013)	Y (survey data)	Y (Table A4)
Nauru	—	Jacob (2000)	Y	
	—	South & Skelton (2000)	Y	
	—	Chin et al. (2011)	Y	
New Caledonia	North Province (Kone, Koumac, Touho, Hienghène)	Virly (2004)	Y (survey data)	Y (Table A3)
	North Eastern Lagoon (Poeubo to Hienghène)	McKenna et al. (2006)	Y (survey data)	Y (Table A3)

Continued

Table A1 (Continued)   List of localities with giant clams (in alphabetical order)

Locality	Locality	Citations	Was data useful for review?	Was data extracted?
New Caledonia	Poum	Vieux (2009)	Y (survey data)	Y (Table A3)
(Continued)	Corne Sud	Wantiez et al. (2007a)	Y (survey data)	Y (Table A3)
	Ile des Pins	Wantiez et al. (2007b)	Y (survey data)	Y (Table A3)
	Bourail	Wantiez et al. (2007c)	Y (survey data)	Y (Table A3)
	Grand Lagon Nord	Wantiez et al. (2008a)	Y (survey data)	Y (Table A3)
	Merlet	Wantiez et al. (2008b)	Y (survey data)	Y (Table A3)
	Ducos Island, Bay of Saint Vincent	Aubert et al. (2009)	Y	
	New Caledonia (50 sites)	Purcell et al. (2009)	Y (survey data)	Y (Table A3)
	Noumea	Chin et al. (2011)	Y	
	New Caledonia	Dumas et al. (2011)	Y (survey data)	Y (Table A3)
	Ioro reef	Schwartzmann et al. (2011)	Y	
	—	Dumas et al. (2013)	Y (survey data)	Y (Table A3)
	and Loyalty Islands	Borsa et al. (2015)	Y	
	Northeastern coast of New Caledonia	Tiavouane & Fauvelot (2016)	Y (DNA)	
	—	Reef Check (1997–1998, 2001, 2003–2011)	Y (survey data)	Y (Table A4)
Niue	—	Dalzell et al. (1993)	Y (survey data)	Y (Table A3)
	—	Vieux et al. (2004)	Y	
	—	Kronen et al. (2008)	Y (survey data)	Y (Table A3)
Northern Mariana Islands	Saipan Island	Flickr	Y	
	Maug Island	Flickr	Y	
Palau	South of Kokor, Western Caroline Islands	Hardy & Hardy (1969)	Y (survey data)	Y (Table A3)
	Helen Reef, Western Caroline Islands	Hester & Jones (1974)	Y (survey data)	Y (Table A3)
	Helen Reef, Western Caroline Islands	Bryan & McConnell (1976)	Y (survey data)	Y (Table A3)
	Helen Reef, Western Caroline Islands	Hirshberger (1980)	Y (survey data)	Y (Table A3)
	—	Isamu (2008)	Y (insufficient data)	
	—	Reef Check (1997, 2000–2003, 2006)	Y (survey data)	Y (Table A4)
Papua New Guinea	Milne Bay Province	Kinch (2001)	Y (survey data)	Y (Table A3)
	Milne Bay Province	Kinch (2002)	Y (survey data)	Y (Table A3)
	Milne Bay Province	Wells & Kinch (2003)	Y	
	Milne Bay Province	Miller & Sweatman (2004)	Y	
	—	Berzunza-Sanchez et al. (2013)	Y (history)	
	Madang and Kavieng	Borsa et al. (2015)	Y	
	Kavieng, New Ireland	Militz et al. (2015)	Y (survey data)	Y (Table A3)
	—	Reef Check (1998–2000, 2002, 2004, 2008–2009)	Y (survey data)	Y (Table A4)

Continued

Table A1 (Continued)   List of localities with giant clams (in alphabetical order)

Locality	Locality	Citations	Was data useful for review?	Was data extracted?
Philippines	Sulu Archipelago	Rosewater (1982)	Y	
	South-Central Philippines	Alcala (1986)	Y (survey data)	Y (Table A3)
	—	Alcala & Alcazar (1987)	Y (insufficient data)	
	Sulu Archipelago and Southern Palawan	Villanoy et al. (1988)	Y	
	—	Gomez & Alcala (1988)	Y (survey data)	see Juinio et al. (1989)
	—	Juinio et al. (1989)	Y (survey data)	Y (Table A3)
	—	Calumpong & Cadiz (1993)	Y (survey data)	Y (Table A3)
	—	Gomez et al. (2000)	Y (insufficient data)	
	—	Calumpong et al. (2002)	Y	
	Tubbataha Reefs Natural Park	Dolorosa & Schoppe (2005)	Y (survey data)	Y (Table A3)
	—	Gomez & Mingoa-Licuanan (2006)	Y (insufficient data)	
	Caniogan Marine Sanctuary, NW Philippines	Cabaitan et al. (2008)	Y (insufficient data)	
	Bolinao Reef System	Dizon et al. (2008)	Y (insufficient data)	
	Tubbataha Reefs Natural Park	Dolorosa (2010)	Y (survey data)	Y (Table A3)
	Tubbataha Reefs Natural Park	Dolorosa & Jontila (2012)	Y (survey data)	Y
	Island of Hadji Panglima Tahil, Sulu	Tabugo et al. (2013)	Y	
	Tubbataha Reefs Natural Park	Dolorosa et al. (2014)	Y	
	Sibulan, Negos, Philippines	Borsa et al. (2015)	Y (DNA)	
	Sabang Reef Fish Sanctuary, Honda Bay	Gonzales et al. (2014a)	Y (survey data)	Y (Table A3)
	Apulit Island, West Sulu Sea, Palawan	Gonzales et al. (2014a)	Y (survey data)	Y (Table A3)
	Apulit Island, Taytay Bay, Palawan	Gonzales et al. (2014b)	Y (survey data)	Y (Table A3)
	Tubbataha Reefs Natural Park	Dolorosa et al. (2015)	Y	
	Tubbataha Reefs Natural Park	Conales et al. (2015)	Y (survey data)	Y (Table A3)
	—	Reef Check (1997–2008, 2010–2014)	Y (survey data)	Y (Table A4)
Pitcairn Islands	—	Paulay (1989)	Y	
	Oeno Atoll	Irving & Dawson (2013)	Y (survey data)	Y (Table A3)
Republic of Kiribati	Fanning Atoll	Kay (1970)	Y	
	—	Taniera (1988)	Y	

Continued

Table A1 (Continued)   List of localities with giant clams (in alphabetical order)

Locality	Locality	Citations	Was data useful for review?	Was data extracted?
Republic of Kiribati	Central Gilbert Islands	Munro (1988)	Y (survey data)	Y (Table A3)
(Continued)	Caroline Atoll (formerly Gilbert Islands)	Kepler & Kepler (1994)	Y (survey data)	Y (Table A3)
	Gilbert Islands	Thomas (2001)	Y	
	Northern Line Islands	Sandin et al. (2008)	Y (survey data)	Y (Table A3)
	Millenium Atoll	Barott et al. (2010)	Y (survey data)	Y (Table A3)
	Northern Line Islands	Williams et al. (2013)	Y	
	Kiritimati, Northern Line Islands	Borsa et al. (2015)	Y	
	—	Thomas (2014)	Y (history)	
Samoa	—	Zann (1989)	Y	
	Upolu, Western Samoa	Zann (1991)	Y	
	Western Samoa	Tacconi & Tisdell (1992) Chapter 13	Y	
	—	Tisdell & Wittenberg (1992)	Y	
	—	South & Skelton (2000)	Y	
	—	Tiitii et al. (2014)	Y	
	—	Flickr	Y	
Saudi Arabia	Jeddah	Hughes (1977)	Y	
	Jeddah	Bodoy (1984)	Y (survey data)	Y (Table A3)
	—	PERSGA (2010)	Y (survey data)	Y (Table A3)
	—	Reef Check (1999,2008–2009)	Y (survey data)	Y (Table A4)
Saya de Malha Banks (currently administered by Mauritius)	—	Sirenko & Scarlato (1991)	Y	
Seychelles	Mahe	Taylor (1968)	Y	
	Seychelle Islands	Selin et al. (1992)	Y (survey data)	Y (Table A3)
	Aride Island Beach	Agombar et al. (2003)	Y (survey data)	Y (Table A3)
	Silhouette Island	Gerlach & Gerlach (2004)	Y	
	—	Reef Check (1997, 2001)	Y (survey data)	Y (Table A4)
Singapore	Singapore	Courtois de Vicose & Chou (1999)	Y (insufficient data)	
	Southern Islands	Guest et al. (2008)	Y (survey data)	Y (Table A3)
	—	Todd & Guest (2008)	Y (insufficient data)	
	—	Soo et al. (2010)	Y (insufficient data)	
	Southern Islands	Neo & Todd (2012a,b)	Y (survey data)	Y (Table A3)
	—	Neo et al. (2013)	Y (insufficient data)	
	—	Neo & Todd (2013)	Y (survey data)	Y (Table A3)

Continued

Table A1 (Continued) List of localities with giant clams (in alphabetical order)

Locality	Locality	Citations	Was data useful for review?	Was data extracted?
Solomon Islands	—	Govan et al. (1988)	Y	
	—	Skewes (1990)	Y (insufficient data)	
	—	Bell et al. (1997)	Y (insufficient data)	
	—	Bell (1999)	Y (insufficient data)	
	Arnavon Marine Conservation Area	Lovell et al. (2004)	Y	
	—	Ramohia (2006)	Y (survey data)	Y (Table A3)
	Bellona (Mungiki) Island	Thaman et al. (2011)	Y	
	—	Borsa et al. (2015)	Y (DNA)	
	—	Reef Check (2005–2012)	Y (survey data)	Y (Table A4)
Somalia	—	Sommer et al. (1996)	Y	
	—	Pilcher & Alsuhaibany (2000)	Y	
South Africa	—	Reef Check (2000–2002, 2005)	Y (survey data)	Y (Table A4)
South China Sea	Xisha Islands (Paracel Islands)	Zhuang (1978)	Y	
	Xisha (Paracel Islands) and Nansha Islands (Spratly Islands)	Bernard et al. (1993)	Y	
	Xisha Islands (Paracel Islands)	Pan & Lan (1998)	Y	
	Pulau Layang Layang (Swallow Reef) (Malaysia)	Sahari et al. (2002)	Y (survey data)	Y (Table A3)
	North Spratly Islands	Van Long et al. (2008)	Y (survey data)	Y (Table A3)
	North Danger Reef and Jackson Atoll	Calumpong et al. (2008)	Y	
	North Danger Reef and Trident Shoal	Lasola & Hoang (2008)	Y (survey data)	Y (Table A3)
	North Danger Reef and Jackson Atoll	Calumpong & Macansantos (2008)	Y (survey data)	Y (Table A3)
	Dongsha Atoll (Pratas Islands) (Taiwan)	Borsa et al. (2015)	Y	
	Taiping Island (Itu Aba Island, Spratly group)	A Frontier in the South China Sea: Biodiversity of Taiping Island, Nansha Islands (2014)	Y	
Sri Lanka	—	Reef Check (2003)	Y (survey data)	Y (Table A4)
Sudan	Harvey reef, Baraja reef, Lighthouse reef, Mersa Towartit	Taylor & Reid (1984)	Y	
	Sanganeb Atoll	CBD Report (N.D.)	Y	
	—	Reef Check (2004, 2009)	Y (survey data)	Y (Table A4)

Continued

Table A1 (Continued)   List of localities with giant clams (in alphabetical order)

Locality	Locality	Citations	Was data useful for review?	Was data extracted?
Taiwan	—	Wu (1999)	Y	
	Northern and Southern Taiwan, Orchid Island, Green Island, Hsiaoliuchiu, Penghu	Tang (2005)	Y	
	Gueishan Island	Huang et al. (2013)	Y	
	Northern and Southern Taiwan, Orchid Island, Green Island, Hsiaoliuchiu, Penghu	Su et al. (2014)	Y	
	—	Reef Check (1998, 2008–2010)	Y (survey data)	Y (Table A4)
Tanzania	Zanzibar	Gossling et al. (2004)	Y	
	Chumbe Island	Daniels (2004)	Y (survey data)	Y (Table A3)
	Kilwa Island, southern Swahili coast	Nakamura (2013)	Y	
	—	Reef Check (1997–1998, 2003–2008)	Y (survey data)	Y (Table A4)
Thailand	Lee-Pae Island, Andaman Seas	Chantrapornsyl et al. (1996)	Y (survey data)	Y (Table A3)
	Surin Islands	Kittiwattanawong (1997)	Y	
	Surin Islands, Andaman Sea and Racha Yai Island, Phuket	Kittiwattanawong (2001)	Y	
	Andaman Seas, Gulf of Thailand	Kittiwattanawong et al. (2001)	Y	
	Surin Islands	Koh et al. (2003)	Y (survey data)	Y (Table A3)
	Surin Islands	Loh et al. (2004)	Y (survey data)	Y (Table A3)
	Mannai Island, Rayong province	Junchompoo et al. (2013)	Y (survey data)	Y (Table A3)
	—	Reef Check (1998–2001, 2003–2015)	Y (survey data)	Y (Table A4)
Tokelau	—	Braley (1989)	Y (survey data)	Y (Table A3)
	—	Tisdell & Wittenberg (1992)	Y	
	—	Vieux et al. (2004)	Y	
Tonga	—	Langi & Hesitoni 'Aloua (1988)	Y (survey data)	Y (Table A3)
	Ha'apai, Vava'u Islands	Lucas et al. (1991)	Y	
	—	Tacconi & Tisdell (1992) Chapter 13	Y	
	—	Tisdell & Wittenberg (1992)	Y	
	—	Chesher (1993): p. 31	Y (survey data)	Y (Table A3)
	Ha'apai, Vava'u Islands	Ledua et al. (1993)	Y (survey data)	Y (Table A3)
	—	Sone & Loto'ahea (1995)	Y	

Continued

Table A1 (Continued)   List of localities with giant clams (in alphabetical order)

Locality	Locality	Citations	Was data useful for review?	Was data extracted?
Tonga (Continued)	Tongatapu Island	Tu'avao et al. (1995)	Y (survey data)	Y (Table A3)
	—	Salvat (2000)	Y	
	—	Reef Check (2002, 2013)	Y (survey data)	Y (Table A4)
Tuvalu	Nukufetau, Nukulaelae, Funafuti	Braley (1988)	Y (survey data)	Y (Table A3)
	Nanumea, Nui	Langi (1990)	Y (survey data)	Y (Table A3)
	—	Lovell et al. (2004)	Y	
	—	Sauni et al. (2008)	Y (survey data)	Y (Table A3)
	—	Job & Ceccarelli (2012)	Y (survey data)	Y (Table A3)
	Funafuti	Siaosi et al. (2012)	Y (survey data)	Y (Table A3)
United States Minor Outlying Islands	Palmyra Atoll	Flickr	Y	
	Kingman Reef National Wildlife Refuge	Flickr	Y	
Vanuatu	—	Zann & Ayling (1988)	Y (survey data)	Y (Table A3)
	—	Bell & Amos (1993)	Y (survey data)	Y [same as Zann & Ayling (1988)]
	—	Lovell et al. (2004)	Y	
	—	Nimoho et al. (2013)	Y (survey data)	Y (Table A3)
	Efate	Borsa et al. (2015)	Y	
	—	Reef Check (2004, 2008, 2011–2012)	Y (survey data)	Y (Table A4)
Viet Nam	An Thoi Archipelago	Latypov (2000)	Y	
	Central Viet Nam	Latypov (2001)	Y	
	Mju and Moon Islands	Latypov (2006)	Y (survey data)	Y (Table A3)
	Con Dao Islands	Selin & Latypov (2011)	Y (survey data)	Y (Table A3)
	Gulf of Siam and South Viet Nam	Latypov & Selin (2011)	Y (survey data)	Y (Table A3)
	Ku Lao Cham Islands	Latypov & Selin (2012a)	Y	
	Cam Ranh Bay	Latypov & Selin (2012b)	Y (survey data)	Y (Table A3)
	—	Latypov (2013)	Y (survey data)	Y (Table A3)
	Khanh Hoa Province	Latypov & Selin (2013)	Y (survey data)	Y
	—	Long & Vo (2013)	Y (survey data)	Cannot be easily retrieved
	—	Reef Check (1998–2006)	Y (survey data)	Y (Table A4)
Wallis and Futuna Islands	Wallis Island	Pollock (1992)	Y	
	Wallis Island	Borsa et al. (2015)	Y	
Yemen	—	PERSGA (2010)	Y (survey data)	Y (Table A3)
	Kamaran Island	Huber & Eschner (2011)	Y	
	—	Reef Check (1999, 2001, 2008)	Y (survey data)	Y (Table A4)

Note: Full reference list in Appendix B.

Table A2 Checklist of giant clam species

Recorded localities	Species	Localities	Reference(s)
Red Sea			
Djibouti	TM	Djibouti	PERSGA (2010)
	TS	Djibouti	PERSGA (2010)
Egypt	TM	Egypt	Wells et al. (1983); PERSGA (2010); Mekawy & Madkour (2012); Mekawy (2014)
		Coral carpets, Northern Bay of Safaga	Zuschin & Pillar (1997)
		Gulf of Aqaba	Kilada (1998); Zuschin & Stachowitsch (2007)
		Abu Sauatir, Northern Red Sea	Ullmann (2013)
	TS	Egypt	Wells et al. (1983); PERSA (2010)
		Gulf of Aqaba	Kilada (1998)
	TSI (previously TCO)	Sinai coast, western Gulf of Aqaba; Northern Red Sea, Egyptian mainland	Richter et al. (2008); Huber & Eschner (2011)
Eritrea	*Tridacna* spp.	No data	Reef Check
Israel	TM	Eilat (southernmost of Israel, Red Sea)	Flickr Eduardo Pszczol (2006)
	TS	Eilat (southernmost of Israel, Red Sea)	Flickr Eduardo Pszczol (2005)
	TSI (previously TCO)	Eilat (southernmost of Israel, Red Sea)	Flickr Eduardo Pszczol (2006)
Jordan	TM	Jordanian coast of Gulf of Aqaba	Roa-Quiaoit (2005); PERSGA (2010)
	TS	Jordanian coast of Gulf of Aqaba	Roa-Quiaoit (2005); PERSGA (2010)
	TSI (previously TCO)	Jordanian Red Sea coast	Richter et al. (2008); Huber & Eschner (2011)
	Tridacna spp.	Jordanian coast of Gulf of Aqaba	Al-Horani et al. (2006)
Saudi Arabia	TM	Jeddah	Hughes (1977); Bodoy (1984)
		Saudi Arabia	Wells et al. (1983); Munro (1989); PERSGA (2010)
	TS	Jeddah	Hughes (1977)
		Saudi Arabia	Wells et al. (1983); PERSGA (2010)
	TSI (previously TCO)	Aqaba, Tabouk	Flickr Magnus Franklin (2010)
Sudan	TM	Harvey reef, Towartit	Taylor & Reid (1984)
		Baraja (patch reef)	Taylor & Reid (1984)
		Sudan	PERSGA (2010)
	TS	Harvey reef, Towartit	Taylor & Reid (1984)
		Sudan	PERSGA (2010)
Yemen	TM	Yemen	PERSGA (2010)
	TS	Yemen	PERSGA (2010)
	TSI (previously TCO)	Kamaran Island	Huber & Eschner (2011)

Continued

Table A2 (Continued)   Checklist of giant clam species

Recorded localities	Species	Localities	Reference(s)
South-East Africa			
Comoros	TM	Itsandra Plongee	Flickr Eric de Troyer
	TS	Itsandra Plongee	Flickr Eric de Troyer
	Tridacna spp.	Nioumachouoi site and Ouenefou reef	Wilkinson (2000)
Kenya	TG?	Kenya	Accordi et al. (2010)
	TM	Kenya	Evans et al. (1977); Wells et al. (1983); Anam & Mostarda (2012)
	TS	Kenya	Wells et al. (1983); Anam & Mostarda (2012)
Madagascar	TG?	Andavadoaka region	Hopkins (2009)
	TM	Madagascar	Wells et al. (1983); Wells (2003), C. Gough (BlueVentures), pers. comm.
	TS	Madagascar	Wells et al. (1983); Wells (2003), C. Gough (BlueVentures), pers. comm.
		Northern Madagascar	Harding & Randriamanantsoa (2008)
		Southwestern Madagascar	Barnes & Rawlinson (2009)
	Tridacna spp.	Andavadoaka region	Harding et al. (2006)
	Giant clams	Andavadoaka region	Nadon et al. (2007)
Mauritius	TG?	Mauritius (Text: "specimen 92 cm long")	Michel et al. (1985)
	TM	Mauritius	Wells et al. (1983); Michel et al. (1985)
		Rodrigues Island	Oliver et al. (2004)
	TS	Mauritius	Wells et al. (1983); Michel et al. (1985)
Mayotte	TM	Mayotte	Blog: Jana around the world; S. Andréfouët, pers. obs.
	TS	Mayotte	Blog: Jana around the world; S. Andréfouët, pers. obs.
Mozambique	TM	Mozambique	Wells et al. (1983)
		Azura Benguerra Island	Unknown
	TS	Mozambique	Wells et al. (1983)
		Quirimba Archipelago	Barnes et al. (1998)
		Paindane Coral Garden	P. Southwood, pers. comm. (2009)
	TSI (previously TCO)	Bazurato Island	Flickr Mark van Malsen (2008)
		Creche, Southern Mozambique	C. Lindeque, pers. comm. (2012)
		Inhambane Province	Flickr Vera & Gordon (2012)
		Mozambique	ReefBuilders.com (2015), N. Helgason, pers. comm. (2015)
La Réunion	TS	La Réunion	Flickr Cedric Peneau (2014)
Seychelles	TC	Coetivy Island	Selin et al. (1992)
	TM	Mahe	Taylor (1968); Selin et al. (1992)
		Seychelles	Wells et al. (1983)
		Aride Island Beach	Agombar et al. (2003)
		Silhouette Island	Gerlach & Gerlach (2004)
	TS	Mahe	Taylor (1968)
		Seychelles	Wells et al. (1983)

Continued

Table A2 (Continued)   Checklist of giant clam species

Recorded localities	Species	Localities	Reference(s)
Seychelles		Aride Island Beach	Agombar et al. (2003)
(Continued)		Silhouette Island	Gerlach & Gerlach (2004)
Somalia	TM	Somalia	Sommer et al. (1996)
	TS	Somalia	Sommer et al. (1996)
	Tridacna spp.	Somalia	Pilcher & Alsuhaibany (2000)
South Africa	TM	South Africa	Wells et al. (1983); Munro (1989)
	TS	South Africa	Munro (1989)
Tanzania	TM	Kilwa Island	Nakamura (2013)
		Chumbe Island	Daniels (2004)
	TS	Zanzibar	Gossling et al. (2004)
		Chumbe Island	Daniels (2004)
Indian Ocean			
Cargados Carajos	TLZ	Cargados Carajos Archipelago	Monsecour (2016)
Archipelago	TM	Cargados Carajos Archipelago	Monsecour (2016)
	TS	Cargados Carajos Archipelago	Monsecour (2016)
Christmas Island	TC	Christmas Island	Sources found in Tan & Low (2014)
	TD	Christmas Island	Sources found in Tan & Low (2014)
	TG	Christmas Island (EXTINCT)	Andrews et al. (1900); Tomlin (1934); Wells & Slack-Smith (2000); Hourston (2010)
	TM	Christmas Island	Tomlin (1934); Wells & Slack-Smith (2000); Hourston (2010)
	TNO	Christmas Island	Neo & Low (2017)
	TS	Christmas Island	Wells & Slack-Smith (2000); Hourston (2010)
Cocos (Keeling)	TC	Cocos (Keeling) Islands	Abbott (1950)
Islands (Australia	TD	Cocos (Keeling) Islands	Maes (1967); Wells et al. (1983); Munro (1989); Wells (1994); Hourston (2010)
Territory)	TG	Cocos (Keeling) Islands (EXTINCT)	Wells (1994); Hender et al. (2001); Hourston (2010)
	TM	Cocos (Keeling) Islands	Maes (1967); Wells (1994); Australian Government (2005); Hourston (2010); Bellchambers & Evans (2013); Evans et al. (2016)
	TS	Cocos (Keeling) Islands (EXTINCT)	Gibson-Hill (1946)
British Indian	TM	Chagos Archipelago	Wells et al. (1983); Sheppard (1984); Chagos Conservation Trust (2014)
Ocean Territory	TS	Chagos Archipelago	Wells et al. (1983); Sheppard (1984); Chagos Conservation Trust (2014)
India	HH	Andaman Islands	Rosewater (1965)
		Nicobar Islands	Rosewater (1965)
	TC	Kavaratti	Namboodiri & Sivadas (1979)
		Andaman Islands	Ramadoss (1983)
		Nicobar Islands	Ramadoss (1983)
	TG	Andaman and Nicobar Islands	Apte et al. (2010)

Continued

Table A2 (Continued) Checklist of giant clam species

Recorded localities	Species	Localities	Reference(s)
India *(Continued)*	TM	Andaman Islands	Ramadoss (1983); Wells et al. (1983); Munro (1989)
		Nicobar Islands	Ramadoss (1983); Munro (1989)
		Laccadives	Munro (1989)
		Lakshadweep Archipelago	George et al. (1986); Apte & Dutta (2010); Apte et al. (2010)
	TS	Andaman Islands	Ramadoss (1983); Munro (1989)
		Nicobar Islands	Ramadoss (1983); Munro (1989)
		Laccadives	Munro (1989)
Maldives	TM	Maldives	Wells et al. (1983)
		Central and northern atolls	Basker (1991)
		Baa Atoll	Andréfouët et al. (2012)
	TS	Maldives	Wells et al. (1983)
		Central and northern atolls	Basker (1991)
		Baa Atoll	Andréfouët et al. (2012)
Saya de Malha Bank	TR	Saya de Malha Bank	Sirenko & Scarlato (1991)
Sri Lanka	TM	Sri Lanka	Wells et al. (1983); Munro (1989)
	TS	Sri Lanka	Munro (1989)
East Asia			
China	TC	Sanya	Qi (2004)
		Xincun	Qi (2004)
	TM	China	Wells et al. (1983)
	TS	Hainan Island (Shalao, Xiaodonghai, Xizhou Islet, Dongzhou Islet, Yezhu Island)	Fiege et al. (1994)
		Sanya	Qi (2004)
		Xincun	Qi (2004)
	Tridacna spp.	Lunya Bay, Hainan Island	Hutchings & Wu (1987)
		Sanya waters	Tadashi et al. (2008)
Hong Kong	TM	Mirs Bay	Morton & Morton (1983)
Japan	HH	Ryukyu	Hirase (1954); Wells et al. (1983)
		Okinawa	Bernard et al. (1993); Okada (1997)
	TC	Amami-oshima	Hirase (1954); Miklos Kazmer, pers. comm. (2015)
		Ishigaki	Kanno et al. (1976)
		Ryukyu	Wells et al. (1983); Munro (1989)
		Okinawa	Bernard et al. (1993); Okada (1997)
		Sesoko Island, Okinawa	Flickr Jin-Yao Ong (2013)
	TG	Yaeyama, Ryukyu	Hirase (1954); Wells et al. (1983)
	TM	Amami-oshima	Hirase (1954)
		Ishigaki	Kanno et al. (1976)
		Japan	Wells et al. (1983)
		Ryukyu (EXTINCT)	Munro (1989)
		Okinawa	Bernard et al. (1993); Okada (1997)

Continued

Table A2 (Continued)   Checklist of giant clam species

Recorded localities	Species	Localities	Reference(s)
Japan *(Continued)*		Ogasawara National Park (Chichi and Haha Islands)	Fujiwara et al. (2000)
		Tokashiki-son, Okinawa	Flickr Nemo's great uncle (2009)
		Clothesline, Okinawa	Flickr chino1138 (2012)
	TNO	Okinawa and Ishigaki Islands	Kubo & Iwai (2007)
	TS	Ryukyu	Hirase (1954); Munro (1989)
		Ishigaki	Kanno et al. (1976)
		Japan	Wells et al. (1983)
		Okinawa	Bernard et al. (1993); Okada (1997)
Taiwan	HH	EXTINCT	Bernard et al. (1993); Munro (1989)
		Hengchun, Lanyu	Wu (1999)
	TC	Taiwan	Bernard et al. (1993)
		Hengchun, Lanyu	Wu (1999)
		Gueishan Island	Huang et al. (2013)
	TD	EXTINCT	Bernard et al. (1993)
		Hengchun, Lanyu	Wu (1999)
	TG	EXTINCT	Bernard et al. (1993); Munro (1989)
		Hengchun, Lanyu	Wu (1999)
		Gueishan Island	Huang et al. (2013)
	TM	Taiwan	Wells et al. (1983); Bernard et al. (1993)
		Taipei, Suao, Daikanko, Kaohsiung, Shaoliuchiu, Penghu, Hengchun, Lanyu	Wu (1999)
		Gueishan Island	Huang et al. (2013)
		Lamay Island	Flickr Dennis Wong (2013)
		Lanyu, Orchid Island	Flickr Blowing Puffer Fish (2015)
	TNO	Northern and Southern Taiwan, Orchid Island, Green Island, Hsiaoliuchiu, Penghu	Tang (2005); Su et al. (2014)
	TS	Taiwan	Bernard et al. (1993)
		Hengchun, Lanyu	Wu (1999)
		Green Island	Flickr Michael Huang (2007); Flickr rcmlee99 (2015)
South China Sea	HH	Xisha Islands (Paracel Islands)	Zhuang (1978); Pan & Lan (1998); Qi (2004)
		Xisha (Paracel Islands) and Nansha Islands (Spratly Islands)	Bernard et al. (1993); Liu (2013)
		Pulau Layang Layang (Swallow Reef)	Sahari et al. (2002)
		North Danger Reef (Spratly Islands)	Calumpong et al. (2008); Calumpong & Macansantos (2008)
		Jackson Atoll (Spratly Islands)	Calumpong et al. (2008)
		Taiping Island (Itu Aba Island)	A Frontier in the SCS (2014)
	TC	Xisha Islands (Paracel Islands)	Zhuang (1978); Qi (2004)
		Nansha Islands (Spratly Islands)	Bernard et al. (1993); Liu (2013)
		Pulau Layang Layang (Swallow Reef)	Sahari et al. (2002)

Continued

Table A2 (Continued) Checklist of giant clam species

Recorded localities	Species	Localities	Reference(s)
South China Sea		North Spratly Islands	Van Long et al. (2008)
(Continued)		North Danger Reef (Spratly Islands)	Calumpong et al. (2008); Lasola & Hoang (2008); Calumpong & Macansantos (2008)
		Jackson Atoll (Spratly Islands)	Calumpong et al. (2008); Calumpong & Macansantos (2008)
		Trident Shoal (Spratly Islands)	Lasola & Hoang (2008)
		Taiping Island (Itu Aba Island)	A Frontier in the SCS (2014)
	TD	Xisha Islands (Paracel Islands)	Zhuang (1978); Qi (2004)
		Xisha (Paracel Islands) and Nansha Islands (Spratly Islands)	Bernard et al. (1993); Liu (2013)
	TG	Xisha Islands (Paracel Islands)	Zhuang (1978); Qi (2004)
		Xisha (Paracel Islands) and Nansha Islands (Spratly Islands)	Bernard et al. (1993); Liu (2013)
		Pulau Layang Layang (Swallow Reef)	Sahari et al. (2002)
	TM	Xisha Islands (Paracel Islands)	Zhuang (1978); Qi (2004)
		Xisha (Paracel Islands) and Nansha Islands (Spratly Islands)	Bernard et al. (1993); Liu (2013)
		Pulau Layang Layang (Swallow Reef)	Sahari et al. (2002)
		North Danger Reef (Spratly Islands)	Calumpong et al. (2008); Calumpong & Macansantos (2008)
		Jackson Atoll (Spratly Islands)	Calumpong et al. (2008); Calumpong & Macansantos (2008)
		Taiping Island (Itu Aba Island)	A Frontier in the SCS (2014)
	TNO	Dongsha Atoll (Pratas Islands)	Borsa et al. (2015)
	TS	Xisha Islands (Paracel Islands)	Zhuang (1978); Qi (2004)
		Xisha (Paracel Islands) and Nansha Islands (Spratly Islands)	Bernard et al. (1993); Liu (2013)
		Pulau Layang Layang (Swallow Reef)	Sahari et al. (2002)
		North Spratly Islands	Van Long et al. (2008)
		North Danger Reef (Spratly Islands)	Calumpong et al. (2008); Lasola & Hoang (2008); Calumpong & Macansantos (2008)
		Jackson Atoll (Spratly Islands)	Calumpong et al. (2008); Calumpong & Macansantos (2008)
		Trident Shoal (Spratly Islands)	Lasola & Hoang (2008)
		Taiping Island (Itu Aba Island)	A Frontier in the SCS (2014)
South-East Asia			
Brunei	*Tridacna* spp.	No data	Reef Check
Cambodia	TG?	Song Saa Private Island, Koh Rong Archipelago	Savage et al. (2013)
		Koh Rong and Koh Koun, Koh Rong Archipelago	Thorne et al. (2015)

Continued

Table A2 (Continued)   Checklist of giant clam species

Recorded localities	Species	Localities	Reference(s)
Cambodia *(Continued)*	TC	Song Saa Private Island, Koh Rong Archipelago	J.M. Savage, pers. comm.
	TS	Song Saa Private Island, Koh Rong Archipelago	J. Wong, pers. comm.
	Tridacna spp.	Cambodia	Chou et al. (2002)
		Koh Kong	Vibol (N.D.); Kim et al. (2004)
		Koh Sdach	Vibol (N.D.); Kim et al. (2004)
		Koh Rong	Vibol (N.D.); Kim et al. (2004); Van Bochove et al. (2011)
		Koh Tang	Vibol (N.D.); Kim et al. (2004)
East Timor	TC	Dili Rock	Flickr Nick Hobgood (2006)
	TG	Dili Rock	Flickr Nick Hobgood (2006)
	TNO	Dili Rock	Flickr Nick Hobgood (2006)
Indonesia	HH	Indonesia	Wells et al. (1983); Pasaribu (1988); Munro (1989)
		Karimun Jawa, Central Indonesia	Brown & Muskanofola (1985)
		Genting Island	Pringgenies et al. (1995)
		Gulf of Tomini, Sulawesi	Wells (2001)
		Rajah Ampat Islands, Papua Province	Wells (2002); Flickr Raja Ampat Biodiversity (2013)
		Pari Island	Eliata et al. (2003); Panggabean (2007)
		Kei Kecil waters, Southeast Maluku	Kusnadi et al. (2008); Hernawan (2010)
		Savu Sea, East Nusa Tenggara Province	Naguit et al. (2012)
	HP	Northeastern Indonesia	Wells et al. (1983); Pasaribu (1988); Munro (1989)
		Seruni Island	Pringgenies et al. (1995)
		Gulf of Tomini, Sulawesi	Wells (2001)
		Rajah Ampat Islands, Papua Province	Wells (2002)
	TC	Indonesia	Wells et al. (1983); Pasaribu (1988); Munro (1989)
		Karimun Jawa, Central Indonesia	Brown & Muskanofola (1985)
		Genting Island	Pringgenies et al. (1995)
		Seruni Island	Pringgenies et al. (1995)
		Sambangan Island	Pringgenies et al. (1995)
		Gulf of Tomini, Sulawesi	Wells (2001)
		Rajah Ampat Islands, Papua Province	Wells (2002)
		Pari Island	Eliata et al. (2003)
		Seribu Islands	Yusuf et al. (2009)
		Manado waters	Yusuf et al. (2009)
		Kei Kecil waters, Southeast Maluku	Kusnadi et al. (2008); Hernawan (2010)
		Savu Sea, East Nusa Tenggara Province	Naguit et al. (2012)

Continued

Table A2 (Continued)   Checklist of giant clam species

Recorded localities	Species	Localities	Reference(s)
Indonesia	TD	Irian Jaya	Wells et al. (1983)
(Continued)		Indonesia	Wells et al. (1983); Pasaribu (1988); Munro (1989)
		Karimun Jawa, Central Indonesia	Brown & Muskanofola (1985)
		Rajah Ampat Islands, Papua Province	Wells (2002)
		Kei Kecil waters, Southeast Maluku	Hernawan (2010)
		Komodo	Flickr yudas_net (2009)
	TG	Indonesia	Wells et al. (1983); Pasaribu (1988); Munro (1989)
		Karimun Jawa, Central Indonesia	Brown & Muskanofola (1985)
		Gulf of Tomini, Sulawesi	Wells (2001)
		Rajah Ampat Islands, Papua Province	Wells (2002)
		Bunaken, Manado waters	Yusuf et al. (2009); Flickr Matt Kieffer (2010)
		Kei Kecil waters, Southeast Maluku	Kusnadi et al. (2008); Hernawan (2010)
		Kri Island, Irian Jaya	Flickr Eric Cheng (2004)
		Current City, Komodo	Flickr Maximilian Hand (2008)
		West Papua	Flickr Paul Cowell (2011)
	TM	Indonesia	Wells et al. (1983); Pasaribu (1988); Munro (1989)
		Karimun Jawa, Central Indonesia	Brown & Muskanofola (1985)
		Genting Island	Pringgenies et al. (1995)
		Seruni Island	Pringgenies et al. (1995)
		Sambangan Island	Pringgenies et al. (1995)
		Gulf of Tomini, Sulawesi	Wells (2001)
		Rajah Ampat Islands, Papua Province	Wells (2002)
		Pari Island	Eliata et al. (2003)
		Anambas and Natuna Islands	Tan & Kastoro (2004); Flickr Fauzan Rizki (2015)
		Seribu Islands	Yusuf et al. (2009)
		Manado waters	Yusuf et al. (2009)
		Kei Kecil waters, Southeast Maluku	Hernawan (2010)
		Savu Sea, East Nusa Tenggara Province	Naguit et al. (2012)
		Komodo National Park, East Nusa Tenggara	Flickr Nick Hobgood (2006)
	TNO	Bunaken and Alor Archipelago	Borsa et al. (2015)
		Doi Island, Molucca Sea	Borsa et al. (2015)
	TS	Indonesia	Wells et al. (1983); Pasaribu (1988); Munro (1989)
		Karimun Jawa, Central Indonesia	Brown & Muskanofola (1985)
		Gulf of Tomini, Sulawesi	Wells (2001)

Continued

Table A2 (Continued)   Checklist of giant clam species

Recorded localities	Species	Localities	Reference(s)
Indonesia *(Continued)*		Rajah Ampat Islands, Papua Province	Wells (2002)
		Pari Island	Eliata et al. (2003)
		Seribu Islands	Yusuf et al. (2009)
		Manado waters	Yusuf et al. (2009)
		Kei Kecil waters, Southeast Maluku	Kusnadi et al. (2008); Hernawan (2010)
		Savu Sea, East Nusa Tenggara Province	Naguit et al. (2012)
		Wakatobi	Flickr Richard Johnson (2007)
		Aceh	Flickr iderq_shai (2010)
		Komodo	Flickr Brandon (2011)
		Cenderawasih Bay	Flickr lcn2012a (2012)
		West Papua	Flickr Sailendivers (2012)
Malaysia	HP	Sabah, East Malaysia	Tan & Zulfigar (2001, 2003)
	HH	North Borneo	Wells et al. (1983)
		Malaysia	Munro (1989); Tan & Zulfigar (2001, 2003)
		Johore Islands	Zulfigar & Tan (2000)
	TC	Western coast of the Malay Peninsula	Wells et al. (1983)
		North Borneo	Wells et al. (1983)
		Malaysia	Munro (1989); Tan & Zulfigar (2001, 2003)
		Pulau Redang (Terengganu)	Mohamed-Pauzi et al. (1994)
		Pulau Tioman (Pahang)	Tan et al. (1998)
		Johore Islands	Zulfigar & Tan (2000)
	TD	Sabah, East Malaysia	Tan & Zulfigar (2003)
	TG	Malaysia	Munro (1989); Tan & Zulfigar (2001, 2003)
		Pulau Redang (Terengganu)	Mohamed-Pauzi et al. (1994)
		Pulau Tioman (Pahang)	Tan et al. (1998)
	TM	Malaysia	Wells et al. (1983); Munro (1989); Tan & Zulfigar (2001, 2003)
		North Borneo	Wells et al. (1983)
		Pulau Redang (Terengganu)	Mohamed-Pauzi et al. (1994)
		Pulau Tioman (Pahang)	Tan et al. (1998)
		Johore Islands	Zulfigar & Tan (2000)
	TS	Malaysia	Wells et al. (1983); Munro (1989); Tan & Zulfigar (2001, 2003)
		North Borneo	Wells et al. (1983)
		Pulau Redang (Terengganu)	Mohamcd-Pauzi et al. (1994)
		Pulau Tioman (Pahang)	Tan et al. (1998)
		Johore Islands	Zulfigar & Tan (2000)
	Tridacna spp.	Tun Sakaran Marine Park, East Sabah	Montagne et al. (2013)

Continued

Table A2 (Continued)   Checklist of giant clam species

Recorded localities	Species	Localities	Reference(s)
Myanmar (Burma)	HH	Burma	Munro (1989); Wells (1997)
	TG	Burma	Munro (1989)
	TM	Burma	Munro (1989)
	TS	Burma	Munro (1989)
Philippines	HH	Philippines	Wells et al. (1983); Gomez & Alcala (1988); Juinio et al. (1989); Munro (1989)
		Central Visayas	Alcala (1986)
		Palawan	Alcala (1986)
		Cagayan	Alcala (1986)
		Sulu Archipelago and Southern Palawan	Villanoy et al. (1988)
		Tubbataha	Dolorosa & Schoppe (2005); Dolorosa (2010); Dolorosa et al. (2015)
		Hadji Panglima Tahil, Sulu	Tabugo et al. (2013)
	HP	Sulu Archipelago, Southern Philippines	Rosewater (1982); Wells et al. (1983); Gomez & Alcala (1988); Juinio et al. (1989); Munro (1989)
		Cagayan	Alcala (1986)
		Tubbataha	Calumpong & Cadiz (1993); Dolorosa (2010); Dolorosa & Jontila (2012); Dolorosa et al. (2014, 2015)
		Sulu Archipelago and Southern Palawan	Villanoy et al. (1988)
	TC	Philippines	Wells et al. (1983); Gomez & Alcala (1988); Juinio et al. (1989); Munro (1989)
		Central Visayas	Alcala (1986)
		Western Visayas	Alcala (1986)
		Palawan	Alcala (1986)
		Cagayan	Alcala (1986); Calumpong & Cadiz (1993)
		Tubbataha	Calumpong & Cadiz (1993); Dolorosa & Schoppe (2005); Dolorosa (2010); Dolorosa & Jontila (2012); Gonzales et al. (2014b); Dolorosa et al. (2015); Conales et al. (2015)
		Sumilon Island	Calumpong & Cadiz (1993)
		Balicasag Island	Calumpong & Cadiz (1993)
		Pamilacan Island	Calumpong & Cadiz (1993)
		Bolisong, Negros Oriental	Calumpong et al. (2002)
	TD	Philippines	Wells et al. (1983); Gomez & Alcala (1988); Juinio et al. (1989); Munro (1989)
		Palawan	Alcala (1986)
		Cagayan	Alcala (1986)

Continued

Table A2 (Continued)   Checklist of giant clam species

Recorded localities	Species	Localities	Reference(s)
Philippines		Tubbataha	Dolorosa et al. (2010, 2015)
(Continued)		Sabang Reef Fish Sanctuary, Honda Bay, Palawan	Gonzales et al. (2014a)
	TG	Philippines	Wells et al. (1983); Gomez & Alcala (1988); Juinio et al. (1989); Munro (1989)
		Palawan	Alcala (1986)
		Cagayan	Alcala (1986)
		Sulu Archipelago and Southern Palawan	Villanoy et al. (1988)
		Sabang Reef Fish Sanctuary, Honda Bay, Palawan	Gonzales et al. (2014a)
		Tubbataha	Dolorosa et al. (2015)
	TM	Philippines	Wells et al. (1983); Gomez & Alcala (1988); Junio et al. (1989); Munro (1989)
		Central Visayas	Alcala (1986)
		Western Visayas	Alcala (1986)
		Palawan	Alcala (1986)
		Cagayan	Alcala (1986); Calumpong & Cadiz (1993)
		Tubbataha	Calumpong & Cadiz (1993); Dolorosa & Schoppe (2005); Dolorosa (2010); Dolorosa & Jontila (2012); Dolorosa et al. (2015)
		Sumilon Island	Calumpong & Cadiz (1993)
		Apo Island	Calumpong et al. (2002)
	TNO	Sibulan, Negros (Lizano & Santos, 2014)	Borsa et al. (2015)
	TS	Philippines	Wells et al. (1983); Gomez & Alcala (1988); Juinio et al. (1989); Munro (1989)
		Central Visayas	Alcala (1986)
		Western Visayas	Alcala (1986)
		Palawan	Alcala (1986)
		Cagayan	Alcala (1986)
		Sulu Archipelago and Southern Palawan	Villanoy et al. (1988)
		Tubbataha	Calumpong & Cadiz (1993); Dolorosa & Schoppe (2005); Dolorosa (2010); Dolorosa & Jontila (2012); Dolorosa et al. (2015)
		Bolisong, Negros Oriental	Calumpong et al. (2002)
		Apo Island	Calumpong et al. (2002)
		Hadji Panglima Tahil, Sulu	Tabugo et al. (2013)
		Sabang Reef Fish Sanctuary, Honda Bay, Palawan	Gonzales et al. (2014a)

Continued

Table A2 (Continued)   Checklist of giant clam species

Recorded localities	Species	Localities	Reference(s)
Singapore	HH	Singapore	Wells et al. (1983); Munro (1989); Neo & Todd (2012a,b, 2013)
	TC	Singapore	Wells et al. (1983); Munro (1989); Guest et al. (2008); Neo & Todd (2012a,b, 2013)
	TG	Singapore	Neo & Todd (2012a,b, 2013)
	TM	Singapore	Munro (1989); Guest et al. (2008); Neo & Todd (2012a,b, 2013)
	TS	Singapore	Munro (1989); Guest et al. (2008); Neo & Todd (2012a,b, 2013)
Thailand	TC	Thailand	Wells et al. (1983); Munro (1989)
		Lee-Pae Island	Chantrapornsyl et al. (1996)
		Surin Islands	Koh et al. (2003); Loh et al. (2004)
		Mannai Island, Rayong Province	Junchompoo et al. (2013)
	TG	Thailand	Munro (1989)
		EXTINCT	Kittiwattanawong (2001)
	TM	Thailand	Wells et al. (1983); Munro (1989)
		Lee-Pae Island	Chantrapornsyl et al. (1996)
		Surin Islands	Kittiwattanawong (1997); Koh et al. (2003)
		Phuket Islands	Kittiwattanawong (1997)
		Adang Rawii Islands	Kittiwattanawong (1997)
	TS	Thailand	Wells et al. (1983); Munro (1989)
		Lee-Pae Island	Chantrapornsyl et al. (1996)
		Surin Islands	Kittiwattanawong et al. (2001); Koh et al. (2003)
		Gulf of Thailand	Kittiwattanawong et al. (2001)
		Mannai Island, Rayong Province	Junchompoo et al. (2013)
Viet Nam	TC	An Thoi Archipelago, South China Sea	Latypov (2000)
		Mju Island	Latypov (2006)
		Hon Bay Canh Island and Hon Cau Island, Con Dao	Latypov & Selin (2011)
		Con Dao Archipelago	Selin & Latypov (2011)
		Khanh Hoa Province	Latypov & Selin (2013)
		Giang Bo Reef	Latypov (2013)
		Re Island	Latypov (2013)
		Bath Long Vi Reef	Latypov (2013)
	TG	Ha Long Bay (shells)	M.L. Neo, pers. obs. (2014)
	TM	Cham Islands, Central Viet Nam	Latypov (2001)
		Tho Chau, Con Dao, Thu Islands	Latypov & Selin (2011)
		Ku Lao Cham Islands	Latypov & Selin (2012a)
		Hon Nai Island, Cam Ranh Bay	Latypov & Selin (2012b)
	TS	An Thoi Archipelago, South China Sea	Latypov (2000)
		Tho Chau, Con Dao, Thu Islands	Latypov & Selin (2011)

Continued

Table A2 (Continued) Checklist of giant clam species

Recorded localities	Species	Localities	Reference(s)
Viet Nam		Khanh Hoa Province	Latypov & Selin (2013)
(Continued)		Giang Bo Reef	Latypov (2013)
		Re Island	Latypov (2013)
		Bath Long Vi Reef	Latypov (2013)
Australia			
Australia	HH	Western Australia	Wells et al. (1983); Rees et al. (2003)
		Queensland	Wells et al. (1983)
		Australia	Munro (1989)
	TC	Orpheus Island, Palm Island Group, Queensland	Hamner & Jones (1976)
		Great Barrier Reef, Queensland	Wells et al. (1983)
		Australia	Munro (1989)
		Western Australia	Rees et al. (2003)
	TD	Australia	Wells et al. (1983); Munro (1989)
		Great Barrier Reef, Queensland	Braley (1987a,b); Alder & Braley (1989); Pearson & Munro (1991)
		Western Australia	Rees et al. (2003)
	TG	Western Australia	Wells et al. (1983); Rees et al. (2003)
		Great Barrier Reef, Queensland	Wells et al. (1983); Braley (1987a,b); Alder & Braley (1989); Pearson & Munro (1991)
		Australia	Munro (1989)
		Heron Island, southern Great Barrier Reef	Strotz et al. (2010)
	TM	One Tree Island, Queensland	McMichael (1974)
		Australia	Wells et al. (1983); Munro (1989)
		Lord Howe Island, New South Wales	Wells et al. (1983)
		Western Australia	Rees et al. (2003)
		Montebello Islands, Western Australia	Wells et al. (2000)
		Solitary Islands Marine Park, northern New South Wales	Smith (2011)
	TMB (previously TT)	No data	Newman & Gomez (2000)
	TNI (now TNO)	Five Finger Reef, south of Coral Bay, WA	Penny & Willan (2014); Borsa et al. (2015)
	TNO	Western Australia (Huelsken et al., 2013)	Borsa et al. (2015)
		Ningaloo Marine Park, WA	Black et al. (2011); Johnson et al. (2016)
	TS	Australia	Wells et al. (1983); Munro (1989)
		Montebello Islands, Western Australia	Wells et al. (2000)
		Western Australia	Rees et al. (2003)

Continued

Table A2 (Continued)   Checklist of giant clam species

Recorded localities	Species	Localities	Reference(s)
Pacific Ocean			
Melanesia			
Fiji	HH	EXTINCT	Wells et al. (1983); Munro (1989); Tisdell & Wittenberg (1992); Seeto et al. (2012)
	TC	Lakeba Island	Vuki et al. (1992)
		Cicia Island	Vuki et al. (1992)
	TD	Fiji	Lewis et al. (1988); Munro (1989); Tacconi & Tisdell (1992b); Tisdell & Wittenberg (1992)
		Lakeba Island	Vuki et al. (1992)
		Tuvuca Island	Vuki et al. (1992)
		Cicia Island	Vuki et al. (1992)
		Balavu Island	Vuki et al. (1992)
		Ono Island	Flickr avipoodle (2010)
	TG	EXTINCT	Lewis et al. (1988); Munro (1989); Tisdell & Wittenberg (1992)
		Fiji	Flickr Jex207 (2007)
	TM	Fiji	Wells et al. (1983); Lewis et al. (1988); Munro (1989); Tacconi & Tisdell (1992b); Tisdell & Wittenberg (1992)
		Lakeba Island	Vuki et al. (1992)
		Tuvuca Island	Vuki et al. (1992)
		Cicia Island	Vuki et al. (1992)
		Balavu Island	Vuki et al. (1992)
	TNO	Moon Reef, Viti Levu	Borsa et al. (2015)
		Bega Lagoon Resort	Flickr CrashDiver (2010); Flickr scuba_dot_com (2014)
	TS	Fiji	Wells et al. (1983); Lewis et al. (1988); Munro (1989); Tacconi & Tisdell (1992b); Tisdell & Wittenberg (1992)
		Lakeba Island	Vuki et al. (1992)
		Tuvuca Island	Vuki et al. (1992)
		Cicia Island	Vuki et al. (1992)
		Balavu Island	Vuki et al. (1992)
		Wakaya, Koro Sea	Flickr Paul & Jill (2011)
	TMB (previously TT)	Eastern Islands (Lau)	Lewis & Ledua (1988); Lucas et al. (1991); Ledua et al. (1993)
		Fiji	Tacconi & Tisdell (1992b)
		Matokana	Ledua et al. (1993)
New Caledonia	HH	New Caledonia	Wells et al. (1983); Munro (1989); Dumas et al. (2011, 2013)
		North Eastern Lagoon (Poeubo to Hienghène)	McKenna et al. (2006)
		Bourail	Wantiez et al. (2007c)
		Grand Lagon Nord	Wantiez et al. (2008a)

Continued

Table A2 (Continued) Checklist of giant clam species

Recorded localities	Species	Localities	Reference(s)
New Caledonia (Continued)		Ducos Island, Bay of Saint Vincent	Aubert et al. (2009)
		Ioro Reef	Schwartzmann et al. (2011)
	TC	North Eastern Lagoon (Poeubo to Hienghène)	McKenna et al. (2006)
		Poum	Vieux (2009)
		Bourail	Wantiez et al. (2007c)
		Grand Lagon Nord	Wantiez et al. (2008a)
		Merlet	Wantiez et al. (2008b)
		New Caledonia	Dumas et al. (2011)
	TD	New Caledonia	Wells et al. (1983); Munro (1989); Purcell et al. (2009); Dumas et al. (2011)
		North Eastern Lagoon (Poeubo to Hienghène)	McKenna et al. (2006)
		Corne Sud	Wantiez et al. (2007a)
		Ile des Pins	Wantiez et al. (2007b)
		Bourail	Wantiez et al. (2007c)
		Grand Lagon Nord	Wantiez et al. (2008a)
		Merlet	Wantiez et al. (2008b)
	TG	FOSSIL TG FOUND / EXTINCT	Munro (1989)
	TM	New Caledonia	Wells et al. (1983); Munro (1989); Purcell et al. (2009); Dumas et al. (2011, 2013)
		North Eastern Lagoon (Poeubo to Hienghène)	McKenna et al. (2006)
		Poum	Vieux (2009)
		Corne Sud	Wantiez et al. (2007a)
		Ile des Pins	Wantiez et al. (2007b)
		Bourail	Wantiez et al. (2007c)
		Grand Lagon Nord	Wantiez et al. (2008a)
		Merlet	Wantiez et al. (2008b)
	TMB (previously TT)	Loyalty Islands	Bouchet et al. (2001); Tiavouane & Fauvelot (2016)
	TNO	Loyalty Islands	Borsa et al. (2015)
		Hienghene, northeastern coast	Borsa et al. (2015)
	TS	New Caledonia	Wells et al. (1983); Munro (1989); Purcell et al. (2009); Dumas et al. (2011, 2013)
		North Eastern Lagoon (Poeubo to Hienghène)	McKenna et al. (2006)
		Poum	Vieux (2009)
		Corne Sud	Wantiez et al. (2007a)
		Ile des Pins	Wantiez et al. (2007b)
		Bourail	Wantiez et al. (2007c)
		Grand Lagon Nord	Wantiez et al. (2008a)
		Merlet	Wantiez et al. (2008b)

Continued

Table A2 (Continued) Checklist of giant clam species

Recorded localities	Species	Localities	Reference(s)
Papua New Guinea	HH	PNG	Wells et al. (1983); Munro (1989)
		Milne Bay Province	Kinch (2001, 2002); Wells & Kinch (2003)
	HP	Milne Bay Province	Kinch (2001, 2002)
	TC	PNG	Wells et al. (1983)
		Milne Bay Province	Kinch (2001, 2002); Wells & Kinch (2003)
	TD	PNG	Wells et al. (1983); Munro (1989)
		Milne Bay Province	Kinch (2002); Wells & Kinch (2003)
	TG	PNG	Wells et al. (1983); Munro (1989)
		Milne Bay Province	Kinch (2001, 2002); Wells & Kinch (2003)
	TM	PNG	Wells et al. (1983); Munro (1989)
		Milne Bay Province	Kinch (2001, 2002); Wells & Kinch (2003)
	TNO	Madang Province, Kavieng	Borsa et al. (2015)
	TS	PNG	Wells et al. (1983); Munro (1989)
		Milne Bay Province	Kinch (2001, 2002); Wells & Kinch (2003)
Solomon Islands	HH	Solomon Islands	Wells et al. (1983); Govan et al. (1988); Munro (1989); Ramohia (2006)
		Bellona (Mungiki) Island	Thaman et al. (2011)
	TC	Solomon Islands	Wells et al. (1983); Govan et al. (1988); Munro (1989); Ramohia (2006)
		Bellona (Mungiki) Island	Thaman et al. (2011)
	TD	Solomon Islands	Govan et al. (1988); Munro (1989); Ramohia (2006)
		Bellona (Mungiki) Island	Thaman et al. (2011)
	TG	Solomon Islands	Wells et al. (1983); Govan et al. (1988); Munro (1989); Ramohia (2006); Flickr Artefacque (2014)
		Bellona (Mungiki) Island	Thaman et al. (2011)
		Arnavon Island	Flickr LMMA Network (2006)
	TM	Solomon Islands	Wells et al. (1983); Govan et al. (1988); Munro (1989); Ramohia (2006)
		Bellona (Mungiki) Island	Thaman et al. (2011)
		Gizo	Flickr Shea Pletz (2011)
	TNO	Solomon Islands (Huelsken et al., 2013)	Borsa et al. (2015)
	TS	Solomon Islands	Govan et al. (1988); Munro (1989); Ramohia (2006)
		Bellona (Mungiki) Island	Thaman et al. (2011)
		Bulo Island, off SE Gatokae Island, Mbatuna	Flickr Jose B (2015)
Vanuatu (and New Hebrides)	HH	Vanuatu (and New Hebrides)	Wells et al. (1983); Munro (1989); Zann & Ayling (1988); Bell & Amos (1993)
	TC	Vanuatu	Zann & Ayling (1988); Bell & Amos (1993)

Continued

Table A2 (Continued) Checklist of giant clam species

Recorded localities	Species	Localities	Reference(s)
Vanuatu (and New Hebrides) *(Continued)*	TD	EXTINCT	Munro (1989); Bell & Amos (1993)
	TG	EXTINCT	Wells et al. (1983); Munro (1989); Bell & Amos (1993)
	TM	Vanuatu (and New Hebrides)	Wells et al. (1983); Munro (1989); Zann & Ayling (1988); Bell & Amos (1993)
	TNO	Efate	Borsa et al. (2015)
	TS	Vanuatu	Munro (1989); Zann & Ayling (1988); Bell & Amos (1993)
Micronesia			
Federated States of Micronesia (FSM)	HH	LOW NUMBERS (REINTRODUCTION)	Munro (1989); Smith (1992)
	TD	NO WILD STOCKS (INTRODUCTION)	Munro (1989); Smith (1992)
	TG	Lamotrek and West Fayu Atolls (relict)	Price & Fagolimul (1988); Munro (1989)
		FSM	Smith (1992)
	TM	FSM	Munro (1989); Smith (1992)
	TNO	Kosrae, part of the Caroline Islands	Borsa et al. (2015)
	TS	FSM	Munro (1989); Smith (1992)
Guam	HH	EXTINCT	Munro (1989); Anonymous (1994); Paulay (2003)
	TD	INTRODUCED FROM PALAU	Wells et al. (1983); Anonymous (1994); Paulay (2003)
	TG	EXTINCT	Munro (1989); Anonymous (1994); Paulay (2003)
	TM	Guam	Munro (1989); Stojkovich (1977); Anonymous (1994); Paulay (2003)
		Mariana Islands	Flickr NOAA Photo Library, David Burdick (2010)
	TS	Guam	Anonymous (1994); Paulay (2003)
		Cocos West Island	Flickr GingrichCrew (2011)
Republic of Kiribati	HH	Gilbert Islands	Wells et al. (1983); Munro (1988, 1989); Thomas (2001)
		Kiribati Islands	Taniera (1988); Thomas (2014)
	TG	Gilbert Islands	Wells et al. (1983); Munro (1988, 1989); Thomas (2001)
		Kiribati Islands	Taniera (1988); Thomas (2014)
	TM	Fanning Island	Kay (1970)
		Gilbert Islands	Wells et al. (1983); Munro (1988, 1989); Kepler & Kepler (1994); Thomas (2001)
		Northern Line Islands	Kay (1970); Wells et al. (1983); Munro (1989); Sandin et al. (2008); Williams et al. (2013)
		Phoenix Islands	Munro (1989)
		Kiribati Islands	Taniera (1988); Thomas (2014)
		Millennium Atoll Lagoon	Barott et al. (2010)

Continued

Table A2 (Continued)   Checklist of giant clam species

Recorded localities	Species	Localities	Reference(s)
Republic of Kiribati	TNO	Kiritimati, Northern Line Islands	Borsa et al. (2015)
(Continued)	TS	Gilbert Islands	Wells et al. (1983); Munro (1988,t 1989); Thomas (2001)
		Kiribati Islands	Taniera (1988)
Marshall Islands	HH	Marshall Islands	Wells et al. (1983); Munro (1989); Pinca & Beger (2002)
		Rongelap Atoll	Pinca & Beger (2003)
		Namu Atoll	Beger et al. (2008)
	TD	INTRODUCED FROM PALAU	Munro (1989); Pinca & Beger (2002)
		Mili Atoll	Pinca & Beger (2003)
		Rongelap Atoll	Pinca & Beger (2003)
		Arno Atoll	Beger et al. (2008)
	TG	Marshall Islands	Wells et al. (1983); Munro (1989); Pinca & Beger (2002)
		Mili Atoll	Pinca & Beger (2003)
		Rongelap Atoll	Pinca & Beger (2003)
		Namu Atoll	Beger et al. (2008)
	TM	Marshall Islands	Wells et al. (1983); Munro (1989); Pinca & Beger (2002)
		Mili Atoll	Pinca & Beger (2003)
		Rongelap Atoll	Pinca & Beger (2003)
		Namu Atoll	Beger et al. (2008)
		Majuro Atoll	Beger et al. (2008)
	TS	Marshall Islands	Wells et al. (1983); Munro (1989); Pinca & Beger (2002)
		Mili Atoll	Pinca & Beger (2003)
		Rongelap Atoll	Pinca & Beger (2003)
		Namu Atoll	Beger et al. (2008)
		Majuro Atoll	Beger et al. (2008)
Nauru	TM	LOCALLY EXTINCT	Jacob (2000); South & Skelton (2000)
Northern Mariana	HH	LOCALLY EXTINCT	Munro (1989)
Islands	TC	POSSIBLY EXTINCT	IUCN Red List
	TD	POSSIBLY EXTINCT	IUCN Red List
	TG	LOCALLY EXTINCT	Munro (1989)
	TM	Mariana Islands	Wells et al. (1983); Munro (1989)
		Maug Island, Marianas Trench Marine National Monument	Flickr lucidlou (2007)
		Saipan Island	Flickr Chris (2008)
	TS	Mariana Islands	Wells et al. (1983)
Palau	HH	Caroline Islands	Hardy & Hardy (1969); Hester & Jones (1974); Bryan & McConnell (1976); Hirshberger (1980); Wells et al. (1983)
		Palau	Wells et al. (1983); Munro (1989)
	HP	Palau	Munro (1989)
	TC	Caroline Islands	Hardy & Hardy (1969); Hester & Jones (1974); Wells et al. (1983)
		Palau	Munro (1989)

Continued

Table A2 (Continued)   Checklist of giant clam species

Recorded localities	Species	Localities	Reference(s)
Palau *(Continued)*	TD	Caroline Islands	Hardy & Hardy (1969); Hester & Jones (1974); Bryan & McConnell (1976); Hirshberger (1980); Wells et al. (1983)
		Palau	Munro (1989)
	TG	Caroline Islands	Hardy & Hardy (1969); Hester & Jones (1974); Bryan & McConnell (1976); Hirshberger (1980); Wells et al. (1983)
		Palau	Munro (1989)
	TM	Caroline Islands	Hardy & Hardy (1969); Hester & Jones (1974); Bryan & McConnell (1976); Hirshberger (1980); Wells et al. (1983)
		Palau	Munro (1989)
	TS	Caroline Islands	Hardy & Hardy (1969); Hester & Jones (1974); Bryan & McConnell (1976); Hirshberger (1980); Wells et al. (1983)
		Palau	Munro (1989)
United States Minor Outlying Islands	TM	Wake Island	Wells et al. (1983)
Polynesia			
American Samoa	HH	LOCALLY EXTINCT and REINTRODUCED	Nagaoka (1993); Craig (2009)
	TD	INTRODUCED	Bell (1993)
	TG	INTRODUCED	Bell (1993)
	TM	Rose Atoll	Radtke (1985); Munro (1989); Green & Craig (1999); Craig (2009)
	TS	American Samoa	Munro (1989); Craig (2009)
Cook Islands	TD	INTRODUCED FROM PALAU	Munro (1989); Tisdell & Wittenberg (1992)
	TG	INTRODUCED	Flickr Richard Mayston (2008); Flickr RDPixelShop (2011)
	TM	Cook Islands	Munro (1989); Tisdell & Wittenberg (1992)
		Suwarrow Atoll	Sims & Howard (1988)
		Manihiki Atoll	Sims & Howard (1988)
		Penrhyn Atoll	Sims & Howard (1988)
		Aitutaki	Paulay (1987); Sims & Howard (1988)
		Tongareva Lagoon	Chambers (2007)
	TNO	Rarotonga	Flickr Richard Mayston (2008)
	TS	Cook Islands	Paulay (1987); Tisdell & Wittenberg (1992)
		Aitutaki (RARE)	Sims & Howard (1988); Munro (1989)
French Polynesia	TM	Takapoto Atoll	Jaubert (1977); Richard (1977); Laurent (2001)
		Tuamotu	Wells et al. (1983)
		Polynesie Francaise	Munro (1989)
		Bora Bora Island	Planes et al. (1993)
		Moorea	Laurent (2001)

Continued

Table A2 (Continued)   Checklist of giant clam species

Recorded localities	Species	Localities	Reference(s)
French Polynesia		Anaa	Laurent (2001)
(Continued)		Tatakoto Atoll	Gilbert et al. (2005, 2006b); Andréfouët et al. (2013)
		Fangatau Atoll	Andréfouët et al. (2005); Gilbert et al. (2006b)
		Reao	Gilbert et al. (2006a)
		Pukarua	Gilbert et al. (2006a)
		Raivavae	Gilbert et al. (2006a); Andréfouët et al. (2009); Van Wynsberge et al. (2013)
		Tubuai, Austral Islands	Larrue (2006); Gilbert et al. (2006b); Van Wynsberge et al. (2013)
	TS	Tuamotu	Wells et al. (1983); Andréfouët et al. (2014)
		Tubuai, Austral Islands	Gilbert et al. (2007); Newman & Gomez (2007)
		Gambier	Andréfouët et al. (2014)
Pitcairn Islands	TM	Henderson Island	Wells et al. (1983); Paulay et al. (1989)
		Oeno Lagoon	Paulay et al. (1989); Irving & Dawson (2013)
		Pitcairn Islands	Palomares et al. (2011)
	TS	Ducie Atoll	Paulay et al. (1989)
		Henderson Island	Paulay et al. (1989)
		Pitcairn Islands	Palomares et al. (2011)
Niue	TM	Niue	Dalzell et al. (1993); Kronen et al. (2008)
	TS	LOCALLY EXTINCT	Dalzell et al. (1993); absent in Kronen et al. (2008) surveys
		Niue	Flickr orbitonline (2009); Flickr Sam & Fanny (2012)
Samoa	HH	LOCALLY EXTINCT	Munro (1989); Zann (1991); Tacconi & Tisdell (1992b); South & Skelton (2000)
		REINTRODUCED	Flickr Richard Mayston (2015)
	TD	INTRODUCED	Fisheries Newsletter (2014)
	TG	INTRODUCED	Fisheries Newsletter (2014)
	TM	Samoa	Wells et al. (1983); Munro (1989); Zann (1989); Tisdell & Wittenberg (1992)
		Upolu, Western Samoa	Zann (1991); Tacconi & Tisdell (1992b)
	TS	Samoa	Wells et al. (1983); Munro (1989); Zann (1989); Tisdell & Wittenberg (1992)
		Upolu, Western Samoa	Zann (1991); Tacconi & Tisdell (1992b)
Tokelau	TM	Tokelau	Munro (1989); Tisdell & Wittenberg (1992)
		Fakaofo	Braley (1989)
		Nukunonu	Braley (1989)
		Atafu	Braley (1989)
	TS	Fakaofo	Braley (1989)
		Nukunonu	Braley (1989)

Continued

Table A2 (Continued)   Checklist of giant clam species

Recorded localities	Species	Localities	Reference(s)
Tonga	HH	LOCALLY EXTINCT	Wells et al. (1983); Langi & Hesitoni 'Aloua (1988); Munro (1989); Tacconi & Tisdell (1992b); Loto'ahea & Sone (1995); Salvat (2000)
		Vava'u (EXTINCT)	Chesher (1993)
	TG	LOCALLY EXTINCT	Salvat (2000)
	TD	Tonga	Langi & Hesitoni 'Aloua (1988); Munro (1989); Tacconi & Tisdell (1992b); Tisdell & Wittenberg (1992); Loto'ahea & Sone (1995)
		Vava'u	Chesher (1993)
		Tongatapu Island	Tu'avao et al. (1995)
	TM	Tonga	Wells et al. (1983); Langi & Hesitoni 'Aloua (1988); Munro (1989); Tacconi & Tisdell (1992b); Tisdell & Wittenberg (1992); Loto'ahea & Sone (1995)
		Vava'u	Chesher (1993)
		Tongatapu Island	Tu'avao et al. (1995)
	TS	Tonga	Wells et al. (1983); Langi & Hesitoni 'Aloua (1988); Munro (1989); Tacconi & Tisdell (1992b); Tisdell & Wittenberg (1992); Loto'ahea & Sone (1995)
		Vava'u	Chesher (1993)
		Tongatapu Island	Tu'avao et al. (1995)
	TMB (previously TT)	Vava'u and Ha'apai Islands	Lucas et al. (1991)
		Main islands of Tonga	Tacconi & Tisdell (1992b); Ledua et al. (1993)
Tuvalu	HH	Tuvalu	Munro (1989)
	TD	INTRODUCED	Job & Ceccarelli (2012)
	TG	Tuvalu	Munro (1989); Tacconi & Tisdell (1992a); Sauni et al. (2008)
	TM	Ellice Islands	Wells et al. (1983)
		Nukufetau	Braley (1988); Sauni et al. (2008)
		Funafuti	Braley (1988); Sauni et al. (2008); Job & Ceccarelli (2012); Siaosi et al. (2012)
		Nukulaelae	Braley (1988); Job & Ceccarelli (2012)
		Tuvalu	Munro (1989)
		Nanumea	Langi (1990)
		Nui	Langi (1990)
		Niutao	Sauni et al. (2008)
		Vaitupu	Sauni et al. (2008)
	TS	Ellice Islands	Wells et al. (1983)
		Nukufetau	Braley (1988); Sauni et al. (2008)

Continued

Table A2 (Continued)   Checklist of giant clam species

Recorded localities	Species	Localities	Reference(s)
Tuvalu *(Continued)*		Funafuti	Braley (1988); Sauni et al. (2008); Job & Ceccarelli (2012); Siaosi et al. (2012)
		Tuvalu	Munro (1989)
Wallis and Futuna Islands	TM	Wallis and Futuna	Pollock (1992)
	TNO	Wallis Island	Borsa et al. (2015)
	TS	Wallis and Futuna	Pollock (1992)

Note: Full reference list found in Appendix B. Tg—*Tridacna gigas*; Td—*T. derasa*; Tmb—*T. mbalavuana* (previously *T. tevoroa*); Ts—*T. squamosa*; Tsi—*T. squamosina* (previously *T. costata*); Tr—*T. rosewateri*; Tm—*T. maxima*; Tlz—*T. lorenzi*; Tno—*T. noae*; Tc—*T. crocea*; Hh—*Hippopus hippopus*; Hp—*H. porcellanus*.

Table A3 Global density patterns of wild giant clam populations

Country	Localities surveyed	Species	Year of survey	Method of survey	Approximate area of surveys (ha)	Approximate area of surveys (m²)	Number of ind.	Density (m⁻²)	Reference (population survey)
American Samoa	Rose Atoll	Tm	1982–1984	Belt transects; 30 × 2 m	392	3,920,000	1,338,700	0.34151	Radtke (1985)
American Samoa	Rose Atoll	Tm	1994–1995	Belt transects; 50 × 2 m	615	6,150,000	27,845	0.00453	Green & Craig (1999)
American Samoa	Rose Atoll	Tm, Ts	1994–1995	Belt transects; 50 × 2 m	21	213,000	2,765	0.01298	Green & Craig (1999)
Australia	One Tree Island, Queensland	Tm	1966	Quadrat; 25 × 28 m	—	450	359	0.79778	McMichael (1974)
Australia	One Tree Island, Queensland	Tm	1968	Quadrat; 25 × 28 m	—	450	345	0.76667	McMichael (1974)
Australia	One Tree Island, Queensland	Tm	1969	Quadrat; 25 × 28 m	—	450	374	0.83111	McMichael (1974)
Australia	Orpheus Island, Queensland	Tc	1974–1975	Belt transects; Coral head surface	—	16	70	4.37500	Hammer & Jones (1976)
Australia	Escape reefs (west reef, north reef, east bommie, south reef)	Td	1981	Census of fixed area	3	33,720	205	0.00608	Braley (1987b)
Australia	Escape reefs (west reef, north reef, east bommie, south reef)	Tg	1981	Census of fixed area	3	33,720	254	0.00753	Braley (1987b)
Australia	Escape reefs (west reef, north reef-a, north reef-b, east bommie)	Td	1982	Census of fixed area	1	10,510	97	0.00923	Braley (1987b)
Australia	Escape reefs (west reef, north reef-a, north reef-b, east bommie)	Tg	1982	Census of fixed area	1	10,510	141	0.01342	Braley (1987b)
Australia	Great Barrier Reef (northern)	Td	1983	Quadrat; 50 × 20 m	—	—	—	0.00029	Braley (1987a)

Continued

Table A3 (Continued) Global density patterns of wild giant clam populations

Country	Localities surveyed	Species	Year of survey	Method of survey	Approximate area of surveys (ha)	Approximate area of surveys (m²)	Number of ind.	Density (m⁻²)	Reference (population survey)
Australia	Great Barrier Reef (northern)	Tg	1983	Quadrat; 50 × 20 m	—	—	—	0.00078	Braley (1987a)
Australia	Great Barrier Reef (southern)	Td	1983	Quadrat; 50 × 20 m	—	—	—	0.00059	Braley (1987a)
Australia	Great Barrier Reef (southern)	Tg	1983	Quadrat; 50 × 20 m	—	—	—	0.00006	Braley (1987a)
Australia	Michaelmas Reef, GBR	Tg	1978	Quadrat; 180 × 150 m	—	27,000	1,166	0.04319	Pearson & Munro (1991)
Australia	Michaelmas Reef, GBR	Tg	1980–1981	Quadrat; 180 × 150 m	—	27,000	1,120	0.04148	Pearson & Munro (1991)
Australia	Michaelmas Reef, GBR	Tg	1985	Quadrat; 180 × 150 m	—	27,000	764	0.02830	Pearson & Munro (1991)
Australia	Michaelmas Reef, GBR	Td	1978	Quadrat; 180 × 150 m	—	27,000	46	0.00170	Pearson & Munro (1991)
Australia	Michaelmas Reef, GBR	Td	1980–1981	Quadrat; 180 × 150 m	—	27,000	46	0.00170	Pearson & Munro (1991)
Australia	Michaelmas Reef, GBR	Td	1985	Quadrat; 180 × 150 m	—	27,000	31	0.00115	Pearson & Munro (1991)
Australia	Ashmore Reef	Tm	2003	Distance swim transects; 500 × 5 m	40	397,500	456	0.00115	Rees et al. (2003)
Australia	Cartier Reef	Tm	2003	Distance swim transects; 500 × 5 m	18	180,000	110	0.00061	Rees et al. (2003)
Australia	Mermaid Reef	Tm	2003	Distance swim transects; 500 × 5 m	23	232,500	793	0.00341	Rees et al. (2003)
Australia	Ashmore Reef	Tg	2003	Distance swim transects; 500 × 5 m	40	397,500	49	0.00012	Rees et al. (2003)
Australia	Cartier Reef	Tg	2003	Distance swim transects; 500 × 5 m	18	180,000	0	0.00000	Rees et al. (2003)

Continued

Table A3 (Continued) Global density patterns of wild giant clam populations

Country	Localities surveyed	Species	Year of survey	Method of survey	Approximate area of surveys (ha)	Approximate area of surveys (m²)	Number of ind.	Density (m⁻²)	Reference (population survey)
Australia	Mermaid Reef	Tg	2003	Distance swim transects; 500 × 5 m	23	232,500	79	0.00034	Rees et al. (2003)
Australia	Ashmore Reef	Hh	2003	Distance swim transects; 500 × 5 m	40	397,500	740	0.00186	Rees et al. (2003)
Australia	Cartier Reef	Hh	2003	Distance swim transects; 500 × 5 m	18	180,000	715	0.00397	Rees et al. (2003)
Australia	Mermaid Reef	Hh	2003	Distance swim transects; 500 × 5 m	23	232,500	46	0.00020	Rees et al. (2003)
Australia	Ningaloo Marine Park, WA	Tm (Tno)	2010	Belt transects; varied lengths (10–38 m) and widths (2–5 m)	—	15,173	3,119	0.20556	Black et al. (2011) [Also see Johnson et al. (2016)]
Australia	Surfers South, Ningaloo Marine Park, WA	Tno	2014	Randomized sampling	—	—	—	0.05000	Johnson et al. (2016)
Australia	Mandu Mandu, Ningaloo Marine Park, WA	Tno	2014	Randomized sampling	—	—	—	0.28000	Johnson et al. (2016)
Cambodia	Koh Kong	T	?	Belt transects	—	—	—	2.00000	Vibol (N.D.); Kim et al. (2004)
Cambodia	Koh Sdach	T	?	Belt transects	—	—	—	1.20000	Vibol (N.D.); Kim et al. (2004)
Cambodia	Koh Rong	T	?	Belt transects	—	—	—	0.30000	Vibol (N.D.); Kim et al. (2004)
Cambodia	Koh Tang	T	?	Belt transects	—	—	—	0.30000	Vibol (N.D.); Kim et al. (2004)
Cambodia	Koh Chann, Koh Sdach group	T	2001	Belt transects; 20 × 5 m	—	500	3	0.00600	Chou et al. (2002)
Cambodia	Koh Poi Jepon, Koh Sdach group	T	2001	Belt transects; 20 × 5 m	—	500	30	0.06000	Chou et al. (2002)
Cambodia	Koh Toutint, Koh Sdach group	T	2001	Belt transects; 20 × 5 m	—	500	3	0.00600	Chou et al. (2002)

Continued

Table A3 (Continued) Global density patterns of wild giant clam populations

Country	Localities surveyed	Species	Year of survey	Method of survey	Approximate area of surveys (ha)	Approximate area of surveys (m^2)	Number of ind.	Density (m^{-2})	Reference (population survey)
Cambodia	Koh Dom Long, Koh Sdach group	T	2001	Belt transects; 20 × 5 m	—	500	7	0.01400	Chou et al. (2002)
Cambodia	Poi Chheng Lek, Koh Sdach group	T	2001	Belt transects; 20 × 5 m	—	500	30	0.06000	Chou et al. (2002)
Cambodia	Koh Sdach (south), Koh Sdach group	T	2002	Belt transects; 20 × 5 m	—	500	25	0.05000	Chou et al. (2002)
Cambodia	Koh Ampil Toch, Koh Sdach group	T	2002	Belt transects; 20 × 5 m	—	500	12	0.02400	Chou et al. (2002)
Cambodia	Koh Kmauch, Koh Sdach group	T	2002	Belt transects; 20 × 5 m	—	500	6	0.01200	Chou et al. (2002)
Cambodia	Koh Rong, Koh Kon, Koh Rong Samloem, Phreah Sihanouk Province	T	2010	Belt transects; 20 × 5 m	—	—	—	0–1.35	Van Bochove et al. (2011)
Cambodia	Song Saa Private Island Resort, Koh Rong Archipelago (MPA)	Tg?	2013	Belt transects; 20 × 5 m	—	—	—	0.72000	Savage et al. (2013)
Cambodia	Song Saa Private Island Resort, Koh Rong Archipelago (Surrounding sites)	Tg?	2013	Belt transects; 20 × 5 m	—	—	—	0.26500	Savage et al. (2013)
Cambodia	Song Saa Private Island Resort, Koh Rong Archipelago (Geographically isolated sites)	Tg?	2013	Belt transects; 20 × 5 m	—	—	—	0.13000	Savage et al. (2013)
Cambodia	Koh Rong Archipelago	T	2010–2014	Belt transects; 20 × 5 m	—	—	—	0.02000	Thorne et al. (2015)
Christmas Island	—	T	2005	Belt transects; 50 × 1 m	—	3,300	29	0.00879	Gilligan et al. (2008)

Continued

Table A3 (Continued) Global density patterns of wild giant clam populations

Country	Localities surveyed	Species	Year of survey	Method of survey	Approximate area of surveys (ha)	Approximate area of surveys (m²)	Number of ind.	Density (m⁻²)	Reference (population survey)
Cocos (Keeling) Islands	—	Td, Tm	?	?	—	—	—	0.03000	Hender et al. (2001)
Cocos (Keeling) Islands	Cabbage Patch 10 m	Tm	1999	Reef Check Survey methods	—	—	—	0.15000	Australian Government (2005)
Cocos (Keeling) Islands	Cabbage Patch 10 m	Tm	2000	Reef Check Survey methods	—	—	—	0.00000	Australian Government (2005)
Cocos (Keeling) Islands	Cabbage Patch 10 m	Tm	2001	Reef Check Survey methods	—	—	—	0.08000	Australian Government (2005)
Cocos (Keeling) Islands	Cabbage Patch 10 m	Tm	2003	Reef Check Survey methods	—	—	—	0.00000	Australian Government (2005)
Cocos (Keeling) Islands	Cabbage Patch 10 m	Tm	2004	Reef Check Survey methods	—	—	—	0.07750	Australian Government (2005)
Cocos (Keeling) Islands	100th Site	Tm	2002	Reef Check Survey methods	—	—	—	0.10500	Australian Government (2005)
Cocos (Keeling) Islands	100th Site	Tm	2004	Reef Check Survey methods	—	—	—	0.20250	Australian Government (2005)
Cocos (Keeling) Islands	Cabbage Patch 3 m	Tm	1997	Reef Check Survey methods	—	—	—	0.28250	Australian Government (2005)
Cocos (Keeling) Islands	Cabbage Patch 3 m	Tm	1999	Reef Check Survey methods	—	—	—	0.28000	Australian Government (2005)
Cocos (Keeling) Islands	Cabbage Patch 3 m	Tm	2001	Reef Check Survey methods	—	—	—	0.08000	Australian Government (2005)
Cocos (Keeling) Islands	Cabbage Patch 3 m	Tm	2002	Reef Check Survey methods	—	—	—	0.15000	Australian Government (2005)
Cocos (Keeling) Islands	Cabbage Patch 3 m	Tm	2003	Reef Check Survey methods	—	—	—	0.17000	Australian Government (2005)
Cocos (Keeling) Islands	Cabbage Patch 3 m	Tm	2004	Reef Check Survey methods	—	—	—	0.24000	Australian Government (2005)

Continued

Table A3 (Continued) Global density patterns of wild giant clam populations

Country	Localities surveyed	Species	Year of survey	Method of survey	Approximate area of surveys (ha)	Approximate area of surveys (m²)	Number of ind.	Density (m⁻²)	Reference (population survey)
Cocos (Keeling) Islands	77 sites in Cocos (Keeling) Islands	Tm	2011	Belt transect; 50 × 2 m	—	—	—	0.05400	Bellchambers & Evans (2013)
Cocos (Keeling) Islands	The Rip (no-take site)	Tm	2011	Belt transect; 50 × 2 m			—	1.05500	Bellchambers & Evans (2013)
Cocos (Keeling) Islands	70 sites in Cocos (Keeling) Islands	Tm	2014	Belt transect; 50 × 2 m	—		—	0.06600	Evans et al. (2016)
Cocos (Keeling) Islands	The Rip (no-take site)	Tm	2014	Belt transect; 50 × 2 m		—	—	0.82500	Evans et al. (2016)
Cook Islands	Aitutaki	Tm	?	Swathe transects	—	80	382	4.77500	Sims & Howard (1988)
Cook Islands	Manihiki	Tm	?	Swathe transects	—	825	216	0.26182	Sims & Howard (1988)
Cook Islands	Suwarrow	Tm	?	Swathe transects	—	2,240	130	0.05804	Sims & Howard (1988)
Cook Islands	Tongareva Lagoon	Tm	2006	Quadrat; 50 × 50 m	—	67,500	28,066	0.41579	Chambers (2007)
Djibouti	—	T	1998	Quadrat 10 × 10 m; and 20-minute time swim	—	3,500	348	0.09943	PERSGA (2000)
Djibouti	Maskali Island, Moucha Island, Tadjoura Bay, 7-Brothers Island	T	2002	Belt transects; 20 × 5 m	—	—	—	0.02550	PERSGA (2010)
Djibouti	Maskali Island, Moucha Island, 7-Brothers Island	Tm, Ts	2008	Belt transects; 20 × 5 m	—	2,800	—	0.01040	PERSGA (2010)
Egypt	Northern Bay of Safaga	Tm	?	Random quadrats; 0.25 m²	—	40	66	1.67266	Zuschin & Pillar (1997)
Egypt	SW of Gulf of Aqaba, between Ras Nosrani and Ras Mohammed	Tm	1994	Belt transects; 30 × 2 m	—	1,440	6,709	4.65903	Kilada et al. (1998)

Continued

Table A3 (Continued) Global density patterns of wild giant clam populations

Country	Localities surveyed	Species	Year of survey	Method of survey	Approximate area of surveys (ha)	Approximate area of surveys (m²)	Number of ind.	Density (m⁻²)	Reference (population survey)
Egypt	SW of Gulf of Aqaba, between Ras Nosrani and Ras Mohammed	Ts	1994	Belt transects; 30 × 2 m	—	1,440	45	0.03125	Kilada et al. (1998)
Egypt	Abu Sauatir, Northern Red Sea	Tm	2012	Random quadrats; 0.25 m²	—	491	159	0.32383	Ullmann (2013)
Egypt	Pharaoh Island/Coral Island, Taba	Tm	?	Belt transects; 50 × 5 m	—	750	4	0.00533	Richter et al. (2008)
Egypt	Pharaoh Island/Coral Island, Taba	Ts	?	Belt transects; 50 × 5 m	—	750	2	0.00320	Richter et al. (2008)
Egypt	Ras Amira/Taba Heights, Taba	Tm	?	Belt transects; 50 × 5 m	—	750	3	0.00333	Richter et al. (2008)
Egypt	Ras Amira/Taba Heights, Taba	Ts	?	Belt transects; 50 × 5 m	—	750	2	0.00220	Richter et al. (2008)
Egypt	Buoy, Nuweiba	Tm	?	Belt transects; 50 × 5 m	—	500	1	0.00200	Richter et al. (2008)
Egypt	Buoy, Nuweiba	Ts	?	Belt transects; 50 × 5 m	—	750	1	0.00150	Richter et al. (2008)
Egypt	Fayrouza, Nuweiba	Tm	?	Belt transects; 50 × 5 m	—	500	22	0.04405	Richter et al. (2008)
Egypt	Fayrouza, Nuweiba	Ts	?	Belt transects; 50 × 5 m	—	500	0	0.00030	Richter et al. (2008)
Egypt	Fayrouza, Nuweiba	Tsi	?	Belt transects; 50 × 5 m	—	500	0	0.00029	Richter et al. (2008)
Egypt	Towers, Nuweiba	Tm	?	Belt transects; 50 × 5 m	—	1,500	8	0.00565	Richter et al. (2008)
Egypt	Towers, Nuweiba	Ts	?	Belt transects; 50 × 5 m	—	1,500	4	0.00248	Richter et al. (2008)
Egypt	Blue Hole, Dahab	Tm	?	Belt transects; 50 × 5 m	—	750	55	0.07397	Richter et al. (2008)

Continued

Table A3 (Continued) Global density patterns of wild giant clam populations

Country	Localities surveyed	Species	Year of survey	Method of survey	Approximate area of surveys (ha)	Approximate area of surveys (m²)	Number of ind.	Density (m⁻²)	Reference (population survey)
Egypt	Blue Hole, Dahab	Ts	?	Belt transects; 50 × 5 m	—	750	6	0.00800	Richter et al. (2008)
Egypt	Blue Hole, Dahab	Tsi	?	Belt transects; 50 × 5 m	—	750	0	0.00047	Richter et al. (2008)
Egypt	InMo, Dahab	Tm	?	Belt transects; 50 × 5 m	—	750	54	0.07207	Richter et al. (2008)
Egypt	InMo, Dahab	Ts	?	Belt transects; 50 × 5 m	—	750	3	0.00453	Richter et al. (2008)
Egypt	Lagona, Dahab	Tm	?	Belt transects; 50 × 5 m	—	250	95	0.38150	Richter et al. (2008)
Egypt	Lagona, Dahab	Ts	?	Belt transects; 50 × 5 m	—	250	1	0.00470	Richter et al. (2008)
Egypt	Lagona, Dahab	Tsi	?	Belt transects; 50 × 5 m	—	250	0	0.00117	Richter et al. (2008)
Egypt	Shark Point, Ras Mohammed	Tm	?	Belt transects; 50 × 5 m	—	500	2	0.00480	Richter et al. (2008)
Egypt	Shark Point, Ras Mohammed	Ts	?	Belt transects; 50 × 5 m	—	500	0	0.00030	Richter et al. (2008)
Egypt	Yolanda Bay, Ras Mohammed	Tm	?	Belt transects; 50 × 5 m	—	250	18	0.07390	Richter et al. (2008)
Egypt	Yolanda Bay, Ras Mohammed	Ts	?	Belt transects; 50 × 5 m	—	250	0	0.00150	Richter et al. (2008)
Egypt	Marsa Abu Kalawa, Hurghada	Tm	?	Belt transects; 50 × 5 m	—	250	29	0.11700	Richter et al. (2008)
Egypt	Marsa Abu Kalawa, Hurghada	Ts	?	Belt transects; 50 × 5 m	—	250	1	0.00250	Richter et al. (2008)
Egypt	Marsa Abu Kalawa, Hurghada	Tsi	?	Belt transects; 50 × 5 m	—	250	2	0.00622	Richter et al. (2008)

Continued

Table A3 (Continued) Global density patterns of wild giant clam populations

Country	Localities surveyed	Species	Year of survey	Method of survey	Approximate area of surveys (ha)	Approximate area of surveys (m²)	Number of ind.	Density (m⁻²)	Reference (population survey)
Egypt	Sachwa, Hurghada	Tm	?	Belt transects; 50 × 5 m	—	500	40	0.07905	Richter et al. (2008)
Egypt	Sachwa, Hurghada	Tsi	?	Belt transects; 50 × 5 m	—	500	1	0.00165	Richter et al. (2008)
Egypt	Shab Abu Nuga, Hurghada	Tm	?	Belt transects; 50 × 5 m	—	250	210	0.84000	Richter et al. (2008)
Egypt	Shab Abu Nuga, Hurghada	Ts	?	Belt transects; 50 × 5 m	—	250	1	0.00250	Richter et al. (2008)
Egypt	Shab El Erg (South), Hurghada	Tm	?	Belt transects; 50 × 5 m	—	750	147	0.19570	Richter et al. (2008)
Egypt	Shab El Erg (South), Hurghada	Ts	?	Belt transects; 50 × 5 m	—	750	3	0.00343	Richter et al. (2008)
Egypt	Shab Shabina, Hurghada	Tm	?	Belt transects; 50 × 5 m	—	250	72	0.28670	Richter et al. (2008)
Egypt	Stone Beach/Hamda, Hurghada	Tm	?	Belt transects; 50 × 5 m	—	250	15	0.06090	Richter et al. (2008)
Egypt	Stone Beach/Hamda, Hurghada	Tsi	?	Belt transects; 50 × 5 m	—	250	1	0.00210	Richter et al. (2008)
Egypt	Dahab, Nabq, Ras Norani, Ras Mohamed, Hurghada, Safaga, Hamrawin, Qusier, Marsa Alam	Tm, Ts	2002	Belt transects; 20 × 5 m	—	—	—	0.02150	PERSGA (2010)
Egypt	Nuweiba, Dahab, Sharm El-Sheikh, Hurghada, Safaga, Qusier, Marsa Alam	Tm, Ts	2008	Belt transects; 20 × 5 m	—	5,600	—	0.02930	PERSGA (2010)
Fiji	Cakau Tabu Reef, Lau	Tmb	1986	SCUBA search (per man hour effort)	—	—	1	0.25 clam per man hour	Ledua et al. (1993)

Continued

Table A3 (Continued) Global density patterns of wild giant clam populations

Country	Localities surveyed	Species	Year of survey	Method of survey	Approximate area of surveys (ha)	Approximate area of surveys (m²)	Number of ind.	Density (m⁻²)	Reference (population survey)
Fiji	Vatoa Island, Lau	Tmb	1989	SCUBA search (per man hour effort)	—	—	6	0.30 clam per man hour	Ledua et al. (1993)
Fiji	Vatoa Island, Lau	Tmb	1990	SCUBA search (per man hour effort)	—	—	5	0.20 clam per man hour	Ledua et al. (1993)
Fiji	Vatoa Island, Lau	Tmb	1991	SCUBA search (per man hour effort)	—	—	2	0.30 clam per man hour	Ledua et al. (1993)
French Polynesia	Orapa, Takapoto Lagoon	Tm	1974–1975	Belt transects; 200 × 2.5 m	—	—	—	0.74000	Richard (1977)
French Polynesia	Village, Takapoto Lagoon	Tm	1975–1975	Belt transects; 262.5 × 2.5 m	—	—	—	1.28000	Richard (1977)
French Polynesia	Vairua, Takapoto Lagoon	Tm	1976–1975	Belt transects; 230 × 2.5 m	—	—	—	0.79000	Richard (1977)
French Polynesia	Gnake, Takapoto Lagoon	Tm	1977–1975	Belt transects; 265 × 2.5 m	—	—	—	0.62000	Richard (1977)
French Polynesia	Moorea, Society Islands	Tm	?	Point Centered Quarter Method	—	20,000	700	0.03500	Laurent (2001)
French Polynesia	Takapoto Lagoon, W Tuamotu Islands	Tm	?	Belt transects	—	1,150	161	0.14000	Laurent (2001)
French Polynesia	Anaa, W Tuamotu Islands	Tm	?	Belt transects	—	2,735	55	0.02000	Laurent (2001)
French Polynesia	Reao, E Tuamotu Islands	Tm	2003	Belt transects	—	3,200	26,080	8.15000	Gilbert et al. (2006a)
French Polynesia	Pukarua, E Tuamotu Islands	Tm	2003	Belt transects	—	1,305	17,043	13.06000	Gilbert et al. (2006a)
French Polynesia	Raivavae, Austral Islands	Tm	2003	Belt transects	—	5,485	7,185	1.31000	Gilbert et al. (2006a)
French Polynesia	Fangatau Atoll	Tm	2004	In situ data	—	99	3,781	38.19192	Andréfouët et al. (2005); Gilbert et al. (2006b)
French Polynesia	Tatakoto Atoll	Tm	2004	In situ data	—	70	6,389	90.94662	Gilbert et al. (2006b)

Continued

Table A3 (Continued) Global density patterns of wild giant clam populations

Country	Localities surveyed	Species	Year of survey	Method of survey	Approximate area of surveys (ha)	Approximate area of surveys (m^2)	Number of ind.	Density (m^{-2})	Reference (population survey)
French Polynesia	Tubuai Island	Tm	2004	*In situ* data	—	82	6,400	78.28746	Gilbert et al. (2006b)
French Polynesia	Tatakoto Atoll, Eastern Tuamotu Archipelago	Tm	2004–2005	Remote sensing	—	—	—	544.00000	Gilbert et al. (2005)
French Polynesia	Raivavae, Austral Islands	Tm	2004–2005	Manta tows	—	—	—	0.00–10.67	Andréfouët et al. (2009)
French Polynesia	Raivavae, Austral Islands	Tm	2004–2005	Broad-scale survey—Reef benthos transect surveys	—	—	—	0.05–2.79	Andréfouët et al. (2009)
French Polynesia	Raivavae, Austral Islands	Tm	2004–2005	Remote sensing	—	—	—	0.06–7.40	Andréfouët et al. (2009)
French Polynesia	Tatakoto Atoll	Tm	2012	*In situ* data	—	—	—	38.00000	Andréfouët et al. (2013)
French Polynesia	Tubuai Island (southern ridge)	Tm	2004	Belt transects	—	—	—	3.70000	Van Wynsberge et al. (2013)
French Polynesia	Tubuai Island (reef flat)	Tm	2004	Belt transects	—	—	—	5.80000	Van Wynsberge et al. (2013)
French Polynesia	Tubuai Island (southern ridge)	Tm	2010	Belt transects	—	—	—	4.60000	Wynsberge et al. (2013)
French Polynesia	Tubuai Island (reef flat)	Tm	2010	Belt transects	—	—	—	3.80000	Wynsberge et al. (2013)
French Polynesia	Raivavae, Austral Islands (SW reef flat)	Tm	2005	Belt transects	—	—	—	7.40000	Van Wynsberge et al. (2013)
French Polynesia	Raivavae, Austral Islands (N reef flat)	Tm	2005	Belt transects	—	—	—	0.17000	Van Wynsberge et al. (2013)
French Polynesia	Raivavae, Austral Islands (SW reef flat)	Tm	2010	Belt transects	—	—	—	6.20000	Van Wynsberge et al. (2013)

Continued

Table A3 (Continued) Global density patterns of wild giant clam populations

Country	Localities surveyed	Species	Year of survey	Method of survey	Approximate area of surveys (ha)	Approximate area of surveys (m²)	Number of ind.	Density (m⁻²)	Reference (population survey)
French Polynesia	Raivavae, Austral Islands (N reef flat)	Tm	2010	Belt transects	—	—	—	0.35000	Van Wynsberge et al. (2013)
India	Diglipur, Andaman and Nicobar Island (I)	Tc	?	?	—	—	—	10.00000	Ramadoss (1983)
India	Diglipur, Andaman and Nicobar Island (S)	Tc	?	?	—	—	—	3.00000	Ramadoss (1983)
India	Diglipur, Andaman and Nicobar Island (S)	Tm	?	?	—	—	—	1.00000	Ramadoss (1983)
India	Diglipur, Andaman and Nicobar Island (S)	Ts	?	?	—	—	—	1.00000	Ramadoss (1983)
India	Mayabunder, Andaman and Nicobar Island (I)	Tc	?	?	—	—	—	2.00000	Ramadoss (1983)
India	Mayabunder, Andaman and Nicobar Island (S)	Tc	?	?	—	—	—	1.00000	Ramadoss (1983)
India	Mayabunder, Andaman and Nicobar Island (S)	Tm	?	?	—	—	—	2.00000	Ramadoss (1983)
India	Mayabunder, Andaman and Nicobar Island (S)	Ts	?	?	—	—	—	0.50000	Ramadoss (1983)
India	Havelock Island, Andaman and Nicobar Island (I)	Tc	?	?	—	—	—	15.00000	Ramadoss (1983)
India	Havelock Island, Andaman and Nicobar Island (S)	Tc	?	?	—	—	—	5.00000	Ramadoss (1983)
India	Havelock Island, Andaman and Nicobar Island (I)	Tm	?	?	—	—	—	1.00000	Ramadoss (1983)

Continued

Table A3 (Continued) Global density patterns of wild giant clam populations

Country	Localities surveyed	Species	Year of survey	Method of survey	Approximate area of surveys (ha)	Approximate area of surveys (m²)	Number of ind.	Density (m⁻²)	Reference (population survey)
India	Havelock Island, Andaman and Nicobar Island (S)	Tm	?	?	—	—	—	2.00000	Ramadoss (1983)
India	Havelock Island, Andaman and Nicobar Island (S)	Ts	?	?	—	—	—	0.10000	Ramadoss (1983)
India	Neill Island, Andaman and Nicobar Island (I)	Tc	?	?	—	—	—	2.00000	Ramadoss (1983)
India	Neill Island, Andaman and Nicobar Island (S)	Tm	?	?	—	—	—	1.00000	Ramadoss (1983)
India	Neill Island, Andaman and Nicobar Island (S)	Ts	?	?	—	—	—	0.50000	Ramadoss (1983)
India	Rangat, Andaman and Nicobar Island (I)	Tc	?	?	—	—	—	1.00000	Ramadoss (1983)
India	Rangat, Andaman and Nicobar Island (I)	Tm	?	?	—	—	—	0.50000	Ramadoss (1983)
India	Long Island, Andaman and Nicobar Island (I)	Tc	?	?	—	—	—	2.00000	Ramadoss (1983)
India	Long Island, Andaman and Nicobar Island (S)	Tc	?	?	—	—	—	0.50000	Ramadoss (1983)
India	Long Island, Andaman and Nicobar Island (S)	Tm	?	?	—	—	—	2.00000	Ramadoss (1983)
India	Long Island, Andaman and Nicobar Island (S)	Ts	?	?	—	—	—	0.50000	Ramadoss (1983)
India	Port Blair, Andaman and Nicobar Island (I)	Tc	?	?	—	—	—	1.00000	Ramadoss (1983)
India	Port Blair, Andaman and Nicobar Island (S)	Tc	?	?	—	—	—	2.00000	Ramadoss (1983)

Continued

Table A3 (Continued) Global density patterns of wild giant clam populations

Country	Localities surveyed	Species	Year of survey	Method of survey	Approximate area of surveys (ha)	Approximate area of surveys (m²)	Number of ind.	Density (m⁻²)	Reference (population survey)
India	Port Blair, Andaman and Nicobar Island (S)	Tm	?	?	—	—	—	1.00000	Ramadoss (1983)
India	Port Blair, Andaman and Nicobar Island (S)	Ts	?	?	—	—	—	1.00000	Ramadoss (1983)
India	Ross Island, Andaman and Nicobar Island (S)	Tc	?	?	—	—	—	3.00000	Ramadoss (1983)
India	Ross Island, Andaman and Nicobar Island (S)	Tm	?	?	—	—	—	1.00000	Ramadoss (1983)
India	Ross Island, Andaman and Nicobar Island (S)	Ts	?	?	—	—	—	1.00000	Ramadoss (1983)
India	Chiriyatapu, Andaman and Nicobar Island (I)	Tc	?	?	—	—	—	3.00000	Ramadoss (1983)
India	Chiriyatapu, Andaman and Nicobar Island (S)	Tc	?	?	—	—	—	1.00000	Ramadoss (1983)
India	Chiriyatapu, Andaman and Nicobar Island (I)	Tm	?	?	—	—	—	1.00000	Ramadoss (1983)
India	Chiriyatapu, Andaman and Nicobar Island (S)	Tm	?	?	—	—	—	2.00000	Ramadoss (1983)
India	Chiriyatapu, Andaman and Nicobar Island (I)	Ts	?	?	—	—	—	1.00000	Ramadoss (1983)
India	Chiriyatapu, Andaman and Nicobar Island (S)	Ts	?	?	—	—	—	1.00000	Ramadoss (1983)
India	Little Andaman, Andaman and Nicobar Island (S)	Tc	?	?	—	—	—	2.00000	Ramadoss (1983)
India	Little Andaman, Andaman and Nicobar Island (S)	Tm	?	?	—	—	—	1.00000	Ramadoss (1983)
India	Little Andaman, Andaman and Nicobar Island (S)	Ts	?	?	—	—	—	0.50000	Ramadoss (1983)

Continued

Table A3 (Continued)   Global density patterns of wild giant clam populations

Country	Localities surveyed	Species	Year of survey	Method of survey	Approximate area of surveys (ha)	Approximate area of surveys (m²)	Number of ind.	Density (m⁻²)	Reference (population survey)
India	Car Nicobar, Andaman and Nicobar Island (I)	Tc	?	?	—	—	—	1.00000	Ramadoss (1983)
India	Car Nicobar, Andaman and Nicobar Island (S)	Tc	?	?	—	—	—	1.00000	Ramadoss (1983)
India	Car Nicobar, Andaman and Nicobar Island (S)	Tm	?	?	—	—	—	0.50000	Ramadoss (1983)
India	Car Nicobar, Andaman and Nicobar Island (S)	Ts	?	?	—	—	—	0.50000	Ramadoss (1983)
India	East Bay (Katchall), Andaman and Nicobar Island (I)	Tc	?	?	—	—	—	1.00000	Ramadoss (1983)
India	East Bay (Katchall), Andaman and Nicobar Island (S)	Tc	?	?	—	—	—	1.00000	Ramadoss (1983)
India	Camorta area, Andaman and Nicobar Island (I)	Tc	?	?	—	—	—	1.00000	Ramadoss (1983)
India	Camorta area, Andaman and Nicobar Island (I)	Tm	?	?	—	—	—	1.00000	Ramadoss (1983)
India	Camorta area, Andaman and Nicobar Island (S)	Tm	?	?	—	—	—	1.00000	Ramadoss (1983)
India	Campbell Bay, Andaman and Nicobar Island (I)	Tc	?	?	—	—	—	1.00000	Ramadoss (1983)
India	Campbell Bay, Andaman and Nicobar Island (S)	Tc	?	?	—	—	—	1.00000	Ramadoss (1983)
India	Campbell Bay, Andaman and Nicobar Island (S)	Tm	?	?	—	—	—	1.00000	Ramadoss (1983)
India	Lakshadweep Archipelago	Tm	2005	Belt transects; 100 × 20 m	—	330,000	2,748	0.00833	Apte et al. (2010)

Continued

Table A3 (Continued) Global density patterns of wild giant clam populations

Country	Localities surveyed	Species	Year of survey	Method of survey	Approximate area of surveys (ha)	Approximate area of surveys (m^2)	Number of ind.	Density (m^{-2})	Reference (population survey)
India	Lakshadweep Archipelago	Tm	2006	Belt transects; 100 × 20 m	—	268,000	1,948	0.00727	Apte et al. (2010)
Indonesia	Gelean, Karimun Java	Tm	1983	Belt transects; 100 × 10 m	—	2,000	2	0.00100	Brown & Muskanofola (1985)
Indonesia	Gelean, Karimun Java	Ts	1983	Belt transects; 100 × 10 m	—	2,000	1	0.00050	Brown & Muskanofola (1985)
Indonesia	Bengkoang, Karimun Java	Tc	1983	Belt transects; 100 × 10 m	—	5,000	9	0.00180	Brown & Muskanofola (1985)
Indonesia	Bengkoang, Karimun Java	Tm	1983	Belt transects; 100 × 10 m	—	5,000	28	0.00560	Brown & Muskanofola (1985)
Indonesia	Bengkoang, Karimun Java	Ts	1983	Belt transects; 100 × 10 m	—	5,000	9	0.00180	Brown & Muskanofola (1985)
Indonesia	Menjangan Kecil, Karimun Java	Tc	1983	Belt transects; 100 × 10 m	—	3,000	93	0.03100	Brown & Muskanofola (1985)
Indonesia	Menjangan Kecil, Karimun Java	Tm	1983	Belt transects; 100 × 10 m	—	3,000	185	0.06167	Brown & Muskanofola (1985)
Indonesia	Menjangan Kecil, Karimun Java	Ts	1983	Belt transects; 100 × 10 m	—	3,000	27	0.00900	Brown & Muskanofola (1985)
Indonesia	Karang Besi, Karimun Java	Tc	1983	Belt transects; 100 × 10 m	—	1,000	5	0.00500	Brown & Muskanofola (1985)

Continued

Table A3 (Continued) Global density patterns of wild giant clam populations

Country	Localities surveyed	Species	Year of survey	Method of survey	Approximate area of surveys (ha)	Approximate area of surveys (m²)	Number of ind.	Density (m⁻²)	Reference (population survey)
Indonesia	Karang Besi, Karimun Java	Td	1983	Belt transects; 100 × 10 m	—	1,000	1	0.00100	Brown & Muskanofola (1985)
Indonesia	Karang Besi, Karimun Java	Tm	1983	Belt transects; 100 × 10 m	—	1,000	28	0.02800	Brown & Muskanofola (1985)
Indonesia	Karang Besi, Karimun Java	Ts	1983	Belt transects; 100 × 10 m	—	1,000	10	0.01000	Brown & Muskanofola (1985)
Indonesia	Katang Island, Karimun Java	Tc	1983	Belt transects; 50 × 10 m	—	500	2	0.00400	Brown & Muskanofola (1985)
Indonesia	Katang Island, Karimun Java	Tm	1983	Belt transects; 50 × 10 m	—	500	11	0.02200	Brown & Muskanofola (1985)
Indonesia	Katang Island, Karimun Java	Ts	1983	Belt transects; 50 × 10 m	—	500	3	0.00600	Brown & Muskanofola (1985)
Indonesia	Cemara Kecil, Karimun Java	Tc	1983	Belt transects; 100 × 10 m	—	3,000	6	0.00200	Brown & Muskanofola (1985)
Indonesia	Cemara Kecil, Karimun Java	Tm	1983	Belt transects; 100 × 10 m	—	3,000	17	0.00567	Brown & Muskanofola (1985)
Indonesia	Cemara Kecil, Karimun Java	Ts	1983	Belt transects; 100 × 10 m	—	3,000	19	0.00633	Brown & Muskanofola (1985)

Continued

Table A3 (Continued)   Global density patterns of wild giant clam populations

Country	Localities surveyed	Species	Year of survey	Method of survey	Approximate area of surveys (ha)	Approximate area of surveys (m²)	Number of ind.	Density (m⁻²)	Reference (population survey)
Indonesia	Pari Island	Hh	1984	Belt transects; variable lengths × 5 m	—	13,036	25	0.00192	Eliata et al. (2003)
Indonesia	Pari Island	Tc	1984	Belt transects; variable lengths × 5 m	—	13,036	53	0.00407	Eliata et al. (2003)
Indonesia	Pari Island	Tm	1984	Belt transects; variable lengths × 5 m	—	13,036	3	0.00023	Eliata et al. (2003)
Indonesia	Pari Island	Ts	1984	Belt transects; variable lengths × 5 m	—	13,036	3	0.00023	Eliata et al. (2003)
Indonesia	Pari Island	Hh	2003	Belt transects; variable lengths × 5 m	—	31,692	5	0.00016	Eliata et al. (2003)
Indonesia	Pari Island	Tc	2003	Belt transects; variable lengths × 5 m	—	31,692	76	0.00240	Eliata et al. (2003)
Indonesia	Pari Island	Tm	2003	Belt transects; variable lengths × 5 m	—	31,692	1	0.00003	Eliata et al. (2003)
Indonesia	Pari Island	Ts	2003	Belt transects; variable lengths × 5 m	—	31,692	1	0.00003	Eliata et al. (2003)
Indonesia	Kepulauan Seribu	Tc	?	Belt transects; 100 × 5 m	—	1,500	41	0.02733	Yusuf et al. (2009)
Indonesia	Kepulauan Seribu	Tm	?	Belt transects; 100 × 5 m	—	1,500	25	0.01667	Yusuf et al. (2009)
Indonesia	Kepulauan Seribu	Ts	?	Belt transects; 100 × 5 m	—	1,500	40	0.02667	Yusuf et al. (2009)
Indonesia	Manado	Tc	?	Belt transects; 100 × 5 m	—	2,000	17	0.00850	Yusuf et al. (2009)
Indonesia	Manado	Tg	?	Belt transects; 100 × 5 m	—	2,000	1	0.00050	Yusuf et al. (2009)
Indonesia	Manado	Tm	?	Belt transects; 100 × 5 m	—	2,000	3	0.00150	Yusuf et al. (2009)

Continued

206

Table A3 (Continued) Global density patterns of wild giant clam populations

Country	Localities surveyed	Species	Year of survey	Method of survey	Approximate area of surveys (ha)	Approximate area of surveys (m²)	Number of ind.	Density (m⁻²)	Reference (population survey)
Indonesia	Manado	Ts	?	Belt transects; 100 × 5 m	—	2,000	40	0.02000	Yusuf et al. (2009)
Indonesia	Kei Kecil waters, Southeast Maluku	Hh	2009	Quadrat-Transect method; within a 50 × 50 m quadrat	—	22,500	25	0.00111	Hernawan (2010)
Indonesia	Kei Kecil waters, Southeast Maluku	Tc	2009	Quadrat-Transect method; within a 50 × 50 m quadrat	—	22,500	227	0.01009	Hernawan (2010)
Indonesia	Kei Kecil waters, Southeast Maluku	Td	2009	Quadrat-Transect method; within a 50 × 50 m quadrat	—	22,500	2	0.00009	Hernawan (2010)
Indonesia	Kei Kecil waters, Southeast Maluku	Tg	2009	Quadrat-Transect method; within a 50 × 50 m quadrat	—	22,500	1	0.00004	Hernawan (2010)
Indonesia	Kei Kecil waters, Southeast Maluku	Tm	2009	Quadrat-Transect method; within a 50 × 50 m quadrat	—	22,500	67	0.00298	Hernawan (2010)
Indonesia	Kei Kecil waters, Southeast Maluku	Ts	2009	Quadrat-Transect method; within a 50 × 50 m quadrat	—	22,500	14	0.00062	Hernawan (2010)
Indonesia	Savu Sea, East Nusa Tenggara	Hh	2010	Belt transects; 50 × 5 m	—	6,750	11	0.00163	Naguit et al. (2012)
Indonesia	Savu Sea, East Nusa Tenggara	Tc	2010	Belt transects; 50 × 5 m	—	6,750	91	0.01348	Naguit et al. (2012)
Indonesia	Savu Sea, East Nusa Tenggara	Tm	2010	Belt transects; 50 × 5 m	—	6,750	42	0.00622	Naguit et al. (2012)
Indonesia	Savu Sea, East Nusa Tenggara	Ts	2010	Belt transects; 50 × 5 m	—	6,750	17	0.00256	Naguit et al. (2012)

Continued

Table A3 (Continued) Global density patterns of wild giant clam populations

Country	Localities surveyed	Species	Year of survey	Method of survey	Approximate area of surveys (ha)	Approximate area of surveys (m²)	Number of ind.	Density (m⁻²)	Reference (population survey)
Jordan	City Beach	Tm	?	Belt transects; 50 × 5 m	—	250	2	0.00800	Roa-Quaoit (2005)
Jordan	City Beach	Ts	?	Belt transects; 50 × 5 m	—	250	1	0.00200	Roa-Quaoit (2005)
Jordan	Clinker	Tm	?	Belt transects; 50 × 5 m	—	750	3	0.00333	Roa-Quaoit (2005)
Jordan	Clinker	Ts	?	Belt transects; 50 × 5 m	—	750	3	0.00367	Roa-Quaoit (2005)
Jordan	MSS Reserve	Tm	?	Belt transects; 50 × 5 m	—	750	5	0.00700	Roa-Quaoit (2005)
Jordan	MSS Reserve	Ts	?	Belt transects; 50 × 5 m	—	750	4	0.00467	Roa-Quaoit (2005)
Jordan	Tourist Camp	Tm	?	Belt transects; 50 × 5 m	—	750	4	0.00533	Roa-Quaoit (2005)
Jordan	Tourist Camp	Ts	?	Belt transects; 50 × 5 m	—	750	1	0.00177	Roa-Quaoit (2005)
Jordan	Japanese Garden	Tm	?	Belt transects; 50 × 5 m	—	750	7	0.00967	Roa-Quaoit (2005)
Jordan	Japanese Garden	Ts	?	Belt transects; 50 × 5 m	—	750	5	0.00663	Roa-Quaoit (2005)
Jordan	Gorgon	Tm	?	Belt transects; 50 × 5 m	—	500	2	0.00450	Roa-Quaoit (2005)
Jordan	Gorgon	Ts	?	Belt transects; 50 × 5 m	—	500	2	0.00395	Roa-Quaoit (2005)
Jordan	Big Bay	Tm	?	Belt transects; 50 × 5 m	—	500	2	0.00450	Roa-Quaoit (2005)
Jordan	Big Bay	Ts	?	Belt transects; 50 × 5 m	—	500	2	0.00400	Roa-Quaoit (2005)

Continued

Table A3 (Continued) Global density patterns of wild giant clam populations

Country	Localities surveyed	Species	Year of survey	Method of survey	Approximate area of surveys (ha)	Approximate area of surveys (m²)	Number of ind.	Density (m⁻²)	Reference (population survey)
Jordan	North Royal Dive	Tm	?	Belt transects; 50 × 5 m	—	750	3	0.00367	Roa-Quaoit (2005)
Jordan	North Royal Dive	Ts	?	Belt transects; 50 × 5 m	—	750	1	0.00077	Roa-Quaoit (2005)
Jordan	Intelligence	Ts	?	Belt transects; 50 × 5 m	—	500	3	0.00515	Roa-Quaoit (2005)
Jordan	Thermal Plant	Tm	?	Belt transects; 50 × 5 m	—	500	3	0.00500	Roa-Quaoit (2005)
Jordan	Thermal Plant	Ts	?	Belt transects; 50 × 5 m	—	500	2	0.00300	Roa-Quaoit (2005)
Jordan	Gas Pipeline	Tm	?	Belt transects; 50 × 5 m	—	500	3	0.00500	Roa-Quaoit (2005)
Jordan	Gas Pipeline	Ts	?	Belt transects; 50 × 5 m	—	500	1	0.00110	Roa-Quaoit (2005)
Jordan	Jordan Fertilizer Complex	Tm	?	Belt transects; 50 × 5 m	—	500	1	0.00100	Roa-Quaoit (2005)
Jordan	Jordan Fertilizer Complex	Ts	?	Belt transects; 50 × 5 m	—	500	1	0.00230	Roa-Quaoit (2005)
Jordan	Saudi Arabia Border	Tm	?	Belt transects; 50 × 5 m	—	750	8	0.01000	Roa-Quaoit (2005)
Jordan	Saudi Arabia Border	Ts	?	Belt transects; 50 × 5 m	—	750	2	0.00310	Roa-Quaoit (2005)
Jordan	Aqaba	Tm, Ts	2008	Belt transects; 20 × 5 m	—	2,400	—	0.00750	PERSGA (2010)
Madagascar	Nosy Fasy, Andavadoaka	T	2006	Belt transects; 50 × 5 m	—	750	7	0.00933	Harding et al. (2006)
Madagascar	Shark Alley, Andavadoaka	T	2006	Belt transects; 50 × 5 m	—	750	3	0.00400	Harding et al. (2006)

Continued

Table A3 (Continued) Global density patterns of wild giant clam populations

Country	Localities surveyed	Species	Year of survey	Method of survey	Approximate area of surveys (ha)	Approximate area of surveys (m²)	Number of ind.	Density (m⁻²)	Reference (population survey)
Madagascar	Valleys, Andavadoaka	T	2006	Belt transects; 50 × 5 m	—	750	2	0.00267	Harding et al. (2006)
Madagascar	THB, Andavadoaka	T	2006	Belt transects; 50 × 5 m	—	500	1	0.00267	Harding et al. (2006)
Madagascar	Coco Beach, Andavadoaka	T	2006	Belt transects; 50 × 5 m	—	750	1	0.00133	Harding et al. (2006)
Madagascar	Andavadoaka Rock, Andavadoaka	T	2006	Belt transects; 50 × 5 m	—	750	4	0.00533	Harding et al. (2006)
Madagascar	Andavadoaka, SW Madagascar	T	2004–2005	Belt transects; 10 × 2 m	—	5,440	32	0.00533	Nadon et al. (2007)
Madagascar	Sahamalaza, Northern Madagascar	T	2005–2006	Belt transects; 50 × 5 m	—	2,000	—	0.00550	Harding & Randriamanantsoa (2008)
Madagascar	Tanjona, Northern Madagascar	T	2005–2006	Belt transects; 50 × 5 m	—	2,000	—	0.00050	Harding & Randriamanantsoa (2008)
Madagascar	Cap Masoala, Northern Madagascar	T	2005–2006	Belt transects; 50 × 5 m	—	2,000	—	0.00400	Harding & Randriamanantsoa (2008)
Madagascar	Tampolo, Northern Madagascar	T	2005–2006	Belt transects; 50 × 5 m	—	2,000	—	0.00800	Harding & Randriamanantsoa (2008)
Madagascar	Mananara, Northern Madagascar	T	2005–2006	Belt transects; 50 × 5 m	—	2,000	—	0.00350	Harding & Randriamanantsoa (2008)
Madagascar	Andavadoaka, SW Madagascar	Tg	2005	Belt transects; 10 × 2 m	—	—	—	0.05000	Hopkins (2009)
Madagascar	Andavadoaka, SW Madagascar	Tg	2006	Belt transects; 10 × 2 m	—	—	—	0.04000	Hopkins (2009)

Continued

210

Table A3 (Continued)   Global density patterns of wild giant clam populations

Country	Localities surveyed	Species	Year of survey	Method of survey	Approximate area of surveys (ha)	Approximate area of surveys (m²)	Number of ind.	Density (m⁻²)	Reference (population survey)
Madagascar	Andavadoaka, SW Madagascar	Tg	2007	Belt transects; 10 × 2 m	—	—	—	0.03500	Hopkins (2009)
Madagascar	Andavadoaka, SW Madagascar	Tg	2008	Belt transects; 10 × 2 m	—	—	—	0.02000	Hopkins (2009)
Malaysia	Pulau Tioman	Tc	1996	Line intercept transects (density presented is # per 100 m)	—	1,410	26	0.01844	Tan et al. (1998); no area provided but Othman et al. (2010) provided a survey area
Malaysia	Pulau Tioman	Tm	1997	Line intercept transects (density presented is # per 100 m)	—	1,410	141	0.10000	Tan et al. (1998); no area provided but Othman et al. (2010) provided a survey area
Malaysia	Pulau Tioman	Ts	1998	Line intercept transects (density presented is # per 100 m)	—	1,410	66	0.04681	Tan et al. (1998); no area provided but Othman et al. (2010) provided a survey area
Malaysia	Gaya West-mesh reef, Tun Sakaran Marine Park, Semporna Islands Park, Sabah	T	2011	Belt transects; 50 × 1 m	—	—	—	0.01000	Montagne et al. (2013)
Malaysia	Gaya West-outer slope, Tun Sakaran Marine Park, Semporna Islands Park, Sabah	T	2011	Belt transects; 50 × 1 m	—	—	—	0.02000	Montagne et al. (2013)

Continued

Table A3 (Continued) Global density patterns of wild giant clam populations

Country	Localities surveyed	Species	Year of survey	Method of survey	Approximate area of surveys (ha)	Approximate area of surveys (m²)	Number of ind.	Density (m⁻²)	Reference (population survey)
Malaysia	Gaya East-reef flat, Tun Sakaran Marine Park, Semporna Islands Park, Sabah	T	2011	Belt transects; 50 × 1 m	—	—	—	0.20000	Montagne et al. (2013)
Malaysia	Gaya East-inner slope, Tun Sakaran Marine Park, Semporna Islands Park, Sabah	T	2011	Belt transects; 50 × 1 m	—	—	—	0.09000	Montagne et al. (2013)
Malaysia	Mantabuan-mesh reef, Tun Sakaran Marine Park, Semporna Islands Park, Sabah	T	2011	Belt transects; 50 × 1 m	—	—	—	0.11000	Montagne et al. (2013)
Malaysia	Mantabuan-outer slope, Tun Sakaran Marine Park, Semporna Islands Park, Sabah	T	2011	Belt transects; 50 × 1 m	—	—	—	0.07000	Montagne et al. (2013)
Maldives	Fished reefs: Raa Atoll (Beriyanfaru, Hurasfaru, Maadhaffaru, Dhigufaru, Maadhunifaru reefs) and Shaviyani Atoll (Bolissafaru reef)	Tm	1991	Manta tows	—	38,700	125	0.00322	Basker (1991)
Maldives	Fished reefs: Raa Atoll (Beriyanfaru, Hurasfaru, Maadhaffaru, Dhigufaru, Maadhunifaru reefs) and Shaviyani Atoll (Bolissafaru reef)	Ts	1991	Manta tows	—	38,700	15	0.00039	Basker (1991)

Continued

Table A3 (Continued)   Global density patterns of wild giant clam populations

Country	Localities surveyed	Species	Year of survey	Method of survey	Approximate area of surveys (ha)	Approximate area of surveys (m²)	Number of ind.	Density (m⁻²)	Reference (population survey)
Maldives	Unfished reefs: Shaviyani Atoll (Hurasfaru, Kilisfaru, Mathikomandoo reefs) and Lhaviyani Atoll (Gaa en faru, Madivaru, Felivaru reefs)	Tm	1991	Manta tows	—	44,050	174	0.00395	Basker (1991)
Maldives	Unfished reefs: Shaviyani Atoll (Hurasfaru, Kilisfaru, Mathikomandoo reefs) and Lhaviyani Atoll (Gaa en faru, Madivaru, Felivaru reefs)	Ts	1991	Manta tows	—	44,050	48	0.00109	Basker (1991)
Maldives	Kaafu Atoll	Tm	1991	Manta tows	—	42,400	78	0.00185	Basker (1991)
Maldives	Kaafu Atoll	Ts	1991	Manta tows	—	42,400	14	0.00033	Basker (1991)
Marshall Islands	Shark Alley, Jaboan Point, Rongelap Atoll	T	2002	Belt transects; 50 × 5 m	—	500	4	0.00800	Pinca & Beger (2002)
New Caledonia	North Province (Kone, Koumac, Touho, Hienghène)	Hh	2004	Belt transects; 40 × 1 m	—	8,640	—	0.00151	Virly (2004) (Also see Dumas et al. 2011)
New Caledonia	North Province (Kone, Koumac, Touho, Hienghène)	Hh	2004	Manta tows; 300 × 2 m	—	83,400	—	0.00007	Virly (2004) (Also see Dumas et al. 2011)
New Caledonia	North Province (Kone, Koumac, Touho, Hienghène)	Td	2004	Belt transects; 40 × 1 m	—	8,640	—	0.00035	Virly (2004) (Also see Dumas et al. 2011)

Continued

Table A3 (Continued) Global density patterns of wild giant clam populations

Country	Localities surveyed	Species	Year of survey	Method of survey	Approximate area of surveys (ha)	Approximate area of surveys (m²)	Number of ind.	Density (m⁻²)	Reference (population survey)
New Caledonia	North Province (Kone, Koumac, Touho, Hienghène)	Td	2004	Manta tows; 300 × 2 m	—	83,400	—	0.00022	Virly (2004) (Also see Dumas et al. 2011)
New Caledonia	North Province (Kone, Koumac, Touho, Hienghène)	Tm	2004	Belt transects; 40 × 1 m	—	8,640	—	0.03981	Virly (2004) (Also see Dumas et al. 2011)
New Caledonia	North Province (Kone, Koumac, Touho, Hienghène)	Tm	2004	Manta tows; 300 × 2 m	—	83,400	—	0.00752	Virly (2004) (Also see Dumas et al. 2011)
New Caledonia	North Province (Kone, Koumac, Touho, Hienghène)	Ts	2004	Belt transects; 40 × 1 m	—	8,640	—	0.00081	Virly (2004) (Also see Dumas et al. 2011)
New Caledonia	North Province (Kone, Koumac, Touho, Hienghène)	Ts	2004	Manta tows; 300 × 2 m	—	83,400	—	0.00018	Virly (2004) (Also see Dumas et al. 2011)
New Caledonia	North Eastern Lagoon (Poeubo to Hierghène)	Hh	2004	Time swim transects	—	165,400	—	0.00003	McKenna et al. (2006) (Also see Dumas et al. 2011)
New Caledonia	North Eastern Lagoon (Poeubo to Hienghène)	Tc	2004	Time swim transects	—	165,400	—	0.00274	McKenna et al. (2006) (Also see Dumas et al. 2011)
New Caledonia	North Eastern Lagoon (Poeubo to Hienghène)	Td	2004	Time swim transects	—	165,400	—	0.00003	McKenna et al. (2006) (Also see Dumas et al. 2011)
New Caledonia	North Eastern Lagoon (Poeubo to Hienghène)	Tm	2004	Time swim transects	—	165,400	—	0.00787	McKenna et al. (2006) (Also see Dumas et al. 2011)

Continued

Table A3 (Continued) Global density patterns of wild giant clam populations

Country	Localities surveyed	Species	Year of survey	Method of survey	Approximate area of surveys (ha)	Approximate area of surveys (m²)	Number of ind.	Density (m⁻²)	Reference (population survey)
New Caledonia	North Eastern Lagoon (Poeubo to Hienghène)	Ts	2004	Time swim transects	—	165,400	—	0.00077	McKenna et al. (2006) (Also see Dumas et al. 2011)
New Caledonia	Poum	Tc	2007	Belt transects; 25 × 5 m	—	13,125	—	0.03733	Vieux (2009) (Also see Dumas et al. 2011)
New Caledonia	Poum	Tm	2007	Belt transects; 25 × 5 m	—	13,125	—	0.02080	Vieux (2009) (Also see Dumas et al. 2011)
New Caledonia	Poum	Ts	2007	Belt transects; 25 × 5 m	—	13,125	—	0.00305	Vieux (2009) (Also see Dumas et al. 2011)
New Caledonia	Corne Sud	Td	2006	Belt transects; 50 × 10 m	—	7,500	—	0.00067	Wantiez et al. (2007a) (Also see Dumas et al. 2011)
New Caledonia	Corne Sud	Tm	2006	Belt transects; 50 × 10 m	—	7,500	—	0.02320	Wantiez et al. (2007a) (Also see Dumas et al. 2011)
New Caledonia	Corne Sud	Ts	2006	Belt transects; 50 × 10 m	—	7,500	—	0.00080	Wantiez et al. (2007a) (Also see Dumas et al. 2011)
New Caledonia	Ile des Pins	Td	2006	Belt transects; 50 × 10 m	—	11,500	—	0.00070	Wantiez et al. (2007b) (Also see Dumas et al. 2011)
New Caledonia	Ile des Pins	Tm	2006	Belt transects; 50 × 10 m	—	11,500	—	0.00983	Wantiez et al. (2007b) (Also see Dumas et al. 2011)

Continued

Table A3 (Continued) Global density patterns of wild giant clam populations

Country	Localities surveyed	Species	Year of survey	Method of survey	Approximate area of surveys (ha)	Approximate area of surveys (m²)	Number of ind.	Density (m⁻²)	Reference (population survey)
New Caledonia	Ile des Pins	Ts	2006	Belt transects; 50 × 10 m	—	11,500	—	0.00009	Wantiez et al. (2007b) (Also see Dumas et al. 2011)
New Caledonia	Bourail	Hh	2007	Belt transects; 50 × 10 m	—	3,750	—	0.00053	Wantiez et al. (2007c) (Also see Dumas et al. 2011)
New Caledonia	Bourail	Tc	2007	Belt transects; 50 × 10 m	—	3,750	—	0.00320	Wantiez et al. (2007c) (Also see Dumas et al. 2011)
New Caledonia	Bourail	Td	2007	Belt transects; 50 × 10 m	—	3,750	—	0.00027	Wantiez et al. (2007c) (Also see Dumas et al. 2011)
New Caledonia	Bourail	Tm	2007	Belt transects; 50 × 10 m	—	3,750	—	0.02933	Wantiez et al. (2007c) (Also see Dumas et al. 2011)
New Caledonia	Bourail	Ts	2007	Belt transects; 50 × 10 m	—	3,750	—	0.00080	Wantiez et al. (2007c) (Also see Dumas et al. 2011)
New Caledonia	Grand Lagon Nord	Hh	2007	Belt transects; 50 × 5 m	—	7,250	—	0.00055	Wantiez et al. (2008a) (Also see Dumas et al. 2011)
New Caledonia	Grand Lagon Nord	Tc	2007	Belt transects; 50 × 5 m	—	7,250	—	0.00083	Wantiez et al. (2008a) (Also see Dumas et al. 2011)
New Caledonia	Grand Lagon Nord	Td	2007	Belt transects; 50 × 5 m	—	7,250	—	0.00028	Wantiez et al. (2008a) (Also see Dumas et al. 2011)

Continued

Table A3 (Continued) Global density patterns of wild giant clam populations

Country	Localities surveyed	Species	Year of survey	Method of survey	Approximate area of surveys (ha)	Approximate area of surveys (m²)	Number of ind.	Density (m⁻²)	Reference (population survey)
New Caledonia	Grand Lagon Nord	Tm	2007	Belt transects; 50 × 5 m	—	7,250	—	0.01779	Wantiez et al. (2008a) (Also see Dumas et al. 2011)
New Caledonia	Grand Lagon Nord	Ts	2007	Belt transects; 50 × 5 m	—	7,250	—	0.00069	Wantiez et al. (2008a) (Also see Dumas et al. 2011)
New Caledonia	Reefs of Noumea	Tm	2007	Belt transects	—	2,580	—	0.04729	Dumas et al. (Unpublished) (Also see Dumas et al. 2011)
New Caledonia	Merlet	Tc	2008	Belt transects; 50 × 5 m	—	5,250	—	0.00152	Wantiez et al. (2008b) (Also see Dumas et al. 2011)
New Caledonia	Merlet	Td	2008	Belt transects; 50 × 5 m	—	5,250	—	0.00076	Wantiez et al. (2008b) (Also see Dumas et al. 2011)
New Caledonia	Merlet	Tm	2008	Belt transects; 50 × 5 m	—	5,250	—	0.01029	Wantiez et al. (2008b) (Also see Dumas et al. 2011)
New Caledonia	Merlet	Ts	2008	Belt transects; 50 × 5 m	—	5,250	—	0.00057	Wantiez et al. (2008b) (Also see Dumas et al. 2011)
New Caledonia	New Caledonia (50 sites)	Hh	2005–2008	Manta tows; 100 × 2 m	—	227,799	—	0.00003	Purcell et al. (2009) (Also see Dumas et al. 2011)
New Caledonia	New Caledonia (50 sites)	Td	2006–2008	Manta tows; 100 × 2 m	—	227,800	—	0.00013	Purcell et al. (2009) (Also see Dumas et al. 2011)

Continued

Table A3 (Continued) Global density patterns of wild giant clam populations

Country	Localities surveyed	Species	Year of survey	Method of survey	Approximate area of surveys (ha)	Approximate area of surveys (m²)	Number of ind.	Density (m⁻²)	Reference (population survey)
New Caledonia	New Caledonia (50 sites)	Tm	2006–2008	Manta tows; 100×2 m	—	227,800	—	0.00170	Purcell et al. (2009) (Also see Dumas et al. 2011)
New Caledonia	New Caledonia (50 sites)	Ts	2006–2008	Manta tows; 100×2 m	—	227,800	—	0.00038	Purcell et al. (2009) (Also see Dumas et al. 2011)
New Caledonia	Noumea	Hh, Tm, Ts	2007–2009	Belt transects; 20×1 m	—	5,000	280	0.05600	Dumas et al. (2013)
New Caledonia	Noumea	Tm	2007–2009	Belt transects; 20×1 m	—	5,000	276	0.05522	Dumas et al. (2013)
New Caledonia	Noumea	Hh, Ts	2007–2009	Belt transects; 20×1 m	—	5,000	4	0.00078	Dumas et al. (2013)
New Caledonia	Corne Sud	Hh	2010	Belt transects; 50×2 m	4.42	44,200	—	0.00018	Dumas & Andréfouët (Unpublished) (Also see Dumas et al. 2011)
New Caledonia	Corne Sud	Tc	2010	Belt transects; 50×2 m	4.42	44,200	—	0.00013	Dumas & Andréfouët (Unpublished) (Also see Dumas et al. 2011)
New Caledonia	Corne Sud	Td	2010	Belt transects; 50×2 m	4.42	44,200	—	0.00091	Dumas & Andréfouët (Unpublished) (Also see Dumas et al. 2011)

Continued

Table A3 (Continued)   Global density patterns of wild giant clam populations

Country	Localities surveyed	Species	Year of survey	Method of survey	Approximate area of surveys (ha)	Approximate area of surveys (m²)	Number of ind.	Density (m⁻²)	Reference (population survey)
New Caledonia	Corne Sud	Tm	2010	Belt transects; 50 × 2 m	4.42	44,200	—	0.02038	Dumas & Andréfouët (Unpublished) (Also see Dumas et al. 2011)
New Caledonia	Corne Sud	Ts	2010	Belt transects; 50 × 2 m	4.42	44,200	—	0.00013	Dumas & Andréfouët (Unpublished) (Also see Dumas et al. 2011)
New Caledonia	Reserve Merlet	Hh	2010	Belt transects; 50 × 2 m	3.79	37,900	—	0.00055	Dumas & Andréfouët (Unpublished) (Also see Dumas et al. 2011)
New Caledonia	Reserve Merlet	Tc	2010	Belt transects; 50 × 2 m	3.79	37,900	—	0.00011	Dumas & Andréfouët (Unpublished) (Also see Dumas et al. 2011)
New Caledonia	Reserve Merlet	Td	2010	Belt transects; 50 × 2 m	3.79	37,900	—	0.00087	Dumas & Andréfouët (Unpublished) (Also see Dumas et al. 2011)
New Caledonia	Reserve Merlet	Tm	2010	Belt transects; 50 × 2 m	3.79	37,900	—	0.01047	Dumas & Andréfouët (Unpublished) (Also see Dumas et al. 2011)

Continued

Table A3 (Continued) Global density patterns of wild giant clam populations

Country	Localities surveyed	Species	Year of survey	Method of survey	Approximate area of surveys (ha)	Approximate area of surveys (m²)	Number of ind.	Density (m⁻²)	Reference (population survey)
New Caledonia	Reserve Merlet	Ts	2010	Belt transects; 50 × 2 m	3.79	37,900	—	0.00062	Dumas & Andréfouët (Unpublished) (Also see Dumas et al. 2011)
New Caledonia	Kone (plateau de Koniene)	Hh	2010	Belt transects; 50 × 2 m	2.51	25,100	—	0.00045	Dumas & Andréfouët (Unpublished) (Also see Dumas et al. 2011)
New Caledonia	Kone (plateau de Koniene)	Tc	2010	Belt transects; 50 × 2 m	2.51	25,100	—	0.00003	Dumas & Andréfouët (Unpublished) (Also see Dumas et al. 2011)
New Caledonia	Kone (plateau de Koniene)	Td	2010	Belt transects; 50 × 2 m	2.51	25,100	—	0.00045	Dumas & Andréfouët (Unpublished) (Also see Dumas et al. 2011)
New Caledonia	Kone (plateau de Koniene)	Tm	2010	Belt transects; 50 × 2 m	2.51	25,100	—	0.04063	Dumas & Andréfouët (Unpublished) (Also see Dumas et al. 2011)
New Caledonia	Kone (plateau de Koniene)	Ts	2010	Belt transects; 50 × 2 m	2.51	25,100	—	0.00036	Dumas & Andréfouët (Unpublished) (Also see Dumas et al. 2011)

Continued

Table A3 (Continued) Global density patterns of wild giant clam populations

Country	Localities surveyed	Species	Year of survey	Method of survey	Approximate area of surveys (ha)	Approximate area of surveys (m²)	Number of ind.	Density (m⁻²)	Reference (population survey)
New Caledonia	Pouebo	Hh	2010	Belt transects; 50 × 2 m	2.86	28,600	—	0.00118	Dumas & Andréfouët (Unpublished) (Also see Dumas et al. 2011)
New Caledonia	Pouebo	Tc	2010	Belt transects; 50 × 2 m	2.86	28,600	—	0.00597	Dumas & Andréfouët (Unpublished) (Also see Dumas et al. 2011)
New Caledonia	Pouebo	Td	2010	Belt transects; 50 × 2 m	2.86	28,600	—	0.00005	Dumas & Andréfouët (Unpublished) (Also see Dumas et al. 2011)
New Caledonia	Pouebo	Tm	2010	Belt transects; 50 × 2 m	2.86	28,600	—	0.02194	Dumas & Andréfouët (Unpublished) (Also see Dumas et al. 2011)
New Caledonia	Pouebo	Ts	2010	Belt transects; 50 × 2 m	2.86	28,600	—	0.00094	Dumas & Andréfouët (Unpublished) (Also see Dumas et al. 2011)
New Caledonia	Hienghene	Hh	2010	Belt transects; 50 × 2 m	2.86	28,600	—	0.00005	Dumas & Andréfouët (Unpublished) (Also see Dumas et al. 2011)

Continued

Table A3 (Continued) Global density patterns of wild giant clam populations

Country	Localities surveyed	Species	Year of survey	Method of survey	Approximate area of surveys (ha)	Approximate area of surveys (m²)	Number of ind.	Density (m⁻²)	Reference (population survey)
New Caledonia	Hienghene	Tc	2010	Belt transects; 50×2 m	2.86	28,600	—	0.00175	Dumas & Andréfouët (Unpublished) (Also see Dumas et al. 2011)
New Caledonia	Hienghene	Tm	2010	Belt transects; 50×2 m	2.86	28,600	—	0.01437	Dumas & Andréfouët (Unpublished) (Also see Dumas et al. 2011)
New Caledonia	Hienghene	Ts	2010	Belt transects; 50×2 m	2.86	28,600	—	0.00047	Dumas & Andréfouët (Unpublished) (Also see Dumas et al. 2011)
New Caledonia	Lagon Sud-Ouest	Hh	2010	Belt transects; 50×2 m	1.58	15,800	—	0.00002	Dumas & Andréfouët (Unpublished) (Also see Dumas et al. 2011)
New Caledonia	Lagon Sud-Ouest	Td	2010	Belt transects; 50×2 m	1.58	15,800	—	0.00007	Dumas & Andréfouët (Unpublished) (Also see Dumas et al. 2011)
New Caledonia	Lagon Sud-Ouest	Tm	2010	Belt transects; 50×2 m	1.58	15,800	—	0.00444	Dumas & Andréfouët (Unpublished) (Also see Dumas et al. 2011)

Continued

Table A3 (Continued) Global density patterns of wild giant clam populations

Country	Localities surveyed	Species	Year of survey	Method of survey	Approximate area of surveys (ha)	Approximate area of surveys (m²)	Number of ind.	Density (m⁻²)	Reference (population survey)
New Caledonia	Lagon Sud-Ouest	Ts	2010	Belt transects; 50 × 2 m	1.58	15,800	—	0.00075	Dumas & Andréfouët (Unpublished) (Also see Dumas et al. 2011)
New Caledonia	Mare	Tc	2010	Belt transects; 50 × 2 m	4.72	47,200	—	0.00002	Dumas & Andréfouët (Unpublished) (Also see Dumas et al. 2011)
New Caledonia	Mare	Td	2010	Belt transects; 50 × 2 m	4.72	47,200	—	0.00008	Dumas & Andréfouët (Unpublished) (Also see Dumas et al. 2011)
New Caledonia	Mare	Tm	2010	Belt transects; 50 × 2 m	4.72	47,200	—	0.00398	Dumas & Andréfouët (Unpublished) (Also see Dumas et al. 2011)
New Caledonia	Mare	Ts	2010	Belt transects; 50 × 2 m	4.72	47,200	—	0.00011	Dumas & Andréfouët (Unpublished) (Also see Dumas et al. 2011)
Niue	Niue	Tm	1990	Manta tows	9.24	92,400	641	0.00694	Dalzell et al. (1993)
Niue	Niue	Ts	1990	Manta tows	9.24	92,400	80	0.00087	Dalzell et al. (1993)
Niue	Niue	Tm	?	Manta tows	—	33,840	72	0.00213	Kronen et al. (2008)
Niue	Niue	Ts	?	Manta tows	—	33,840	0	0.00000	Kronen et al. (2008)

Continued

Table A3 (Continued) Global density patterns of wild giant clam populations

Country	Localities surveyed	Species	Year of survey	Method of survey	Approximate area of surveys (ha)	Approximate area of surveys (m²)	Number of ind.	Density (m⁻²)	Reference (population survey)
Palau	South of Koror	Hh	1968	Belt transects	—	1,100	4	0.00364	Hardy & Hardy (1969)
Palau	South of Koror	Tc	1968	Belt transects	—	1,100	153	0.13909	Hardy & Hardy (1969)
Palau	South of Koror	Td	1968	Belt transects	—	1,100	6	0.00545	Hardy & Hardy (1969)
Palau	South of Koror	Tg	1968	Belt transects	—	1,100	2	0.00182	Hardy & Hardy (1969)
Palau	South of Koror	Tm	1968	Belt transects	—	1,100	6	0.00545	Hardy & Hardy (1969)
Palau	South of Koror	Ts	1968	Belt transects	—	1,100	7	0.00636	Hardy & Hardy (1969)
Palau	Helen Reef, Western Caroline Islands	Hh	1972	Line transects, Areal, Drift transects, Towing, Power tows	—	43,800	58	0.00132	Hester & Jones (1974)
Palau	Helen Reef, Western Caroline Islands	Tc	1972	Line transects, Areal, Drift transects, Towing, Power tows	—	43,800	Ubiquitous	Ubiquitous	Hester & Jones (1974)
Palau	Helen Reef, Western Caroline Islands	Td	1972	Line transects, Areal, Drift transects, Towing, Power tows	—	43,800	101	0.00231	Hester & Jones (1974)
Palau	Helen Reef, Western Caroline Islands	Tg	1972	Line transects, Areal, Drift transects, Towing, Power tows	—	43,800	82	0.00187	Hester & Jones (1974)
Palau	Helen Reef, Western Caroline Islands	Tm	1972	Line transects, Areal, Drift transects, Towing, Power tows	—	43,800	Ubiquitous	Ubiquitous	Hester & Jones (1974)

Continued

Table A3 (Continued) Global density patterns of wild giant clam populations

Country	Localities surveyed	Species	Year of survey	Method of survey	Approximate area of surveys (ha)	Approximate area of surveys (m²)	Number of ind.	Density (m⁻²)	Reference (population survey)
Palau	Helen Reef, Western Caroline Islands	Ts	1972	Line transects, Areal, Drift transects, Towing, Power tows	—	43,800	1	0.00002	Hester & Jones (1974)
Palau	Helen Reef, Western Caroline Islands	Hh	1975	Line transects, Areal tows	—	24,800	22	0.00089	Bryan & McConnell (1976)
Palau	Helen Reef, Western Caroline Islands	Tc	1975	Line transects, Areal tows	—	24,800	Ubiquitous	Ubiquitous	Bryan & McConnell (1976)
Palau	Helen Reef, Western Caroline Islands	Td	1975	Line transects, Areal tows	—	24,800	6	0.00024	Bryan & McConnell (1976)
Palau	Helen Reef, Western Caroline Islands	Tg	1975	Line transects, Areal tows	—	24,800	4	0.00016	Bryan & McConnell (1976)
Palau	Helen Reef, Western Caroline Islands	Tm	1975	Line transects, Areal tows	—	24,800	629	0.02536	Bryan & McConnell (1976)
Palau	Helen Reef, Western Caroline Islands	Ts	1975	Line transects, Areal tows	—	24,800	2	0.00008	Bryan & McConnell (1976)
Palau	Helen Reef, Western Caroline Islands	Hh	1976	Line transects, Areal tows	—	15,470	63	0.00407	Hirschberger (1980)
Palau	Helen Reef, Western Caroline Islands	Tc	1976	Line transects, Areal tows	—	15,470	Ubiquitous	Ubiquitous	Hirschberger (1980)
Palau	Helen Reef, Western Caroline Islands	Td	1976	Line transects, Areal tows	—	15,470	7	0.00045	Hirschberger (1980)
Palau	Helen Reef, Western Caroline Islands	Tg	1976	Line transects, Areal tows	—	15,470	4	0.00026	Hirschberger (1980)
Palau	Helen Reef, Western Caroline Islands	Tm	1976	Line transects, Areal tows	—	15,470	312	0.02017	Hirschberger (1980)
Palau	Helen Reef, Western Caroline Islands	Ts	1976	Line transects, Areal tows	—	15,470	3	0.00019	Hirschberger (1980)
Papua New Guinea	Longman/Kosmann reef	Tg	1980	?	—	—	—	0.00090	Chesher (1980)

Continued

Table A3 (Continued) Global density patterns of wild giant clam populations

Country	Localities surveyed	Species	Year of survey	Method of survey	Approximate area of surveys (ha)	Approximate area of surveys (m²)	Number of ind.	Density (m⁻²)	Reference (population survey)
Papua New Guinea	Siata reef, Nuakata	Tg	?	?	—	—	—	0.00100	Tarnasky (1980)
Papua New Guinea	Engineer and Conflict Group islands	Hh	1996	?	—	—	—	0.00201	Kinch (2001)
Papua New Guinea	Engineer and Conflict Group islands	Hp	1996	?	—	—	—	0.00003	Kinch (2001)
Papua New Guinea	Engineer and Conflict Group islands	Tc	1996	?	—	—	—	0.00119	Kinch (2001)
Papua New Guinea	Engineer and Conflict Group islands	Td	1996	?	—	—	—	0.00053	Kinch (2001)
Papua New Guinea	Engineer and Conflict Group islands	Tg	1996	?	—	—	—	0.00004	Kinch (2001)
Papua New Guinea	Engineer and Conflict Group islands	Tm	1997	?	—	—	—	0.00179	Kinch (2001)
Papua New Guinea	Engineer and Conflict Group islands	Ts	1998	?	—	—	—	0.00058	Kinch (2001)
Papua New Guinea	Milne Bay Province	Hh	2001	1126 sites were surveyed	—	—	—	0.00004	Kinch (2002)
Papua New Guinea	Milne Bay Province	Tc	2001	1126 sites were surveyed	—	—	—	0.00149	Kinch (2002)
Papua New Guinea	Milne Bay Province	Td	2001	1126 sites were surveyed	—	—	—	0.00003	Kinch (2002)
Papua New Guinea	Milne Bay Province	Tg	2001	1126 sites were surveyed	—	—	—	0.00008	Kinch (2002)
Papua New Guinea	Milne Bay Province	Tm	2001	1126 sites were surveyed	—	—	—	0.00018	Kinch (2002)
Papua New Guinea	Milne Bay Province	Ts	2001	1126 sites were surveyed	—	—	—	0.00014	Kinch (2002)

Continued

Table A3 (Continued)   Global density patterns of wild giant clam populations

Country	Localities surveyed	Species	Year of survey	Method of survey	Approximate area of surveys (ha)	Approximate area of surveys (m²)	Number of ind.	Density (m⁻²)	Reference (population survey)
Papua New Guinea	Kavieng lagoonal system, New Ireland Province	Tm	2015	Belt transects; 50 × 8 m	—	48,000	181	0.00377	Militz et al. (2015)
Papua New Guinea	Kavieng lagoonal system, New Ireland Province	Tno	2015	Belt transects; 50 × 8 m	—	48,000	131	0.00273	Militz et al. (2015)
Philippines	Central Visayas, Visayas	Tc	1984–1985	Quadrat	3	30,000	49	0.00163	Alcala (1986)
Philippines	Central Visayas, Visayas	Tm	1984–1985	Quadrat	3	30,000	24	0.00080	Alcala (1986)
Philippines	Central Visayas, Visayas	Ts	1984–1985	Quadrat	3	30,000	20	0.00067	Alcala (1986)
Philippines	West Visayas, Visayas	Tc	1984–1985	Flowmeter method	0.7	7,000	16	0.00229	Alcala (1986)
Philippines	West Visayas, Visayas	Tm	1984–1985	Flowmeter method	0.7	7,000	21	0.00300	Alcala (1986)
Philippines	West Visayas, Visayas	Ts	1984–1985	Flowmeter method	0.7	7,000	92	0.01314	Alcala (1986)
Philippines	Cagayan, Sulu Seas	Tc	1984–1985	Flowmeter method	0.5645	5,645	102	0.01807	Alcala (1986)
Philippines	Cagayan, Sulu Seas	Tm	1984–1985	Flowmeter method	0.5645	5,645	144	0.02551	Alcala (1986)
Philippines	Cagayan, Sulu Seas	Ts	1984–1985	Flowmeter method	0.5645	5,645	7	0.00124	Alcala (1986)
Philippines	Palawan	Hh	1984–1985	Flowmeter method	2.1	21,000	29	0.00138	Alcala (1986)
Philippines	Palawan	Tc	1984–1985	Flowmeter method	2.1	21,000	6,901	0.32862	Alcala (1986)
Philippines	Palawan	Td	1984–1985	Flowmeter method	2.1	21,000	8	0.00038	Alcala (1986)
Philippines	Palawan	Tm	1984–1985	Flowmeter method	2.1	21,000	56	0.00267	Alcala (1986)
Philippines	Palawan	Ts	1984–1985	Flowmeter method	2.1	21,000	57	0.00271	Alcala (1986)
Philippines	Western Pangasinan, Luzon	Tc	1984–1986	Belt transects; 100 × 5 m	5.3	53,000	39	0.00074	Juinio et al. (1989)
Philippines	Western Pangasinan, Luzon	Tm	1984–1986	Belt transects; 100 × 5 m	5.3	53,000	6	0.00011	Juinio et al. (1989)
Philippines	Western Pangasinan, Luzon	Ts	1984–1986	Belt transects; 100 × 5 m	5.3	53,000	17	0.00032	Juinio et al. (1989)
Philippines	Polillo, Quezon, Luzon	Hh	1984–1986	Belt transects; 100 × 5 m	2.1	21,000	5	0.00024	Juinio et al. (1989)
Philippines	Polillo, Quezon, Luzon	Tc	1984–1986	Belt transects; 100 × 5 m	2.1	21,000	7,138	0.33990	Juinio et al. (1989)

Continued

Table A3 (Continued) Global density patterns of wild giant clam populations

Country	Localities surveyed	Species	Year of survey	Method of survey	Approximate area of surveys (ha)	Approximate area of surveys (m²)	Number of ind.	Density (m⁻²)	Reference (population survey)
Philippines	Polillo, Quezon, Luzon	Td	1984–1986	Belt transects; 100 × 5 m	2.1	21,000	6	0.00029	Juinio et al. (1989)
Philippines	Polillo, Quezon, Luzon	Tg	1984–1986	Belt transects; 100 × 5 m	2.1	21,000	2	0.00010	Juinio et al. (1989)
Philippines	Polillo, Quezon, Luzon	Tm	1984–1986	Belt transects; 100 × 5 m	2.1	21,000	112	0.00533	Juinio et al. (1989)
Philippines	Polillo, Quezon, Luzon	Ts	1984–1986	Belt transects; 100 × 5 m	2.1	21,000	147	0.00700	Juinio et al. (1989)
Philippines	Zambales, Luzon	Tc	1984–1986	Belt transects; 100 × 5 m	1.04	10,400	19	0.00183	Juinio et al. (1989)
Philippines	Zambales, Luzon	Tm	1984–1986	Belt transects; 100 × 5 m	1.04	10,400	7	0.00067	Juinio et al. (1989)
Philippines	Zambales, Luzon	Ts	1984–1986	Belt transects; 100 × 5 m	1.04	10,400	1	0.00010	Juinio et al. (1989)
Philippines	Albay, Luzon	Tc	1984–1986	Belt transects; 100 × 5 m	1.45	14,500	119	0.00821	Juinio et al. (1989)
Philippines	Albay, Luzon	Tm	1984–1986	Belt transects; 100 × 5 m	1.45	14,500	102	0.00703	Juinio et al. (1989)
Philippines	Albay, Luzon	Ts	1984–1986	Belt transects; 100 × 5 m	1.45	14,500	60	0.00414	Juinio et al. (1989)
Philippines	Sorsogon, Luzon	Tc	1984–1986	Belt transects; 100 × 5 m	1.48	14,800	46	0.00311	Juinio et al. (1989)
Philippines	Sorsogon, Luzon	Tm	1984–1986	Belt transects; 100 × 5 m	1.48	14,800	121	0.00818	Juinio et al. (1989)
Philippines	Sorsogon, Luzon	Ts	1984–1986	Belt transects; 100 × 5 m	1.48	14,800	4	0.00027	Juinio et al. (1989)
Philippines	Calatagan, Luzon	Tc	1984–1986	Belt transects; 100 × 5 m	1.11	11,100	14	0.00126	Juinio et al. (1989)

Continued

Table A3 (Continued) Global density patterns of wild giant clam populations

Country	Localities surveyed	Species	Year of survey	Method of survey	Approximate area of surveys (ha)	Approximate area of surveys (m²)	Number of ind.	Density (m⁻²)	Reference (population survey)
Philippines	Calatagan, Luzon	Tm	1984–1986	Belt transects; 100 × 5 m	1.11	11,100	12	0.00108	Juinio et al. (1989)
Philippines	Calatagan, Luzon	Ts	1984–1986	Belt transects; 100 × 5 m	1.11	11,100	29	0.00261	Juinio et al. (1989)
Philippines	Lubang Island, Luzon	Hh	1984–1986	Belt transects; 100 × 5 m	1.49	14,900	1	0.00007	Juinio et al. (1989)
Philippines	Lubang Island, Luzon	Tc	1984–1986	Belt transects; 100 × 5 m	1.49	14,900	84	0.00564	Juinio et al. (1989)
Philippines	Lubang Island, Luzon	Tm	1984–1986	Belt transects; 100 × 5 m	1.49	14,900	22	0.00148	Juinio et al. (1989)
Philippines	Lubang Island, Luzon	Ts	1984–1986	Belt transects; 100 × 5 m	1.49	14,900	20	0.00134	Juinio et al. (1989)
Philippines	Ambil Island, Luzon	Hh	1984–1986	Belt transects; 100 × 5 m	2.5	25,000	2	0.00008	Juinio et al. (1989)
Philippines	Ambil Island, Luzon	Tc	1984–1986	Belt transects; 100 × 5 m	2.5	25,000	67	0.00268	Juinio et al. (1989)
Philippines	Ambil Island, Luzon	Td	1984–1986	Belt transects; 100 × 5 m	2.5	25,000	9	0.00036	Juinio et al. (1989)
Philippines	Ambil Island, Luzon	Tg	1984–1986	Belt transects; 100 × 5 m	2.5	25,000	1	0.00004	Juinio et al. (1989)
Philippines	Ambil Island, Luzon	Tm	1984–1986	Belt transects; 100 × 5 m	2.5	25,000	112	0.00448	Juinio et al. (1989)
Philippines	Ambil Island, Luzon	Ts	1984–1986	Belt transects; 100 × 5 m	2.5	25,000	82	0.00328	Juinio et al. (1989)
Philippines	Apo Reef, Luzon	Tc	1984–1986	Belt transects; 100 × 5 m	0.88	8,800	26	0.00295	Juinio et al. (1989)
Philippines	Apo Reef, Luzon	Td	1984–1986	Belt transects; 100 × 5 m	0.88	8,800	1	0.00011	Juinio et al. (1989)

Continued

Table A3 (Continued) Global density patterns of wild giant clam populations

Country	Localities surveyed	Species	Year of survey	Method of survey	Approximate area of surveys (ha)	Approximate area of surveys (m²)	Number of ind.	Density (m⁻²)	Reference (population survey)
Philippines	Apo Reef, Luzon	Tm	1984–1986	Belt transects; 100 × 5 m	0.88	8,800	83	0.00943	Juinio et al. (1989)
Philippines	Apo Reef, Luzon	Ts	1984–1986	Belt transects; 100 × 5 m	0.88	8,800	1	0.00011	Juinio et al. (1989)
Philippines	Puerto Galera, Luzon	Tc	1984–1986	Belt transects; 100 × 5 m	1.46	14,600	4	0.00027	Juinio et al. (1989)
Philippines	Puerto Galera, Luzon	Tm	1984–1986	Belt transects; 100 × 5 m	1.46	14,600	14	0.00096	Juinio et al. (1989)
Philippines	Puerto Galera, Luzon	Ts	1984–1986	Belt transects; 100 × 5 m	1.46	14,600	14	0.00096	Juinio et al. (1989)
Philippines	NE Negros, Visayas	Tc	1984–1986	Belt transects; 100 × 5 m	0.29	2,900	2	0.00069	Juinio et al. (1989)
Philippines	NE Negros, Visayas	Tm	1984–1986	Belt transects; 100 × 5 m	0.29	2,900	1	0.00034	Juinio et al. (1989)
Philippines	NE Negros, Visayas	Ts	1984–1986	Belt transects; 100 × 5 m	0.29	2,900	1	0.00034	Juinio et al. (1989)
Philippines	El Nido, Palawan	Hh	1984–1986	Belt transects; 100 × 5 m	2.55	25,500	12	0.00047	Juinio et al. (1989)
Philippines	El Nido, Palawan	Hp	1984–1986	Belt transects; 100 × 5 m	2.55	25,500	1	0.00004	Juinio et al. (1989)
Philippines	El Nido, Palawan	Tc	1984–1986	Belt transects; 100 × 5 m	2.55	25,500	280	0.01098	Juinio et al. (1989)
Philippines	El Nido, Palawan	Tm	1984–1986	Belt transects; 100 × 5 m	2.55	25,500	23	0.00090	Juinio et al. (1989)
Philippines	El Nido, Palawan	Ts	1984–1986	Belt transects; 100 × 5 m	2.55	25,500	125	0.00490	Juinio et al. (1989)
Philippines	Inagauan-Aborlan, Palawan	Td	1984–1986	Belt transects; 100 × 5 m	0.45	4,500	1	0.00022	Juinio et al. (1989)

Continued

Table A3 (Continued) Global density patterns of wild giant clam populations

Country	Localities surveyed	Species	Year of survey	Method of survey	Approximate area of surveys (ha)	Approximate area of surveys (m²)	Number of ind.	Density (m⁻²)	Reference (population survey)
Philippines	Inagauan-Aborlan, Palawan	Tm	1984–1986	Belt transects; 100 × 5 m	0.45	4,500	3	0.00067	Juinio et al. (1989)
Philippines	Inagauan-Aborlan, Palawan	Ts	1984–1986	Belt transects; 100 × 5 m	0.45	4,500	1	0.00022	Juinio et al. (1989)
Philippines	Sombrero Island, Palawan	Hh	1984–1986	Belt transects; 100 × 5 m	0.2	2,000	1	0.00050	Juinio et al. (1989)
Philippines	Sombrero Island, Palawan	Tc	1984–1986	Belt transects; 100 × 5 m	0.2	2,000	50	0.02500	Juinio et al. (1989)
Philippines	Sombrero Island, Palawan	Tm	1984–1986	Belt transects; 100 × 5 m	0.2	2,000	13	0.00650	Juinio et al. (1989)
Philippines	Sombrero Island, Palawan	Ts	1984–1986	Belt transects; 100 × 5 m	0.2	2,000	2	0.00100	Juinio et al. (1989)
Philippines	Cagayan Island, Palawan	Hh	1984–1986	Belt transects; 100 × 5 m	0.64	6,400	5	0.00078	Juinio et al. (1989)
Philippines	Cagayan Island, Palawan	Tc	1984–1986	Belt transects; 100 × 5 m	0.64	6,400	33	0.00516	Juinio et al. (1989)
Philippines	Cagayan Island, Palawan	Tg	1984–1986	Belt transects; 100 × 5 m	0.64	6,400	1	0.00016	Juinio et al. (1989)
Philippines	Cagayan Island, Palawan	Tm	1984–1986	Belt transects; 100 × 5 m	0.64	6,400	167	0.02609	Juinio et al. (1989)
Philippines	Cagayan Island, Palawan	Ts	1984–1986	Belt transects; 100 × 5 m	0.64	6,400	3	0.00047	Juinio et al. (1989)
Philippines	Camiguin Island, Mindanao	Tc	1984–1986	Belt transects; 100 × 5 m	2.13	21,300	24	0.00113	Juinio et al. (1989)
Philippines	Camiguin Island, Mindanao	Tm	1984–1986	Belt transects; 100 × 5 m	2.13	21,300	66	0.00310	Juinio et al. (1989)
Philippines	Camiguin Island, Mindanao	Ts	1984–1986	Belt transects; 100 × 5 m	2.13	21,300	33	0.00155	Juinio et al. (1989)

Continued

Table A3 (Continued) Global density patterns of wild giant clam populations

Country	Localities surveyed	Species	Year of survey	Method of survey	Approximate area of surveys (ha)	Approximate area of surveys (m²)	Number of ind.	Density (m⁻²)	Reference (population survey)
Philippines	Central Visayas, Visayas (Sumilon Island, Balicasag Island, Pamilacan Island)	Tc	1992	Belt transects with 1×1 m quadrats	—	10	41	4.10000	Calumpong & Cadiz (1993)
Philippines	Central Visayas, Visayas (Sumilon Island, Balicasag Island, Pamilacan Island)	Tm	1992	Belt transects with 1×1 m quadrats	—	10	2	0.20000	Calumpong & Cadiz (1993)
Philippines	Cagayan Island, Palawan	Tc	1992	Belt transects with 1×1 m quadrats	—	10	7	0.70000	Calumpong & Cadiz (1993)
Philippines	Cagayan Island, Palawan	Tm	1992	Belt transects with 1×1 m quadrats	—	10	5	0.50000	Calumpong & Cadiz (1993)
Philippines	Tubbataha reefs	Tc	1992	Belt transects with 1×1 m quadrats	—	10	36	3.60000	Calumpong & Cadiz (1993)
Philippines	Tubbataha reefs	Tm	1992	Belt transects with 1×1 m quadrats	—	10	7	0.70000	Calumpong & Cadiz (1993)
Philippines	Tubbataha reefs	Hh	2005	Belt transects; 150×2 m	—	4,500	10	0.00222	Dolorosa & Schoppe (2005)
Philippines	Tubbataha reefs	Tc	2005	Belt transects; 150×2 m	—	4,500	104	0.02311	Dolorosa & Schoppe (2005)
Philippines	Tubbataha reefs	Tm	2005	Belt transects; 150×2 m	—	4,500	29	0.00644	Dolorosa & Schoppe (2005)
Philippines	Tubbataha reefs	Ts	2005	Belt transects; 150×2 m	—	4,500	2	0.00044	Dolorosa & Schoppe (2005)
Philippines	Tubbataha reefs	H spp.	2009–2010	Belt transects; 20×2 m	—	8,320	26	0.00313	Dolorosa (2010)
Philippines	Tubbataha reefs	Tc	2010–2010	Belt transects; 20×2 m	—	8,320	541	0.06502	Dolorosa (2010)

Continued

Table A3 (Continued) Global density patterns of wild giant clam populations

Country	Localities surveyed	Species	Year of survey	Method of survey	Approximate area of surveys (ha)	Approximate area of surveys (m²)	Number of ind.	Density (m⁻²)	Reference (population survey)
Philippines	Tubbataha reefs	Tm	2011–2010	Belt transects; 20 × 2 m	—	8,320	32	0.00385	Dolorosa (2010)
Philippines	Tubbataha reefs	Hp	2008	Belt transects; 100 × 2 m	—	4,200	41	0.00976	Dolorosa & Jontila (2012)
Philippines	Tubbataha reefs	Tc	2008	Belt transects; 100 × 2 m	—	4,200	287	0.06833	Dolorosa & Jontila (2012)
Philippines	Tubbataha reefs	Tm	2008	Belt transects; 100 × 2 m	—	4,200	23	0.00548	Dolorosa & Jontila (2012)
Philippines	Sabang Reef Fish Sanctuary (inside), Honda Bay, Puerto Princesa City, Palawan	Td	2004	Belt transects; 100 × 2 m and two permanent quadrats; 5 × 20 m	—	—	—	0.01000	Gonzales et al. (2014)
Philippines	Sabang Reef Fish Sanctuary (inside), Honda Bay, Puerto Princesa City, Palawan	Ts	2004	Belt transects; 100 × 2 m and two permanent quadrats; 5 × 20 m	—	—	—	0.01500	Gonzales et al. (2014a)
Philippines	Meara Island	Td	2004	Belt transect; 100 × 2 m	—	—	—	0.02500	Gonzales et al. (2014a)
Philippines	Meara Island	Tg	2004	Belt transect; 100 × 2 m	—	—	—	0.01500	Gonzales et al. (2014a)
Philippines	Meara Island	Ts	2004	Belt transect; 100 × 2 m	—	—	—	0.05500	Gonzales et al. (2014a)
Philippines	Apulit Island, Taytay Bay, Palawan	Tc	2006	Belt transect; 100 × ? m	—	—	—	0.06810	Gonzales et al. (2014b)
Philippines	Ranger Station, Tubbataha Reefs Natural Park, Cagayancillo, Palawan	Tc	2009	Belt transects; 20 × 2 m	—	1,600.00	—	0.39250	Conales et al. (2015) [Dolorosa unpublished data]

Continued

233

Table A3 (Continued)   Global density patterns of wild giant clam populations

Country	Localities surveyed	Species	Year of survey	Method of survey	Approximate area of surveys (ha)	Approximate area of surveys (m²)	Number of ind.	Density (m⁻²)	Reference (population survey)
Philippines	Ranger Station, Tubbataha Reefs Natural Park, Cagayancillo, Palawan	Tc	2010	Coral head surveys (n=10)	—	40.75	236	5.79141	Conales et al. (2015)
Pitcairn Islands	Oeno Atoll	Tm	?	—	—	—	—	8 to 10	Irving & Dawson (2013)
Republic of Kiribati	Abemama Atoll, Central Gilbert Islands group	Hh	1985	Manta tows	2860	28,600,000	10,050	0.00035	Munro (1988)
Republic of Kiribati	Abemama Atoll, Central Gilbert Islands group	Tg	1985	Manta tows	2860	28,600,000	6,592	0.00023	Munro (1988)
Republic of Kiribati	Abemama Atoll, Central Gilbert Islands group	Ts	1985	Manta tows	2860	28,600,000	137	0.00000	Munro (1988)
Republic of Kiribati	Abiang Atoll, Central Gilbert Islands group	Hh	1985	Manta tows	8990	89,900,000	19,846	0.00005	Munro (1988)
Republic of Kiribati	Abiang Atoll, Central Gilbert Islands group	Tg	1985	Manta tows	8990	89,900,000	4,931	0.00005	Munro (1988)
Republic of Kiribati	Abiang Atoll, Central Gilbert Islands group	Ts	1985	Manta tows	8990	89,900,000	5,319	0.00006	Munro (1988)
Republic of Kiribati	Maiana Atoll, Central Gilbert Islands group	Hh	1985	Manta tows	2800	28,000,000	1,600	0.00006	Munro (1988)
Republic of Kiribati	Maiana Atoll, Central Gilbert Islands group	Tg	1985	Manta tows	2800	28,000,000	2,150	0.00008	Munro (1988)
Republic of Kiribati	Maiana Atoll, Central Gilbert Islands group	Ts	1985	Manta tows	2800	28,000,000	2,580	0.00009	Munro (1988)
Republic of Kiribati	Tarawa Atoll, Central Gilbert Islands group	Hh	1985	Manta tows	2960	29,600,000	500	0.00002	Munro (1988)
Republic of Kiribati	Tarawa Atoll, Central Gilbert Islands group	Tg	1985	Manta tows	2960	29,600,000	560	0.00002	Munro (1988)
Republic of Kiribati	Tarawa Atoll, Central Gilbert Islands group	Ts	1985	Manta tows	2960	29,600,000	780	0.00003	Munro (1988)

Continued

Table A3 (Continued) Global density patterns of wild giant clam populations

Country	Localities surveyed	Species	Year of survey	Method of survey	Approximate area of surveys (ha)	Approximate area of surveys (m²)	Number of ind.	Density (m⁻²)	Reference (population survey)
Republic of Kiribati	Caroline Atoll (formerly Gilbert Islands)	Tm	?	—	—	—	—	35.00000	Kepler & Kepler (1994)
Republic of Kiribati	Kingman Atoll, Northern Line Islands	Tm	2005	Belt transects; 60 × 2 m	—	1,200	—	0.00750	Sandin et al. (2008)
Republic of Kiribati	Palmyra Atoll, Northern Line Islands	Tm	2005	Belt transects; 60 × 2 m	—	1,200	—	0.00080	Sandin et al. (2008)
Republic of Kiribati	Kiritimati Atoll, Northern Line Islands	Tm	2005	Belt transects; 60 × 2 m	—	600	—	0.00450	Sandin et al. (2008)
Republic of Kiribati	Millennium Atoll (Caroline Atoll)	Tm	2009	Belt transects; 25 × 1 m	174	—	—	1.50000	Barott et al. (2010)
Samoan Archipelago	Upolu, Tutuila, Aunu'u, Ofu-Olosega, Ta'u	Tm, Ts	1994–1995	Belt transects; 50 × 2 m	29	292,000	88	0.00030	Green & Craig (1999)
Saudi Arabia	Tuwwal, Jeddah	Tm	?	Belt transects; 10 × 5 m (varied number of plots)	—	300	63	0.21000	Bodoy (1984)
Saudi Arabia	Shoiba, South of Jeddah	Tm	?	Belt transects; 10 × 5 m (varied number of plots)	—	500	19	0.03800	Bodoy (1984)
Saudi Arabia	North of Sharm-el-Abhur, Jeddah	Tm	?	Belt transects; 10 × 5 m (varied number of plots)	—	300	40	0.13333	Bodoy (1984)
Saudi Arabia	Jeddah northern Corniche	Tm	?	Belt transects; 10 × 5 m (varied number of plots)	—	250	6	0.02400	Bodoy (1984)
Saudi Arabia	Al-Wajh, Jeddah, Farasan Islands	Tm, Ts	2002	Belt transects; 20 × 5 m	—	—	—	0.01410	PERSGA (2010)

Continued

Table A3 (Continued) Global density patterns of wild giant clam populations

Country	Localities surveyed	Species	Year of survey	Method of survey	Approximate area of surveys (ha)	Approximate area of surveys (m²)	Number of ind.	Density (m⁻²)	Reference (population survey)
Saudi Arabia	Haql, Maqna, Duba, Umm Lajj, Mastura, Jeddah, Al Lith, Assir, Farasan	Tm, Ts	2008	Belt transects; 20 × 5 m	—	6,400	—	0.03850	PERSGA (2010)
Seychelles	Seychelles Islands	Tc	1989	Belt transects	—	—	—	1 to 10	Selin et al. (1992)
Seychelles	Aride Island Beach	Tm	2001–2002	Daily 30-min walk at low tide	—	—	—	3—Occasional (9 to 20 specimens)	Agombar et al. (2003)
Seychelles	Aride Island Beach	Ts	2001–2002	Daily 30-min walk at low tide	—	—	—	4—Fairly common (21 to 30 specimens)	Agombar et al. (2003)
Singapore	Southern Islands (7 sites)	Tc	2003	Belt transects; 2 m wide	—	9,670	7	0.00072	Guest et al. (2008)
Singapore	Southern Islands (7 sites)	Tm	2003	Belt transects; 2 m wide	—	9,670	1	0.00010	Guest et al. (2008)
Singapore	Southern Islands (7 sites)	Ts	2003	Belt transects; 2 m wide	—	9,670	15	0.00155	Guest et al. (2008)
Singapore	Southern Islands (29 sites)	Tc	2009–2010	Belt transects; 6 m wide and quadrats; 10 × 10 to 20 × 20 m²	—	87,515	31	0.00035	Neo & Todd (2012, 2013)
Singapore	Southern Islands (29 sites)	Ts	2009–2010	Belt transects; 6 m wide and quadrats; 10 × 10 to 20 × 20 m²	—	87,515	28	0.00032	Neo & Todd (2012, 2013)
Solomon Islands	Solomon Islands	Hh	2004	Belt transects; 300 × 2 m (shallow) and 250 × 50 m (deep)	—	118,350	4	0.00003	Ramohia (2006)

Continued

Table A3 (Continued) Global density patterns of wild giant clam populations

Country	Localities surveyed	Species	Year of survey	Method of survey	Approximate area of surveys (ha)	Approximate area of surveys (m²)	Number of ind.	Density (m⁻²)	Reference (population survey)
Solomon Islands	Solomon Islands	Tc	2004	Belt transects; 300 × 2 m (shallow) and 250 × 50 m (deep)	—	118,350	60	0.00051	Ramohia (2006)
Solomon Islands	Solomon Islands	Td	2004	Belt transects; 300 × 2 m (shallow) and 250 × 50 m (deep)	—	118,350	17	0.00014	Ramohia (2006)
Solomon Islands	Solomon Islands	Tg	2004	Belt transects; 300 × 2 m (shallow) and 250 × 50 m (deep)	—	118,350	12	0.00010	Ramohia (2006)
Solomon Islands	Solomon Islands	Tm	2004	Belt transects; 300 × 2 m (shallow) and 250 × 50 m (deep)	—	118,350	115	0.00097	Ramohia (2006)
Solomon Islands	Solomon Islands	Ts	2004	Belt transects; 300 × 2 m (shallow) and 250 × 50 m (deep)	—	118,350	95	0.00080	Ramohia (2006)
South China Sea (Malaysia)	Pulau Layang Layang, Sabah	Hh	2002	Timed Roving Diver technique; 1 hour; 9 sampling sites	—	—	1	?	Sahari et al. (2002)
South China Sea (Malaysia)	Pulau Layang Layang, Sabah	Tc	2002	Timed Roving Diver technique; 1 hour; 9 sampling sites	—	—	71	?	Sahari et al. (2002)
South China Sea (Malaysia)	Pulau Layang Layang, Sabah	Tg	2002	Timed Roving Diver technique; 1 hour; 9 sampling sites	—	—	6	?	Sahari et al. (2002)

Continued

Table A3 (Continued) Global density patterns of wild giant clam populations

Country	Localities surveyed	Species	Year of survey	Method of survey	Approximate area of surveys (ha)	Approximate area of surveys (m²)	Number of ind.	Density (m⁻²)	Reference (population survey)
South China Sea (Malaysia)	Pulau Layang Layang, Sabah	Tm	2002	Timed Roving Diver technique; 1 hour; 9 sampling sites	—	—	8	?	Sahari et al. (2002)
South China Sea (Malaysia)	Pulau Layang Layang, Sabah	Ts	2002	Timed Roving Diver technique; 1 hour; 9 sampling sites	—	—	37	?	Sahari et al. (2002)
South China Sea (North Spratly Islands)	Trident—JOMSRE III	Tc	2005	Belt transects; 20 × 10 m	—	800	4	0.00500	Van Long et al. (2008)
South China Sea (North Spratly Islands)	Trident—JOMSRE III	Ts	2005	Belt transects; 20 × 10 m	—	800	10	0.01250	Van Long et al. (2008)
South China Sea (North Spratly Islands)	NE North East Cay	Tc	2005	Belt transects; 20 × 10 m	—	800	5	0.00625	Van Long et al. (2008)
South China Sea (North Spratly Islands)	NE North East Cay	Ts	2005	Belt transects; 20 × 10 m	—	800	8	0.01000	Van Long et al. (2008)
South China Sea (North Spratly Islands)	E North East Cay	Tc	2005	Belt transects; 20 × 10 m	—	800	7	0.00875	Van Long et al. (2008)
South China Sea (North Spratly Islands)	E North East Cay	Ts	2005	Belt transects; 20 × 10 m	—	800	6	0.00750	Van Long et al. (2008)
South China Sea (North Spratly Islands)	SW North East Cay	Tc	2005	Belt transects; 20 × 10 m	—	800	3	0.00375	Van Long et al. (2008)

Continued

Table A3 (Continued) Global density patterns of wild giant clam populations

Country	Localities surveyed	Species	Year of survey	Method of survey	Approximate area of surveys (ha)	Approximate area of surveys (m²)	Number of ind.	Density (m⁻²)	Reference (population survey)
South China Sea (North Spratly Islands)	SW North East Cay	Ts	2005	Belt transects; 20 × 10 m	—	800	9	0.01125	Van Long et al. (2008)
South China Sea (North Spratly Islands)	NE South West Cay	Tc	2005	Belt transects; 20 × 10 m	—	800	4	0.00500	Van Long et al. (2008)
South China Sea (North Spratly Islands)	NE South West Cay	Ts	2005	Belt transects; 20 × 10 m	—	800	11	0.01375	Van Long et al. (2008)
South China Sea (North Spratly Islands)	SW South West Cay	Tc	2005	Belt transects; 20 × 10 m	—	800	4	0.00500	Van Long et al. (2008)
South China Sea (North Spratly Islands)	SW South West Cay	Ts	2005	Belt transects; 20 × 10 m	—	800	6	0.00750	Van Long et al. (2008)
South China Sea (Spratly Islands)	Trident Shoal; Station 1	Tc	2005	Belt transects; 20 × 10 m (10 m depth)	—	800	2	0.00250	Lasola & Hoang (2008)
South China Sea (Spratly Islands)	Trident Shoal; Station 1	Ts	2005	Belt transects; 20 × 10 m (10 m depth)	—	800	5	0.00625	Lasola & Hoang (2008)
South China Sea (Spratly Islands)	South West Cay; Station 2	Tc	2005	Belt transects; 20 × 10 m (10 m depth)	—	800	4	0.00500	Lasola & Hoang (2008)
South China Sea (Spratly Islands)	South West Cay; Station 2	Ts	2005	Belt transects; 20 × 10 m (10 m depth)	—	800	6	0.00750	Lasola & Hoang (2008)
South China Sea (Spratly Islands)	North East Cay; Station 3	Tc	2005	Belt transects; 20 × 10 m (10 m depth)	—	800	5	0.00625	Lasola & Hoang (2008)

Continued

Table A3 (Continued) Global density patterns of wild giant clam populations

Country	Localities surveyed	Species	Year of survey	Method of survey	Approximate area of surveys (ha)	Approximate area of surveys (m²)	Number of ind.	Density (m⁻²)	Reference (population survey)
South China Sea (Spratly Islands)	North East Cay; Station 3	Ts	2005	Belt transects; 20 × 10 m (10 m depth)	—	800	8	0.01000	Lasola & Hoang (2008)
South China Sea (Spratly Islands)	North East Cay; Station 4	Tc	2005	Belt transects; 20 × 10 m (10 m depth)	—	800	7	0.00875	Lasola & Hoang (2008)
South China Sea (Spratly Islands)	North East Cay; Station 4	Ts	2005	Belt transects; 20 × 10 m (10 m depth)	—	800	6	0.00750	Lasola & Hoang (2008)
South China Sea (Spratly Islands)	South West Cay; Station 5	Tc	2005	Belt transects; 20 × 10 m (10 m depth)	—	800	4	0.00500	Lasola & Hoang (2008)
South China Sea (Spratly Islands)	South West Cay; Station 5	Ts	2005	Belt transects; 20 × 10 m (10 m depth)	—	800	11	0.01375	Lasola & Hoang (2008)
South China Sea (Spratly Islands)	North East Cay; Station 6	Tc	2005	Belt transects; 20 × 10 m (10 m depth)	—	800	3	0.00375	Lasola & Hoang (2008)
South China Sea (Spratly Islands)	North East Cay; Station 6	Ts	2005	Belt transects; 20 × 10 m (10 m depth)	—	800	9	0.01125	Lasola & Hoang (2008)
South China Sea (Spratly Islands)	NE Cay; North Danger Reef	Hh	2007	Belt transects; 500 × 1 m	—	1,500	—	0.00060	Calumpong & Macansantos (2008)
South China Sea (Spratly Islands)	NE Cay; North Danger Reef	Tc	2007	Belt transects; 500 × 1 m	—	1,500	—	0.01000	Calumpong & Macansantos (2008)

Continued

Table A3 (Continued) Global density patterns of wild giant clam populations

Country	Localities surveyed	Species	Year of survey	Method of survey	Approximate area of surveys (ha)	Approximate area of surveys (m²)	Number of ind.	Density (m⁻²)	Reference (population survey)
South China Sea (Spratly Islands)	NE Cay; North Danger Reef	Tm	2007	Belt transects; 500 × 1 m	—	1,500	—	0.00400	Calumpong & Macansantos (2008)
South China Sea (Spratly Islands)	NE Cay; North Danger Reef	Ts	2007	Belt transects; 500 × 1 m	—	1,500	—	0.00060	Calumpong & Macansantos (2008)
South China Sea (Spratly Islands)	S Reef; North Danger Reef	Tm	2007	Belt transects; 500 × 1 m	—	1,000	—	0.01200	Calumpong & Macansantos (2008)
South China Sea (Spratly Islands)	S Reef; North Danger Reef	Ts	2007	Belt transects; 500 × 1 m	—	1,000	—	0.00200	Calumpong & Macansantos (2008)
South China Sea (Spratly Islands)	N Reef; North Danger Reef	Hh	2007	Belt transects; 500 × 1 m	—	1,000	—	0.00100	Calumpong & Macansantos (2008)
South China Sea (Spratly Islands)	N Reef; North Danger Reef	Tc	2007	Belt transects; 500 × 1 m	—	1,000	—	0.00300	Calumpong & Macansantos (2008)
South China Sea (Spratly Islands)	N Reef; North Danger Reef	Tm	2007	Belt transects; 500 × 1 m	—	1,000	—	0.01300	Calumpong & Macansantos (2008)
South China Sea (Spratly Islands)	N Reef; North Danger Reef	Ts	2007	Belt transects; 500 × 1 m	—	1,000	—	0.00100	Calumpong & Macansantos (2008)
South China Sea (Spratly Islands)	Jenkins Reef; North Danger Reef	Tc	2007	Belt transects; 500 × 1 m	—	500	11	0.02200	Calumpong & Macansantos (2008)

Continued

Table A3 (Continued) Global density patterns of wild giant clam populations

Country	Localities surveyed	Species	Year of survey	Method of survey	Approximate area of surveys (ha)	Approximate area of surveys (m²)	Number of ind.	Density (m⁻²)	Reference (population survey)
South China Sea (Spratly Islands)	Jenkins Reef; North Danger Reef	Tm	2007	Belt transects; 500 × 1 m	—	500	23	0.04600	Calumpong & Macansantos (2008)
South China Sea (Spratly Islands)	Jenkins Reef; North Danger Reef	Ts	2007	Belt transects; 500 × 1 m	—	500	2	0.00400	Calumpong & Macansantos (2008)
South China Sea (Spratly Islands)	Dickinson Reef; Jackson Atoll	Tm	2007	Belt transects; 500 × 1 m	—	500	3	0.00600	Calumpong & Macansantos (2008)
South China Sea (Spratly Islands)	Hoare Reef; Jackson Atoll	Tc	2007	Belt transects; 500 × 1 m	—	500	1	0.00200	Calumpong & Macansantos (2008)
South China Sea (Spratly Islands)	Hoare Reef; Jackson Atoll	Tm	2007	Belt transects; 500 × 1 m	—	500	1	0.00200	Calumpong & Macansantos (2008)
South China Sea (Spratly Islands)	Danger Reef; Jackson Atoll	Tc	2007	Belt transects; 500 × 1 m	—	500	2	0.00400	Calumpong & Macansantos (2008)
South China Sea (Spratly Islands)	Danger Reef; Jackson Atoll	Tm	2007	Belt transects; 500 × 1 m	—	500	4	0.00800	Calumpong & Macansantos (2008)
South China Sea (Spratly Islands)	Patch Reef; Jackson Atoll	Ts	2007	Belt transects; 500 × 1 m	—	500	1	0.00200	Calumpong & Macansantos (2008)
Sudan	Wingate, Sanganeb, Tawartit, Suakin, Tala Tala Saghir	Tm, Ts	2002	Belt transects; 20 × 5 m	—	—	—	0.01480	PERSGA (2010)
Sudan	O'Seif, Arkiyai, Port-Sudan, Suakin	Tm, Ts	2008	Belt transects; 20 × 5 m	—	3,200	—	0.03250	PERSGA (2010)

Continued

Table A3 (Continued) Global density patterns of wild giant clam populations

Country	Localities surveyed	Species	Year of survey	Method of survey	Approximate area of surveys (ha)	Approximate area of surveys (m²)	Number of ind.	Density (m⁻²)	Reference (population survey)
Tanzania	North; Chumbe's reef sanctuary	T	2004	Belt transects; 20 × 5 m	—	400	6	0.01500	Daniels (2004)
Tanzania	Middle; Chumbe's reef sanctuary	T	2004	Belt transects; 20 × 5 m	—	400	5	0.01250	Daniels (2004)
Tanzania	South; Chumbe's reef sanctuary	T	2004	Belt transects; 20 × 5 m	—	400	14	0.03500	Daniels (2004)
Thailand	Lee-Pae Island, Andaman Sea	Tc	?	Belt transects; 100 × 4 m	—	6,400	1,562	0.24406	Chantrapornsyl et al. (1996)
Thailand	Lee-Pae Island, Andaman Sea	Tm	?	Belt transects; 100 × 4 m	—	6,400	403	0.06297	Chantrapornsyl et al. (1996)
Thailand	Lee-Pae Island, Andaman Sea	Ts	?	Belt transects; 100 × 4 m	—	6,400	1	0.00016	Chantrapornsyl et al. (1996)
Thailand	Surin Islands (11 sites)	mostly Tc; Tm, Ts scarce	2003	Belt transects; 20 × 5 m	—	6,000	210	0.03500	Koh et al. (2003)
Thailand	Surin Islands (16 sites)	mostly Tc; Tm, Ts scarce	2004	Belt transects; 20 × 5 m	—	10,400	154	0.01481	Loh et al. (2004)
Thailand	Mannai Island, Rayong Province	Tc	2009–2010	Belt transect; 100 × 2 m	—	200	117	0.58500	Junchompoo et al. (2013)
Thailand	Mannai Island, Rayong Province	Ts	2010–2010	Belt transect; 100 × 2 m	—	200	12	0.06000	Junchompoo et al. (2013)
Tokelau	Fakaofo Atoll	Tm	1989	Surface tow and reef flat transects	21.44	214,400	34,312	0.16004	Braley (1989)
Tokelau	Nukunonu Atoll	Tm	1989	Surface tow and reef flat transects	19.67	196,700	44,318	0.22531	Braley (1989)
Tokelau	Atafu Atoll	Tm	1989	Surface tow and reef flat transects	11.1	111,000	11,048	0.09953	Braley (1989)

Continued

Table A3 (Continued) Global density patterns of wild giant clam populations

Country	Localities surveyed	Species	Year of survey	Method of survey	Approximate area of surveys (ha)	Approximate area of surveys (m²)	Number of ind.	Density (m⁻²)	Reference (population survey)
Tokelau	Fakaofo Atoll	Ts	1989	Surface tow and reef flat transects	21.44	214,400	25	0.00012	Braley (1989)
Tokelau	Nukunonu Atoll	Ts	1989	Surface tow and reef flat transects	19.67	196,700	206	0.00105	Braley (1989)
Tokelau	Atafu Atoll	Ts	1989	Surface tow and reef flat transects	11.1	111,000	0	0.00000	Braley (1989)
Tonga	East Malinoa Island	T	1978–1979	Snorkelling and SCUBA (time-based surveys)	—	—	—	6.5 clam per man 0.5 hour	Langi & Hesitoni 'Aloua (1988)
Tonga	West Malinoa Island	T	1979–1979	Snorkelling and SCUBA (time-based surveys)	—	—	—	4 clam per man 0.5 hour	Langi & Hesitoni 'Aloua (1988)
Tonga	NW Fafa Island	T	1980–1979	Snorkelling and SCUBA (time-based surveys)	—	—	—	2 clam per man 0.5 hour	Langi & Hesitoni 'Aloua (1988)
Tonga	SW Fafa Island	T	1981–1979	Snorkelling and SCUBA (time-based surveys)	—	—	—	1.5 clam per man 0.5 hour	Langi & Hesitoni 'Aloua (1988)
Tonga	NW Makaha'a Island	T	1982–1979	Snorkelling and SCUBA (time-based surveys)	—	—	—	3.5 clam per man 0.5 hour	Langi & Hesitoni 'Aloua (1988)
Tonga	Hakau Mamao 1	T	1983–1979	Snorkelling and SCUBA (time-based surveys)	—	—	—	9.5 clam per man 0.5 hour	Langi & Hesitoni 'Aloua (1988)
Tonga	Hakau Mamao 2	T	1984–1979	Snorkelling and SCUBA (time-based surveys)	—	—	—	7 clam per man 0.5 hour	Langi & Hesitoni 'Aloua (1988)

Continued

Table A3 (Continued) Global density patterns of wild giant clam populations

Country	Localities surveyed	Species	Year of survey	Method of survey	Approximate area of surveys (ha)	Approximate area of surveys (m²)	Number of ind.	Density (m⁻²)	Reference (population survey)
Tonga	East Malinoa Island	Tm, Ts	1987	Snorkelling and SCUBA (time-based surveys)	—	—	—	1 clam per man 0.5 hour	Langi & Hesitoni 'Aloua (1988)
Tonga	West Malinoa Island	Tm, Ts	1988	Snorkelling and SCUBA (time-based surveys)	—	—	—	0.7 clam per man 0.5 hour	Langi & Hesitoni 'Aloua (1988)
Tonga	NW Fafa Island	Tm, Ts	1989	Snorkelling and SCUBA (time-based surveys)	—	—	—	4.4 clam per man 0.5 hour	Langi & Hesitoni 'Aloua (1988)
Tonga	NW Makaha'a Island	Tm, Ts	1990	Snorkelling and SCUBA (time-based surveys)	—	—	—	3.3 clam per man 0.5 hour	Langi & Hesitoni 'Aloua (1988)
Tonga	Hakau Mamao 1	Tm, Ts	1991	Snorkelling and SCUBA (time-based surveys)	—	—	—	19 clam per man 0.5 hour	Langi & Hesitoni 'Aloua (1988)
Tonga	Hakau Mamao 2	Tm, Ts	1992	Snorkelling and SCUBA (time-based surveys)	—	—	—	22.8 clam per man 0.5 hour	Langi & Hesitoni 'Aloua (1988)
Tonga	Vava'u Island Group	Td	1987	Timed surveys (64.35 h)	—	—	0	0 clam per man hour	Chesher (1993)
Tonga	Vava'u Island Group	Tm	1987	Timed surveys (64.35 h)	—	—	1,183	18.4 clam per man hour	Chesher (1993)
Tonga	Vava'u Island Group	Ts	1987	Timed surveys (64.35 h)	—	—	132	2.1 clam per man hour	Chesher (1993)
Tonga	Vava'u Island Group	Td	1988	Timed surveys (69.92 h)	—	—	2	0.03 clam per man hour	Chesher (1993)
Tonga	Vava'u Island Group	Tm	1988	Timed surveys (69.92 h)	—	—	1,032	14.8 clam per man hour	Chesher (1993)
Tonga	Vava'u Island Group	Ts	1988	Timed surveys (69.92 h)	—	—	99	1.4 clam per man hour	Chesher (1993)

Continued

245

Table A3 (Continued)   Global density patterns of wild giant clam populations

Country	Localities surveyed	Species	Year of survey	Method of survey	Approximate area of surveys (ha)	Approximate area of surveys (m²)	Number of ind.	Density (m⁻²)	Reference (population survey)
Tonga	Vava'u Island Group	Td	1989	Timed surveys (64.75 h)	—	—	45	0.7 clam per man hour	Chesher (1993)
Tonga	Vava'u Island Group	Tm	1989	Timed surveys (64.75 h)	—	—	1,336	20.6 clam per man hour	Chesher (1993)
Tonga	Vava'u Island Group	Ts	1989	Timed surveys (64.75 h)	—	—	161	2.5 clam per man hour	Chesher (1993)
Tonga	Vava'u Island Group	Td	1990	Timed surveys (55.37 h)	—	—	82	1.5 clam per man hour	Chesher (1993)
Tonga	Vava'u Island Group	Tm	1990	Timed surveys (55.37 h)	—	—	1,044	18.9 clam per man hour	Chesher (1993)
Tonga	Vava'u Island Group	Ts	1990	Timed surveys (55.37 h)	—	—	266	4.8 clam per man hour	Chesher (1993)
Tonga	Lofanga, Ha'apai	Tmb	1989	SCUBA search (per man hour effort)	—	—	1	1 clam per man hour	Ledua et al. (1993)
Tonga	Auhangamea channel, Uiha Island, Ha'apai	Tmb	1989	SCUBA search (per man hour effort)	—	—	12	2.5 clam per man hour	Ledua et al. (1993)
Tonga	Kahefahefa Island, Vava'u	Tmb	1990	SCUBA search (per man hour effort)	—	—	5	0.35 clam per man hour	Ledua et al. (1993)
Tonga	Kahefahefa Island, Vava'u	Tmb	1991	SCUBA search (per man hour effort)	—	—	6	0.04 clam per man hour	Ledua et al. (1993)
Tonga	Kahefahefa Island, Vava'u	Tmb	1991	SCUBA search (per man hour effort)	—	—	1	2 clam per man hour	Ledua et al. (1993)
Tonga	Faka'osi Reef, Pangai, Ha'apai	Tmb	1992	SCUBA search (per man hour effort)	—	—	16	0.33 clam per man hour	Ledua et al. (1993)
Tonga	Luahoko Island, Ha'apai	Tmb	1992	SCUBA search (per man hour effort)	—	—	21	9.1 clam per man hour	Ledua et al. (1993)
Tonga	Atata Island, Tongatapu Island Group	Tm	1993	Towing, Free swimming, SCUBA (90 minutes)	—	—	12	4 clam per man 0.5 hour	Tu'avao et al. (1995)

Continued

Table A3 (Continued) Global density patterns of wild giant clam populations

Country	Localities surveyed	Species	Year of survey	Method of survey	Approximate area of surveys (ha)	Approximate area of surveys (m²)	Number of ind.	Density (m⁻²)	Reference (population survey)
Tonga	Atata Island, Tongatapu Island Group	Ts	1993	Free swimming, SCUBA (60 minutes)	—	—	4	2 clam per man 0.5 hour	Tu'avao et al. (1995)
Tonga	NW Fafa Island, Tongatapu Island Group	Td	1993	Towing (40 minutes)	—	—	1	0.75 clam per man 0.5 hour	Tu'avao et al. (1995)
Tonga	NW Fafa Island, Tongatapu Island Group	Tm	1993	Towing (40 minutes)	—	—	4	3 clam per man 0.5 hour	Tu'avao et al. (1995)
Tonga	NW Fafa Island, Tongatapu Island Group	Ts	1993	Towing (40 minutes)	—	—	2	1.5 clam per man 0.5 hour	Tu'avao et al. (1995)
Tonga	Hakau Mamao Reef, Tongatapu Island Group	Td	1993	Towing (30 minutes)	—	—	2	2 clam per man 0.5 hour	Tu'avao et al. (1995)
Tonga	Hakau Mamao Reef, Tongatapu Island Group	Tm	1993	Towing (90 minutes)	—	—	9	3 clam per man 0.5 hour	Tu'avao et al. (1995)
Tonga	Hakau Mamao Reef, Tongatapu Island Group	Ts	1993	Towing (60 minutes)	—	—	1	0.5 clam per man 0.5 hour	Tu'avao et al. (1995)
Tonga	Niutoua, Tongatapu Island Group	Tm	1993	Towing (60 minutes)	—	—	0.5	0.25 clam per man 0.5 hour	Tu'avao et al. (1995)
Tonga	Haveluiku, Tongatapu Island Group	Tm	1993	Towing (60 minutes)	—	—	1.5	0.75 clam per man 0.5 hour	Tu'avao et al. (1995)
Tonga	Haveluiku, Tongatapu Island Group	Ts	1993	Towing (60 minutes)	—	—	0.5	0.25 clam per man 0.5 hour	Tu'avao et al. (1995)
Tonga	Monotapu, Tongatapu Island Group	Td	1993	Towing (60 minutes)	—	—	3	1.5 clam per man 0.5 hour	Tu'avao et al. (1995)
Tonga	Monotapu, Tongatapu Island Group	Tm	1993	Towing (60 minutes)	—	—	4	2 clam per man 0.5 hour	Tu'avao et al. (1995)
Tonga	Ha'atafu, Tongatapu Island Group	Tm	1993	Towing (60 minutes)	—	—	31.5	15.75 clam per man 0.5 hour	Tu'avao et al. (1995)
Tonga	Ha'atafu, Tongatapu Island Group	Ts	1993	Towing (60 minutes)	—	—	1	0.5 clam per man 0.5 hour	Tu'avao et al. (1995)

Continued

247

Table A3 (Continued) Global density patterns of wild giant clam populations

Country	Localities surveyed	Species	Year of survey	Method of survey	Approximate area of surveys (ha)	Approximate area of surveys (m²)	Number of ind.	Density (m⁻²)	Reference (population survey)
Tonga	Hakauiki Reef, Tongatapu Island Group	Tm	1993	Towing (60 minutes)	—	—	33.5	16.75 clam per man 0.5 hour	Tu'avao et al. (1995)
Tonga	Malinoa Island, Tongatapu Island Group	Tm	1994	Free swimming (180 minutes)	—	—	8	1.33 clam per man 0.5 hour	Tu'avao et al. (1995)
Tuvalu	Nukufetau Atoll	Tm	?	?	—	—	—	0.00630	Braley (1988)
Tuvalu	Funafuti Atoll	Tm	?	?	—	—	—	0.01010	Braley (1988)
Tuvalu	Nukulaelae Atoll	Tm	?	?	—	—	—	0.00031	Braley (1988)
Tuvalu	Nukufetau Atoll	Ts	?	?	—	—	—	0.00007	Braley (1988)
Tuvalu	Funafuti Atoll	Ts	?	?	—	—	—	0.00014	Braley (1988)
Tuvalu	Nukulaelae Atoll	Ts	?	?	—	—	—	0.00000	Braley (1988)
Tuvalu	Nanumea Atoll	Tm	?	?	—	—	—	0.00006	Langi (1990)
Tuvalu	Nui Atoll	Tm	?	?	—	—	—	0.00027	Langi (1990)
Tuvalu	Funafuti Atoll	Tm	2004	Belt transects; 300 × 2 m and 40 × 1 m	—	54,120	164	0.00303	Sauni et al. (2008)
Tuvalu	Funafuti Atoll	Ts	2004	Belt transects; 300 × 2 m and 40 × 1 m	—	54,120	16	0.00030	Sauni et al. (2008)
Tuvalu	Nukufetau Atoll	Tm	2004	Belt transects; 300 × 2 m and 40 × 1 m	—	46,320	125	0.00271	Sauni et al. (2008)
Tuvalu	Nukufetau Atoll	Ts	2004	Belt transects; 300 × 2 m	—	43,200	5	0.00012	Sauni et al. (2008)
Tuvalu	Vaitupu Islands (central group)	Tm	2005	Belt transects; 300 × 2 m	—	43,200	37	0.00086	Sauni et al. (2008)
Tuvalu	Niutao Islands (northern group)	Tm	2005	Belt transects; 300 × 2 m and 40 × 1 m	—	17,280	3	0.00017	Sauni et al. (2008)

Continued

Table A3 (Continued)   Global density patterns of wild giant clam populations

Country	Localities surveyed	Species	Year of survey	Method of survey	Approximate area of surveys (ha)	Approximate area of surveys (m²)	Number of ind.	Density (m⁻²)	Reference (population survey)
Tuvalu	Nanumea Atoll	Td, Tm, Ts	2010	Belt transects; 50 × 4 m	—	5,400	0	0.00000	Job & Ceccarelli (2012)
Tuvalu	Nukulaelae Atoll	Td, Tm, Ts	2010	Belt transects; 50 × 4 m	—	6,000	0	0.00000	Job & Ceccarelli (2012)
Tuvalu	Funafuti Atoll	Td, Tm, Ts	2010	Belt transects; 25 × 4 m	—	12,600	114	0.00904	Job & Ceccarelli (2012)
Tuvalu	Funafuti Conservation Area (FCA), Funafuti Atoll	Tm	2011	Manta tows; 300 × 6 m	—	21,600	—	0.00238	Siaosi et al. (2012)
Tuvalu	Funafuti Conservation Area (FCA), Funafuti Atoll	Ts	2011	Manta tows; 300 × 6 m	—	21,600	—	0.00046	Siaosi et al. (2012)
Tuvalu	Fongafale, Funafuti Atoll	Tm	2011	Belt transects; 40 × 6 m	—	2,400	—	0.00500	Siaosi et al. (2012)
Tuvalu	Fongafale, Funafuti Atoll	Ts	2011	Belt transects; 40 × 6 m	—	2,400	—	0.00167	Siaosi et al. (2012)
Tuvalu	Funafuti Conservation Area (FCA), Funafuti Atoll	Tm	2011	Belt transects; 40 × 6 m	—	1,440	—	0.01250	Siaosi et al. (2012)
Vanuatu	Inyeug Island, Anatom (Lagoon patch reef)	Hh	1988	Spot dives, manta tows, or belt transects	—	—	—	0.00100	Zann & Ayling (1988)
Vanuatu	Moso Island, Efate	Hh	1988	Spot dives, manta tows, or belt transects	—	—	—	0.00030	Zann & Ayling (1988)
Vanuatu	Cook's Reef, Efate (Lagoon)	Hh	1988	Spot dives, manta tows, or belt transects	—	—	—	0.00250	Zann & Ayling (1988)
Vanuatu	Cook's Reef, Efate (Slope)	Hh	1988	Spot dives, manta tows, or belt transects	—	—	—	0.00010	Zann & Ayling (1988)
Vanuatu	SE Reef, Pentecost	Hh	1988	Spot dives, manta tows, or belt transects	—	—	—	0.00090	Zann & Ayling (1988)

Continued

Table A3 (Continued) Global density patterns of wild giant clam populations

Country	Localities surveyed	Species	Year of survey	Method of survey	Approximate area of surveys (ha)	Approximate area of surveys (m²)	Number of ind.	Density (m⁻²)	Reference (population survey)
Vanuatu	Lesalav Bay, Pentecost	Hh	1988	Spot dives, manta tows, or belt transects	—	—	—	0.00010	Zann & Ayling (1988)
Vanuatu	Reef Islands, Pentecost	Hh	1988	Spot dives, manta tows, or belt transects	—	—	—	0.00230	Zann & Ayling (1988)
Vanuatu	Hog Bay, Espiritu Santo	Hh	1988	Spot dives, manta tows, or belt transects	—	—	—	0.00020	Zann & Ayling (1988)
Vanuatu	Moso Island, Efate	Tc	1988	Spot dives, manta tows, or belt transects	—	—	—	0.00030	Zann & Ayling (1988)
Vanuatu	Port Anatom, Anatom	Tm	1988	Spot dives, manta tows, or belt transects	—	—	—	0.00160	Zann & Ayling (1988)
Vanuatu	Inyeug Island, Anatom (Reef slope)	Tm	1988	Spot dives, manta tows, or belt transects	—	—	—	0.00500	Zann & Ayling (1988)
Vanuatu	Inyeug Island, Anatom (Lagoon patch reef)	Tm	1988	Spot dives, manta tows, or belt transects	—	—	—	0.00200	Zann & Ayling (1988)
Vanuatu	Port Patrick, Anatom	Tm	1988	Spot dives, manta tows, or belt transects	—	—	—	0.00160	Zann & Ayling (1988)
Vanuatu	Lakariata, Tanna	Tm	1988	Spot dives, manta tows, or belt transects	—	—	—	0.00050	Zann & Ayling (1988)
Vanuatu	Lelepa, Efate	Tm	1988	Spot dives, manta tows, or belt transects	—	—	—	0.00030	Zann & Ayling (1988)
Vanuatu	Moso Island, Efate	Tm	1988	Spot dives, manta tows, or belt transects	—	—	—	0.00070	Zann & Ayling (1988)
Vanuatu	Cook's Reef, Efate (Lagoon)	Tm	1988	Spot dives, manta tows, or belt transects	—	—	—	0.00100	Zann & Ayling (1988)
Vanuatu	Cook's Reef, Efate (Slope)	Tm	1988	Spot dives, manta tows, or belt transects	—	—	—	0.00050	Zann & Ayling (1988)
Vanuatu	SE Reef, Pentecost	Tm	1988	Spot dives, manta tows, or belt transects	—	—	—	0.00060	Zann & Ayling (1988)

Continued

Table A3 (Continued) Global density patterns of wild giant clam populations

Country	Localities surveyed	Species	Year of survey	Method of survey	Approximate area of surveys (ha)	Approximate area of surveys (m²)	Number of ind.	Density (m⁻²)	Reference (population survey)
Vanuatu	Loltong Bay, Pentecost	Tm	1988	Spot dives, manta tows, or belt transects	—	—	—	0.00200	Zann & Ayling (1988)
Vanuatu	Lesalav Bay, Pentecost	Tm	1988	Spot dives, manta tows, or belt transects	—	—	—	0.00090	Zann & Ayling (1988)
Vanuatu	Reef Islands, Pentecost	Tm	1988	Spot dives, manta tows, or belt transects	—	—	—	0.00130	Zann & Ayling (1988)
Vanuatu	Hog Bay, Espiritu Santo	Tm	1988	Spot dives, manta tows, or belt transects	—	—	—	0.00020	Zann & Ayling (1988)
Vanuatu	Maskelynes, Malekula Group	Tc	1988	Belt transects; 50 × 5 m	—	18,750	7	0.00075	Zann & Ayling (1988)
Vanuatu	Maskelynes, Malekula Group	Tm	1988	Belt transects; 50 × 5 m	—	18,750	14	0.00075	Zann & Ayling (1988)
Vanuatu	Atchin Island, Malekula Group	Tm	1988	Belt transects; 50 × 5 m	—	3,750	2	0.00053	Zann & Ayling (1988)
Vanuatu	Malecula, Malekula Group	Tm	1988	Belt transects; 50 × 5 m	—	7,500	2	0.00027	Zann & Ayling (1988)
Vanuatu	Maskelynes, Malekula Group	Ts	1988	Belt transects; 50 × 5 m	—	18,750	4	0.00021	Zann & Ayling (1988)
Vanuatu	Malecula, Malekula Group	Ts	1988	Belt transects; 50 × 5 m	—	7,500	1	0.00013	Zann & Ayling (1988)
Vanuatu	Inside taboo area; Analcauhat, Aneityum	Tm	2011–2012	—	—	—	—	0.00733	Nimoho et al. (2013)
Vanuatu	Outside taboo area; Analcauhat, Aneityum	Tm	2011–2012	—	—	—	—	0.00275	Nimoho et al. (2013)
Vanuatu	Inside taboo area; Mangaliliu, Efate	Tm	2011–2012	—	—	—	—	0.01214	Nimoho et al. (2013)
Vanuatu	Outside taboo area; Mangaliliu, Efate	Tm	2011–2012	—	—	—	—	0.01412	Nimoho et al. (2013)

Continued

Table A3 (Continued) Global density patterns of wild giant clam populations

Country	Localities surveyed	Species	Year of survey	Method of survey	Approximate area of surveys (ha)	Approximate area of surveys (m²)	Number of ind.	Density (m⁻²)	Reference (population survey)
Viet Nam	Mju Island, Nha Trang Bay, Khanh Hoa Province	Tc	1981	—	—	—	—	0.50000	Latypov (2006)
Viet Nam	Hon Bay Canh Island and Hon Cau Island, Con Dao Islands (Lagoons)	Tc	2010	—	—	—	—	15–20 clams per m²	Latypov & Selin (2011)
Viet Nam	Tho Chau, Con Dao, and Thu Islands (Reef slope)	Tm	2010	—	—	—	—	0.08–0.1 clams per m²	Latypov & Selin (2011)
Viet Nam	Tho Chau, Con Dao, and Thu Islands (Reef flat)	Ts	2010	—	—	—	—	0.10000	Latypov & Selin (2011)
Viet Nam	Tho Chau, Con Dao, and Thu Islands (Reef slope)	Ts	2010	—	—	—	—	0.2–0.5 clams per m²	Latypov & Selin (2011)
Viet Nam	Bay Canh Island, Con Dao Archipelago	Tc	2010	Belt transects	—	—	—	23.00000	Selin & Latypov (2011)
Viet Nam	Cau Island, Con Dao Archipelago	Tc	2011	Belt transects	—	—	—	25.00000	Selin & Latypov (2011)
Viet Nam	Hon Nai Island, Cam Ranh Bay, southern Viet Nam	Tm	?	1 m² quadrats along 100 m transect	—	—	—	0.20000	Latypov & Selin (2012b)
Viet Nam	Giang Bo Reef	Tc	2004–2007	1 m² quadrats along 100–200 m transect	—	—	—	2.00000	Latypov (2013)
Viet Nam	Giang Bo Reef	Ts	2004–2007	1 m² quadrats along 100–200 m transect	—	—	—	0.10000	Latypov (2013)
Viet Nam	Mju Island, Nha Trang, Khanh Hoa Province	Tc	2004–2005	1 m² quadrats along 100 m transect	—	5	—	0.50000	Latypov & Selin (2013)

Continued

Table A3 (Continued) Global density patterns of wild giant clam populations

Country	Localities surveyed	Species	Year of survey	Method of survey	Approximate area of surveys (ha)	Approximate area of surveys (m²)	Number of ind.	Density (m⁻²)	Reference (population survey)
Viet Nam	Mju Island, Nha Trang Bay, Khanh Hoa Province	Ts	2005–2005	1 m² quadrats along 100 m transect	—	5	—	0.10000	Latypov & Selin (2013)
Yemen	Tiqfash Island, Shalatem Island, Myyun Island, Shaqraa coast, Sikha Island, Macroqha Island, Socotra Island	Tm, Ts	2008	Belt transects; 20 × 5 m	—	4,800	—	0.00020	PERSGA (2010)

Notes: Full reference list in Appendix B. ? denotes information is unknown or unverified. Hh — *Hipposus hipposus*; Hp — *H. porcellanus*; T —*Tridacna*; Tc — *T. costata*; Td — *T. derasa*; Tg — *T. gigas*; Tm — *T. maxima*; Tmb — *T. mbalavuana* (previously *T. tevoroa*); Tno — *T. noae*; Ts — *T. squamosa*; Tsi — *T. squamosina* (previously *T. costata*). Original density figures were erroneous and corrected in this table: Black et al. 2011 (Tno), Brown & Muskanofola 1985 (Tc, Tm, Ts), Junchompoo et al. 2013 (Tc), Munro 1988 (Hh, Tg, Ts), Tan et al. 1998 (Tc, Tm, Ts), Yusuf et al. 2009 (Tg) Density figures computed based on average of all densities from individual surveys: Barott et al. 2010 (Tm), Bellchambers & Evans 2013 (Tm), Braley 1987a (Td, Tg), Braley 1988 (Tm, Ts), Calumpong & Macansantos 2008 (Hh, Tc, Tm, Ts), Dumas & Andréfouët 2011 (Hh, Tc, Td, Tm, Ts), Dumas et al. 2013 (Hh, Tm, Ts), Evans et al. 2006 (Tm), Gonzales et al. 2014b (Tc), Harding & Randriamanantsoa 2008 (T), Hender et al. 2001 (Td, Tm), Hopkins 2009 (Tg), Kepler & Kepler 1994 (Tm), Langi 1990 (Tm), McKenna et al. 2006 (Hh, Tc, Td, Tm, Ts), Montagne et al. 2013 (T), PERSGA 2010 (Tm, Ts), Purcell et al. 2009 (Hh, Td, Tm, Ts), Sandin et al. 2008 (Tm), Siaosi et al. 2012 (Tm, Ts), Thorne et al. 2015 (T), Vieux 2009 (Tc, Tm, Ts), Virly 2004 (Hh, Td, Tm, Ts), Wantiez et al. 2007a,b,c, 2008a,b (Hh, Tc, Td, Tm, Ts)

Table A4 Global distribution of giant clams (Reef Check)

Country	Reef Site	Monitoring years (total clam density = number of individuals per 100 m²)																		
		1997	1998	1999	2000	2001	2002	2003	2004	2005	2006	2007	2008	2009	2010	2011	2012	2013	2014	2015
American Samoa	Amalau Bay	1.0	—	—	—	—	—	—	—	—	—	—	—	—	—	—	—	—	—	—
American Samoa	Bait Reef—The Trench (Back reef crest) Site 1	—	—	—	—	—	—	1.0	—	—	—	—	—	—	—	—	—	—	—	—
Australia	Agincourt Reef—12 Apostles (Back reef slope) Site 1	—	—	—	—	—	—	1.5	—	—	—	—	—	—	—	—	—	—	—	—
Australia	Agincourt Reef—Agincourt 2D (Pontoon) (Back reef slope) Site 1	—	—	—	—	—	—	—	—	—	—	—	1.5	—	—	—	—	—	—	—
Australia	Agincourt Reef—Agincourt 3D (Pontoon) (Back reef slope) Site 1	—	—	—	—	—	—	9.75	3.75	—	—	—	—	—	3.25	1.88	0.5	0.75	—	—
Australia	Agincourt Reef—Agincourt 3D (Pontoon) (Back reef slope) Site 2	—	—	—	—	—	—	—	2.75	—	—	—	—	4.25	4.25	3.25	2.5	2.5	—	—

Continued

Table A4 (Continued) Global distribution of giant clams (Reef Check)

Country	Reef Site	Monitoring years (total clam density = number of individuals per 100 m^2)																		
		1997	1998	1999	2000	2001	2002	2003	2004	2005	2006	2007	2008	2009	2010	2011	2012	2013	2014	2015
Australia	Agincourt Reef—Agincourt 3D (Pontoon) (Back reef slope) Site 3	—	—	—	—	—	—	—	4.75	6.0	5.25	—	—	—	—	—	—	—	—	—
Australia	Agincourt Reef—Barracuda Bommie (Back reef crest) Site 1	—	—	—	—	—	—	4.0	—	—	—	4.0	—	—	—	—	—	—	—	—
Australia	Agincourt Reef—End of the World (Back reef slope) Site 1	—	—	—	—	—	4.5	—	—	—	—	—	—	—	—	—	—	—	—	—
Australia	Agincourt Reef—Harry's Bommie (Back reef slope) Site 1	—	—	—	—	—	—	3.5	—	—	—	—	—	3.25	—	—	—	—	—	—
Australia	Agincourt Reef—Phil's Reef (Back reef slope) Site 1	—	—	—	—	—	—	0.75	—	—	—	—	—	0.25	—	—	—	—	—	—
Australia	Agincourt Reef—Phil's Reef (Back reef slope) Site2	—	—	—	—	—	—	—	—	—	—	—	—	1.25	—	—	—	—	—	—

Continued

Table A4 (Continued) Global distribution of giant clams (Reef Check)

Country	Reef Site	Monitoring years (total clam density = number of individuals per 100 m²)																		
		1997	1998	1999	2000	2001	2002	2003	2004	2005	2006	2007	2008	2009	2010	2011	2012	2013	2014	2015
Australia	Agincourt Reef—Playground (Back reef wall) Site 1	—	—	—	—	—	—	—	—	—	0.75	1.5	—	—	—	—	—	—	—	—
Australia	Agincourt Reef—The Point (Back reef slope) Site 1	—	—	—	—	—	—	—	3.25	—	—	12.0	—	3.25	—	—	—	—	—	—
Australia	Agincourt Reef—Turtle Bay (Back reef wall) Site 1	—	—	—	—	—	—	—	—	—	—	—	—	0.75	0.25	—	—	—	—	—
Australia	Arlington Reef	—	—	—	—	—	5.5	—	—	—	—	—	—	—	—	—	—	—	—	—
Australia	Bait Reef—The Clusters (Back reef crest) Site 1	—	—	—	—	—	—	3.0	—	—	—	—	—	—	—	—	—	—	—	—
Australia	Barolin Rocks Reef—Barolin Rocks (Woongarra Marine Park) (Fringing reef seaward) Site 2	—	—	—	—	—	—	—	—	—	—	—	—	—	—	—	—	0.5	—	—
Australia	Bashful Bommie	—	—	—	—	—	—	1.0	—	3.75	2.0	—	—	—	—	—	—	—	—	—
Australia	Bashful Bommie Haba	—	—	—	—	—	—	—	—	—	—	1.25	—	—	—	—	—	—	—	—

Continued

Table A4 (Continued)   Global distribution of giant clams (Reef Check)

Country	Reef Site	Monitoring years (total clam density = number of individuals per 100 m^2)																		
		1997	1998	1999	2000	2001	2002	2003	2004	2005	2006	2007	2008	2009	2010	2011	2012	2013	2014	2015
Australia	Bashful Bommie Site 1	—	—	—	—	—	—	—	—	—	—	2.5	—	—	—	—	—	—	—	—
Australia	Blue Buoy Bashful Bommie	—	—	—	—	—	—	—	—	1.88	1.0	—	—	—	—	—	—	—	—	—
Australia	Blue Pearl Bay rocks	—	—	—	—	—	—	1.5	—	—	—	5.5	—	—	—	—	—	—	—	—
Australia	Blue Pearl Bay Site 2	—	—	—	—	—	—	—	—	47.25	19.75	—	—	—	—	—	—	—	—	—
Australia	Briggs Reef—Briggs Reef (Back reef slope) Site 1	—	—	—	—	—	—	2.5	—	—	—	—	—	—	—	—	—	—	—	—
Australia	Briggs Reef—Fish Bowl (Back reef slope) Site 1	—	—	—	—	—	—	—	—	—	—	—	1.75	—	—	—	—	—	—	—
Australia	Bundegi	0.25	—	—	—	—	—	—	—	—	—	—	—	—	—	—	—	—	—	—
Australia	Cod Hole Ribbon Reef #10	—	—	—	—	—	—	—	—	—	—	—	—	—	—	—	—	—	—	—
Australia	Coral Cay Beach	—	—	—	1.25	—	—	—	—	—	—	—	—	—	—	—	—	—	—	—
Australia	Currimundi Reef—Currimundi Reef (Back reef slope) Site 2	—	—	—	—	—	—	—	—	—	—	—	—	0.25	—	—	—	—	—	—
Australia	Davies Reef	—	—	—	—	—	—	—	—	7.25	—	—	—	—	—	—	—	—	—	—

Continued

257

Table A4 (Continued) Global distribution of giant clams (Reef Check)

Country	Reef Site	Monitoring years (total clam density = number of individuals per 100 m^2)																		
		1997	1998	1999	2000	2001	2002	2003	2004	2005	2006	2007	2008	2009	2010	2011	2012	2013	2014	2015
Australia	Davies Reef—The Lost World (Back reef wall) Site 1	—	—	—	—	—	—	—	—	—	7.0	3.25	—	—	—	—	—	—	—	—
Australia	Fantasea ReefWorld Pantoon (Hardy Reef)	—	—	—	—	—	—	7.0	—	—	—	1.5	—	—	—	—	—	—	—	—
Australia	Fitzroy Beach Dive	—	—	—	—	—	4.25	—	—	—	—	—	—	—	—	—	—	—	—	—
Australia	Fitzroy Sandy Patches Dive	—	—	—	—	—	7.25	—	—	—	—	—	—	—	—	—	—	—	—	—
Australia	Flat Rock Island—Shark Gulley (Fringing reef seaward) Site 1	—	—	—	—	—	—	—	—	—	—	—	—	0.25	—	1.0	—	—	—	—
Australia	Flat Rock Island—The Nursery (Fringing reef leeward) Site 1	—	—	—	—	—	—	—	—	—	—	—	—	—	—	0.25	0.25	—	—	—
Australia	Flinders Reef—Aladdin's Cave (Fringing reef seaward) Site 1	—	—	—	—	—	—	—	—	—	—	—	—	0.5	0.5	0.75	—	—	—	—

Continued

258

Table A4 (Continued)   Global distribution of giant clams (Reef Check)

Country	Reef Site	Monitoring years (total clam density = number of individuals per 100 m²)																		
		1997	1998	1999	2000	2001	2002	2003	2004	2005	2006	2007	2008	2009	2010	2011	2012	2013	2014	2015
Australia	Flinders Reef—Aladdin's Cave (Other) Site 1	—	—	—	—	—	—	—	—	—	—	—	—	—	—	—	—	0.25	—	—
Australia	Flinders Reef—Nursery (Back reef slope) Site 1	—	—	—	—	—	—	—	—	—	—	—	—	—	—	1.25	0.5	—	—	—
Australia	Flinders Reef—Nursery (Back reef slope) Site 3	—	—	—	—	—	—	—	—	—	—	—	—	—	—	—	0.75	—	—	—
Australia	Flinders Reef—Nursery (Fringing reef leeward) Site 1	—	—	—	—	—	—	—	—	—	—	—	0.25	0.5	0.5	—	—	—	—	—
Australia	Flinders Reef—Nursery (Fringing reef leeward) Site 3	—	—	—	—	—	—	—	—	—	—	—	—	0.25	—	—	—	—	—	—
Australia	Flynns Reef—Gordon's Mooring (Missing Habitat) Site 1	—	—	—	—	—	—	—	—	—	2.0	—	—	—	—	—	—	—	—	—

Continued

Table A4 (Continued) Global distribution of giant clams (Reef Check)

Country	Reef Site	Monitoring years (total clam density = number of individuals per 100 m^2)																		
		1997	1998	1999	2000	2001	2002	2003	2004	2005	2006	2007	2008	2009	2010	2011	2012	2013	2014	2015
Australia	Flynns Reef—Yellow Mooring (aka Fish bowl) (Back reef slope) Site 1	—	—	—	—	—	—	—	—	—	—	—	—	0.75	—	—	—	—	—	—
Australia	Hardy Reef—Hardy Reef (Back reef wall) Site 1	—	—	—	—	—	3.75	—	—	—	—	—	—	5.0	4.5	—	—	—	—	—
Australia	Hardy Reef—Hardy Reef (Back reef wall) Site 2	—	—	—	—	—	—	—	—	—	—	—	—	6.5	3.0	—	—	—	—	—
Australia	Hardy Reef—Hardy Reef (Back reef wall) Site 3	—	—	—	—	—	—	—	—	—	—	—	—	1.25	8.0	1.25	—	6.25	—	—
Australia	Hardy Reef, Fantasea Reef World, Whitsundays	—	—	—	—	—	—	—	—	19.75	5.75	—	—	—	—	—	—	—	—	—
Australia	Hastings Reef—North Hastings A (Back reef wall) Site 1	—	—	—	—	—	—	—	—	—	—	—	—	0.75	0.5	—	—	—	—	—

Continued

Table A4 (Continued) Global distribution of giant clams (Reef Check)

Country	Reef Site	Monitoring years (total clam density = number of individuals per 100 m²)																		
		1997	1998	1999	2000	2001	2002	2003	2004	2005	2006	2007	2008	2009	2010	2011	2012	2013	2014	2015
Australia	Hastings Reef—North Hastings A (Lagoon) Site 1	—	—	—	—	—	—	1.25	—	—	—	—	—	—	1.0	—	—	—	—	—
Australia	Hastings Reef—North Hastings B (Back reef wall) Site 1	—	—	—	—	—	—	—	—	—	—	—	3.25	0.5	—	—	—	—	—	—
Australia	Hastings Reef—South Hastings (Back reef wall) Site 1	—	—	—	—	—	—	—	—	—	0.25	—	—	—	—	—	—	—	—	—
Australia	Hastings Reef (North) Down Under Dive Site 1	—	—	—	—	—	—	—	—	—	3.75	—	—	—	—	—	—	—	—	—
Australia	Hastings Reef (North) Down Under Dive Site 3	—	—	—	—	—	—	—	—	—	1.0	—	—	—	—	—	—	—	—	—
Australia	Hayman Island Reefs—Blue Pearl Bay (Fringing reef leeward) Site 1	—	—	—	—	2.0	—	—	—	—	—	—	—	2.0	12.75	—	—	—	—	—

Continued

261

Table A4 (Continued)   Global distribution of giant clams (Reef Check)

Country	Reef Site	Monitoring years (total clam density = number of individuals per 100 m^2)																		
		1997	1998	1999	2000	2001	2002	2003	2004	2005	2006	2007	2008	2009	2010	2011	2012	2013	2014	2015
Australia	Hayman Island Reefs—Blue Pearl Bay (Fringing reef leeward) Site 3	—	—	—	—	—	—	—	—	—	13.5	—	—	—	7.5	—	—	—	—	—
Australia	Herald Cay	6.75	—	—	—	—	—	—	—	—	—	—	—	—	—	—	—	—	—	—
Australia	Heron Island	2.75	—	—	—	—	—	—	—	—	—	—	—	—	—	—	—	—	—	—
Australia	Heron Reef—Canyons (Back reef slope) Site 1	—	—	—	—	—	—	—	—	—	—	—	—	—	—	—	—	2.25	—	—
Australia	Heron Reef—Cappuccino Express (Reef flat) Site 1	—	—	—	—	—	—	—	—	—	—	—	—	—	—	—	—	1.25	—	—
Australia	Heron Reef—Coral Garden (Fringing reef seaward) Site 1	—	—	—	—	—	—	—	—	—	—	—	—	—	—	—	—	0.5	—	—
Australia	Heron Reef—Coral Grotto (Back reef slope) Site 1	—	—	—	—	—	—	—	—	—	—	—	—	—	—	1.75	—	1.0	—	—

Continued

262

Table A4 (Continued) Global distribution of giant clams (Reef Check)

Country	Reef Site	Monitoring years (total clam density = number of individuals per 100 m^2)																		
		1997	1998	1999	2000	2001	2002	2003	2004	2005	2006	2007	2008	2009	2010	2011	2012	2013	2014	2015
Australia	Heron Reef—Harry's Bommie (Fringing reef seaward) Site 1	—	—	—	—	—	—	—	—	—	—	—	—	—	—	0.25	—	0.5	—	—
Australia	Heron Reef—Last Resort (Reef flat) Site 1	—	—	—	—	—	—	—	—	—	—	—	—	—	—	—	—	1.25	—	—
Australia	Heron Reef- Heron Bommie (Fringing reef seaward) Site 1	—	—	—	—	—	—	—	—	—	—	—	—	—	—	1.75	—	0.5	—	—
Australia	Heron Reef- Jetty Flat (Reef flate) Site 1	—	—	—	—	—	—	—	—	—	—	—	—	—	—	2.5	1.0	0.25	—	—
Australia	Heron Reef- Libby's Lair (Back reef slope) Site 1	—	—	—	—	—	—	—	—	—	—	—	—	—	—	1.25	—	0.5	—	—
Australia	Heron Reef- North Bommie (Back reef slope) Site 1	—	—	—	—	—	—	—	—	—	—	—	—	—	—	1.0	—	0.25	—	—

Continued

Table A4 (Continued) Global distribution of giant clams (Reef Check)

Country	Reef Site	Monitoring years (total clam density = number of individuals per 100 m²)																		
		1997	1998	1999	2000	2001	2002	2003	2004	2005	2006	2007	2008	2009	2010	2011	2012	2013	2014	2015
Australia	Heron Reef- Research Zone (Reef flat) Site 1	—	—	—	—	—	—	—	—	—	—	—	—	—	—	1.25	—	0.75	—	—
Australia	Heron Reef- Shark Bay (Reef flat) Site 1	—	—	—	—	—	—	—	—	—	—	—	—	—	—	1.75	0.5	1.5	—	—
Australia	Hook Island—Luncheon Bay (Fringing reef leeward) Site 1	—	—	—	—	—	—	—	—	—	—	—	—	—	—	—	—	6.5	—	—
Australia	Inner Gneerings—The Caves (Other) Site 2	—	—	—	—	—	—	—	—	—	—	—	—	—	0.25	—	—	0.25	—	—
Australia	Inner Gneerings—The Caves (Reef flat) Site 1	—	—	—	—	—	—	—	—	—	—	—	—	0.25	0.25	0.25	—	0.25	—	—
Australia	John Brewer Reef—John Brewer (Back reef crest) Site 1	—	—	—	—	—	—	—	18.5	6.25	1.75	—	—	—	—	—	—	—	—	—

Continued

Table A4 (Continued) Global distribution of giant clams (Reef Check)

Country	Reef Site	Monitoring years (total clam density = number of individuals per 100 m²)																		
		1997	1998	1999	2000	2001	2002	2003	2004	2005	2006	2007	2008	2009	2010	2011	2012	2013	2014	2015
Australia	John Brewer Reef—John Brewer (Back reef crest) Site 2	—	—	—	—	—	—	—	10.75	4.0	—	—	—	—	—	—	—	—	—	—
Australia	John Brewer Reef site 3	—	—	—	—	—	—	—	13.25	—	—	—	—	—	—	—	—	—	—	—
Australia	Keeper Reef—Keeper Reef (Back reef slope) Site 1	—	—	—	—	—	—	0.5	2.5	—	4.25	—	—	—	—	—	—	—	—	—
Australia	Kelso Reef	8.0	—	—	—	—	15.0	—	—	—	—	—	—	—	—	—	—	—	—	—
Australia	Knuckle Reef—Knuckle Reef (Back reef slope) Site 1	—	—	—	—	—	—	—	—	—	22.75	—	12.25	7.5	11.75	6.5	—	—	—	—
Australia	Knuckle Reef—Knuckle Reef (Back reef slope) Site 2	—	—	—	—	—	—	—	—	—	22.0	—	—	6.0	10.5	—	—	—	—	—
Australia	Lady Elliot Island—Coral Gardens (Fringing reef seaward) Site 1	—	—	—	—	—	—	—	—	—	—	—	—	—	—	—	—	1.0	—	—
Australia	Lady Elliot Island—Lady Elliot Reef Lagoon 2 (Reef flat) Site 1	—	—	—	—	—	—	—	—	—	—	—	—	—	—	—	—	2.75	—	—

Continued

Table A4 (Continued) Global distribution of giant clams (Reef Check)

Country	Reef Site	Monitoring years (total clam density = number of individuals per 100 m²)																		
		1997	1998	1999	2000	2001	2002	2003	2004	2005	2006	2007	2008	2009	2010	2011	2012	2013	2014	2015
Australia	Lady Elliot Island—Reefy Seconds (Fringing reef seaward) Site 1	—	—	—	—	—	—	—	—	—	—	—	—	—	—	—	—	0.25	—	—
Australia	Lady Elliot Island- Lady Elliot Reef Lagoon (Lagoon) Site 1	—	—	—	—	—	—	—	—	—	—	—	—	—	—	3.25	—	11.5	—	—
Australia	Lodestone Reef	0.25	—	—	—	—	—	—	—	—	—	—	—	—	—	—	—	—	—	—
Australia	Low Isles	1.0	—	3.5	—	—	—	—	—	—	—	—	—	—	—	—	—	—	—	—
Australia	Low Isles "Lagoon West"	—	—	—	—	11.0	—	—	—	—	—	—	—	—	—	—	—	—	—	—
Australia	Low Isles Reef—Low Isles (Fringing reef leeward) Site 1	—	—	—	—	—	2.75	—	1.0	15.25	8.25	—	11.75	3.5	1.0	—	—	8.25	—	—
Australia	Low Isles Reef—Low Isles (Fringing reef leeward) Site 2	—	—	—	—	—	—	—	—	39.5	26.75	—	—	3.75	0.25	—	—	—	—	—
Australia	Magnetic Island Reefs—Alma Bay (Fringing reef leeward) Site 2	—	—	—	—	—	—	—	—	—	—	—	—	—	—	—	—	0.25	—	—

Continued

Table A4 (Continued) Global distribution of giant clams (Reef Check)

Country	Reef Site	Monitoring years (total clam density = number of individuals per 100 m²)																		
		1997	1998	1999	2000	2001	2002	2003	2004	2005	2006	2007	2008	2009	2010	2011	2012	2013	2014	2015
Australia	Magnetic Island Reefs—Florence Bay (Fringing reef leeward) Site 2	—	—	—	—	—	—	—	—	—	—	—	—	—	—	—	0.25	—	—	—
Australia	Magnetic Island Reefs—Middle Reef (Fringing reef seaward) Site 3	—	—	—	—	—	—	—	—	0.25	—	—	—	—	—	—	—	—	—	—
Australia	Magnetic Island Reefs—Middle Reef (Reef flat) Site 2	—	—	—	—	—	—	—	—	—	0.25	—	—	—	—	—	—	—	—	—
Australia	Magnetic Island Reefs—Nelly Bay (Fringing reef leeward) Site 1	—	—	—	—	—	—	—	—	—	—	—	0.25	—	—	—	—	—	—	—
Australia	Magnetic Island Reefs—Picnic Reef (Fringing reef leeward) Site 1	—	—	—	—	—	—	—	—	0.25	0.12	—	—	—	—	—	—	—	—	—
Australia	Magnetic Island Reefs—Picnic Reef (Fringing reef leeward) Site 2	—	—	—	—	—	—	—	—	—	0.12	—	—	—	—	—	—	—	—	—

Continued

267

Table A4 (Continued) Global distribution of giant clams (Reef Check)

Country	Reef Site	Monitoring years (total clam density = number of individuals per 100 m²)																		
		1997	1998	1999	2000	2001	2002	2003	2004	2005	2006	2007	2008	2009	2010	2011	2012	2013	2014	2015
Australia	Magnetic Island Reefs—Picnic Reef (Fringing reef leeward) Site 3	—	—	—	—	—	—	—	—	—	0.38	—	—	—	—	—	—	—	—	—
Australia	Maureens Cove	4.0	—	—	—	—	—	—	—											
Australia	Michaelmas Cay	—	—	—	—	—	5.5	—	—	—	—	—	—	—	—	—	—	—	—	—
Australia	Michaelmas Reef—Breaking Patches (Back reef wall) Site 1	—	—	—	—	—	—	—	—	—	—	—	—	0.25	—	—	—	—	—	—
Australia	Michaelmas Reef—Breaking Patches (Reef flat) Site 1	—			—			—	—		—	—	—	0.5	—	—	—	—	—	—
Australia	Michaelmas Reef—Breaking Patches (Reef flat) Site 2	—	—		—			—	—		—	—	—	0.5	—	—	—	—	—	—
Australia	Michaelmas Reef—Long Bommie (Reef flat) Site 1	—	—		—			—	—		—	—	—	0.25	—	—	—	—	—	—

Continued

268

Table A4 (Continued) Global distribution of giant clams (Reef Check)

Country	Reef Site	Monitoring years (total clam density = number of individuals per 100 m²)																		
		1997	1998	1999	2000	2001	2002	2003	2004	2005	2006	2007	2008	2009	2010	2011	2012	2013	2014	2015
Australia	Milne Reef—Swimming Pool (Back reef slope) Site 1	—	—	—	—	—	—	—	—	—	0.5	—	—	0.5	—	—	—	—	—	—
Australia	Moore Reef—Reef Magic Pontoon (Back reef slope) Site 1	—	—	—	—	—	—	—	—	—	—	—	—	0.75	—	—	—	—	—	—
Australia	Moore Reef—Reef Magic Pontoon (Back reef wall) Site 1	—	—	—	—	—	—	—	—	—	—	—	0.25	—	0.5	—	—	—	—	—
Australia	Moore Reef—Reef Magic Pontoon (Back reef wall) Site 2	—	—	—	—	—	—	—	—	—	—	—	—	0.5	—	—	—	—	—	—
Australia	Moore Reef—Sunlover cruises pontoon (Back reef slope) Site 1	—	—	—	—	—	—	—	—	—	—	—	—	—	1.25	—	—	—	—	—

Continued

Table A4 (Continued)   Global distribution of giant clams (Reef Check)

Country	Reef Site	Monitoring years (total clam density = number of individuals per 100 m^2)																		
		1997	1998	1999	2000	2001	2002	2003	2004	2005	2006	2007	2008	2009	2010	2011	2012	2013	2014	2015
Australia	Moore Reef—Sunlover cruises pontoon (Back reef slope) Site 2	—	—	—	—	—	—	—	—	—	—	—	—	—	3.75	—	—	—	—	—
Australia	Moore Reef (Reef Magic) Site 1	—	—	—	—	—	—	—	—	1.25	—	—	—	—	—	—	—	—	—	—
Australia	Moore Reef (Reef Magic) Site 1	—	—	—	—	—	—	—	—	—	0.25	—	—	—	—	—	—	—	—	—
Australia	Moore Reef (Reef Magic) Site 3	—	—	—	—	—	—	—	—	—	1.75	—	—	—	—	—	—	—	—	—
Australia	Moore Reef Site 2	—	—	—	—	—	—	—	—	0.25	—	—	—	—	—	—	—	—	—	—
Australia	Mudjimba (Old Woman) Island—Mudjimba Island (Fringing reef leeward) Site 1	—	—	—	—	—	—	—	—	—	—	—	—	—	—	0.75	—	—	—	—

Continued

Table A4 (Continued) Global distribution of giant clams (Reef Check)

Country	Reef Site	Monitoring years (total clam density = number of individuals per 100 m²)																		
		1997	1998	1999	2000	2001	2002	2003	2004	2005	2006	2007	2008	2009	2010	2011	2012	2013	2014	2015
Australia	Mudjimba (Old Woman) Island—The Ledge (Mudjimba Island) (Fringing reef leeward) Site 1	—	—	—	—	—	—	—	—	—	—	—	—	—	—	—	—	0.25	—	—
Australia	Norman Reef—Norman Reef North (Back reef slope) Site 1	—	—	—	—	—	—	—	—	—	—	—	—	1.0	—	—	—	—	—	—
Australia	Norman Reef—Norman Reef North (Back reef slope) Site 2	—	—	—	—	—	—	—	—	—	—	—	—	0.75	—	—	—	—	—	—
Australia	Norman Reef—Norman Reef North (Back reef wall) Site 1	—	—	—	—	—	2.5	—	—	—	—	—	—	—	—	—	—	—	—	—
Australia	Norman Reef—Norman Reef South (Back reef slope) Site 1	—	—	—	—	—	—	—	—	—	0.75	—	—	—	2.5	—	—	—	—	—
Australia	Normandy Island	—	—	—	—	—	1.25	—	—	—	—	—	—	—	—	—	—	—	—	—

Continued

Table A4 (Continued) Global distribution of giant clams (Reef Check)

Country	Reef Site	Monitoring years (total clam density = number of individuals per 100 m²)																		
		1997	1998	1999	2000	2001	2002	2003	2004	2005	2006	2007	2008	2009	2010	2011	2012	2013	2014	2015
Australia	North Hastings	—	—	—	—	—	—	1.5	—	—	—	—	—	—	—	—	—	—	—	—
Australia	North Horn-Osprey Reef	—	—	2.25	—	—	—	—	—	—	—	—	—	—	—	—	—	—	—	—
Australia	Opal Reef	—	—	1.0	—	0.5	—	—	—	—	—	—	—	—	—	—	—	—	—	—
Australia	Opal Reef—Bashful Bommie (Back reef slope) Site 1	—	—	—	—	—	1.5	—	—	—	—	—	1.38	0.62	3.25	4.25	3.25	5.75	—	—
Australia	Opal Reef—Bashful Bommie (Back reef slope) Site 2	—	—	—	—	—	—	—	3.25	—	—	—	—	—	3.0	—	—	—	—	—
Australia	Opal Reef—Cathedrals (Back reef slope) Site 1	—	—	—	—	—	—	—	—	—	1.25	—	—	1.0	—	—	—	—	—	—
Australia	Opal Reef—SNO (South North Opal) (Back reef crest) Site 1	—	—	—	—	—	—	—	—	—	—	—	—	0.75	1.0	1.75	—	—	—	—
Australia	Opal Reef—SNO (South North Opal) (Back reef slope) Site 2	—	—	—	—	—	—	—	—	—	—	—	—	—	—	—	—	1.0	—	—

Continued

Table A4 (Continued) Global distribution of giant clams (Reef Check)

		Monitoring years (total clam density = number of individuals per 100 m²)																	
Country	Reef Site	1997	1998	1999	2000	2001	2002	2003	2004	2005	2006	2007	2008	2009	2010	2011	2012	2013	2014	2015
Australia	Opal Reef—SNO (South North Opal) (Reef flat) Site 1	—	—	—	—	—	—	—	—	—	—	—	—	—	1.25	—	—	—	—	—
Australia	Opal Reef—Split Bommie (Back reef wall) Site1	—	—	—	—	—	—	—	1.0	1.5	1.0	0.25	0.5	1.25	0.38	—	—	—	—	—
Australia	Opal Reef—The Wedge (Back reef slope) Site 1	—	—	—	—	—	—	—	1.5	1.25	1.0	1.5	1.25	0.75	2.5	—	2.5	1.0	—	—
Australia	Opal Reef—Two Tone (Reef flat) Site 1	—	—	—	—	—	—	2.5	3.5	—	—	4.75	—	2.0	1.75	—	2.5	2.25	—	—
Australia	Osprey Reef—Admiralty Anchor (Back reef wall) Site 1	17.5	—	—	—	—	9.25	16.0	16.0	23.5	8.0	14.5	13.5	—	—					
Australia	Osprey Reef—North Horn (Back reef wall) Site 1	—	—	—	—	1.25	1.75	1.5	1.75	8.75	—	—	4.25	—	—					

Continued

273

Table A4 (Continued) Global distribution of giant clams (Reef Check)

		Monitoring years (total clam density = number of individuals per 100 m^2)																			
Country	Reef Site	1997	1998	1999	2000	2001	2002	2003	2004	2005	2006	2007	2008	2009	2010	2011	2012	2013	2014	2015
Australia	Oyster Stacks—Oyster Stacks North (Fringing reef seaward) Site 1	—	—	—	—	—	—	—	—	—	—	—	—	—	—	—	—	2.0	—	—
Australia	Oyster Stacks—Oyster Stacks South (Fringing reef seaward) Site 1	—	—	—	—	—	—	—	—	—	—	—	—	—	—	—	—	1.5	—	—
Australia	Palm Beach Reef—Palm Beach Reef (Fringing reef seaward) Site 1	—	—	—	—	—	—	—	—	—	—	—	—	—	—	—	—	0.75	—	—
Australia	Palm Island Reefs—Cattle Bay (Fringing reef leeward) Site 1	—	—	—	—	—	—	—	—	94.5	18.5	—	—	—	7.75	—	—	—	—	—
Australia	Palm Island Reefs—Cattle Bay (Fringing reef leeward) Site 2	—	—	—	—	—	—	—	—	84.75	77.5	—	—	—	—	—	—	—	—	—
Australia	Palm Island Reefs—Cattle Bay (Fringing reef leeward) Site3	—	—	—	—	—	—	—	—	164.5	29.25	—	—	—	—	—	—	—	—	—

Continued

Table A4 (Continued) Global distribution of giant clams (Reef Check)

Country	Reef Site	Monitoring years (total clam density = number of individuals per 100 m^2)																		
		1997	1998	1999	2000	2001	2002	2003	2004	2005	2006	2007	2008	2009	2010	2011	2012	2013	2014	2015
Australia	Palm Island Reefs—Curacoa Island (Fringing reef leeward) Site 1	—	—	—	—	—	—	—	—	3.62	—	—	—	—	—	—	—	—	—	—
Australia	Palm Island Reefs—Curacoa Island (Fringing reef leeward) Site 2	—	—	—	—	—	—	—	—	8.75	—	—	—	—	—	—	—	—	—	—
Australia	Palm Island Reefs—Curacoa Island (Fringing reef leeward) Site 3	—	—	—	—	—	—	—	—	2.75	0.25	—	—	—	—	—	—	—	—	—
Australia	Palm Island Reefs—Fantome (Fringing reef seaward) Site 1	—	—	—	—	—	—	—	—	9.5	8.5	—	—	—	—	—	—	—	—	—
Australia	Palm Island Reefs—Fantome (Fringing reef seaward) Site 2	—	—	—	—	—	—	—	—	15.75	4.0	—	—	—	—	—	—	—	—	—
Australia	Palm Island Reefs—Fantome (Fringing reef seaward) Site 3	—	—	—	—	—	—	—	—	39.5	5.0	—	—	—	—	—	—	—	—	—

Continued

Table A4 (Continued) Global distribution of giant clams (Reef Check)

Country	Reef Site	Monitoring years (total clam density = number of individuals per 100 m²)																		
		1997	1998	1999	2000	2001	2002	2003	2004	2005	2006	2007	2008	2009	2010	2011	2012	2013	2014	2015
Australia	Palm Island Reefs—Juno Bay (Fringing reef leeward) Site 1	—	—	—	—	—	—	—	—	119.75	191.75	—	—	—	—	—	—	—	—	—
Australia	Palm Island Reefs—Juno Bay (Fringing reef leeward) Site 2	—	—	—	—	—	—	—	—	91.75	121.75	—	—	—	—	—	—	—	—	—
Australia	Palm Island Reefs—Juno Bay (Fringing reef leeward) Site 3	—	—	—	—	—	—	—	—	45.75	244.75	—	—	—	—	—	—	—	—	—
Australia	Palm Island Reefs—Pelorus (Fringing reef leeward) Site 1	—	—	—	—	—	—	—	—	6.25	24.5	—	—	—	—	—	—	—	—	—
Australia	Palm Island Reefs—Pelorus (Fringing reef leeward) Site 2	—	—	—	—	—	—	—	—	6.5	8.75	—	—	—	—	—	—	—	—	—
Australia	Palm Island Reefs—Pelorus (Fringing reef leeward) Site 3	—	—	—	—	—	—	—	—	19.5	22.0	—	—	—	—	—	—	—	—	—

Continued

Table A4 (Continued) Global distribution of giant clams (Reef Check)

Country	Reef Site	Monitoring years (total clam density = number of individuals per 100 m²)																		
		1997	1998	1999	2000	2001	2002	2003	2004	2005	2006	2007	2008	2009	2010	2011	2012	2013	2014	2015
Australia	Peel Island-South Peel (Fringing reef seaward) Site 1	—			—	—	—	—	—	—	—	—	—	—	—	—	—	0.75	—	—
Australia	Phil's Bommie	0.75	—	—	—	—	—	—	—	—	—	—	—	—	—	—	—	—	—	—
Australia	Radical Bay	—	—	—	—	—	—	—	—	—	—	—	—	—	—	—	—	—	—	—
Australia	Rainbow Reef Keeper Reef	—	—	—	—	—	—	1.25	—	—	—	—	—	—	—	—	—	—	—	—
Australia	Rat Island	—	—	—	—	—	—	—	—	—	—	—	—	—	—	—	—	—	—	—
Australia	Reef	3.5	—	—	—	—	—	—	—	—	—	—	—	—	—	—	—	—	—	—
Australia	Ribbon Reef #5 southern patch	—	—	0.25	—	—	—	—	—	—	—	—	—	—	—	—	—	—	—	—
Australia	Ribbon Reef 10—Challenger Bay (Back reef crest) Site 1	—	3.75	2.5	—	—	4.75	—	8.38	12.5	—	—	3.5	—	—	—	—	—	—	—
Australia	Ribbon Reef 10—Challenger Bay (Back reef crest) Site 2	—						1.75												
Australia	Ribbon Reef 10—No Name Reef (Back reef slope) Site 1	—	3.25	2.0	—						—				—				—	

Continued

277

Table A4 (Continued) Global distribution of giant clams (Reef Check)

Country	Reef Site	Monitoring years (total clam density = number of individuals per 100 m^2)																		
		1997	1998	1999	2000	2001	2002	2003	2004	2005	2006	2007	2008	2009	2010	2011	2012	2013	2014	2015
Australia	Ribbon Reef 10—Pixie Gardens (Back reef wall) Site 1	—	—	—	—	—	—	—	1.5	1.5	—	—	1.5	—	—	—	—	—	—	—
Australia	Ribbon Reef 3—Clam Beds (Back reef slope) Site 1	—	—	—	—	—	2.5	2.75	—	4.25	5.25	—	4.0	—	—	—	—	—	—	—
Australia	Ribbon Reef 3—Flare Point (Back reef slope) Site 1	—	—	—	—	—	—	—	1.5	—	—	—	—	—	—	—	—	—	—	—
Australia	Ribbon Reef 3—Tracey's Wonderland (Joanies Joy) (Back reef slope) Site 1	—	—	—	—	—	—	—	—	3.75	1.25	—	2.0	—	—	—	—	—	—	—
Australia	Sandy Island	—	—	—	—	—	—	—	—	—	—	—	—	—	—	—	—	—	—	—
Australia	Saxon Reef—Saxon Reef (Back reef slope) Site 1	—	—	—	—	—	—	2.25	—	—	—	—	—	1.0	0.75	—	—	0.5	—	—
Australia	Saxon Reef—Saxon Reef (Back reef slope) Site 2	—	—	—	—	—	2.75	—	—	—	—	—	—	—	1.0	—	—	—	—	—

Continued

Table A4 (Continued) Global distribution of giant clams (Reef Check)

Country	Reef Site	1997	1998	1999	2000	2001	2002	2003	2004	2005	2006	2007	2008	2009	2010	2011	2012	2013	2014	2015
		Monitoring years (total clam density = number of individuals per 100 m²)																		
Australia	Shag Rock Island—Shag Rock North (Fringing reef seaward) Site 2	—	—	—	—	—	—	—	—	—	—	—	—	1.0	—	0.25	—	—	—	—
Australia	Shag Rock Island—Shag Rock South (Back reef crest) Site 1	—	—	—	—	—	—	—	—	—	—	—	—	0.25	—	—	—	—	—	—
Australia	Shag Rock Island—Shag Rock South (Fringing reef seaward) Site 1	—	—	—	—	—	—	—	—	—	—	—	—	—	0.5	0.5	—	—	—	—
Australia	South Mandu Reef—South Mandu Reef 1 (Fringing reef seaward) Site 1	—	—	—	—	—	—	—	—	—	—	—	—	—	—	—	—	0.25	—	—
Australia	South Mandu Reef—South Mandu Reef 2 (Fringing reef seaward) Site 1	—	—	—	—	—	—	—	—	—	—	—	—	—	—	—	—	0.5	—	—
Australia	St Crispin Reef—North Point (Back reef slope) Site 1	—	—	—	—	—	—	—	—	—	2.25	4.25	—	—	—	—	—	—	—	—

Continued

279

Table A4 (Continued) Global distribution of giant clams (Reef Check)

Country	Reef Site	Monitoring years (total clam density = number of individuals per 100 m²)																		
		1997	1998	1999	2000	2001	2002	2003	2004	2005	2006	2007	2008	2009	2010	2011	2012	2013	2014	2015
Australia	Tantabiddi	—	—	—	—	—	—	—	—	—	—	—	—	—	—	—	—	—	—	—
Australia	Tantabiddi Reef—Tantabiddi Sanctuary Zone (Fringing reef seaward) Site 1	—	0.75	—	—	—	—	—	—	—	—	—	—	—	—	—	—	0.5	—	—
Australia	Thetford Reef—Thetford Reef (Back reef slope) Site 1	—	—	—	—	—	—	—	—	—	—	—	—	0.25	—	—	—	—	—	—
Australia	Thetford Reef—Thetford Reef (Reef flat) Site 1	—	—	—	—	—	—	—	—	—	—	—	—	3.25	—	—	—	—	—	—
Australia	Upulo Cay	—	—	—	—	—	3.75	—	—	—	—	—	—	—	—	—	—	—	—	—
Australia	Vlasoff Reef	—	—	—	—	—	1.0	—	—	—	—	—	—	—	—	—	—	—	—	—
Australia	Wheeler Reef—Students Bommie (Back reef slope) Site 1	—	—	—	—	—	—	—	—	—	—	6.5	—	—	—	—	—	—	—	—
Australia	Wheeler Reef—The Mooring (Back reef slope) Site 1	—	—	—	—	—	—	1.25	3.0	4.25	3.25	—	—	—	1.75	—	0.5	—	—	—
Australia	Zodiac Rock/Groote Eylandt	—	—	—	2.5	5.0	—	—	—	—	—	—	—	—	—	—	—	—	—	—
Cambodia	Fishing Bay	—	—	—	—	—	—	—	—	—	—	—	—	0.69	—	—	—	—	—	—
Cambodia	KK01A	—	—	—	—	—	—	—	—	—	—	—	—	0.25	1.25	—	—	—	—	—

Continued

Table A4 (Continued) Global distribution of giant clams (Reef Check)

Country	Reef Site	Monitoring years (total clam density = number of individuals per 100 m²)																		
		1997	1998	1999	2000	2001	2002	2003	2004	2005	2006	2007	2008	2009	2010	2011	2012	2013	2014	2015
Cambodia	KK03	—	—	—	—	—	—	—	—	—	—	—	—	—	0.5	—	—	—	—	—
Cambodia	KK03A	—	—	—	—	—	—	—	—	—	—	—	—	1.75		—	—	—	—	—
Cambodia	KK03B	—	—	—	—	—	—	—	—	—	—	—	—	0.25	—	—	—	—	—	—
Cambodia	KK03C	—	—	—	—	—	—	—	—	—	—	—	—	0.25	—	—	—	—	—	—
Cambodia	KK04A	—	—	—	—	—	—	—	—	—	—	—	—	0.5	—	—	—	—	—	—
Cambodia	KK06A	—	—	—	—	—	—	—	—	—	—	—	—	0.25	—	—	—	—	—	—
Cambodia	Koh Krosa Kandal	—	—	—	—	—	—	2.25	—	—	—	—	—	—		—	—	—	—	—
Cambodia	Koh Mano (channel)	—	—	—	—	—	—	3.25	—	—	—	—	—	—		—	—	—	—	—
Cambodia	Koh Poah (Site 7)	—	—	—	—	—	—	—	—	—	—	—	—	0.25	—	—	—	—	—	—
Cambodia	Koh Rong Samlem/Koh Samlem Straits	—	0.25	—	—	2.0	—	—	—											
Cambodia	KR02A	—	—	—	—	—	—	—	—	—	—	—	—	5.0	—	—	—	—	—	—
Cambodia	KR05C	—	—	—	—	—	—	—	—	—	—	—	—	0.75	—	—	—	—	—	—
Cambodia	KR06A	—	—	—	—	—	—	—	—	—	—	—	—	2.25	—	—	—	—	—	—
Cambodia	KR06B	—	—	—	—	—	—	—	—	—	—	—	—	4.0	—	—	—	—	—	—
Cambodia	KR17A	—	—	—	—	—	—	—	—	—	—	—	—	0.25	1.25	—	—	—	—	—
Cambodia	KR18A	—	—	—	—	—	—	—	—	—	—	—	—	1.5	—	—	—	—	—	—
Cambodia	KR22A	—	—	—	—	—	—	—	—	—	—	—	—	2.75	—	—	—	—	—	—
Cambodia	KS02	—	—	—	—	—	—	—	—	—	—	—	—	—	0.25	—	—	—	—	—
Cambodia	KS03A	—	—	—	—	—	—	—	—	—	—	—	—	1.0	—	—	—	—	—	—
Cambodia	KS03B	—	—	—	—	—	—	—	—	—	—	—	—	1.25	—	—	—	—	—	—
Cambodia	KS04A	—	—	—	—	—	—	—	—	—	—	—	—	0.5	—	—	—	—	—	—
Cambodia	KS05A	—	—	—	—	—	—	—	—	—	—	—	—	0.25	—	—	—	—	—	—
Cambodia	KS12A	—	—	—	—	—	—	—	—	—	—	—	—	1.25	—	—	—	—	—	—

Continued

281

Table A4 (Continued) Global distribution of giant clams (Reef Check)

Country	Reef Site	Monitoring years (total clam density = number of individuals per 100 m²)																		
		1997	1998	1999	2000	2001	2002	2003	2004	2005	2006	2007	2008	2009	2010	2011	2012	2013	2014	2015
Cambodia	KS12B	—	—	—	—	—	—	—	—	—	—	—	—	2.0	—	—	—	—	—	—
Cambodia	Tuear Khang Cherng	—	—	—	—	—	—	—	—	—	—	—	—	1.5	—	—	—	—	—	—
Cambodia	Vietnamese Bay	—	—	—	—	—	—	—	—	—	—	—	—	0.75	—	—	—	—	—	—
China	Dadonghai 1	—	—	—	1.25	—	—	—	—	—	—	—	—	—	—	—	—	—	—	—
China	Dadonghai 2	—	—	—	0.25	—	—	—	—	—	—	—	—	—	—	—	—	—	—	—
China	Dadonghai 3	—	—	—	0.75	—	—	—	—	—	—	—	—	—	—	—	—	—	—	—
China	Dadonghai 4	—	—	—	0.25	—	—	—	—	—	—	—	—	—	—	—	—	—	—	—
China	Xiao Dong Hai	—	—	—	—	—	0.12	—	—	—	—	—	—	—	—	—	—	—	—	—
Christmas Island	Chicken Farm	—	—	—	—	—	—	5.5	6.0	4.0	1.75	6.25	—	—	—	—	—	—	—	—
Christmas Island	Flying Fish Cove	—	—	—	—	—	—	9.5	10.0	8.0	3.5	3.38	—	—	—	—	—	—	—	—
Cocos (Keeling) Islands	100th	—	—	—	—	—	10.5	19.25	20.25	—	—	21.5	—	—	—	—	—	—	—	—
Cocos (Keeling) Islands	Banyak Coral	—	—	—	—	—	—	—	—	—	—	0.25	—	—	—	—	—	—	—	—
Cocos (Keeling) Islands	Banyak Coral—Pulu Keeling National Park	—	—	—	—	—	—	—	—	—	—	—	0.5	—	—	—	—	—	—	—
Cocos (Keeling) Islands	Cabbage Patch	38.25	—	—	—	—	—	—	—	—	—	—	—	—	—	—	—	—	—	—
Cocos (Keeling) Islands	Cabbage Patch (10m)	—	9.25	15.0	—	8.0	2.25	—	7.75	—	—	8.25	8.0	—	—	—	—	—	—	—

Continued

Table A4 (Continued) Global distribution of giant clams (Reef Check)

Country	Reef Site		Monitoring years (total clam density = number of individuals per 100 m^2)																		
		1997	1998	1999	2000	2001	2002	2003	2004	2005	2006	2007	2008	2009	2010	2011	2012	2013	2014	2015
Cocos (Keeling) Islands	Cabbage Patch (3m)	—	—	27.5	—	13.75	14.25	17.25	23.5	—	—	27.5	34.5	—	—	—	—	—	—	—
Cocos (Keeling) Islands	Cologne Gardens	—	—			—	1.5	1.25	—	1.5	—	2.0	—	—	—	—	—	—	—	—
Cocos (Keeling) Islands	Horsburgh Island North	—	—			0.25	—	—	—			—	—	—	—	—	—	—	—	—
Cocos (Keeling) Islands	North Point	—	—			—	—	0.25	—			0.25	—	—	—	—	—	—	—	—
Cocos (Keeling) Islands	Prison Gardens	—	—			—	1.25	1.0	2.25	—		4.5	—	—	—	—	—	—	—	—
Cocos (Keeling) Islands	Pulu Chepelok	—	—			—	1.0	—	1.5	—		1.75	—	—	—	—	—	—	—	—
Cocos (Keeling) Islands	Soft Coral Garden	—	—			—	—	—	—	—		0.25	—	—	—	—	—	—	—	—
Cocos (Keeling) Islands	Two Trees	—	—			—	—	0.25	—	0.25	—	0.5	—	—	—	—	—	—	—	—
Cook Islands	Amuri 2	—	—				—	—	—	0.25	—	—	—	—	—	—	—	—	—	—
Cook Islands	Atuatane	—	—				—	—	—	1.0	—	—	—	—	—	—	—	—	—	—
Cook Islands	Maina	—	—				—	—	—	0.75	—	—	—	—	—	—	—	—	—	—
Cook Islands	North of pass 2—outreef	—	—				—	—	—	1.0	—	—	—	—	—	—	—	—	—	—

Continued

Table A4 (Continued) Global distribution of giant clams (Reef Check)

Country	Reef Site	Monitoring years (total clam density = number of individuals per 100 m²)																		
		1997	1998	1999	2000	2001	2002	2003	2004	2005	2006	2007	2008	2009	2010	2011	2012	2013	2014	2015
Cook Islands	Northwest Corner	—	—	—	—	—	—	—	—	5.5	—	—	—	—	—	—	—	—	—	—
Cook Islands	South of pass—outreef	—	—	—	—	—	—	—	—	0.25	—	—	—	—	—	—	—	—	—	—
Cook Islands	Southwest Manuae	—	—	—	—	—	—	—	—	1.0	—	—	—	—	—	—	—	—	—	—
Cook Islands	Tongarutu	—	—	—	—	—	—	—	—	1.25	—	—	—	—	—	—	—	—	—	—
East Timor	Acrema	—	—	—	—	—	—	—	1.25	—	—	—	—	—	—	—	—	—	—	—
East Timor	K41	—	—	—	—	—	—	—	—	—	—	—	0.12	—	—	—	—	—	—	—
East Timor	North of Tanjung Reta	—	—	—	—	—	—	—	0.75	—	—	—	—	—	—	—	—	—	—	—
East Timor	South of Barstool	—	—	—	—	—	—	—	1.0	—	—	—	—	—	—	—	—	—	—	—
East Timor	South of Tanjung Reta	—	—	—	—	—	—	—	2.75	—	—	—	—	—	—	—	—	—	—	—
Egypt	3 islands	—	—	—	—	—	—	—	18.25	—	—	—	—	—	—	—	—	—	—	—
Egypt	Abu Hashish	2.5	—	—	—	—	—	—	—	—	—	—	—	—	—	—	—	—	—	—
Egypt	Abu Hashish South	—	—	—	—	5.5	—	—	—	—	—	—	—	—	—	—	—	—	—	—
Egypt	Abu Helal	—	—	—	—	—	—	—	—	—	—	2.57	—	3.0	—	4.0	—	4.38	5.92	—
Egypt	Abu Lakkany	1.75	—	—	—	—	—	—	—	—	—	—	—	—	—	—	—	—	—	—
Egypt	Abu Muchadi	7.5	—	—	—	—	—	—	—	—	—	—	—	—	—	—	—	—	—	—
Egypt	Abu Talha	—	—	—	—	—	—	4.75	2.0	7.0	3.25	—	—	—	—	—	—	—	—	—
Egypt	Amphoras	7.13	—	—	11.38	—	—	—	—	—	—	—	—	—	—	—	—	—	—	—
Egypt	Blue Hole	—	—	—	—	—	—	—	—	—	—	8.25	—	9.5	7.63	3.5	6.0	10.0	7.88	—
Egypt	Canyon North	—	—	—	—	—	—	—	—	—	—	2.25	—	5.63	3.25	3.5	—	—	5.63	—
Egypt	Canyon South-Coral Garden	—	—	—	—	—	—	—	3.0	18.5	—	3.75	—	3.0	0.5	—	—	—	2.75	—

Continued

Table A4 (Continued)   Global distribution of giant clams (Reef Check)

Country	Reef Site	Monitoring years (total clam density = number of individuals per 100 m²)																		
		1997	1998	1999	2000	2001	2002	2003	2004	2005	2006	2007	2008	2009	2010	2011	2012	2013	2014	2015
Egypt	Castle Beach-Ras Shitan	—	—	—	—	—	—	—	—	—	—	—	2.5	—	—	—	—	—	—	—
Egypt	Checkpoint	—	—	—	—	—	—	2.25	—	—	—	1.0	—	—	—	—	—	3.5	—	—
Egypt	Dahab Lighthouse	—	—	—	—	—	3.0	—	—	5.25	6.25	—	—	11.5	8.88	7.25	9.0	—	5.58	3.75
Egypt	Dahab Moray Garden	—	—	—	—	—	2.0	—	6.2	28.0	5.58	—	—	8.63	14.88	8.15	11.75	16.94	18.75	25.88
Egypt	Disha Outside	0.75	—	—	—	5.5	—	—	—	—	—	—	2.0	—	—	—	—	—	—	—
Egypt	Eel Garden North	—	—	—	—	—	—	—	9.0	—	—	—	—	—	—	—	—	—	—	—
Egypt	El Quadim Bay inside the bay	—	—	—	—	—	—	—	—	8.22	—	—	—	—	—	—	—	—	—	—
Egypt	El Quadim Bay outside the bay	—	—	—	—	—	—	—	—	3.38	—	—	—	—	—	—	—	—	—	—
Egypt	Falfulea	4.5	—	—	—	25.63	—	—	—	—	14.0	—	—	—	—	—	—	—	—	—
Egypt	Fanadir	0.88	—	—	—	3.25	—	—	—	—	—	—	0.75	0.25	—	—	—	—	—	—
Egypt	Far Garden	12.25	—	—	4.25	—	—	3.75	—	—	—	—	—	—	—	—	—	—	—	—
Egypt	Gabr el Bint	—	—	—	—	—	—	—	—	—	—	—	—	6.38	7.63	6.5	10.25	—	8.75	—
Egypt	Gabr el Bint North	—	—	—	—	—	—	—	16.75	7.25	5.38	10.63	—	—	—	—	—	—	—	—
Egypt	Gabr el Bint South	—	—	—	—	—	—	—	19.13	—	—	—	—	—	—	—	—	—	—	—
Egypt	Gamul Kebir	1.75	—	—	—	—	—	—	—	—	—	—	—	—	—	—	—	—	—	—
Egypt	Gamul Soraya	4.75	—	—	—	—	—	—	—	—	—	—	—	—	—	—	—	—	—	—
Egypt	Gebel el Rosas North	—	—	—	—	—	—	—	—	93.5	—	—	—	—	—	—	—	—	—	—
Egypt	Godda Abu Ramada	1.25	—	—	—	—	—	—	—	—	—	—	—	—	—	—	—	—	—	—

Continued

Table A4 (Continued) Global distribution of giant clams (Reef Check)

Country	Reef Site	Monitoring years (total clam density = number of individuals per 100 m^2)																		
		1997	1998	1999	2000	2001	2002	2003	2004	2005	2006	2007	2008	2009	2010	2011	2012	2013	2014	2015
Egypt	Gordon Reef Tiran	7.5	—	—	—	—	—	—	—	—	—	4.0	—	4.13	—	—	—	—	—	—
Egypt	Gota Abu Ramada	—	—	—	—	—	—	—	—	—	—	—	3.5	—	—	—	—	—	—	—
Egypt	Gotta Nakari	—	—	—	—	—	—	—	—	2.38	0.5	—	—	—	—	—	—	—	—	—
Egypt	Islands	—	—	—	—	—	—	—	20.08	—	—	—	10.75	—	—	—	—	—	—	—
Egypt	Islands North	—	—	—	—	—	—	—	—	—	—	—	—	6.75	5.0	6.0	12.25	11.25	—	—
Egypt	Islands South	—	—	—	—	—	—	—	—	—	—	—	—	4.25	6.25	8.25	—	11.75	8.67	—
Egypt	Jackfish Alley	6.75	—	—	5.75	—	7.25	—	—	—	—	—	—	—	—	—	—	—	—	—
Egypt	Jackson Reef Tiran	4.38	—	—	—	2.63	—	1.13	2.5	—	—	0.25	—	3.13	—	—	—	—	—	—
Egypt	Jolanda Reef	0.75	—	—	—	—	—	—	—	—	—	—	—	—	—	—	—	—	—	—
Egypt	Kalawy A	—	—	—	—	—	—	—	—	—	—	—	10.13	10.88	—	—	—	—	—	—
Egypt	Kalawy B	—	—	—	—	—	—	—	—	—	—	—	3.88	2.75	—	—	—	—	—	—
Egypt	Maagana	2.88	—	—	—	—	—	—	—	—	—	—	—	—	—	—	—	—	—	—
Egypt	Markaz Ratz	—	—	—	—	—	—	40.0	—	1.25	—	2.0	—	—	—	—	—	—	—	—
Egypt	Markaz Ratz South	—	—	—	—	—	—	—	—	—	—	7.75	—	—	—	—	—	—	—	—
Egypt	Marsa Abu Dabab North	—	—	—	—	13.75	—	—	14.38	—	—	—	—	—	—	—	—	—	—	—
Egypt	Marsa Assalaya North	—	—	—	—	—	—	—	7.0	11.75	—	5.0	—	—	—	—	—	—	—	4.63
Egypt	Marsa Egla North	—	—	—	—	—	—	27.0	6.63	15.0	—	10.25	—	—	5.63	—	10.19	—	—	—
Egypt	Marsa Gabel El Rosas	—	—	—	—	—	—	—	—	—	—	—	—	—	5.38	—	7.25	—	—	—
Egypt	Marsa Nakari North	—	—	—	—	—	—	—	—	—	—	—	—	—	1.88	4.38	—	4.25	—	8.63

Continued

Table A4 (Continued)   Global distribution of giant clams (Reef Check)

Country	Reef Site	Monitoring years (total clam density = number of individuals per 100 m²)																		
		1997	1998	1999	2000	2001	2002	2003	2004	2005	2006	2007	2008	2009	2010	2011	2012	2013	2014	2015
Egypt	Marsa Nakari South	—	—	—	—	—	—	—	—	—	—	—	—	—	3.5	5.0	—	5.88	—	10.5
Egypt	Marsa Samadai North	—	—	—	—	—	—	15.13	11.25	7.75	—	8.25	—	9.0	—	—	14.0	3.5	—	—
Egypt	Marsa Shagra North Reef	—	—	—	—	—	—	—	—	—	—	—	—	5.0	—	5.5	—	7.13	—	8.75
Egypt	Marsa Shagra South Reef	—	—	—	—	—	—	—	—	—	—	—	—	15.88	—	5.0	—	24.33	—	19.5
Egypt	Marsa Shaqraa-Hosam Helmy Camp	—	—	—	—	—	—	—	—	—	—	—	2.5	—	—	—	—	—	—	—
Egypt	Marsa Shuni North	—	—	—	—	21.75	—	—	3.25	—	—	—	—	—	—	—	—	—	—	—
Egypt	Marsa Tondoba (3 Sisters)	—	—	—	—	—	—	—	—	49.5	—	66.5	—	—	—	—	56.5	—	—	—
Egypt	Marsa Tondoba North	—	—	—	—	—	—	32.75	43.38	49.5	37.25	40.75	—	48.75	—	—	63.63	56.5	—	—
Egypt	Marsa Um Tondoba North Entrance	—	—	—	—	—	—	—	—	—	—	—	—	—	—	—	—	146.5	—	—
Egypt	Marsa Wizr-Mangrove Bay Resort Beach	—	—	—	—	—	—	—	—	—	—	—	2.75	—	—	—	—	—	—	—
Egypt	Mashraba-Nesima	—	—	—	—	—	—	—	—	—	—	8.25	—	—	—	—	—	—	—	—
Egypt	Melia Sinai Beach	—	—	—	—	—	—	—	—	—	—	—	3.25	—	—	—	—	—	—	—
Egypt	Middle Garden	21.25	—	—	17.0	12.75	4.0	—	—	—	—	—	—	—	—	—	—	—	—	—

Continued

287

Table A4 (Continued) Global distribution of giant clams (Reef Check)

Country	Reef Site	Monitoring years (total clam density = number of individuals per 100 m²)																		
		1997	1998	1999	2000	2001	2002	2003	2004	2005	2006	2007	2008	2009	2010	2011	2012	2013	2014	2015
Egypt	Middle-far Garden	15.13	—	—	—	—	—	—	—	—	—	—	—	—	—	—	—	—	—	—
Egypt	Moray Garden	—	—	—	—	—	—	—	10.0	—	—	—	—	—	—	—	—	—	—	—
Egypt	Nuweiba Hilton	0.25	—	—	—	—	—	—	—	—	—	—	—	—	—	—	—	—	—	—
Egypt	Panorama Reef	—	—	—	—	—	—	—	—	—	—	—	—	13.0	—	—	—	—	—	—
Egypt	Pharoah's Island	3.13	—	—	—	—	—	—	—	—	—	—	—	—	—	—	—	—	—	—
Egypt	Ras Abre El Bint	3.75	—	—	—	—	—	—	—	—	—	—	—	—	—	—	—	—	—	—
Egypt	Ras Abu Soma	0.69	—	—	—	—	—	—	—	—	—	—	2.13	—	—	—	—	—	—	—
Egypt	Ras Bob	48.13	—	—	—	—	—	10.5	—	—	—	—	—	—	—	—	—	—	—	—
Egypt	Ras Gamila	—	—	—	—	—	—	3.0	—	—	—	—	—	—	—	—	—	—	—	—
Egypt	Ras Ghaloum North	—	—	—	—	—	—	—	5.5	—	—	—	—	—	—	—	—	—	—	—
Egypt	Ras Ghaloum South	—	—	—	—	—	—	—	10.25	—	—	—	—	—	—	—	—	—	—	—
Egypt	Ras Ghozlani	—	—	—	—	8.0	—	5.75	—	—	—	—	—	—	—	—	—	—	—	—
Egypt	Ras Gumilla	9.25	—	—	—	—	—	—	—	—	—	—	—	—	—	—	—	—	—	—
Egypt	Ras Katy	1.75	—	—	—	—	—	6.5	—	—	—	—	—	—	—	—	—	—	—	—
Egypt	Ras Mohamed-Anemone City	—	—	—	—	—	—	—	—	—	—	—	0.88	—	—	—	—	—	—	—
Egypt	Ras Nusrani	8.0	—	—	19.75	—	6.25	—	—	—	—	—	—	—	—	—	—	—	—	—
Egypt	Rick's Reef	—	—	—	—	—	—	—	11.67	23.0	12.38	—	—	7.88	5.38	5.88	13.33	13.5	16.94	—
Egypt	Samaka Mogeha (North Marsa Nakari)	—	—	—	—	—	—	3.5	—	6.75	—	—	—	—	—	—	—	—	—	—
Egypt	Sha'ab Abu Danab	—	—	—	—	—	—	—	—	—	—	—	—	2.25	—	—	—	—	—	—
Egypt	Shaab Shagra "Elphinstone"	—	—	—	—	—	—	—	—	—	—	—	—	—	2.38	—	2.88	—	—	—
Egypt	Shaab Claude	3.75	—	—	—	—	—	—	—	—	—	—	—	—	—	—	—	—	—	—

Continued

Table A4 (Continued) Global distribution of giant clams (Reef Check)

Country	Reef Site	Monitoring years (total clam density = number of individuals per 100 m²)																		
		1997	1998	1999	2000	2001	2002	2003	2004	2005	2006	2007	2008	2009	2010	2011	2012	2013	2014	2015
Egypt	Shaab Samadai-Pinacle Kebir	—									1.5									—
Egypt	Shaab Shear	4.17	—	—	—	—			—		—	—	—	—	—	—	—	—	—	—
Egypt	Shabrul	0.75	—	—	—	—			—		—	—	—	—	—	—	—	—	—	—
Egypt	Shark Bay	7.88	—	—	—	8.75			—		—	—	—	—	—	—	—	—	—	—
Egypt	Sharm Abu Dabab South				—									18.25		23.88		34.63		
Egypt	Sharm Fukeri North	—							1.38	2.25		6.75					5.25			—
Egypt	Southern Oasis	—							—		—				3.0				—	—
Egypt	Stone Beach	3.13	—		—	8.0			—		—								—	—
Egypt	Temple	5.5	—		—	9.0	31.0		—		—								—	—
Egypt	Thomas Reef	7.5	—		—		3.38		—		—								—	—
Egypt	Three Pools	—	—	—	—				—	25.5	26.0	—			—			—	—	—
Egypt	Tobia Gebir	—							—		—			2.75	—			—	—	—
Egypt	Tondoba Bay, outer reef North	—							66.75	18.25	—						62.75	150.25		
Egypt	Torfa Fanous East	—											4.33	—			—			
Egypt	Torfa Heaven	—		—		—			21.75	10.38	26.0	—		—		—	—			
Egypt	Torfa Mekki	—	—	—				12.0	—							—	—			
Egypt	Tower	9.88			9.25		6.5		—											
Egypt	Tubya Kebir	2.75				—			—							4.13				
Egypt	Umm Sid	—									2.5	3.25		5.63	7.5					
Egypt	Wadi Lahami House Reef 1	—									2.5			2.5		—	12.0		9.13	—

Continued

Table A4 (Continued) Global distribution of giant clams (Reef Check)

Country	Reef Site	Monitoring years (total clam density = number of individuals per 100 m²)																		
		1997	1998	1999	2000	2001	2002	2003	2004	2005	2006	2007	2008	2009	2010	2011	2012	2013	2014	2015
Egypt	Wadi Lahami House Reef 2	—	—	—	—	—	—	—	—	—	—	—	—	2.88	—	—	4.67	—	—	—
Egypt	White Knight	21.5	—	—	—	4.5	—	—	—	—	—	—	—	—	—	—	—	—	—	—
Eritrea	Dur Gaam Island	—	—	—	0.13	—	—	—	—	—	—	—	—	—	—	—	—	—	—	—
Federated States of Micronesia	Buoy 16 (Inpuspusa)	—	—	—	0.75	—	0.5	0.25	0.25	0.25	—	—	—	—	—	—	—	—	—	—
Federated States of Micronesia	Buoy 18	—	—	—	—	0.75	—	—	—	—	—	—	—	—	—	—	—	—	—	—
Federated States of Micronesia	Buoy 23	—	—	—	0.25	—	—	—	—	—	—	—	—	—	—	—	—	—	—	—
Federated States of Micronesia	Buoy 27 (Tukunsru N.)	—	—	—	2.0	0.5	1.5	—	0.75	0.25	—	—	—	—	—	—	—	—	—	—
Federated States of Micronesia	Buoy 29 Sroac	—	—	—	—	—	—	—	—	0.5	—	—	—	—	—	—	—	—	—	—
Federated States of Micronesia	Buoy 39 (Shark Island)	—	—	—	—	0.25	—	—	—	0.25	0.25	—	0.25	—	—	—	—	—	—	—
Federated States of Micronesia	EMB #31 Panyacah	—	—	—	—	0.75	—	—	—	—	1.25	—	0.75	—	—	—	—	—	—	—
Federated States of Micronesia	EMB #34 Molsron Mwot West	—	—	—	—	0.25	0.5	—	—	3.25	—	—	—	—	—	—	—	—	—	—

Continued

Table A4 (Continued)   Global distribution of giant clams (Reef Check)

Country	Reef Site	Monitoring years (total clam density = number of individuals per 100 m²)																		
		1997	1998	1999	2000	2001	2002	2003	2004	2005	2006	2007	2008	2009	2010	2011	2012	2013	2014	2015
Federated States of Micronesia	EMB #43 Tafunsak Srisrik	—	—	—	—	—	1.25	—	—	1.25	—	0.5	—	—	—	—	—	—	—	—
Federated States of Micronesia	EMB #47 Kisacs	—	—	—	—	—	—	—	—	—	—	—	0.25	—	—	—	—	—	—	—
Federated States of Micronesia	EMB #49 Inkaratoah	—	—	—	—	—	—	—	—	0.25	—	—	—	—	—	—	—	—	—	—
Federated States of Micronesia	EMB #53 Metais	—	—	—	—	1.0	0.5	—	—	—	—	—	—	—	—	—	—	—	—	—
Federated States of Micronesia	EMB 16 Inpuspusa	—	—	—	—	—	—	—	—	0.25	—	—	—	—	—	—	—	—	—	—
Federated States of Micronesia	EMB 16 Inpuspusa	—	—	—	—	—	—	—	—	—	0.25	—	0.5	—	—	—	—	—	—	—
Federated States of Micronesia	Kisacs EMB 47	—	—	—	—	—	—	—	—	—	—	—	—	—	—	—	—	—	—	—
Federated States of Micronesia	Molsron Malem EMB#8	—	—	—	—	0.25	—	—	—	—	0.25	0.25	—	—	—	—	—	—	—	—
Federated States of Micronesia	North Tukunsruh EMB 27	—	—	—	—	—	—	—	—	—	0.5	—	—	—	—	—	—	—	—	—
Federated States of Micronesia	Sunrise Reef	—	—	—	—	—	0.62	—	—	—	—	—	—	—	—	—	—	—	—	—

Continued

291

Table A4 (Continued) Global distribution of giant clams (Reef Check)

Country	Reef Site	1997	1998	1999	2000	2001	2002	2003	2004	2005	2006	2007	2008	2009	2010	2011	2012	2013	2014	2015
										Monitoring years (total clam density = number of individuals per 100 m²)									
Federated States of Micronesia	Trochus Sanctuary	—	—	—	0.75	0.5	0.62	0.25	—	0.38	—	—	0.12	—	—	—	—	—	—	—
Federated States of Micronesia	Wanyan	—	—	—	—	—	2.25	—	—	—	—	—	—	—	—	—	—	—	—	—
Fiji	2 Thumbs Up Arch Bommie	—	—	—	—	—	—	—	—	—	—	—	—	—	—	0.75	—	—	—	—
Fiji	Aanuya Reef Edge	—	—	—	—	—	—	—	0.5	—	—	—	—	—	—	—	—	—	—	—
Fiji	Alice Reef	—	—	—	—	—	—	0.25	—	0.25	—	—	—	—	—	—	—	—	—	—
Fiji	Angel Reef	0.25	—	—	—	—	0.75	—	—	—	—	—	—	—	—	—	—	—	—	—
Fiji	Anthias Avenue	—	—	—	—	—	—	—	—	—	—	—	—	—	—	0.25	—	—	—	—
Fiji	Aquaventure Dive Shop	—	—	—	—	—	0.5	—	—	—	—	—	—	—	—	—	—	—	—	—
Fiji	Back Reef, Magic Mushrooms, South Save-a-Tack	—	—	—	—	—	—	—	—	—	0.25	—	—	—	—	—	—	—	—	—
Fiji	Barracuda Point	—	—	—	—	—	—	—	—	—	0.5	—	—	—	—	—	—	—	—	—
Fiji	Bella's Reef Nadogo	—	—	—	—	—	—	—	0.25	—	—	—	—	—	—	—	—	—	—	—
Fiji	Big Blue 3	—	—	—	0.5	—	—	—	—	—	—	—	—	—	—	—	—	—	—	—
Fiji	Big Blue 6	—	—	—	0.25	—	—	—	—	—	—	—	—	—	—	—	—	—	—	—
Fiji	Bird Rock	—	—	—	—	—	—	—	1.5	—	—	—	—	—	—	—	—	—	—	—
Fiji	Black Magic Mountain	—	—	—	—	—	0.25	—	—	—	—	—	—	—	—	—	—	—	—	—

Continued

Table A4 (Continued)   Global distribution of giant clams (Reef Check)

Country	Reef Site	Monitoring years (total clam density = number of individuals per 100 m^2)																		
		1997	1998	1999	2000	2001	2002	2003	2004	2005	2006	2007	2008	2009	2010	2011	2012	2013	2014	2015
Fiji	Blue Ribbon Eel Reef	—	—	—	—	—	0.5	—	0.75	0.25	—	—	—	—	—	—	—	—	—	—
Fiji	Blue Wall	—	—	—	—	—	—	—	—	—	—	—	—	—	—	—	—	—	—	—
Fiji	Buca Bay Channel Marker	0.5	—	—	—	—	1.0	—	—	—	—	—	—	—	—	—	—	—	—	—
Fiji	Castaway House Reef	—	—	—	—	4.5	—	—	3.0	—	—	—	—	—	—	—	—	—	—	—
Fiji	Castaway Pinnacles	—	—	—	—	—	0.25	—	—	—	—	—	—	—	—	—	—	—	—	—
Fiji	Castaway Resort House Reef	—	—	—	—	—	—	2.75	—	—	—	—	—	—	—	—	—	—	—	—
Fiji	Cat's Meow Shoal	—	—	—	—	—	—	—	—	—	—	0.25	—	—	—	—	—	—	—	—
Fiji	Cousteau Jetty shallow	—	—	—	—	—	0.25	0.25	—	—	0.75	0.75	—	—	—	—	—	—	—	—
Fiji	Cousteau Lighthouse	—	—	—	—	—	—	—	0.12	—	0.25	0.5	0.12	—	—	—	—	—	—	—
Fiji	Cowrie Crawl	—	—	—	—	—	—	—	—	—	—	0.25	—	—	—	—	—	—	—	—
Fiji	Cukini/Nadogo Mangrove Island	—	—	—	—	—	—	—	0.25	—	—	—	—	—	—	—	—	—	—	—
Fiji	Dennis Patch	—	—	—	—	—	—	0.25	—	—	—	—	—	—	—	—	—	—	—	—
Fiji	Dreadlocks	—	—	—	—	—	—	—	—	—	—	0.25	—	—	—	—	—	—	—	—
Fiji	E6	—	—	—	—	—	—	—	—	—	—	0.75	0.5	—	—	—	—	—	—	—
Fiji	Fish Factory	—	—	—	—	—	0.75	—	—	—	—	—	—	—	—	—	—	—	—	—
Fiji	Fragle Rock	—	—	—	—	—	—	—	0.5	—	—	—	—	—	—	—	—	—	—	—
Fiji	G-6 Reef	—	—	—	—	—	0.25	—	—	—	—	—	—	—	—	—	—	—	—	—

Continued

293

Table A4 (Continued) Global distribution of giant clams (Reef Check)

Country	Reef Site	Monitoring years (total clam density = number of individuals per 100 m²)																		
		1997	1998	1999	2000	2001	2002	2003	2004	2005	2006	2007	2008	2009	2010	2011	2012	2013	2014	2015
Fiji	Garden of Eden	—	—	—	—	—	0.12	—	—	—	—	—	—	—	—	—	—	—	—	—
Fiji	Golden Nuggets	—	—	—	—	—	0.25	0.38	0.25	0.25	—	—	—	—	—	—	—	—	—	—
Fiji	Golden Nuggets Deep	—	—	—	—	—	—	0.38	—	—	—	—	—	—	—	—	—	—	—	—
Fiji	Golden Reef	—	—	—	1.5	0.25	—	—	—	—	—	—	—	—	—	—	—	—	—	—
Fiji	Great White Wall	—	—	—	—	—	0.25	—	0.5	—	—	—	—	0.25	0.25	0.25	—	—	—	—
Fiji	Honeymoon	—	—	—	—	—	—	0.25	—	—	—	—	—	—	—	—	—	—	—	—
Fiji	Honeymoon Island	—	—	—	—	—	—	—	0.08	—	—	—	—	—	—	—	—	—	—	—
Fiji	House Reef Raviniake	—	—	—	—	—	—	4.25	0.5	—	—	—	—	—	—	—	—	—	—	—
Fiji	House Reef Raviniake 2	—	—	—	—	—	—	0.5	—	—	—	—	—	—	—	—	—	—	—	—
Fiji	Inner Barrier SW Qalito Island	—	—	—	—	—	—	—	0.5	—	—	—	—	—	—	—	—	—	—	—
Fiji	Instant Replay	—	—	—	—	—	0.25	—	—	—	—	—	—	—	—	—	—	—	—	—
Fiji	Islet off Kia	—	—	—	—	—	—	—	0.75	—	—	—	—	—	—	—	—	—	—	—
Fiji	Jacky's Reef	—	—	—	—	—	3.25	—	—	—	—	—	—	—	—	—	—	—	—	—
Fiji	Jerry's Jelly	—	—	—	—	—	—	—	—	—	1.25	—	—	—	—	—	—	—	—	—
Fiji	Jerry's Jelly/ Blue Ribbon Eel Reef	—	—	—	—	—	—	—	—	—	—	—	—	—	—	0.25	—	—	—	—
Fiji	Jimmy's Reef	—	—	—	—	—	2.25	—	—	—	—	—	—	—	—	—	—	—	—	—
Fiji	Korovou	—	—	—	—	—	0.62	—	—	—	—	—	—	—	—	—	—	—	—	—
Fiji	Kosova Point	—	—	—	—	—	—	—	0.25	—	—	—	—	—	—	—	—	—	—	—
Fiji	Lawaki Beach House	—	—	—	—	—	—	—	—	—	—	—	1.0	—	—	—	—	—	—	—

Continued

Table A4 (Continued) Global distribution of giant clams (Reef Check)

Country	Reef Site	Monitoring years (total clam density = number of individuals per 100 m²)																		
		1997	1998	1999	2000	2001	2002	2003	2004	2005	2006	2007	2008	2009	2010	2011	2012	2013	2014	2015
Fiji	Lawaki Beach Resort (Deeper Reef Edge)	—	—	—	—	—	—	—	0.25	—	—	—	—	—	—	—	—	—	—	—
Fiji	Lawaki Beach Resort House Reef North	—	—	—	—	—	—	—	0.5	—	—	—	—	—	—	—	—	—	—	—
Fiji	Lawaki Beach Resort House Reef South	—	—	—	—	—	—	0.25	—	—	—	—	—	—	—	—	—	—	—	—
Fiji	Lighthouse	—	—	—	—	0.75	—	—	—	—	—	—	—	—	—	—	—	—	—	—
Fiji	Likuliku 1	—	—	—	—	—	—	—	0.25	—	—	—	—	—	—	—	—	—	—	—
Fiji	Likuliku 4	—	—	—	—	—	—	—	0.25	—	—	—	—	—	—	—	—	—	—	—
Fiji	Likuri Pass South outer wall	—	—	—	—	—	—	0.25	—	—	—	—	—	—	—	—	—	—	—	—
Fiji	Lomanisue North	—	—	—	—	—	—	—	0.25	—	—	—	—	—	—	—	—	—	—	—
Fiji	Maccadame Reef	—	—	0.5	—	—	—	—	—	—	—	—	—	—	—	—	—	—	—	—
Fiji	Magic Mushrooms	—	—	—	—	—	—	—	0.25	—	—	—	—	—	—	—	—	—	—	—
Fiji	Makamaka Point	—	—	—	—	—	0.38	0.25	—	—	—	—	—	—	—	—	—	—	—	—
Fiji	Malamala	—	—	—	—	—	—	1.0	—	—	—	—	—	—	—	—	—	—	—	—
Fiji	Mali Passage	—	—	—	—	—	—	—	0.25	—	—	—	—	—	—	—	—	—	—	—
Fiji	Malolo 2	—	—	—	—	—	—	0.75	—	—	—	—	—	—	—	—	—	—	—	—
Fiji	Manta Ray Point	—	—	—	—	—	0.25	0.33	—	—	—	—	—	—	—	—	—	—	—	—
Fiji	Motuli Bawa	—	—	—	—	—	—	—	1.5	—	—	—	—	—	—	—	—	—	—	—
Fiji	Mount Mutiny	—	—	—	—	—	—	—	—	—	—	0.25	—	—	—	0.25	—	—	—	—

Continued

295

Table A4 (Continued) Global distribution of giant clams (Reef Check)

Country	Reef Site	Monitoring years (total clam density = number of individuals per 100 m²)																		
		1997	1998	1999	2000	2001	2002	2003	2004	2005	2006	2007	2008	2009	2010	2011	2012	2013	2014	2015
Fiji	Muiavuso Flats	—	—	—	—	—	—	0.5	0.25	—	—	—	—	—	—	—	—	—	—	—
Fiji	Mystery Reef	0.75	—	—	—	—	—	—	—	0.25	0.12	0.12	—	—	—	—	—	—	—	—
Fiji	Naivua Keraquma Reef	—	—	—	—	1.25	—	—	—	—	—	—	—	—	—	—	—	—	—	—
Fiji	Naiyacayaca	—	—	—	—	—	0.12	0.92	—	—	—	—	—	—	—	—	—	—	—	—
Fiji	Nakubu Reef	—	—	—	—	—	—	0.75	—	—	—	—	—	—	—	—	—	—	—	—
Fiji	Nakubu Reef site A	—	—	—	—	—	1.5	—	—	—	—	—	—	—	—	—	—	—	—	—
Fiji	Namoa Reef	—	—	—	—	—	—	—	5.75	—	—	—	—	—	—	—	—	—	—	—
Fiji	Navini Southwest-subsurface beachcomber	—	—	—	—	—	—	—	0.25	—	—	—	—	—	—	—	—	—	—	—
Fiji	North Castaway	—	—	—	—	—	—	2.75	1.25	—	—	—	—	—	—	—	—	—	—	—
Fiji	North Reef	—	—	—	—	—	—	—	0.75	—	—	—	—	—	—	—	—	—	—	—
Fiji	Oarmans Bay	—	—	—	—	—	—	—	—	—	—	0.25	—	1.25	—	—	—	—	—	—
Fiji	OJ's	—	—	—	—	—	0.25	0.44	—	—	—	—	—	—	—	—	—	—	—	—
Fiji	Outer side of Inner Barrier	—	—	—	—	—	—	0.5	—	—	—	—	—	—	—	—	—	—	—	—
Fiji	Outer wall near south Kaibu opposite Pres Vale	—	—	—	—	—	—	—	—	—	—	—	—	—	1.0	—	—	—	—	—
Fiji	Outer wall South Kaibu Island	—	—	—	—	—	—	—	—	—	—	—	—	—	0.5	—	—	—	—	—
Fiji	Ovulavula Reef	—	—	—	—	—	1.5	—	—	—	—	—	—	—	—	—	—	—	—	—
Fiji	Pinnacle	—	—	—	—	—	1.0	0.25	—	—	—	—	—	—	—	—	—	—	—	—
Fiji	Pleasure Point	—	—	—	—	—	1.5	—	—	—	—	—	—	—	—	—	—	—	—	—

Continued

296

Table A4 (Continued) Global distribution of giant clams (Reef Check)

Country	Reef Site	Monitoring years (total clam density = number of individuals per 100 m^2)																		
		1997	1998	1999	2000	2001	2002	2003	2004	2005	2006	2007	2008	2009	2010	2011	2012	2013	2014	2015
Fiji	Purple Haze Reef	—	—	—	—	—	1.25	—	—	—	—	—	—	—	—	—	—	—	—	—
Fiji	Rainbow Reef: The Corner	—	—	—	—	—	—	—	—	—	—	—	—	—	0.25	—	—	—	—	—
Fiji	Ravanaki House Reef	—	—	—	—	—	—	—	0.25	—	—	—	—	—	—	—	—	—	—	—
Fiji	Raviravi Naku Pass Inner Reef	—	—	—	—	—	—	—	0.25	—	—	—	—	—	—	—	—	—	—	—
Fiji	RCA1	—	—	—	—	0.5	—	—	—	—	—	—	—	—	—	—	—	—	—	—
Fiji	RCA4	—	—	—	—	1.75	—	—	—	—	—	—	—	—	—	—	—	—	—	—
Fiji	Rocky Bay	—	—	—	—	—	—	1.0	—	—	—	—	—	—	—	—	—	—	—	—
Fiji	Ronnie's Bommies	—	—	—	—	—	—	—	0.75	—	—	—	—	—	—	—	—	—	—	—
Fiji	Runners Bay	—	—	—	—	0.25	—	—	—	—	—	—	—	—	—	—	—	—	—	—
Fiji	Sea Fan, Storm Island	—	—	—	—	—	—	—	—	—	—	0.25	—	—	—	—	—	—	—	—
Fiji	Sem's Point (Blue Ribbon Eel)	—	—	—	—	—	—	0.5	0.5	—	—	—	—	—	—	—	—	—	—	—
Fiji	Shark Alley	—	—	—	—	—	0.44	0.19	—	—	—	—	—	—	—	—	—	—	—	—
Fiji	Small White Wall	—	—	—	—	—	—	0.5	0.5	—	—	—	—	—	—	—	—	—	—	—
Fiji	Sunflower	—	—	—	—	—	—	0.12	0.38	—	—	—	—	—	0.25	—	—	—	—	—
Fiji	Supermarket	—	—	—	—	—	—	0.5	0.5	—	—	—	—	—	—	—	—	—	—	—
Fiji	Talailau	—	—	—	—	—	—	—	0.25	—	—	—	—	—	—	—	—	—	—	—
Fiji	Tavewa Island	—	—	—	—	—	0.5	—	—	—	—	—	—	—	—	—	—	—	—	—
Fiji	Tetons	—	—	—	—	—	—	—	—	—	—	1.0	0.25	—	—	0.75	—	—	—	—
Fiji	The Corner	—	—	—	—	—	—	—	0.5	—	—	—	—	—	—	—	—	—	—	—

Continued

297

Table A4 (Continued) Global distribution of giant clams (Reef Check)

Country	Reef Site	Monitoring years (total clam density = number of individuals per 100 m²)																		
		1997	1998	1999	2000	2001	2002	2003	2004	2005	2006	2007	2008	2009	2010	2011	2012	2013	2014	2015
Fiji	The End of the World	—	—	—	—	—	—	—	0.25	—	—	—	—	—	—	—	—	—	—	—
Fiji	Timeless	—	—	—	—	—	—	—	0.75	—	—	—	—	—	—	—	—	—	—	—
Fiji	Tokoriki Wall	—	—	—	—	—	—	—	0.5	—	—	—	—	—	—	—	—	—	—	—
Fiji	Treasure Island	—	—	—	—	—	—	1.0	—	—	—	—	—	—	—	—	—	—	—	—
Fiji	Tukituki, Vatukarasa Reef	—	—	—	—	—	—	—	—	—	—	—	0.25	—	—	—	—	—	—	—
Fiji	Vatuka Island	—	—	—	—	—	—	—	0.5	—	—	—	—	—	—	—	—	—	—	—
Fiji	Vesi Bay	—	—	—	—	—	—	—	—	—	0.5	0.25	—	—	—	0.25	—	—	—	—
Fiji	Vuna Qiliqili	—	—	—	—	—	2.25	0.25	—	—	—	—	—	—	—	—	—	—	—	—
Fiji	Waidigi	—	—	—	—	—	—	0.25	—	—	—	—	—	—	—	—	—	—	—	—
Fiji	Wainalovo East	—	—	—	—	—	—	—	0.25	—	—	—	—	—	—	—	—	—	—	—
Fiji	Wainalovo West	—	—	—	—	—	—	—	0.25	—	—	—	—	—	—	—	—	—	—	—
Fiji	Wainimaloro Bay/Safari Lodge Reef Slope	—	—	—	—	—	—	2.25	—	0.5	—	—	—	—	—	—	—	—	—	—
Fiji	Waitabu Cut	—	—	—	—	—	—	—	—	—	—	0.25	—	—	—	—	—	—	—	—
Fiji	Waitabu Fishing Grounds	—	—	—	—	—	—	—	—	—	—	—	—	—	—	—	—	—	—	—
Fiji	Waitabu MPA reef flat	—	—	—	—	—	—	—	0.5	—	—	—	—	—	—	—	—	—	—	—
Fiji	Waitabu MPA Slope	—	—	—	—	—	—	—	—	—	—	—	—	—	—	0.25	—	—	—	—
Fiji	Wakaya Lion's Den	—	—	—	—	—	—	—	—	—	—	0.5	0.25	—	—	—	—	—	—	—

Continued

Table A4 (Continued) Global distribution of giant clams (Reef Check)

Country	Reef Site	Monitoring years (total clam density = number of individuals per 100 m²)																		
		1997	1998	1999	2000	2001	2002	2003	2004	2005	2006	2007	2008	2009	2010	2011	2012	2013	2014	2015
Fiji	Watu Express Reef (destroyed)	—	—	—	—	—	0.5	—	—	—	—	—	—	—	—	—	—	—	—	—
Fiji	Whiskey Reef (off western shore of Macatawa Levu Island)	—	—	—	—	—	—	—	—	—	—	0.25	—	—	—	—	—	—	—	—
Fiji	Wilkes Passage	—	—	—	—	—	—	2.75	0.25	—	—	—	—	—	—	—	—	—	—	—
Fiji	Yanu Somila	—	—	—	—	—	—	—	0.75	—	—	—	—	—	—	—	—	—	—	—
French Polynesia	12 Apotres	—	—	—	—	—	—	—	—	—	—	0.5	0.5	0.25	0.25	—	—	—	—	—
French Polynesia	Aquarium	—	—	—	—	44.0	30.5	28.88	45.25	63.5	96.75	99.0	130.38	277.5	301.25	—	225.25	—	325.0	—
French Polynesia	Bora Bora Nui Resort (fringing reef)	—	—	—	—	—	—	—	—	19.75	26.75	34.5	85.25	83.5	58.62	71.5	66.25	—	43.0	—
French Polynesia	Bora Bora Resort (fringing reef)	—	—	—	—	—	—	—	—	1.0	26.5	34.0	27.5	38.0	40.5	40.88	35.5	—	31.0	—
French Polynesia	Club Med (fringing reef)	—	—	—	—	—	—	—	—	3.0	4.5	2.75	9.75	11.25	13.5	13.25	13.5	—	6.0	—
French Polynesia	College	—	—	—	—	28.0	20.0	—	—	—	—	—	—	—	—	—	—	—	—	—
French Polynesia	Eboulement	—	—	—	—	—	—	—	—	—	0.5	—	—	—	—	—	—	—	—	—
French Polynesia	Faaa/La Faille St Etienne	—	—	—	—	—	—	—	—	—	0.5	—	—	—	—	—	—	—	—	—
French Polynesia	Fakarava Atoll	—	—	0.5	—	—	—	—	—	—	—	—	—	—	—	—	—	—	—	—

Continued

Table A4 (Continued) Global distribution of giant clams (Reef Check)

Country	Reef Site	Monitoring years (total clam density = number of individuals per 100 m²)																		
		1997	1998	1999	2000	2001	2002	2003	2004	2005	2006	2007	2008	2009	2010	2011	2012	2013	2014	2015
French Polynesia	Fausse Passe (Teraea)	—	—	—	—	—	—	—	—	—	—	31.0	21.75	21.25	—	—	—	—	—	—
French Polynesia	Haapiti	—	—	—	—	—	—	—	—	—	7.75	—	—	—	—	—	—	—	—	—
French Polynesia	Haramea	—	—	—	—	—	—	—	—	—	138.75	—	—	—	—	—	—	—	—	—
French Polynesia	Hart	—	—	—	—	—	—	—	—	—	—	17.0	55.0	50.5	—	—	—	—	—	—
French Polynesia	Hawaiki Nui at Raiatea	—	—	—	—	—	—	—	—	—	—	—	10.25	—	—	—	—	—	—	—
French Polynesia	Hotel Bora Bora	—	—	—	—	—	—	—	—	1.75	4.5	5.75	14.0	19.5	8.25	11.5	10.25	—	—	—
French Polynesia	Huahine Pearl Farm	—	—	—	—	—	—	—	—	—	1.0	—	—	—	—	—	—	—	—	—
French Polynesia	Jardin de Corail (Coral Garden)	—	—	—	—	—	32.25	173.25	319.88	366.25	387.5	305.0	247.5	172.5	125.62	88.12	61.25	—	58.5	—
French Polynesia	Jardin de Fitii	—	—	—	—	—	—	—	—	—	0.5	—	—	—	—	—	—	—	—	—
French Polynesia	Kopuapiro	—	—	—	—	—	—	—	—	36.0	—	—	—	—	—	—	—	—	—	—
French Polynesia	Le Meridien/Manta (Pinacle)	—	—	—	—	—	—	—	—	3.25	3.0	3.0	8.0	10.75	8.38	9.25	12.0	—	15.75	—
French Polynesia	Maharepa	—	—	—	7.0	3.0	8.5	6.25	—	12.5	15.25	—	—	26.75	—	—	—	—	—	—
French Polynesia	Mahu	—	—	—	—	—	—	—	—	—	153.25	291.75	268.0	263.5	—	—	—	—	—	—
French Polynesia	Matira	—	—	—	—	—	—	190.12	433.0	424.0	583.75	463.75	225.0	182.5	61.88	43.5	21.5	—	19.25	—

Continued

Table A4 (Continued) Global distribution of giant clams (Reef Check)

Country	Reef Site	Monitoring years (total clam density = number of individuals per 100 m²)																		
		1997	1998	1999	2000	2001	2002	2003	2004	2005	2006	2007	2008	2009	2010	2011	2012	2013	2014	2015
French Polynesia	Mohio	—	—	—	—	—	133.0	116.12	240.88	123.5	133.75	176.25	196.25	170.25	167.0	172.25	177.0	—	92.5	—
French Polynesia	Motu Haapiti	—	—	—	—	—	—	—	—	—	—	11.5	136.75	151.25	230.38	249.12	254.0	—	233.25	—
French Polynesia	Motu Ome	—	—	—	—	—	—	—	—	—	—	3.75	28.25	30.5	—	—	—	—	—	—
French Polynesia	Motu Tane	—	—	—	—	—	—	—	—	—	—	54.0	166.0	139.25	128.0	143.25	151.25	—	153.5	—
French Polynesia	Motu Tapu	—	—	—	—	—	—	—	—	—	—	5.0	67.0	78.5	64.0	71.25	74.5	—	74.5	—
French Polynesia	Motu Te Avapiti	—	—	—	—	—	—	—	—	—	—	—	0.5	—	—	—	—	—	—	—
French Polynesia	Motu Vahapiapia	—	—	—	—	—	—	—	—	66.25	—	—	—	—	—	—	—	—	—	—
French Polynesia	Napoleon	—	—	—	—	—	—	—	—	—	—	—	1.0	—	—	—	—	—	—	—
French Polynesia	Nuhinuhi	—	—	—	—	—	—	—	—	—	—	3.5	3.5	3.0	2.5	—	—	—	—	—
French Polynesia	Otaha	—	—	—	—	—	—	—	—	—	71.75	44.0	60.0	60.0	96.5	—	—	—	—	—
French Polynesia	Paea	—	—	—	—	—	—	—	—	—	22.5	16.75	—	—	—	—	—	—	—	—
French Polynesia	Papa Mahuea	—	—	—	—	—	—	—	—	—	—	81.75	91.0	89.0	—	—	—	—	—	—
French Polynesia	Papetoai/ Christian	—	—	—	—	—	—	—	1.25	0.25	1.0	—	—	—	—	—	—	—	—	—
French Polynesia	Patito	—	—	—	—	—	—	—	—	—	—	3.75	6.75	6.25	—	—	—	—	—	—

Continued

Table A4 (Continued) Global distribution of giant clams (Reef Check)

Country	Reef Site	Monitoring years (total clam density = number of individuals per 100 m²)																		
		1997	1998	1999	2000	2001	2002	2003	2004	2005	2006	2007	2008	2009	2010	2011	2012	2013	2014	2015
French Polynesia	Pinacle Marara	—	—	—	—	—	2.5	1.5	2.38	—	1.0	2.0	2.25	1.25	4.75	4.5	4.75	—	4.5	—
French Polynesia	Povai	—	—	—	—	—	24.75	14.12	11.12	11.25	4.75	7.75	18.0	26.5	17.12	11.5	14.25	—	13.0	—
French Polynesia	Pufana	—	—	—	—	—	—	—	—	23.5	—	—	—	—	—	—	—	—	—	—
French Polynesia	Pukatoa	—	—	—	—	—	—	—	—	1.25	—	—	—	—	—	—	—	—	—	—
French Polynesia	Requin de Feu	—	—	—	—	—	—	—	—	—	—	—	0.25	—	—	—	—	—	—	—
French Polynesia	Revatua (fringing reef)	—	—	—	—	—	—	—	—	2.75	3.0	1.5	9.5	12.0	18.88	26.5	24.75	—	20.25	—
French Polynesia	Sofitel Marara	—	—	—	—	—	—	—	—	0.5	—	0.25	1.0	1.25	1.25	1.38	1.75	—	1.0	—
French Polynesia	Sofitel Motu	—	—	—	—	—	0.5	0.62	1.38	—	0.5	0.5	6.75	8.5	13.75	17.75	20.5	—	35.5	—
French Polynesia	Tapu	—	—	—	—	—	5.0	10.0	—	—	—	—	—	—	—	—	—	—	—	—
French Polynesia	Tapu (external slope)	—	—	—	—	—	—	—	—	—	1.5	2.0	3.25	8.5	5.62	5.38	7.0	—	5.75	—
French Polynesia	Te Tamanu	—	—	—	—	—	—	—	—	—	—	—	1.25	0.75	0.75	—	—	—	—	—
French Polynesia	Temae	—	—	—	—	—	—	—	—	—	1.25	1.75	—	—	—	—	—	—	—	—
French Polynesia	The Aquarium	—	—	—	—	—	—	—	—	—	—	—	—	—	—	—	—	5.25	—	—
French Polynesia	Tiahura	—	—	—	—	—	—	—	—	—	—	1.75	—	—	—	—	—	—	—	—
French Polynesia	Tiger Shark	—	—	—	—	—	—	—	—	0.25	—	—	—	—	—	—	—	—	—	—

Continued

302

Table A4 (Continued) Global distribution of giant clams (Reef Check)

Country	Reef Site	Monitoring years (total clam density = number of individuals per 100 m²)																		
		1997	1998	1999	2000	2001	2002	2003	2004	2005	2006	2007	2008	2009	2010	2011	2012	2013	2014	2015
French Polynesia	Toau North West outreef	—	—	—	—	—	—	—	—	0.25	—	—	—	—	—	—	—	—	—	—
French Polynesia	Toau South Lagoon	—	—	—	—	—	—	—	—	2.5	—	—	—	—	—	—	—	—	—	—
French Polynesia	Togamaitu-i-uta	—	—	—	—	—	—	—	—	92.25	—	—	—	—	—	—	—	—	—	—
French Polynesia	Tohea NE	—	—	—	—	—	—	—	—	98.0	—	—	—	—	—	—	—	—	—	—
French Polynesia	Tohea SE	—	—	—	—	—	—	—	—	11.0	—	—	—	—	—	—	—	—	—	—
French Polynesia	Top Dive (fringing reef)	—	—	—	—	—	—	—	—	15.0	9.25	19.5	20.75	25.0	35.0	34.88	30.25	—	18.25	—
French Polynesia	Turiroa	—	—	—	—	—	—	102.12	291.38	205.75	—	280.0	64.5	47.5	34.12	—	32.5	—	45.0	—
French Polynesia	Vaioue	—	—	—	—	—	—	—	—	0.25	—	—	—	—	—	—	—	—	—	—
French Polynesia	Vavaratea	—	—	—	—	—	—	—	—	—	20.25	—	—	—	—	—	—	—	—	—
Guam	Double Reef	—	1.0	4.0	—	3.5	—	—	1.38	—	—	—	—	—	—	—	—	—	—	—
Guam	Facpi Point	—	1.0	0.5	—	1.25	—	—	—	—	—	—	—	—	—	—	—	—	—	—
Hong Kong	Crescent Island East (Ngo Mei Chau)	—	—	—	—	—	—	—	—	—	0.5	—	—	—	—	—	—	—	—	—
Hong Kong	Double Island	—	—	—	—	—	—	1.25	—	—	—	—	—	—	—	—	—	—	—	—
Hong Kong	Siu Long Kei	—	—	—	—	—	—	—	—	—	—	—	—	—	—	0.5	—	—	—	—
India	Kadmat	—	0.5	—	—	—	—	—	—	—	—	—	—	—	—	—	—	—	—	—
Indonesia	Ahe Dive Resort Housereef	—	—	—	—	—	—	—	—	—	—	—	—	—	0.25	—	—	—	—	—
Indonesia	Air Karang	—	—	—	—	—	—	—	—	1.0	—	—	—	1.25	—	—	—	—	—	—

Continued

Table A4 (Continued) Global distribution of giant clams (Reef Check)

Country	Reef Site	Monitoring years (total clam density = number of individuals per 100 m²)																		
		1997	1998	1999	2000	2001	2002	2003	2004	2005	2006	2007	2008	2009	2010	2011	2012	2013	2014	2015
Indonesia	Air Tejun, Mursala Island	—	—	—	—	—	—	—	—	1.0	—	—	—	—	—	—	—	—	—	—
Indonesia	Amed	—	—	—	—	1.25	—	—	—	—	—	—	—	—	—	—	—	—	—	—
Indonesia	Ampalas	1.25	—	—	—	—	—	—	—	—	—	—	—	—	—	—	—	—	—	—
Indonesia	Ampana Tete (Tete B)	—	—	—	—	—	—	4.5	—	—	—	—	—	—	—	—	—	—	—	—
Indonesia	Asu	—	—	—	—	0.5	—	—	—	—	—	—	—	—	—	—	—	—	—	—
Indonesia	Baiya	—	—	—	—	—	—	—	—	—	—	—	2.5	—	—	—	—	—	—	—
Indonesia	Bama	—	—	—	—	—	0.5	—	—	—	—	—	—	2.5	—	—	—	—	—	—
Indonesia	Bangkaru	—	—	—	—	—	—	—	—	0.75	—	—	—	—	—	—	—	—	—	—
Indonesia	Bangkaru 2	—	—	—	—	—	—	—	—	1.0	—	—	—	—	—	—	—	—	—	—
Indonesia	Barranglompo (Site 1) Tenggara	—	—	—	—	—	—	—	2.0	0.5	—	—	—	—	—	—	—	—	—	—
Indonesia	Barranglompo (Site 2) Barat	—	—	—	—	—	—	—	—	2.25	—	—	—	—	—	—	—	—	—	—
Indonesia	Barrier Reef	—	—	—	—	—	—	1.0	—	—	—	—	—	—	—	—	—	—	—	—
Indonesia	Batee Gla	—	—	—	—	—	—	—	0.75	—	—	—	—	—	—	—	—	—	—	—
Indonesia	Benteng	—	—	—	—	—	—	—	—	—	—	—	—	1.25	—	—	—	—	—	—
Indonesia	Benteng Reef 1	—	—	10.0	—	—	—	—	—	—	—	—	—	—	—	—	—	—	—	—
Indonesia	Benteng Reef 2	—	—	12.25	—	—	—	—	—	—	—	—	—	—	—	—	—	—	—	—
Indonesia	Berakit	—	—	—	—	—	—	—	0.12	0.25	—	—	—	—	—	—	—	—	—	—
Indonesia	Bida Dari	—	—	—	—	—	—	—	—	—	—	—	—	1.75	—	—	—	—	2.25	—
Indonesia	Bilik	—	—	—	—	—	0.25	—	—	1.0	—	—	—	—	—	—	—	—	—	—
Indonesia	Bingin Bondalem	—	—	—	—	—	—	—	—	—	—	—	—	2.75	—	—	—	—	—	—
Indonesia	Binongko Island (site 13)	—	—	—	4.25	—	—	—	—	—	—	—	—	—	—	—	—	—	—	—
Indonesia	Bisabora	—	—	—	—	—	—	—	—	—	—	—	—	—	—	0.5	—	—	—	—

Continued

304

Table A4 (Continued)   Global distribution of giant clams (Reef Check)

Country	Reef Site	Monitoring years (total clam density = number of individuals per 100 m²)																		
		1997	1998	1999	2000	2001	2002	2003	2004	2005	2006	2007	2008	2009	2010	2011	2012	2013	2014	2015
Indonesia	Blue Coral	—	—	—	—	—	—	0.75	—	—	—	—	—	—	—	—	—	—	—	—
Indonesia	Budo	—	—	—	—	—	3.0	—	—	—	—	—	—	—	—	—	—	—	—	—
Indonesia	Bulan Madu Gili Air	—	—	—	—	—	—	0.25	—	—	—	—	—	—	—	—	—	—	—	—
Indonesia	Buoy 4 Hoga Island	—	—	—	—	5.0	—	1.75	—	—	—	—	—	—	—	—	—	—	—	—
Indonesia	Buoy 5	—	—	—	—	—	—	1.5	—	—	—	—	—	—	—	—	—	—	—	—
Indonesia	Burung Island	—	—	0.5	4.5	1.0	—	1.5	—	—	—	—	—	—	—	—	—	—	—	—
Indonesia	Cemara Besar	—	—	—	—	—	—	—	2.75	1.0	0.5	—	—	—	—	—	—	—	—	—
Indonesia	Cemara Kecil Island	1.75	—	—	4.75	1.75	—	3.5	—	—	—	—	—	—	—	—	—	—	—	—
Indonesia	Coast	1.0	—	—	—	—	—	—	—	—	—	—	—	—	—	—	—	—	—	—
Indonesia	Coral Eye House Reef	—	—	—	—	—	—	—	—	—	—	—	—	—	—	0.5	—	—	—	—
Indonesia	Coral Garden	—	—	—	—	6.5	—	—	—	—	—	—	—	—	—	—	—	—	—	—
Indonesia	Coral Meadow	—	—	—	—	—	—	—	0.75	—	—	—	—	—	—	—	—	—	—	—
Indonesia	East Bauluang Island	—	—	—	—	—	—	2.0	—	—	—	—	—	—	—	—	—	—	—	—
Indonesia	East Kapoposang	—	—	—	0.75	—	—	—	—	—	—	—	—	—	—	—	—	—	—	—
Indonesia	Fan Garden	—	—	—	—	—	—	—	—	—	—	—	—	—	0.5	—	—	—	—	—
Indonesia	Fukui	—	—	—	—	—	0.25	—	—	—	—	—	—	—	—	—	—	—	—	—
Indonesia	Gangga Jetty (Gangga Island)	—	—	—	—	—	—	—	—	—	—	—	—	—	—	4.75	—	0.25	0.5	—
Indonesia	Gangga Village (Gangga Island)	—	—	—	—	—	—	—	—	—	—	—	—	—	—	1.0	—	0.25	—	—
Indonesia	Gapang Beach	—	—	—	—	—	—	—	13.75	—	—	—	—	—	—	—	—	—	—	—

Continued

Table A4 (Continued)   Global distribution of giant clams (Reef Check)

Country	Reef Site	Monitoring years (total clam density = number of individuals per 100 m²)																		
		1997	1998	1999	2000	2001	2002	2003	2004	2005	2006	2007	2008	2009	2010	2011	2012	2013	2014	2015
Indonesia	Garden Eel	—	—	1.0	—	—	2.5	1.0	—	—	—	—	—	—	—	—	—	—	—	—
Indonesia	Geleang Island	—	—	1.25	1.5	—	—	2.75	—	—	—	—	—	—	—	—	—	—	—	—
Indonesia	Gili Sulat	—	—	—	1.0	—	—	—	—	—	—	—	—	—	—	—	—	—	—	—
Indonesia	Gosong Bira Besar	—	—	—	—	—	—	—	—	—	—	5.0	—	—	—	—	—	—	—	—
Indonesia	Gosong Kapas Reef	0.25	—	—	—	—	—	—	—	—	—	—	—	—	—	—	—	—	—	—
Indonesia	Gosong Sawo	—	—	—	—	—	—	0.5	—	—	—	—	—	—	—	—	—	—	—	—
Indonesia	Grand Ika Gili Air	—	—	—	—	—	—	0.5	—	—	—	—	—	—	—	—	—	—	—	—
Indonesia	Grubby's	—	—	—	—	1.5	—	—	—	—	—	—	—	—	—	—	—	—	—	—
Indonesia	Gusung Tarojaya, Salemo Island	—	—	—	—	—	—	—	3.25	—	—	—	—	—	—	—	—	—	—	—
Indonesia	H. Kasim	—	—	—	—	—	—	1.25	—	—	—	—	—	—	—	—	—	—	—	—
Indonesia	Hans Reef	—	—	—	—	—	—	—	—	—	—	—	—	—	—	—	—	—	1.0	—
Indonesia	Halik	—	—	—	—	—	—	—	—	—	—	—	—	—	—	—	1.5	—	1.0	—
Indonesia	Hidden Reef	—	—	—	—	—	—	—	—	—	—	—	—	—	—	—	—	—	0.5	—
Indonesia	Hoga Buoy 3	—	—	—	—	2.75	—	1.75	—	—	—	—	—	—	—	—	—	—	—	—
Indonesia	Hoga Buoy 4	—	—	—	—	—	—	—	—	0.75	—	—	—	—	—	—	—	—	—	—
Indonesia	Hoga Island Buoy 2	—	—	—	7.5	8.25	—	0.5	—	—	—	—	—	—	—	—	—	—	—	—
Indonesia	Horuo Reef	—	—	—	—	—	—	—	—	0.75	—	—	—	—	—	—	—	—	—	—
Indonesia	Ilona Bondalem	—	—	—	—	—	—	—	—	—	—	—	—	1.0	—	—	—	—	—	—
Indonesia	Indaa Atoll	—	—	—	—	—	—	—	—	—	—	—	—	—	1.5	—	—	—	—	—
Indonesia	Indah Reef	—	—	—	—	—	—	2.0	—	—	—	—	—	—	—	—	—	—	—	—
Indonesia	Jeladi Wilis	—	—	—	—	—	—	0.75	—	—	—	—	—	—	—	—	—	—	—	—
Indonesia	Jepun	—	—	—	—	0.25	—	—	—	—	—	—	—	—	—	—	—	—	—	—

Continued

Table A4 (Continued)   Global distribution of giant clams (Reef Check)

Country	Reef Site	Monitoring years (total clam density = number of individuals per 100 m²)																		
		1997	1998	1999	2000	2001	2002	2003	2004	2005	2006	2007	2008	2009	2010	2011	2012	2013	2014	2015
Indonesia	Kahuku (Bangka Island)	—	—	—	—	—	—	—	—	—	—	—	—	—	—	0.25	—	0.25	0.13	—
Indonesia	Kajang	—	—	—	—	—	—	—	—	—	—	—	—	1.5	—	—	—	—	—	—
Indonesia	Kakatu	—	—	—	—	—	—	0.12	—	—	—	—	—	—	—	—	—	—	—	—
Indonesia	Kaledupa 2	—	—	—	—	2.0	—	1.0	—	—	—	—	—	—	—	—	—	—	—	—
Indonesia	Kaledupa Buoy 1	—	—	—	—	—	—	1.75	1.75	—	—	—	—	—	—	—	—	—	—	—
Indonesia	Kaledupa Double Spur	—	—	—	—	2.5	—	0.25	1.0	—	—	—	—	—	—	—	—	—	—	—
Indonesia	Kaledupa Double Spur (site 18)	—	—	—	1.75	—	—	—	—	—	—	—	—	—	—	—	—	—	—	—
Indonesia	Kaledupa near Matingola Village (site 12)	—	—	—	3.75	—	—	—	—	—	—	—	—	—	—	—	—	—	—	—
Indonesia	Kaledupa North Coast	—	—	—	—	0.75	—	—	—	—	—	—	—	—	—	—	—	—	—	—
Indonesia	Kaledupa opposite Hoga (site 33)	—	—	—	—	0.75	—	—	—	—	—	—	—	—	—	—	—	—	—	—
Indonesia	Kaledupa SW tip (site 16)	—	—	—	1.25	—	—	—	—	—	—	—	—	—	—	—	—	—	—	—
Indonesia	Kaledupa West (site 11)	—	—	—	2.25	—	—	—	—	—	—	—	—	—	—	—	—	—	—	—
Indonesia	Kanawa Island	—	—	—	—	—	—	—	—	—	—	—	—	—	—	—	—	—	4.5	—
Indonesia	Kapaenaue (reef on right)	—	—	—	—	0.5	—	—	—	—	—	—	—	—	—	—	—	—	—	—
Indonesia	Kapota 2	—	—	—	—	—	—	—	—	—	—	—	—	—	0.75	—	—	—	—	—
Indonesia	Kapota Ollo	—	—	—	—	—	—	—	—	—	—	—	—	—	1.0	—	—	—	—	—

Continued

307

Table A4 (Continued) Global distribution of giant clams (Reef Check)

Country	Reef Site		Monitoring years (total clam density = number of individuals per 100 m²)																	
		1997	1998	1999	2000	2001	2002	2003	2004	2005	2006	2007	2008	2009	2010	2011	2012	2013	2014	2015
Indonesia	Kapote (reef on right)	—	—	—	—	0.25	—	—	—	—	—	—	—	—	—	—	—	—	—	—
Indonesia	Kapote Island (reef on right)	—	—	—	—	0.75	—	—	—	—	—	—	—	—	—	—	—	—	—	—
Indonesia	Karang Kaledupa	—	—	—	—	2.5	—	—	—	—	—	—	—	—	—	—	—	—	—	—
Indonesia	Karang Kaledupa (Atoll)	—	—	—	6.5	—	—	—	—	—	—	—	—	—	—	—	—	—	—	—
Indonesia	Karang Kaledupa (reef on right)	—	—	—	—	2.5	—	—	—	—	—	—	—	—	—	—	—	—	—	—
Indonesia	Karang Kaledupa (west-reef on right)	—	—	—	—	0.5	—	—	—	—	—	—	—	—	—	—	—	—	—	—
Indonesia	Karang Kaledupa atoll (site 6)	—	—	—	5.75	—	—	—	—	—	—	—	—	—	—	—	—	—	—	—
Indonesia	Karang Kapote SW (site 21)	—	—	—	1.25	—	—	—	—	—	—	—	—	—	—	—	—	—	—	—
Indonesia	Karang Kasih	—	—	—	—	—	—	—	—	1.0	—	—	—	—	—	—	—	—	—	—
Indonesia	Karang Mayit	—	—	—	—	—	—	—	—	—	—	—	—	1.0	—	—	—	—	—	—
Indonesia	Katiet Patch Reef	—	—	—	—	—	0.5	—	—	—	—	—	—	—	—	—	—	—	—	—
Indonesia	Kayu Duwi Tejakula	—	—	—	—	—	—	—	—	—	—	—	—	0.25	—	—	—	—	—	—
Indonesia	Kayunyole	—	—	—	—	—	—	—	—	—	—	—	2.25	—	—	—	—	—	—	—
Indonesia	Kecil Island	0.25	—	—	—	—	—	—	—	—	—	—	—	—	—	—	—	—	—	—
Indonesia	Kollozoa Reef	—	—	—	—	—	—	—	—	2.5	—	—	—	—	—	—	—	—	—	—

Continued

Table A4 (Continued) Global distribution of giant clams (Reef Check)

Country	Reef Site	Monitoring years (total clam density = number of individuals per 100 m²)																		
		1997	1998	1999	2000	2001	2002	2003	2004	2005	2006	2007	2008	2009	2010	2011	2012	2013	2014	2015
Indonesia	Kondang Buntung	—	—	—	—	—	—	—	—	0.25	—	—	—	—	—	—	—	—	—	—
Indonesia	Kuburan Cina	—	—	—	—	—	—	—	—	—	—	—	—	—	0.25	—	—	—	—	—
Indonesia	Kulati Wreck	—	—	—	—	—	—	—	—	4.0	—	—	—	—	—	—	—	—	—	—
Indonesia	Labuana 1	—	—	—	—	—	5.25	2.0	—	—	—	—	—	—	—	—	—	—	—	—
Indonesia	Labuana 2	—	—	—	—	—	2.75	3.0	—	—	—	—	—	—	—	—	—	—	—	—
Indonesia	Labuana 3	—	—	—	—	—	0.5	—	0.75	—	—	—	—	—	—	—	—	—	—	—
Indonesia	Labuhan Kapal, Mursala Island	—	—	—	—	—	—	—	—	3.5	—	—	—	—	—	—	—	—	—	—
Indonesia	Latondu	0.5	—	—	—	—	—	—	—	—	—	—	—	—	—	—	—	—	—	—
Indonesia	Legon Waru-Sangyang	—	—	—	—	—	2.0	—	—	—	—	—	—	—	—	—	—	—	—	—
Indonesia	Lekuan 3	—	—	—	0.75	1.5	0.25	—	—	—	—	—	—	—	—	—	—	—	—	—
Indonesia	Lempuyang	—	—	—	—	—	—	—	—	—	—	—	—	2.0	—	—	—	—	—	—
Indonesia	Lhok Me	—	—	—	—	—	—	—	—	—	—	—	—	1.75	—	—	—	—	—	—
Indonesia	Lighthouse Point	0.38	—	—	—	—	—	—	—	—	—	—	—	—	—	—	—	—	—	—
Indonesia	Linggan	—	—	—	—	—	—	—	—	0.25	—	—	—	—	—	—	—	—	—	—
Indonesia	Lintea Atoll (reef on right)	—	—	—	—	1.25	—	—	—	—	—	—	—	—	—	—	—	—	—	—
Indonesia	Lintea Atoll, reef on left	—	—	—	—	2.0	—	—	—	—	—	—	—	—	—	—	—	—	—	—
Indonesia	Lintea Kaledupa	—	—	—	—	2.5	—	—	—	—	—	—	—	—	—	—	—	—	—	—
Indonesia	Luna Park	—	—	—	—	0.25	—	—	—	—	—	—	—	—	—	—	—	—	—	—
Indonesia	Makmur	—	—	—	—	—	—	0.25	—	—	—	—	—	—	—	—	—	—	—	—
Indonesia	Malenge Reef 2	—	—	6.0	—	—	—	—	—	—	—	—	—	—	—	—	—	—	—	—
Indonesia	Malotong	—	—	—	—	—	—	—	—	—	—	—	1.5	—	—	—	—	—	—	—
Indonesia	Manga Tasik	1.25	—	—	—	—	—	—	—	—	—	—	—	—	—	—	—	—	—	—
Indonesia	Mangkasa Point	—	—	—	—	—	—	—	—	—	—	11.0	—	—	—	—	—	—	—	—
Indonesia	Manta Avenue	—	—	—	—	0.25	—	—	—	—	—	—	—	—	—	—	—	—	—	—

Continued

309

Table A4 (Continued) Global distribution of giant clams (Reef Check)

Country	Reef Site	Monitoring years (total clam density = number of individuals per 100 m^2)																		
		1997	1998	1999	2000	2001	2002	2003	2004	2005	2006	2007	2008	2009	2010	2011	2012	2013	2014	2015
Indonesia	Manta Point	—	—	—	—	—	—	—	—	—	—	—	—	—	—	—	—	0.88	—	—
Indonesia	Mari Mabuk Reef	—	—	—	—	—	—	—	—	3.75	—	—	—	—	1.0	—	—	—	—	—
Indonesia	Matingola Village	—	—	—	—	2.0	—	0.5	—	—	—	—	—	—	—	—	—	—	—	—
Indonesia	Mendati Village (reef on left)	—	—	—	—	1.0	—	—	—	—	—	—	—	—	—	—	—	—	—	—
Indonesia	Menjangan Besar Island	—	—	0.5	0.25	—	—	—	8.25	1.5	—	—	0.75	—	—	—	—	—	—	—
Indonesia	Menjangan Kecil Island	0.25	—	3.5	1.0	—	—	0.75	2.5	—	—	—	—	—	—	—	—	—	—	—
Indonesia	Menyawakan	—	—	1.25	2.75	0.25	—	1.0	—	—	0.12	—	—	—	—	—	—	—	—	—
Indonesia	Meras	—	—	—	—	—	0.75	—	—	—	—	—	—	—	—	—	—	—	—	—
Indonesia	Moor	—	—	—	—	—	—	—	—	—	—	—	—	—	0.5	—	—	—	—	—
Indonesia	Negeri Lima	0.75	—	—	—	—	—	—	—	—	—	—	—	—	—	—	—	—	—	—
Indonesia	North Barang Caddi	—	—	—	—	—	—	—	—	0.75	—	—	2.5	—	—	—	—	—	—	—
Indonesia	North side of Pulau Rondo	—	—	—	—	—	—	—	—	0.25	—	—	—	—	—	—	—	—	—	—
Indonesia	Nusa Penida	0.25	—	—	—	—	—	—	—	—	—	—	—	—	—	—	—	—	—	—
Indonesia	Nusa Tiga	—	—	—	—	—	—	0.75	—	—	—	—	—	—	—	—	—	—	—	—
Indonesia	P. Kumbur	4.25	—	—	—	—	—	—	—	—	—	—	—	—	—	—	—	—	—	—
Indonesia	P. Nuburi	3.25	—	—	—	—	—	—	—	—	—	—	—	—	—	—	—	—	—	—
Indonesia	P. Pari (selatan)	—	—	—	—	—	—	—	—	—	—	—	—	—	—	—	—	—	—	—
Indonesia	P. Pepaya	7.75	—	—	—	—	—	—	—	—	—	—	—	—	—	—	—	—	—	—
Indonesia	P. Pramuka (timur)	—	—	—	—	0.25	—	—	—	—	—	—	—	—	—	—	—	—	—	—
Indonesia	P. Rajuni Kecil	—	—	—	—	1.5	—	—	—	—	—	—	—	—	—	—	—	—	—	—

Continued

Table A4 (Continued) Global distribution of giant clams (Reef Check)

Country	Reef Site	1997	1998	1999	2000	2001	2002	2003	2004	2005	2006	2007	2008	2009	2010	2011	2012	2013	2014	2015
													Monitoring years (total clam density = number of individuals per 100 m^2)						
Indonesia	P. Rajuni Kecil (selatan)	—	—	—	—	0.75	—	—	—	—	—	—	—	—	—	—	—	—	—	—
Indonesia	P. Rajuni Kecil (timur)	—	—	—	—	1.25	—	—	—	—	—	—	—	—	—	—	—	—	—	—
Indonesia	Pagang Island	—	—	—	—	—	—	—	—	0.75	—	—	—	—	—	—	—	—	—	—
Indonesia	Pahawang Island	—	—	—	—	—	0.5	—	—	—	—	—	—	—	—	—	—	—	—	—
Indonesia	Pak Kasims	—	—	—	—	3.0	—	0.75	1.5	0.75	—	—	—	—	—	—	—	—	—	—
Indonesia	Palau Langkai	—	—	—	—	0.25	—	—	—	—	—	—	—	—	—	—	—	—	—	—
Indonesia	Palau Lanyukang	—	—	—	—	0.25	—	—	—	—	—	—	—	—	—	—	—	—	—	—
Indonesia	Pangempa Reef	—	—	10.75	—	—	—	—	—	—	—	—	—	—	—	—	—	—	—	—
Indonesia	Panjang Island (North site)	—	—	—	—	—	—	—	—	—	—	—	—	0.75	—	—	—	—	—	—
Indonesia	Panjang Island (South site)	—	—	—	—	—	—	—	—	—	—	—	—	1.0	—	—	—	—	—	—
Indonesia	Pantai Gapang	—	—	—	—	—	—	11.75	—	—	—	—	—	—	—	—	—	—	—	—
Indonesia	Pasir Putih	—	—	—	—	—	—	—	—	1.25	—	—	—	—	—	—	—	—	—	—
Indonesia	Pasoso 1	—	—	—	—	4.0	2.38	2.0	—	—	—	—	—	—	—	—	—	—	—	—
Indonesia	Pasoso 2	—	—	—	—	2.75	1.12	1.5	—	—	—	—	—	—	—	—	—	—	—	—
Indonesia	Pasoso 3	—	—	—	—	1.5	2.12	2.0	—	—	—	—	—	—	—	—	—	—	—	—
Indonesia	Pasoso 4	—	—	—	—	—	1.25	0.5	—	—	—	—	—	—	—	—	—	—	—	—
Indonesia	Pastel	—	—	—	—	—	—	—	1.25	—	—	—	—	—	—	—	—	—	—	—
Indonesia	Pelabuhan Lahewa	—	—	—	—	—	—	—	—	2.0	—	—	—	—	—	—	—	—	—	—
Indonesia	Pemuteran (Pertemuan Dekat)	—	—	—	0.5	—	—	—	—	—	—	—	—	—	—	—	—	—	—	—
Indonesia	Poncan Gadang Island	—	—	—	—	—	—	0.25	—	—	—	—	—	—	—	—	—	—	—	—

Continued

Table A4 (Continued) Global distribution of giant clams (Reef Check)

Country	Reef Site	Monitoring years (total clam density = number of individuals per 100 m²)																		
		1997	1998	1999	2000	2001	2002	2003	2004	2005	2006	2007	2008	2009	2010	2011	2012	2013	2014	2015
Indonesia	Pos 1	—	—	2.0	—	—	—	—	—	—	—	—	—	—	—	—	—	—	—	—
Indonesia	Pulau Bagu	—	—	—	—	—	—	—	—	2.75	—	—	—	—	—	—	—	—	—	—
Indonesia	Pulau Belanda	—	—	—	—	—	—	—	—	—	—	1.25	—	—	—	—	—	—	—	—
Indonesia	Pulau Bira Besar	—	—	—	—	—	—	—	—	—	—	0.5	—	—	—	—	—	—	—	—
Indonesia	Pulau Buro	—	—	—	—	—	—	—	—	0.25	—	—	—	—	—	—	—	—	—	—
Indonesia	Pulau Kalong 1	—	—	—	—	—	—	0.5	—	—	—	—	—	—	—	—	—	—	—	—
Indonesia	Pulau Kalong 2	—	—	—	—	—	—	1.0	—	—	—	—	—	—	—	—	—	—	—	—
Indonesia	Pulau Kayu Angin Bira	—	—	—	—	—	—	—	—	—	—	0.25	—	—	—	—	—	—	—	—
Indonesia	Pulau Panjang 1	—	—	—	—	—	—	0.25	—	—	—	—	—	—	—	—	—	—	—	—
Indonesia	Pulau Panjang 2	—	—	—	—	—	—	1.25	—	—	—	—	—	—	—	—	—	—	—	—
Indonesia	Pulo Rondo II	—	—	—	—	—	—	—	—	0.25	—	—	—	—	—	—	—	—	—	—
Indonesia	Putih Island	—	—	—	—	—	—	2.5	—	—	—	—	—	—	—	—	—	—	—	—
Indonesia	Ridge 1 Hoga Island	—	—	—	—	0.75	—	1.5	—	—	—	—	—	—	—	—	—	—	—	—
Indonesia	Roine Selatan (South)	—	—	—	—	—	—	—	—	—	—	—	—	—	0.25	—	—	—	—	—
Indonesia	Rubiah Pier	—	—	—	—	—	—	1.25	—	—	—	—	—	—	—	—	—	—	—	—
Indonesia	Rumah Merah	—	—	—	—	—	—	—	28.0	—	—	—	—	—	1.0	—	—	—	—	—
Indonesia	Saboloh Besar	—	—	—	0.5	—	—	—	—	—	—	—	—	—	—	—	—	—	—	—
Indonesia	Saboloh Kecil	—	—	—	4.25	—	—	—	—	—	—	—	—	—	—	—	—	—	0.88	—
Indonesia	Sahaong	—	—	—	—	—	—	—	—	—	—	—	—	—	—	2.75	—	0.25	—	—
Indonesia	Samalona Island	—	—	0.5	—	—	—	—	—	—	—	—	—	—	—	—	—	—	—	—
Indonesia	Sambangan Island	—	—	—	—	—	—	—	—	—	0.75	—	—	—	—	—	—	—	—	—
Indonesia	Sampela 3	—	—	—	—	—	—	—	—	3.25	—	—	—	—	—	—	—	—	—	—
Indonesia	Sampela 4	—	—	—	—	—	—	—	—	1.25	—	—	—	—	—	—	—	—	—	—

Continued

Table A4 (Continued) Global distribution of giant clams (Reef Check)

Country	Reef Site	Monitoring years (total clam density = number of individuals per 100 m²)																		
		1997	1998	1999	2000	2001	2002	2003	2004	2005	2006	2007	2008	2009	2010	2011	2012	2013	2014	2015
Indonesia	Sampela Buoy 1	—	—	—	—	—	—	—	1.25	—	—	—	—	—	—	—	—	—	—	—
Indonesia	Sampela Buoy 2 (reef on right)	—	—	—	—	—	—	1.25	—	—	—	—	—	—	—	—	—	—	—	—
Indonesia	Sampela Outer	—	—	—	—	—	—	—	—	2.5	—	—	—	—	—	—	—	—	—	—
Indonesia	Sampela Village	—	—	—	0.25	—	—	—	—	—	—	—	—	—	—	—	—	—	—	—
Indonesia	Sangyang Island	0.25	—	—	—	—	—	—	—	—	—	—	—	—	—	—	—	—	—	—
Indonesia	Sanur	—	—	—	—	0.25	1.0	—	—	—	—	—	—	—	—	—	—	—	—	—
Indonesia	Saponda Laut 3	—	—	—	—	—	—	—	—	—	1.75	—	—	—	—	—	—	—	—	—
Indonesia	Saponda Laut 4	—	—	—	—	—	—	—	—	—	1.25	—	—	—	—	—	—	—	—	—
Indonesia	Sawah (West of Tokobao Island)	—	—	—	—	—	—	—	—	2.0	—	—	—	—	—	—	—	—	—	—
Indonesia	SE Barrang Lompo	—	—	—	—	—	—	—	—	—	—	—	1.0	—	—	—	—	—	—	—
Indonesia	Sea Garden Point	4.0	—	—	—	—	—	—	—	—	—	—	—	—	—	—	—	—	—	—
Indonesia	Sebayor Kecil	—	—	—	—	—	—	—	—	—	—	—	—	—	—	—	—	—	0.25	—
Indonesia	Semak Daun Island (east)	—	—	—	—	—	0.75	—	—	—	—	—	—	—	—	—	—	—	—	—
Indonesia	Semak Daun Island (south)	—	—	—	—	—	1.25	—	—	—	—	—	—	—	—	—	—	—	—	—
Indonesia	Semak Daun Island (west)	—	—	—	—	—	1.0	—	—	—	—	—	—	—	—	—	—	—	—	—
Indonesia	Shark Point	—	—	—	—	—	—	—	—	—	—	—	0.25	1.25	—	0.5	0.5	1.0	—	—
Indonesia	Sianas	—	—	—	—	—	—	1.75	—	—	—	—	—	—	—	—	—	—	—	—
Indonesia	Simacan	—	—	—	—	—	—	—	—	0.75	—	—	—	2.0	—	—	—	—	—	—
Indonesia	Sintok Island	1.75	—	—	—	—	—	—	—	—	—	—	—	—	—	—	—	—	—	—
Indonesia	Sironjong Island	—	—	—	7.75	—	—	—	—	5.5	—	—	—	—	—	—	—	—	—	—

Continued

Table A4 (Continued) Global distribution of giant clams (Reef Check)

Country	Reef Site	Monitoring years (total clam density = number of individuals per 100 m²)																		
		1997	1998	1999	2000	2001	2002	2003	2004	2005	2006	2007	2008	2009	2010	2011	2012	2013	2014	2015
Indonesia	Site 7 Tokabao Island	—	—	—	5.5	—	—	—	—	—	—	—	—	—	—	—	—	—	—	—
Indonesia	Sombano Reef	—	—	—	—	4.0	—	1.0	4.0	1.5	—	—	—	—	—	—	—	—	—	—
Indonesia	Sombu	—	—	—	—	—	—	—	0.5	—	—	—	—	—	0.75	—	—	—	—	—
Indonesia	South Barrang Caddi	—	—	—	—	—	—	—	—	1.0	—	—	1.75	—	—	—	—	—	—	—
Indonesia	South Bauluang Island	—	—	—	—	—	—	1.5	—	—	—	—	—	—	—	—	—	—	—	—
Indonesia	South Kapoposang	—	—	—	0.5	—	—	—	—	—	—	—	—	—	—	—	—	—	—	—
Indonesia	Stasiun 4 (Tanjung Besar)	—	—	—	—	1.5	—	—	—	—	—	—	—	—	—	—	—	—	—	—
Indonesia	Stasiun 3 (Tanjung Kecil)	—	—	—	—	0.75	—	—	—	—	—	—	—	—	—	—	—	—	—	—
Indonesia	Stasiun II Bokori	—	—	—	—	—	—	—	—	0.25	—	—	—	—	—	—	—	—	—	—
Indonesia	Stasiun III Bokori	—	—	—	—	—	—	—	—	0.5	—	—	—	—	—	—	—	—	—	—
Indonesia	Statsiun I Bokori	—	—	—	—	—	—	—	—	1.0	—	—	—	—	—	—	—	—	—	—
Indonesia	Sumanga Island	—	—	—	—	2.25	—	—	—	—	—	—	—	—	—	—	—	—	—	—
Indonesia	Sumpat	—	—	—	—	—	0.25	0.12	0.38	—	—	—	—	—	—	—	—	—	—	—
Indonesia	Sumpat Island	0.25	—	—	—	2.25	—	—	—	—	—	—	—	—	—	—	—	—	—	—
Indonesia	SW Kaledupa	—	—	—	—	2.25	—	—	—	—	—	—	—	—	—	—	—	—	—	—
Indonesia	Table Coral City	—	—	—	1.75	0.5	—	—	—	—	—	—	—	—	—	—	—	—	—	—
Indonesia	Tahu	—	—	—	—	—	—	1.25	—	—	—	—	—	—	—	—	—	—	—	—
Indonesia	Taka Malang	—	—	—	—	—	—	—	—	—	—	—	1.0	0.75	—	—	—	—	—	—
Indonesia	Tambu Bay 1—Pulau Katupat	—	—	—	—	—	—	—	—	—	—	4.0	—	—	—	—	—	—	—	—

Continued

Table A4 (Continued) Global distribution of giant clams (Reef Check)

Country	Reef Site	Monitoring years (total clam density = number of individuals per 100 m^2)																		
		1997	1998	1999	2000	2001	2002	2003	2004	2005	2006	2007	2008	2009	2010	2011	2012	2013	2014	2015
Indonesia	Tambu Bay 3—Awesang	—	—	—	—	—	—	—	—	—	—	1.0	—	—	—	—	—	—	—	—
Indonesia	Tambu Bay 4—Palau (Santigi)	—	—	—	—	—	—	—	—	—	—	2.0	—	—	—	—	—	—	—	—
Indonesia	Tambu Bay 5—Sibualong	—	—	—	—	—	—	—	—	—	—	0.5	—	—	—	—	—	—	—	—
Indonesia	Tanjong Berakit	—	1.0	—	—	—	—	—	—	—	—	—	—	—	—	—	—	—	—	—
Indonesia	Tanjung Api 1	—	—	—	—	—	3.0	2.25	—	—	—	—	—	—	—	—	—	—	—	—
Indonesia	Tanjung Api 2	—	—	—	—	—	3.25	2.0	—	—	—	—	—	—	—	—	—	—	—	—
Indonesia	Tanjung Api 4	—	—	—	—	—	—	2.25	—	—	—	—	—	—	—	—	—	—	—	—
Indonesia	Tanjung Gelam	—	—	—	—	—	—	—	1.5	2.25	—	—	—	—	—	—	—	—	—	—
Indonesia	Tanjung Husi II (Bangka Island)	—	—	—	—	—	—	—	—	—	—	—	—	—	—	—	—	—	0.13	—
Indonesia	Tanjung Patok	—	—	—	—	—	—	—	—	—	—	—	—	—	1.25	—	—	—	—	—
Indonesia	Tanjung Pisok	—	—	—	—	—	0.25	—	—	—	0.5	—	—	—	—	—	—	—	—	—
Indonesia	Tawaeli Talise	—	—	—	—	—	1.25	1.25	—	—	0.5	—	—	—	—	—	—	—	—	—
Indonesia	Teluk Krueng Raya	—	—	—	—	—	—	—	—	0.5	—	—	—	—	—	—	—	—	—	—
Indonesia	Tengah Island	3.75	—	—	—	—	—	—	—	—	—	—	—	—	—	—	—	—	—	—
Indonesia	Tenggara P. Putri	—	—	—	—	—	—	—	—	—	—	—	—	1.0	—	—	—	—	—	—
Indonesia	Tete B	—	—	—	—	—	—	—	—	—	—	—	1.0	—	—	—	—	—	—	—
Indonesia	Tokabao Island	—	—	—	16.5	1.0	—	—	—	—	—	—	—	—	—	—	—	—	—	—
Indonesia	Tokabao Island (site 15)	—	—	—	0.5	—	—	—	—	—	—	—	—	—	—	—	—	—	—	—
Indonesia	Tokobao Atoll	—	—	—	—	—	—	0.5	0.5	—	—	—	—	—	—	—	—	—	—	—
Indonesia	Tolandano	—	—	—	—	1.25	—	—	—	—	—	—	—	—	—	—	—	—	—	—

Continued

Table A4 (Continued) Global distribution of giant clams (Reef Check)

Country	Reef Site	Monitoring years (total clam density = number of individuals per 100 m²)																		
		1997	1998	1999	2000	2001	2002	2003	2004	2005	2006	2007	2008	2009	2010	2011	2012	2013	2014	2015
Indonesia	Tolandano Island (reef on right)	—	—	—	—	0.5	—	—	—	—	—	—	—	—	—	—	—	—	—	—
Indonesia	Tomea Island (near boat mooring)	—	—	—	—	2.25	—	—	—	—	—	—	—	—	—	—	—	—	—	—
Indonesia	Tomea Island (north side)	—	—	—	—	—	—	—	2.0	—	—	—	—	—	—	—	—	—	—	—
Indonesia	Tomea island (waha village)	—	—	—	2.5	—	—	—	—	—	—	—	—	—	—	—	—	—	—	—
Indonesia	Tomea near Waha (reef on right)	—	—	—	—	1.0	—	0.75	—	—	—	—	—	—	—	—	—	—	—	—
Indonesia	Tomea Village (site 9)	—	—	—	2.5	—	—	—	—	—	—	—	—	—	—	—	—	—	—	—
Indonesia	Toyopakeh Pontoon	—	—	—	—	—	0.5	—	—	—	—	—	—	—	—	—	—	—	—	—
Indonesia	Tulamben	—	—	—	—	1.75	—	—	—	—	—	—	—	—	—	—	—	—	—	—
Indonesia	Tulamben House Reef	—	—	—	—	—	—	—	—	—	—	—	—	1.0	—	—	—	—	—	—
Indonesia	Tunang Reef	—	—	—	—	—	—	—	—	—	—	—	—	—	—	—	—	—	0.5	—
Indonesia	Turtle Street	0.75	—	—	—	—	—	—	—	—	—	—	—	—	—	—	—	—	—	—
Indonesia	Uebone	—	—	—	—	—	—	—	—	—	—	—	1.75	—	—	—	—	—	—	—
Indonesia	Ujung Aramanyang	—	—	—	—	—	—	—	—	—	—	—	—	2.5	—	—	—	—	—	—
Indonesia	Ujung Batu Kapal 2	—	—	—	—	—	—	—	—	0.75	—	—	—	—	—	—	—	—	—	—
Indonesia	Ulasa Island	0.25	—	—	—	—	—	—	—	—	—	—	—	—	—	—	—	—	—	—
Indonesia	Waha Jetty	—	—	—	—	2.75	—	—	—	—	—	—	—	—	—	—	—	—	—	—

Continued

Table A4 (Continued) Global distribution of giant clams (Reef Check)

Country	Reef Site	Monitoring years (total clam density = number of individuals per 100 m²)																		
		1997	1998	1999	2000	2001	2002	2003	2004	2005	2006	2007	2008	2009	2010	2011	2012	2013	2014	2015
Indonesia	Waha Jetty Tomea	—	—	—	—	—	—	0.25	—	—	—	—	—	—	—	—	—	—	—	—
Indonesia	Waha Jetty Tomea (site 4)	—	—	—	7.0	—	—	—	—	—	—	—	—	—	—	—	—	—	—	—
Indonesia	Waha Pinnacle	—	—	—	—	—	—	—	1.5	—	—	—	—	—	—	—	—	—	—	—
Indonesia	Waha Selatan	—	—	—	—	—	—	—	—	—	—	—	—	—	0.75	—	—	—	—	—
Indonesia	Waha Wanci	—	—	—	—	—	—	0.38	—	—	—	—	—	—	—	—	—	—	—	—
Indonesia	Wanci Sombu Village (reef on right)	—	—	—	—	3.0	—	—	—	—	—	—	—	—	—	—	—	—	—	—
Indonesia	Wandoka	—	—	—	—	—	—	—	—	—	—	—	—	—	0.5	—	—	—	—	—
Indonesia	Wandoka Wanci (reef on left)	—	—	—	—	0.25	—	—	—	—	—	—	—	—	—	—	—	—	—	—
Indonesia	Watu Lawang	—	—	—	—	—	—	—	—	—	—	—	—	0.5	—	—	—	—	—	—
Indonesia	Watuno	—	—	—	—	—	—	—	—	—	—	—	—	—	1.5	—	—	—	—	—
Indonesia	Waworaha Beach site 1	—	—	—	—	—	—	—	1.0	—	—	—	—	—	—	—	—	—	—	—
Indonesia	Waworaha Beach site 2	—	—	—	—	—	—	—	1.25	—	—	—	—	—	—	—	—	—	—	—
Indonesia	Waworaha Beach Site 3	—	—	—	—	—	—	—	1.5	—	—	—	—	—	—	—	—	—	—	—
Indonesia	Wayag	—	—	—	—	—	—	—	—	—	—	—	—	—	—	—	—	3.5	—	—
Indonesia	West Barrang Lompo	—	—	—	—	—	—	—	—	—	—	—	1.0	—	—	—	—	—	—	—
Indonesia	West Bauluang Island	—	—	—	—	—	—	1.5	—	—	—	—	—	—	—	—	—	—	—	—
Indonesia	West Samalona	—	—	—	1.25	—	—	—	—	—	—	—	—	—	—	—	—	—	—	—
Israel	Coral Beach Nature Reserve	0.13	0.5	—	—	—	—	—	—	—	—	—	—	—	—	—	—	—	—	—

Continued

Table A4 (Continued) Global distribution of giant clams (Reef Check)

Country	Reef Site	Monitoring years (total clam density = number of individuals per 100 m²)																		
		1997	1998	1999	2000	2001	2002	2003	2004	2005	2006	2007	2008	2009	2010	2011	2012	2013	2014	2015
Israel	Nature Reserve	—	—	—	—	0.63	—	—	—	—	—	—	—	—	—	—	—	—	—	—
Israel	North Princess Hotel	—	1.75	—	—	—	—	—	—	—	—	—	—	—	—	—	—	—	—	—
Israel	South Princess Hotel	—	2.0	—	—	—	—	—	—	—	—	—	—	—	—	—	—	—	—	—
Japan	Airport North	—	—	—	—	3.0	—	—	—	—	—	—	—	—	—	—	—	—	—	—
Japan	Airport North (10 m)	—	—	—	—	1.0	—	—	—	—	—	—	—	—	—	—	—	—	—	—
Japan	Akashita	—	—	1.25	—	—	—	—	—	—	—	—	—	—	—	—	—	—	—	—
Japan	Akazumijuki, Yabiji	—	—	—	—	—	—	—	—	—	—	—	—	—	—	—	—	—	0.5	—
Japan	Anadomari-oki	—	—	—	—	—	1.25	—	—	—	—	—	—	—	—	—	—	—	—	—
Japan	Ankyaba	—	—	—	—	7.75	4.5	—	—	—	—	—	—	—	—	—	—	—	—	—
Japan	Arakawasita	—	—	2.75	—	—	—	—	—	—	—	—	—	—	—	—	—	—	—	—
Japan	Byobudani, Chichijima Is	—	—	—	—	—	—	—	—	—	—	—	—	—	—	—	0.5	—	—	—
Japan	Fugausa	—	—	—	—	—	—	0.5	—	—	—	0.25	1.0	0.5	—	—	—	—	—	—
Japan	Gahi	—	—	—	—	—	—	—	—	—	—	2.25	—	—	—	—	—	—	—	—
Japan	Hirashima	—	0.5	—	0.25	0.25	0.25	—	—	—	—	—	—	—	—	—	—	—	—	—
Japan	Miyako Island	—	—	—	0.5	—	—	—	—	—	—	—	—	—	—	—	—	—	—	—
Japan	Nakanose	—	—	7.25	—	2.5	3.5	2.5	2.12	2.5	1.0	—	0.75	0.5	1.75	1.5	1.75	—	0.75	—
Japan	Nakanose Kanokawa Bay	—	—	—	2.25	—	—	—	—	—	—	—	—	—	—	—	—	—	—	—
Japan	Nishi-hama	—	—	5.5	8.75	6.0	3.25	—	—	—	2.5	—	—	—	—	—	—	—	—	—
Japan	Nishiumi	—	—	—	—	—	—	—	—	—	0.12	—	—	—	—	—	—	—	—	—
Japan	Nita-hama	—	—	1.5	0.5	—	—	—	—	—	—	—	—	—	—	—	—	—	—	—

Continued

Table A4 (Continued) Global distribution of giant clams (Reef Check)

Country	Reef Site	Monitoring years (total clam density = number of individuals per 100 m²)																		
		1997	1998	1999	2000	2001	2002	2003	2004	2005	2006	2007	2008	2009	2010	2011	2012	2013	2014	2015
Japan	North reef of Kohama Island	—	—	—	0.75	1.0	1.75	1.5	—	0.75	—	0.5	0.75	0.25	—	—	1.75	—	0.38	—
Japan	North west offing of Chabana	—	—	—	—	—	4.75	—	—	—	—	—	—	2.75	3.5	—	—	—	—	—
Japan	Northern west of Doo-Reef	—	—	—	2.5	—	—	—	—	—	—	—	—	—	—	—	—	—	—	—
Japan	Offing of Ara Beach	—	—	—	—	1.5	—	—	—	—	—	—	—	—	—	—	—	—	—	—
Japan	Offing of Tomori New Harbor	—	—	—	—	5.5	7.75	—	—	—	—	—	—	—	—	—	—	—	—	—
Japan	Oganzaki Ishigaki Island	—	—	0.25	0.25	—	—	—	—	—	—	—	—	—	—	—	—	—	—	—
Japan	Oganzaki Toudaishita	—	0.25	—	—	—	—	—	—	—	—	—	—	—	—	—	—	—	—	—
Japan	Ohgamijima Northwest	—	—	—	—	—	—	—	0.5	—	—	—	—	—	—	—	—	—	—	—
Japan	Oodo	—	—	—	—	—	1.5	—	—	—	—	—	—	—	—	—	—	—	—	—
Japan	Sakieda	0.5	—	2.5	1.25	—	—	—	—	—	—	—	—	—	—	—	—	—	—	—
Japan	Sakuraguchi	—	—	—	—	—	—	0.75	0.75	—	—	—	—	—	—	—	—	—	—	—
Japan	South of Fukapanari	—	—	—	—	—	—	4.5	6.5	—	—	—	—	—	—	—	—	—	—	—
Japan	South of Futami-iwa	—	—	—	—	—	—	0.25	—	—	—	—	—	—	—	—	—	—	—	—
Japan	Southern offing of Hatenohama	—	—	—	—	0.75	1.0	—	—	—	—	—	—	—	—	—	—	—	—	—
Japan	Sunabe	—	—	—	—	—	—	—	—	5.5	10.12	—	—	—	—	—	—	—	—	—
Japan	Tamaruru Point	—	1.5	—	—	—	—	—	—	—	—	—	—	—	—	—	—	—	—	—

Continued

319

Table A4 (Continued) Global distribution of giant clams (Reef Check)

Country	Reef Site	Monitoring years (total clam density = number of individuals per 100 m²)																		
		1997	1998	1999	2000	2001	2002	2003	2004	2005	2006	2007	2008	2009	2010	2011	2012	2013	2014	2015
Japan	Tomori (Kumanomi Paradise) group A data	—	—	—	3.0	—	—	—	—	—	—	—	—	—	—	—	—	—	—	—
Japan	Tomori (Kumanomi Paradise) Group B data	—	—	—	5.0	—	—	—	—	—	—	—	—	—	—	—	—	—	—	—
Japan	Uentoro	—	—	—	1.25	1.75	1.5	—	—	—	—	—	—	—	—	—	—	—	—	—
Japan	Yabiji	—	—	—	—	0.5	—	—	—	—	—	—	—	—	—	—	—	—	—	—
Japan	Yonasone of Iriomote Island	1.0	—	0.5	—	1.0	2.0	—	1.25	—	—	—	—	—	—	—	—	—	—	—
Jordan	Aquarium	—	—	—	—	—	—	—	—	—	—	1.25	—	—	—	—	—	—	—	—
Jordan	First Bay	—	—	—	—	—	—	—	—	—	—	0.5	—	—	—	—	—	—	—	—
Jordan	Japanese Garden	—	—	—	—	—	—	—	—	—	—	0.5	—	—	—	—	—	—	—	—
Kenya	Lobster Malindi	—	—	—	—	—	—	—	0.75	—	—	—	—	—	—	—	—	—	—	—
Kenya	Malindi Barracuda Channel	—	—	—	—	—	—	0.75	—	—	—	—	—	—	—	—	—	—	—	—
Kenya	Malindi Coral Garden	—	—	—	—	—	—	0.25	—	—	—	—	—	—	—	—	—	—	—	—
Kenya	Malindi North Reef	—	—	—	—	—	—	0.12	—	—	—	—	—	—	—	—	—	—	—	—
Kenya	Navy Malindi	—	—	—	—	—	—	—	0.5	—	—	—	—	—	—	—	—	—	—	—
La Réunion	Bleu Marine	—	—	—	—	—	—	—	—	—	—	—	0.25	—	—	—	—	—	—	—
La Réunion	Boucan Canot Lagon	—	—	—	—	—	—	—	—	—	—	—	—	1.25	1.75	—	0.88	—	—	—

Continued

320

Table A4 (Continued) Global distribution of giant clams (Reef Check)

Country	Reef Site	Monitoring years (total clam density = number of individuals per 100 m²)																		
		1997	1998	1999	2000	2001	2002	2003	2004	2005	2006	2007	2008	2009	2010	2011	2012	2013	2014	2015
La Réunion	Boucan Canot PE	—	—	—	—	—	—	—	—	—	—	—	—	1.25	—	0.75	0.75	—	—	—
La Réunion	Cap la Houssaye	—	—	—	—	—	—	—	—	—	—	1.25	0.5	—	0.5	—	0.25	—	—	—
La Réunion	Ermitage PE	—	—	—	—	—	—	—	—	—	—	—	0.5	0.5	1.0	—	0.5	1.25	—	—
La Réunion	Etang Sale Lagon	—	—	—	—	—	—	—	—	—	—	—	—	—	12.0	—	2.75	1.25	—	—
La Réunion	Etang Sale Sud	—	—	—	—	—	—	—	—	—	—	—	2.0	—	—	—	—	—	—	—
La Réunion	Grand Fond	—	—	—	—	—	—	—	—	—	—	—	3.0	—	—	—	—	—	—	—
La Réunion	Hermitage Lagon	—	—	—	—	—	—	—	—	—	—	—	—	0.75	0.25	—	—	—	—	—
La Réunion	Livingstone Lagon	—	—	—	—	—	—	—	—	—	—	—	—	—	1.75	—	0.25	—	—	—
La Réunion	Livingstone PE	—	—	—	—	—	—	—	—	—	—	—	—	—	1.0	—	0.62	—	—	—
La Réunion	Novotel	—	—	—	—	—	—	—	—	—	—	—	1.75	—	—	—	—	—	—	—
La Réunion	Plage Saint Leu	—	—	—	—	—	—	—	—	—	—	—	—	—	1.0	0.75	—	—	—	—
La Réunion	Roches Noires Lagon	—	—	—	—	—	—	—	—	—	—	—	—	0.75	—	1.25	0.62	—	—	—
La Réunion	Roches Noires PE	—	—	—	—	—	—	—	—	—	—	—	—	0.25	0.25	—	0.75	—	—	—
La Réunion	Saline Nord Lagon	—	—	—	—	—	—	—	—	—	—	—	—	1.25	—	0.25	1.0	—	—	—
La Réunion	Saline Nord PE	—	—	—	—	—	—	—	—	—	—	—	—	0.25	0.25	—	—	—	—	—
La Réunion	Spot de Saint-Leu	—	—	—	—	—	—	0.75	—	—	0.25	1.25	0.75	0.5	—	—	—	—	—	—
La Réunion	Spot del'Hermitage	—	—	—	—	—	—	—	0.75	0.25	0.5	1.0	—	—	—	—	—	—	—	—
La Réunion	Spot Etang Sale	—	—	—	—	—	—	0.5	1.5	0.5	0.5	0.5	2.75	0.75	—	0.75	1.0	0.25	—	—
La Réunion	Spot Perroquet	—	—	—	—	—	—	—	—	—	—	—	1.0	—	—	—	0.25	—	—	—

Continued

Table A4 (Continued) Global distribution of giant clams (Reef Check)

Country	Reef Site	Monitoring years (total clam density = number of individuals per 100 m²)																		
		1997	1998	1999	2000	2001	2002	2003	2004	2005	2006	2007	2008	2009	2010	2011	2012	2013	2014	2015
La Réunion	Tessier PE	—	—	—	—	—	—	—	—	—	—	—	—	—	—	—	1.5	—	—	—
La Réunion	Trou d'eau Lagon	—	—	—	—	—	—	—	—	—	—	—	—	0.5	—	—	0.5	—	—	—
La Réunion	Trou d'eau PE	—	—	—	—	—	—	—	—	—	—	—	—	0.75	—	0.5	0.62	—	—	—
Madagascar	Ankarea	—	—	—	—	—	—	—	—	—	—	—	—	—	—	6.0	—	—	—	—
Madagascar	Chesterfield Island	—	—	—	—	1.25	—	—	—	—	—	—	—	—	—	—	—	—	—	—
Madagascar	Coral Garden	—	—	—	—	—	—	—	—	—	—	—	—	0.88	0.25	—	—	—	—	—
Madagascar	Coral Garden, Bay of Ranobe	—	—	—	—	—	—	—	—	—	—	—	—	0.88	—	—	—	—	—	—
Madagascar	East Nosy Fasy	—	—	—	—	—	—	0.25	—	—	—	—	—	—	—	—	—	—	—	—
Madagascar	East Nosy Hao	—	—	—	—	—	—	0.5	—	—	—	—	—	—	—	—	—	—	—	—
Madagascar	Fred's Reef	—	—	—	—	—	—	—	0.25	—	—	—	—	—	—	—	—	—	—	—
Madagascar	La Piscine	—	—	—	—	—	—	—	—	—	—	2.25	—	—	—	—	—	—	—	—
Madagascar	North Nosy Fasy	—	—	—	—	—	—	0.5	—	—	—	—	—	—	—	—	—	—	—	—
Madagascar	North Nosy Hao	—	—	—	—	—	—	0.5	—	—	—	—	—	—	—	—	—	—	—	—
Madagascar	Olaf's Reef	—	—	—	—	—	—	—	2.25	—	—	2.25	—	—	—	—	—	—	—	—
Madagascar	Recruitment Complex	—	—	—	—	—	—	—	—	0.25	—	—	—	—	—	—	—	—	—	—
Madagascar	Seven Little Sharks	—	—	—	—	—	—	—	0.75	—	—	—	—	—	—	—	—	—	—	—
Madagascar	Tanikely 1	—	—	—	—	—	—	—	—	—	—	—	—	—	—	0.75	—	—	—	—
Madagascar	Tanikely 2	—	—	—	—	—	—	—	—	—	—	—	—	—	—	0.75	—	—	—	—
Madagascar	Tsara 1	—	—	—	—	—	—	—	—	—	—	—	—	—	—	5.5	—	—	—	—
Madagascar	Tsara 2	—	—	—	—	—	—	—	—	—	—	—	—	—	—	2.75	—	—	—	—
Madagascar	West Nosy Fasy	—	—	—	—	—	—	0.25	—	—	—	—	—	—	—	—	—	—	—	—
Malaysia	Abalone	—	—	—	—	—	—	—	—	—	—	—	—	—	—	—	0.25	—	—	—
Malaysia	Abect House Reef	—	—	—	—	—	—	—	—	—	—	—	—	—	—	—	—	—	1.5	—

Continued

Table A4 (Continued) Global distribution of giant clams (Reef Check)

Country	Reef Site	Monitoring years (total clam density = number of individuals per 100 m^2)																		
		1997	1998	1999	2000	2001	2002	2003	2004	2005	2006	2007	2008	2009	2010	2011	2012	2013	2014	2015
Malaysia	Adam's Point	—	—	—	—	—	—	—	—	—	—	—	—	—	—	—	—	—	1.25	—
Malaysia	Ali Baba Rock	—	—	—	—	—	—	—	—	—	—	—	—	4.75	—	—	—	—	—	—
Malaysia	Anemone Centre	—	—	—	—	—	—	—	—	—	—	—	—	—	0.25	—	0.5	—	—	—
Malaysia	Anemone Garden	—	—	—	—	—	—	—	—	—	0.25	—	—	—	—	—	—	—	—	—
Malaysia	Atlantis Bay House Reef	—	—	—	—	—	—	—	—	—	—	—	0.25	5.0	—	—	—	—	—	—
Malaysia	Banggi Outer Northeast Reef 1	—	—	—	—	—	—	0.75	—	—	—	—	—	—	—	—	—	—	—	—
Malaysia	Banggi Outer Northeast Reef 2	—	—	—	—	—	—	0.75	—	—	—	—	—	—	—	—	—	—	—	—
Malaysia	Bankawan 1	—	—	—	—	—	—	8.0	—	—	—	—	—	—	—	—	—	—	—	—
Malaysia	Bankawan 2	—	—	—	—	—	—	0.25	—	—	—	—	—	—	—	—	—	—	—	—
Malaysia	Bankawan East	—	—	—	—	—	—	11.0	—	—	—	—	—	—	—	—	—	—	—	—
Malaysia	Bankawan Reef (SW)	—	—	—	—	—	—	0.25	—	—	—	—	—	—	—	—	—	—	—	—
Malaysia	Bankawan Reef 3	—	—	—	—	—	—	13.75	—	—	—	—	—	—	—	—	—	—	—	—
Malaysia	Bankawan South	—	—	—	—	—	—	16.0	—	—	—	—	—	—	—	—	—	—	—	—
Malaysia	Baratua	—	—	—	—	—	—	—	—	—	—	—	—	—	—	—	—	—	1.0	—
Malaysia	Base Camp	—	—	—	—	—	—	—	—	—	—	—	—	—	—	—	2.0	—	0.25	—
Malaysia	Batik	—	—	—	—	—	—	—	—	—	—	—	—	—	—	—	—	—	1.5	—
Malaysia	Batu Layar	—	—	—	—	—	—	—	—	—	—	—	0.25	1.75	—	1.0	1.25	—	1.0	—
Malaysia	Batu Malang	—	—	—	—	—	—	—	—	—	—	—	—	—	1.5	2.12	1.38	—	—	—
Malaysia	Batu Nisan	—	—	—	—	—	—	—	—	—	—	44.0	37.75	30.25	31.25	47.25	31.75	—	20.25	—
Malaysia	Batu Tabir	—	—	—	—	—	—	—	—	—	—	—	1.0	1.5	—	3.75	0.25	—	5.25	—
Malaysia	Beach 3	—	—	—	—	—	—	—	—	—	—	—	—	—	—	—	—	—	1.25	—

Continued

323

Table A4 (Continued) Global distribution of giant clams (Reef Check)

Country	Reef Site	Monitoring years (total clam density = number of individuals per 100 m²)																		
		1997	1998	1999	2000	2001	2002	2003	2004	2005	2006	2007	2008	2009	2010	2011	2012	2013	2014	2015
Malaysia	Bimbo Rock	—	—	—	—	—	—	—	—	—	—	—	—	—	—	—	—	—	—	—
Malaysia	Black Coral Garden	—	—	—	—	—	—	0.75	—	—	—	—	—	—	0.25	—	0.25	—	—	—
Malaysia	Bodgaya Dead End Channel	—	2.5	5.5	—	—	—	—	—	—	—	—	—	—	—	—	—	—	—	—
Malaysia	Bodgaya South Rim	—	5.5	—	—	—	—	—	—	—	—	—	—	—	—	—	—	—	—	—
Malaysia	Bodgaya South Rim Outer Reef	—	—	—	—	—	—	—	1.75	—	—	—	—	—	—	—	—	—	—	—
Malaysia	Bohayan Island	—	1.25	—	—	—	—	—	—	—	—	—	—	—	—	—	—	—	0.75	—
Malaysia	Bugis Bay	—	—	—	—	—	—	—	—	—	—	—	—	—	—	—	1.0	—	—	—
Malaysia	Bumphead Bay	—	—	—	—	—	—	—	—	—	—	—	—	—	—	—	3.75	—	0.5	—
Malaysia	Burn-Burn	—	—	—	—	—	—	—	—	—	—	—	—	—	—	—	—	—	0.75	—
Malaysia	Cabbage Reef	—	—	—	—	—	—	—	—	—	—	—	—	—	—	—	—	—	1.25	—
Malaysia	Cahaya Way, Bohayan Island	—	—	—	—	—	—	—	—	—	—	—	—	—	0.75	0.25	0.5	—	1.5	—
Malaysia	Chagar Hutan	8.5	—	—	—	—	—	—	—	—	—	6.25	—	—	—	—	—	—	—	—
Malaysia	Chagar Hutang (R2)	—	—	—	2.75	—	—	—	—	—	—	—	—	—	—	—	—	—	1.5	—
Malaysia	Chagar Hutang East	—	—	—	—	—	—	—	—	—	—	—	—	0.25	—	—	—	—	0.25	—
Malaysia	Chebah	—	—	—	—	—	—	—	—	—	—	—	—	—	—	—	0.5	—	—	—
Malaysia	Chebeh	—	—	—	—	—	—	—	—	—	—	—	1.25	1.25	—	1.25	2.0	—	4.5	—
Malaysia	Cliff Hanger	—	—	—	—	—	—	—	—	—	—	—	—	—	—	0.5	—	—	—	—
Malaysia	Coral Bay	—	—	—	—	—	—	—	—	—	—	—	—	—	—	—	5.5	—	—	—
Malaysia	Coral Garden 1, Kapas	—	—	—	—	—	—	—	—	—	—	—	—	—	—	0.25	1.75	—	—	—
Malaysia	Coral Garden 3, Kapas	—	—	—	—	—	—	—	—	—	—	—	—	—	0.25	0.5	—	—	—	—

Continued

Table A4 (Continued) Global distribution of giant clams (Reef Check)

Country	Reef Site	Monitoring years (total clam density = number of individuals per 100 m²)																		
		1997	1998	1999	2000	2001	2002	2003	2004	2005	2006	2007	2008	2009	2010	2011	2012	2013	2014	2015
Malaysia	Coral Garden, Mataking Besar	—	—	—	—	—	—	—	—	—	—	—	—	—	—	—	1.75	—	2.75	—
Malaysia	Coral Garden, Mataking Island	—	—	—	—	—	—	—	—	—	—	—	—	—	3.0	2.75	—	—	—	—
Malaysia	Coral Heaven	0.25	—	—	—	—	—	—	—	—	—	—	—	—	—	—	—	—	—	—
Malaysia	Coral Redang House Reef North	—	—	—	—	—	—	5.25	—	—	—	—	—	—	—	—	—	—	—	—
Malaysia	Coral Resort House Reef, Redang Island	—	—	—	—	—	—	—	5.25	—	—	—	—	—	—	—	—	—	—	—
Malaysia	Coral View Reef	—	—	—	—	—	—	—	—	—	—	5.75	—	—	—	—	2.5	—	—	—
Malaysia	Coral Reef	—	—	—	—	—	—	—	—	—	—	—	—	—	—	—	—	—	0.5	—
Malaysia	Danawan Reef, Siamil, Kapalai	—	—	—	—	—	—	—	—	—	—	—	—	—	36.5	—	—	—	—	—
Malaysia	Denawan	—	—	—	—	—	—	—	—	—	—	—	—	—	—	—	—	—	0.25	—
Malaysia	Dead End Channel	—	—	—	—	—	—	—	4.5	—	—	—	—	—	—	—	—	—	3.0	—
Malaysia	Diver's Lodge House Reef	—	—	—	—	—	—	—	—	—	—	—	—	0.25	—	—	—	—	—	—
Malaysia	Drop-off	—	—	—	—	—	—	—	—	—	—	—	—	—	—	—	—	—	0.5	—
Malaysia	East Palau Pinang	—	—	—	—	—	—	—	0.25	—	—	—	—	—	—	—	—	—	—	—
Malaysia	Edwin Rock	—	—	—	—	—	—	—	—	—	—	—	—	—	—	25.75	9.75	—	0.25	—
Malaysia	Eve's Garden	—	—	—	—	—	—	—	—	—	—	—	—	—	0.08	—	—	—	—	—
Malaysia	Fan Canyon	—	—	—	—	—	—	—	—	—	—	—	—	—	—	3.25	1.75	—	3.5	—
Malaysia	Fly Rock	—	—	—	—	—	—	—	—	—	—	—	—	1.75	—	—	—	—	—	—

Continued

Table A4 (Continued) Global distribution of giant clams (Reef Check)

Country	Reef Site	Monitoring years (total clam density = number of individuals per 100 m^2)																		
		1997	1998	1999	2000	2001	2002	2003	2004	2005	2006	2007	2008	2009	2010	2011	2012	2013	2014	2015
Malaysia	Fish Eye	—	—	—	—	—	—	—	—	—	—	—	—	—	—	—	—	—	0.25	—
Malaysia	Fresh Water Bay, Tenggol	—	—	—	—	—	—	—	—	—	—	—	—	0.75	—	0.75	0.5	—	0.5	—
Malaysia	Fringe Reef NE Patanunan	—	—	—	0.25	—	—	—	—	—	—	—	—	—	—	—	—	—	—	—
Malaysia	Fringing Reef S of Karakit	—	—	—	0.62	—	—	—	—	—	—	—	—	—	—	—	—	—	—	—
Malaysia	Fringing Reef S. Molleangan Besar	—	—	—	0.5	—	—	—	—	—	—	—	—	—	—	—	—	—	—	—
Malaysia	Fringing Reef SE Balak	—	—	—	1.0	—	—	—	—	—	—	—	—	—	—	—	—	—	—	—
Malaysia	Fringing Reef SE side of Balak	—	—	—	1.12	—	—	—	—	—	—	—	—	—	—	—	—	—	—	—
Malaysia	Fringing Reef SW Balak	—	—	—	1.5	—	—	—	—	—	—	—	—	—	—	—	—	—	—	—
Malaysia	Froggie Fort	—	—	—	—	—	—	—	—	—	—	—	—	0.5	0.25	2.0	0.25	—	0.5	—
Malaysia	Goby Rock	—	—	—	—	—	—	—	—	—	—	—	—	—	—	—	—	—	0.25	—
Malaysia	Great Wall, Kapalai	—	—	—	—	—	—	—	—	—	—	—	—	—	0.5	—	—	—	—	—
Malaysia	Gua Rajawali	—	—	—	—	—	—	—	—	—	—	—	—	1.5	—	3.25	1.25	—	1.75	—
Malaysia	Gua Sumbang	—	—	—	—	—	—	—	—	—	—	—	—	—	—	—	0.75	—	—	—
Malaysia	Gusung-gusung	—	—	—	—	—	—	—	—	—	—	—	—	0.5	—	—	—	—	—	—
Malaysia	Hanging Garden	—	—	—	—	—	—	—	—	—	—	—	—	—	—	—	—	—	0.75	—
Malaysia	Heritage Row (P. Bidong)	—	—	—	—	—	—	—	—	—	—	—	—	—	—	—	2.0	—	0.5	—

Continued

Table A4 (Continued) Global distribution of giant clams (Reef Check)

Country	Reef Site	Monitoring years (total clam density = number of individuals per 100 m^2)																		
		1997	1998	1999	2000	2001	2002	2003	2004	2005	2006	2007	2008	2009	2010	2011	2012	2013	2014	2015
Malaysia	House Reef, Mataking Besar Island	—	—	—	—	—	—	—	—	—	—	—	—	—	—	—	0.75	—	0.75	—
Malaysia	Italian Place	—	—	—	—	—	—	—	—	—	—	—	—	—	—	—	0.5	—	0.5	—
Malaysia	Jahat North	—	—	—	—	—	—	—	—	—	—	—	—	—	—	—	6.75	—	—	—
Malaysia	Jahat East	—	—	—	—	—	—	—	—	—	—	—	—	—	—	—	—	—	1.75	—
Malaysia	Japanese Garden P. Payar	—	—	—	—	—	—	—	—	—	—	—	—	—	—	—	0.25	—	—	—
Malaysia	Jawfish	—	—	—	—	—	—	—	—	—	—	—	—	0.5	0.75	—	—	—	—	—
Malaysia	Juara Rocks	—	—	—	—	—	—	—	—	—	—	—	—	—	—	—	10.0	—	—	—
Malaysia	Juara South	—	—	—	—	—	—	—	—	—	—	—	—	—	—	—	1.25	—	—	—
Malaysia	Kador Bay/ Teluk Kador	—	—	—	—	—	—	—	—	—	—	9.25	5.88	4.75	—	5.0	2.25	—	6.25	—
Malaysia	Kampong Dogoton (Pulau Banggi)	—	—	—	—	—	—	16.25	—	—	—	—	—	—	—	—	—	—	—	—
Malaysia	Kapalai Rock, Kapalai Island	—	—	—	—	—	—	—	—	—	—	—	0.25	—	—	—	—	—	—	—
Malaysia	Kapikan NE	—	0.25	—	—	—	—	—	0.25	—	—	—	—	—	—	—	—	—	0.63	—
Malaysia	Karakit Reef	—	—	0.5	—	—	—	—	—	—	—	—	—	—	—	—	—	—	—	—
Malaysia	Ken Point	2.5	—	—	—	—	—	—	—	—	—	—	—	—	—	—	—	—	—	—
Malaysia	Ken's Rock	—	—	—	—	—	—	—	—	—	—	—	—	0.5	0.25	1.25	0.25	—	0.75	—
Malaysia	Kerengga Kecil North West	—	—	—	—	—	—	—	—	—	—	—	—	5.75	—	7.5	—	—	—	—
Malaysia	Labas	—	—	—	—	—	—	—	—	—	—	—	—	7.25	—	1.75	2.12	—	2.5	—
Malaysia	Lam's Point	—	—	—	—	—	—	—	—	—	—	—	—	—	—	—	—	—	0.25	—
Malaysia	Lighthouse Front	—	—	—	—	—	—	—	—	—	—	—	2.25	2.0	—	0.5	—	—	—	—

Continued

327

Table A4 (Continued) Global distribution of giant clams (Reef Check)

Country	Reef Site	Monitoring years (total clam density = number of individuals per 100 m^2)																		
		1997	1998	1999	2000	2001	2002	2003	2004	2005	2006	2007	2008	2009	2010	2011	2012	2013	2014	2015
Malaysia	Limau Jambongan	—	—	—	—	—	—	—	—	—	—	—	—	0.5	—	—	—	—	—	—
Malaysia	Linggisan	—	—	—	—	—	—	—	—	—	—	—	—	—	—	—	0.5	—	3.0	—
Malaysia	Lobster Bay	—	—	—	—	—	—	—	—	—	—	—	—	—	—	—	—	—	0.25	—
Malaysia	Lobster Rock, Kapalai	—	—	—	—	—	—	—	—	—	—	—	—	—	4.0	—	—	—	—	—
Malaysia	Lubani Reef	—	—	—	—	—	—	—	—	—	—	—	—	0.5	—	—	—	—	—	—
Malaysia	Lycia Garden	—	—	—	—	—	—	—	—	—	—	—	—	1.5	0.25	1.25	1.5	—	0.5	—
Malaysia	Macromania Baturua	—	—	—	—	—	—	—	—	—	—	—	—	—	—	—	—	—	0.75	—
Malaysia	Madidarah South	—	—	—	—	—	—	—	—	—	—	—	—	1.5	—	—	—	—	—	—
Malaysia	Maganting Island	—	2.5	—	—	—	—	—	—	—	—	—	—	—	—	—	—	—	—	—
Malaysia	Mak Simpan	—	—	—	—	—	—	—	—	—	—	—	—	—	—	—	—	—	3.0	—
Malaysia	Malang Rock	—	—	—	—	—	—	—	—	—	—	2.12	1.62	2.5	—	—	—	—	0.5	—
Malaysia	Mamutik Reef	—	—	—	—	—	—	—	—	—	—	—	—	—	—	—	0.75	—	0.5	—
Malaysia	Mandarin Valley, Kapalai	—	—	—	—	—	—	—	—	—	—	—	—	—	8.75	—	—	—	—	—
Malaysia	Mandidarah East	—	—	—	—	—	—	—	—	—	—	—	—	1.5	—	—	—	—	—	—
Malaysia	Manimpan	—	—	—	—	—	—	—	—	—	—	—	—	0.5	—	—	—	—	—	—
Malaysia	Mantabuan Channel	—	—	—	—	—	—	—	—	—	—	—	—	0.75	—	—	—	—	—	—
Malaysia	Mantabuan North-East	—	7.75	3.5	—	—	—	—	—	—	—	—	—	—	—	—	—	—	1.5	—
Malaysia	Manukan West	—	—	—	—	—	—	—	—	—	—	—	—	—	—	—	2.75	—	3.25	—
Malaysia	Mari-Mari House Reef	—	—	—	—	—	—	—	—	—	—	—	—	—	—	—	0.25	—	—	—

Continued

Table A4 (Continued) Global distribution of giant clams (Reef Check)

Country	Reef Site	Monitoring years (total clam density = number of individuals per 100 m^2)																		
		1997	1998	1999	2000	2001	2002	2003	2004	2005	2006	2007	2008	2009	2010	2011	2012	2013	2014	2015
Malaysia	Mataking House Reef	—	—	—	—	—	—	—	—	—	—	—	—	1.5	2.0	2.0	—	—	0.75	—
Malaysia	Mel's Rock	—	—	—	—	—	—	—	—	—	—	—	—	—	0.25	—	0.25	—	—	—
Malaysia	Melina Undisturbed	—	—	—	—	—	—	—	—	—	—	—	—	—	—	—	—	—	4.63	—
Malaysia	Merrangis Reef	—	—	—	—	—	—	—	—	—	—	—	—	—	—	—	—	—	0.5	—
Malaysia	Mid Reef (left)	—	—	—	—	—	—	—	—	—	—	—	—	—	—	—	2.0	—	0.17	—
Malaysia	Mid Reef (right)	—	—	—	—	—	—	—	—	—	—	—	—	—	—	—	0.75	—	—	—
Malaysia	Mid Rock, Roach Reef	—	—	—	—	—	—	—	—	—	—	—	0.25	—	—	—	—	—	—	—
Malaysia	Moray Reef	—	—	—	—	—	—	—	—	—	—	—	—	—	—	—	0.5	—	—	—
Malaysia	Munjor	—	—	—	—	—	—	—	—	—	—	—	—	—	—	—	—	—	0.75	—
Malaysia	North Point, Pulau Sipadan	—	—	—	—	—	—	—	0.25	—	—	—	—	—	—	—	—	—	—	—
Malaysia	Northern Valley	—	—	—	—	—	—	—	—	—	—	—	—	—	—	0.25	—	—	—	—
Malaysia	Nyak (Tioman East)	—	—	—	—	—	—	—	—	—	—	—	—	—	—	—	—	—	0.75	—
Malaysia	Old Man of the Sea	—	—	—	—	—	—	—	—	—	—	—	—	—	—	—	6.0	—	—	—
Malaysia	P. Kerengga Kecil	—	—	—	—	—	—	—	—	—	—	—	—	—	—	—	5.75	—	7.0	—
Malaysia	P. Nanga	—	—	—	—	—	—	—	—	—	—	—	—	—	—	—	—	—	0.25	—
Malaysia	P. Rawa	—	—	—	—	—	—	—	—	—	—	—	—	—	—	—	0.75	—	—	—
Malaysia	P. Tinggi/ Tanjung Gua Sumbang	—	—	—	—	—	—	—	—	—	—	—	—	—	—	—	—	—	0.25	—
Malaysia	Pandan-Pandan	—	—	—	—	—	—	—	—	—	—	—	—	—	—	—	—	—	0.25	—
Malaysia	Pandanan Bay, Pandanan Island	—	—	—	—	—	—	—	—	—	—	—	—	—	0.75	0.75	0.25	—	—	—

Continued

Table A4 (Continued) Global distribution of giant clams (Reef Check)

Country	Reef Site	Monitoring years (total clam density = number of individuals per 100 m²)																		
		1997	1998	1999	2000	2001	2002	2003	2004	2005	2006	2007	2008	2009	2010	2011	2012	2013	2014	2015
Malaysia	Panglima 1	—	—	—	1.0	—	—	—	—	—	—	—	—	—	—	—	—	—	—	—
Malaysia	Paradise	3.25	—	—	—	—	—	—	—	—	—	—	—	—	—	—	—	—	0.75	—
Malaysia	Paradise 2, Mabul	—	—	—	—	—	—	—	—	—	—	—	1.25	—	—	2.75	—	—	—	—
Malaysia	Pasir Akar	—	—	—	—	—	—	—	—	—	—	—	—	—	—	0.5	0.5	—	0.25	—
Malaysia	Pasir Tenggara	—	—	—	—	—	—	—	—	—	—	—	—	—	—	—	0.5	—	2.0	—
Malaysia	Pasir Tenggara (P. Bidong)	—	—	—	—	—	—	—	—	—	—	—	—	—	—	—	1.0	—	0.5	—
Malaysia	Patch Reef 2km SW Balak	—	—	—	1.5	—	—	—	—	—	—	—	—	—	—	—	—	—	—	—
Malaysia	Patch Reef b/w Balak and Panukaran	—	—	—	1.25	—	—	—	—	—	—	—	—	—	—	—	—	—	—	—
Malaysia	Patch Reef NE Surundang Reef	—	—	—	0.75	—	—	—	—	—	—	—	—	—	—	—	—	—	—	—
Malaysia	Pegaso Reef	—	—	—	—	—	—	—	—	—	—	—	—	—	—	—	0.25	—	0.5	—
Malaysia	Pelangi House Reef South	—	—	—	—	—	—	7.75	—	—	—	—	—	—	—	—	—	—	—	—
Malaysia	Penut (Tioman East Side)	—	—	—	—	—	—	—	—	—	—	—	—	—	—	—	—	—	0.25	—
Malaysia	Pertigi Bay, Redang Island	—	—	—	—	—	—	—	0.75	—	—	—	—	—	—	—	—	—	—	—
Malaysia	Pinang	—	—	—	—	—	—	—	—	—	—	—	2.0	2.0	—	—	—	—	—	—
Malaysia	Pinnacle 3	—	—	1.5	—	—	—	—	—	—	—	—	—	—	—	—	—	—	—	—
Malaysia	Pirates Reef	—	—	—	—	—	—	—	1.5	—	4.5	4.58	6.62	2.5	4.25	1.25	0.88	—	0.25	—
Malaysia	Police Beach	—	—	—	—	—	—	—	—	—	—	—	—	—	—	—	1.0	—	0.25	—
Malaysia	Police Gate	—	—	—	—	—	—	—	—	—	—	—	—	—	—	—	0.5	—	0.25	—

Continued

Table A4 (Continued) Global distribution of giant clams (Reef Check)

Country	Reef Site	Monitoring years (total clam density = number of individuals per 100 m²)																		
		1997	1998	1999	2000	2001	2002	2003	2004	2005	2006	2007	2008	2009	2010	2011	2012	2013	2014	2015
Malaysia	Pom Pom Jetty	—	—	—	—	—	—	—	—	—	—	—	—	—	—	1.25	1.75	—	0.25	—
Malaysia	Pu Manatbuan NE	—	—	—	—	—	—	—	9.0	—	—	—	—	—	—	—	—	—	—	—
Malaysia	Pulau Bohayan	—	1.25	—	—	—	—	—	—	—	—	—	—	—	—	—	—	—	—	—
Malaysia	Pulau Burung	—	—	—	—	—	—	—	—	—	—	—	—	—	—	—	—	—	0.25	—
Malaysia	Pulau Guhan	—	—	—	—	—	—	0.25	—	—	—	—	—	—	—	—	—	—	—	—
Malaysia	Pulau Kalangkaman 1	—	—	—	—	—	—	1.5	—	—	—	—	—	—	—	—	—	—	—	—
Malaysia	Pulau Kalangkaman 2	—	—	—	—	—	—	0.25	—	—	—	—	—	—	—	—	—	—	—	—
Malaysia	Pulau Karah	—	—	—	—	—	—	—	—	—	—	—	—	—	—	—	0.75	—	1.0	—
Malaysia	Pulau Kerengga Besar	—	—	—	—	—	—	—	—	—	—	—	—	—	—	27.5	37.5	—	22.75	—
Malaysia	Pulau Kerengga East	—	—	—	—	—	—	—	—	—	—	—	43.75	44.25	—	—	—	—	—	—
Malaysia	Pulau Kerengga West	—	—	—	—	—	—	—	—	—	—	—	1.25	2.25	—	—	—	—	—	—
Malaysia	Pulau Laila	—	—	—	—	—	—	—	—	—	—	—	—	—	—	—	—	—	0.75	—
Malaysia	Pulau Lang (off Pulau Aur)	—	—	—	—	—	—	—	—	—	1.62	—	1.5	—	13.5	—	—	—	—	—
Malaysia	Pulau Lang Tengah	—	—	—	11.5	—	—	—	—	—	—	—	—	—	—	—	—	—	—	—
Malaysia	Pulau Latoan (Bankawan Reef)	—	—	—	—	—	—	13.0	—	—	—	—	—	—	—	—	—	—	—	—
Malaysia	Pulau Lima	2.5	1.25	—	—	—	—	—	—	—	—	—	—	—	—	—	—	—	—	—
Malaysia	Pulau Lima Southern Tip	—	—	—	—	—	—	—	—	—	—	0.12	0.25	—	—	0.75	—	—	—	—
Malaysia	Pulau Lima, R4	—	—	—	0.5	—	—	—	—	—	—	—	—	—	—	—	—	—	—	—

Continued

Table A4 (Continued) Global distribution of giant clams (Reef Check)

Country	Reef Site	Monitoring years (total clam density = number of individuals per 100 m²)																		
		1997	1998	1999	2000	2001	2002	2003	2004	2005	2006	2007	2008	2009	2010	2011	2012	2013	2014	2015
Malaysia	Pulau Lima, Southern Tip	—	—	—	—	—	—	0.25	—	—	—	—	—	—	—	—	—	—	0.25	—
Malaysia	Pulau Ling	—	—	—	2.0	—	—	—	—	—	—	—	—	—	—	—	—	—	—	—
Malaysia	Pulau Ling, R3A	—	—	—	2.5	—	—	—	—	—	—	—	—	—	—	—	—	—	—	—
Malaysia	Pulau Maganting	—	2.5	—	—	—	—	—	—	—	—	—	—	—	—	—	—	—	—	—
Malaysia	Pulau Paku Besar	—	—	—	—	—	—	—	—	—	—	—	0.25	0.25	—	2.5	0.25	—	2.0	—
Malaysia	Pulau Paku Kecil	—	—	—	—	—	—	7.5	—	—	—	—	—	—	—	—	—	—	—	—
Malaysia	Pulau Paku Kecil SW	—	—	—	—	—	—	—	—	—	—	0.38	0.75	0.25	—	1.25	1.25	—	0.5	—
Malaysia	Pulau Perhentian Kecil/ D'Lagoon	—	—	—	—	—	—	32.0	—	—	—	19.25	34.5	22.0	—	31.0	9.0	—	11.75	—
Malaysia	Pulau Pinang Marine Park	—	—	—	—	—	—	—	—	—	—	0.75	1.0	—	—	1.5	2.0	—	1.75	—
Malaysia	Pulau Rawa, Coral Garden	—	—	—	—	—	—	—	—	—	—	—	—	1.25	—	0.75	—	—	—	—
Malaysia	Pulau Silumpat	—	2.0	—	—	—	—	—	—	—	—	—	—	—	—	—	—	—	—	—
Malaysia	Pulau Susu Dara Besar	—	—	—	—	—	—	5.75	—	—	—	—	—	—	—	—	—	—	—	—
Malaysia	Pulau Tabawan	—	1.0	—	—	—	—	—	—	—	—	—	—	—	—	—	—	—	—	—
Malaysia	Pulau Tabun	—	—	—	—	—	—	—	—	—	—	—	—	—	—	—	—	—	0.25	—
Malaysia	Pulau Tengkorak	—	—	—	—	—	—	—	—	—	—	—	—	—	—	—	2.75	—	2.0	—
Malaysia	Pulau Yu Besar	—	—	—	—	—	—	—	—	—	—	—	—	—	—	—	0.75	—	0.5	—
Malaysia	Pulau Yu Kecil	—	—	—	—	—	—	—	—	—	—	—	—	—	—	—	1.0	—	1.0	—
Malaysia	Pygmy Rock, Siamil, Kapalai	—	—	—	—	—	—	—	—	—	—	—	—	—	2.5	—	—	—	—	—

Continued

Table A4 (Continued)   Global distribution of giant clams (Reef Check)

Country	Reef Site	Monitoring years (total clam density = number of individuals per 100 m²)																		
		1997	1998	1999	2000	2001	2002	2003	2004	2005	2006	2007	2008	2009	2010	2011	2012	2013	2014	2015
Malaysia	Rajawali Reef	—	—	—	—	—	—	—	—	—	—	—	—	3.0	—	—	—	—	—	—
Malaysia	Rayner's Rock	—	—	—	—	—	—	—	—	—	—	—	—	0.25	—	—	—	—	—	—
Malaysia	Redang Kalong House Reef	—	—	—	—	—	—	3.75	—	—	—	—	—	1.5	—	2.25	4.25	—	8.75	—
Malaysia	Reef 38	—	—	—	—	—	—	—	—	—	—	—	—	—	—	—	0.25	—	0.5	—
Malaysia	Reef 77	—	—	—	—	—	—	—	—	—	—	—	—	2.0	1.75	—	1.75	—	0.25	—
Malaysia	Renggis Island North Side	—	—	—	—	—	—	—	—	—	0.5	3.12	0.92	—	—	0.38	0.38	—	0.75	—
Malaysia	Renggis Island South Side	—	—	—	—	—	—	—	—	—	—	1.62	—	—	—	—	—	—	1.25	—
Malaysia	Renggis West	—	—	—	—	—	—	—	—	—	—	—	—	—	—	—	0.5	—	—	—
Malaysia	Ribbon Reef	—	—	—	—	—	—	—	—	—	—	—	—	—	—	—	—	—	4.08	—
Malaysia	Ribbon Valley	0.75	—	—	—	—	—	—	—	—	—	—	—	—	—	—	—	—	—	—
Malaysia	Rizal/Riza Garden	—	—	—	—	—	—	—	—	—	—	—	—	—	—	—	0.5	—	0.25	—
Malaysia	Roach Reef	—	—	—	—	—	—	—	—	—	—	—	—	—	—	—	—	—	0.25	—
Malaysia	Rock 'n' Roll Bay	—	—	—	—	—	—	—	—	—	—	—	—	—	—	—	3.0	—	—	—
Malaysia	S1-D2 Pulau Lang	—	—	—	—	—	—	—	—	—	—	—	—	—	—	—	0.5	—	—	—
Malaysia	Sahara	—	—	—	—	—	—	—	—	—	—	—	—	—	—	—	0.5	—	0.25	—
Malaysia	Sandbar North	—	—	—	—	—	—	—	—	—	—	—	—	—	—	29.25	—	—	—	—
Malaysia	Sandbar South	—	—	—	—	—	—	—	—	—	—	—	—	—	4.25	6.25	2.5	—	4.25	—
Malaysia	Sapi Reef	—	—	—	—	—	—	—	—	—	—	—	—	—	—	—	1.25	—	1.25	—
Malaysia	Scuba Junkie House Reef	—	—	—	—	—	—	—	—	—	—	—	—	—	—	3.0	—	—	—	—
Malaysia	Scubasa Reef	0.62	—	—	—	—	—	—	—	—	—	—	—	1.0	0.75	0.5	—	—	—	—
Malaysia	Sea Bell	—	—	—	—	—	—	—	—	—	—	0.5	1.75	—	—	0.5	1.5	—	0.25	—

Continued

333

Table A4 (Continued) Global distribution of giant clams (Reef Check)

Country	Reef Site	Monitoring years (total clam density = number of individuals per 100 m²)																		
		1997	1998	1999	2000	2001	2002	2003	2004	2005	2006	2007	2008	2009	2010	2011	2012	2013	2014	2015
Malaysia	Semaggot	—	—	—	—	—	—	—	—	—	—	—	—	0.75	—	—	—	—	—	—
Malaysia	Sepoi	—	—	—	—	—	—	—	—	—	—	—	—	—	1.25	1.5	1.75	—	1.5	—
Malaysia	Sepoi Island	—	—	—	—	—	—	—	—	—	—	—	—	—	—	—	1.5	—	—	—
Malaysia	Shark Point	—	—	—	—	—	—	—	—	—	—	—	—	—	—	—	—	—	0.25	—
Malaysia	Si Amil	—	—	—	—	—	—	—	—	—	—	—	—	—	—	—	—	—	0.5	—
Malaysia	Sibuang Point	—	—	—	9.0	—	—	—	—	—	—	—	—	—	—	—	—	—	—	—
Malaysia	Silent Reef, Kapas	—	—	—	—	—	—	—	—	—	—	—	—	—	0.25	0.25	1.25	—	—	—
Malaysia	Silumpat Island	—	2.0	—	—	—	—	—	—	—	—	—	—	—	—	—	—	—	—	—
Malaysia	Sipindung Reef	—	—	—	—	—	—	—	—	—	—	—	—	2.25	—	—	—	—	—	—
Malaysia	Siwa	—	—	—	—	—	—	—	—	—	—	—	—	0.12	—	—	—	—	—	—
Malaysia	Siwa 4	—	—	—	—	—	—	—	—	—	—	—	—	—	0.5	0.75	—	—	0.25	—
Malaysia	Siwa Penyu	—	—	—	—	—	—	—	—	0.25	—	—	—	0.25	0.12	—	0.5	—	0.25	—
Malaysia	Siwa Sunday	—	—	—	—	—	—	—	—	—	—	—	—	—	—	0.25	—	—	—	—
Malaysia	Slasher Beach	—	—	—	—	—	—	—	—	—	—	—	—	—	—	—	0.5	—	—	—
Malaysia	Small Reef	—	—	—	—	—	—	—	—	—	—	—	—	—	—	—	—	—	0.25	—
Malaysia	South Lanting	—	—	—	—	—	—	—	—	—	—	—	—	—	—	—	0.5	—	—	—
Malaysia	South Pinang	—	—	—	—	—	—	—	1.0	—	—	—	—	—	—	—	—	—	—	—
Malaysia	South Rim	—	—	—	—	—	—	—	—	—	—	—	—	—	—	—	—	—	0.38	—
Malaysia	Soyak Island	—	—	—	—	—	—	—	—	—	—	2.25	4.0	5.75	4.25	10.58	—	—	—	—
Malaysia	Soyak Island South	—	—	—	—	—	—	—	—	—	—	3.75	—	1.75	4.25	13.25	—	—	—	—
Malaysia	Soyak North/ Tridacna Bay	—	—	—	—	—	—	—	—	—	—	—	—	—	—	—	2.0	—	1.25	—
Malaysia	Soyak South	—	—	—	—	—	—	—	—	—	—	—	—	—	—	—	6.75	—	4.25	—
Malaysia	Sting Ray City, Kapalai	—	—	—	—	—	—	—	—	—	—	—	—	—	0.5	—	—	—	—	—

Continued

334

Table A4 (Continued) Global distribution of giant clams (Reef Check)

Country	Reef Site	Monitoring years (total clam density = number of individuals per 100 m^2)																		
		1997	1998	1999	2000	2001	2002	2003	2004	2005	2006	2007	2008	2009	2010	2011	2012	2013	2014	2015
Malaysia	Stingray City, Timba—Timba Island	—	—	—	—	—	—	—	—	—	—	—	—	—	5.75	—	3.0	—	6.5	—
Malaysia	Sulug	—	—	—	—	—	—	—	—	—	—	—	—	—	—	—	2.0	—	1.5	—
Malaysia	SW corner of palau balak	—	—	—	1.0	—	—	—	—	—	—	—	—	—	—	—	—	—	—	—
Malaysia	Sweetlips Rock, Mataking Island	—	—	—	—	—	—	—	—	—	—	—	—	—	3.75	1.25	—	—	6.5	—
Malaysia	Sweetlips Rock, Mataking Kecil Island	—	—	—	—	—	—	—	—	—	—	—	—	—	—	—	2.75	—	—	—
Malaysia	Tabawan Island	—	1.08	—	—	—	—	—	—	—	—	—	—	—	—	—	—	—	—	—
Malaysia	Tahingan	—	—	—	—	—	—	—	—	—	—	—	—	0.75	—	—	—	—	—	—
Malaysia	Takon	—	0.5	—	—	—	—	—	—	—	—	—	—	—	—	—	—	—	—	—
Malaysia	Takun	—	0.5	—	—	—	—	—	—	—	—	—	—	—	—	—	—	—	—	—
Malaysia	Talang Besar East	—	—	—	—	—	—	—	—	—	—	—	—	—	—	0.25	—	—	—	—
Malaysia	Tanjung Besi	—	—	—	—	—	—	—	—	—	—	—	3.75	3.0	1.5	0.75	—	—	1.75	—
Malaysia	Tanjung Kenangan	—	—	—	—	—	—	—	—	—	—	—	—	—	—	—	—	—	0.5	—
Malaysia	Tanjung Wokong	—	—	—	—	—	—	—	—	—	—	—	—	—	—	—	—	—	0.25	—
Malaysia	Telok Dalam	—	—	—	—	—	—	—	—	—	—	—	—	—	—	—	5.25	—	—	—
Malaysia	Teluk Gadung	—	—	—	—	—	—	—	—	—	—	—	—	1.25	—	—	—	—	—	—
Malaysia	Teluk Jawa, Kapas	—	—	—	—	—	—	—	—	—	—	—	—	—	0.75	0.5	—	—	—	—
Malaysia	Teluk Nakhoda	—	—	—	—	—	—	—	—	—	—	—	—	2.5	—	—	—	—	—	—
Malaysia	Teluk Rajawali	—	—	—	—	—	—	—	—	—	—	—	—	—	—	—	2.5	—	1.75	—
Malaysia	Teluran	—	—	—	—	—	—	—	—	—	—	—	—	0.75	—	—	—	—	—	—

Continued

Table A4 (Continued) Global distribution of giant clams (Reef Check)

Country	Reef Site	Monitoring years (total clam density = number of individuals per 100 m²)																		
		1997	1998	1999	2000	2001	2002	2003	2004	2005	2006	2007	2008	2009	2010	2011	2012	2013	2014	2015
Malaysia	Tekek House Reef	—	—	—	—	—	—	—	—	—	—	10.25	9.62	—	—	10.25	—	—	3.63	—
Malaysia	w	—	—	—	0.25	—	—	—	—	—	—	—	—	—	—	—	—	—	—	—
Malaysia	Terumbu Kili	0.25	0.5	—	—	—	—	—	—	—	—	—	—	—	—	0.5	—	—	0.5	—
Malaysia	Tg Tengah Southside, Pasir Panjang, Redang Island	—	—	—	—	—	—	—	0.5	—	—	—	—	—	—	—	—	—	—	—
Malaysia	Tiga Ruang Reef	—	—	—	—	—	—	—	—	—	—	—	—	2.88	0.5	1.25	1.75	—	3.75	—
Malaysia	Timba Timba	—	—	—	—	—	—	—	—	—	—	—	—	—	—	3.5	—	—	0.5	—
Malaysia	Tk Miyang	—	—	—	—	—	—	—	—	—	—	—	—	—	—	—	0.5	—	—	—
Malaysia	Tk. Jawa, Dayang	—	—	—	—	—	—	—	—	—	—	—	—	—	—	—	0.25	—	—	—
Malaysia	Toby Reef	—	—	2.25	1.0	—	—	—	—	—	—	—	—	—	—	—	—	—	—	—
Malaysia	Tokong Burung	—	—	—	—	—	—	4.25	—	—	—	—	—	—	—	—	—	—	—	—
Malaysia	Tomok	—	—	—	—	—	—	—	—	—	—	—	1.75	2.75	1.0	2.75	1.0	—	0.75	—
Malaysia	Treasure Hunt, Pandanan Island	—	—	—	—	—	—	—	—	—	—	—	0.25	—	—	—	—	—	1.13	—
Malaysia	Tukas Laut	—	—	—	—	—	—	—	—	—	—	—	1.0	11.75	—	2.5	—	—	—	—
Malaysia	Turtle Bay, Tenggol	—	—	—	—	—	—	—	—	—	—	1.0	—	—	—	1.75	—	—	—	—
Malaysia	Turtle Point	—	—	—	—	—	—	—	—	—	—	—	—	—	—	—	3.25	—	1.0	—
Malaysia	Veron/Veron Fan Garden	—	—	—	—	—	—	—	—	—	—	—	—	8.0	1.0	2.75	3.75	—	0.5	—
Malaysia	West End of Serundang Reef	—	—	—	2.25	—	—	—	—	—	—	—	—	—	—	—	—	—	—	—
Malaysia	Yoshi Point	1.0	—	—	—	—	—	—	—	—	—	—	—	—	—	—	—	—	—	—

Continued

Table A4 (Continued)   Global distribution of giant clams (Reef Check)

Country	Reef Site	Monitoring years (total clam density = number of individuals per 100 m^2)																		
		1997	1998	1999	2000	2001	2002	2003	2004	2005	2006	2007	2008	2009	2010	2011	2012	2013	2014	2015
Malaysia	Zorro	—	—	—	—	—	—	—	—	—	—	—	—	—	—	—	3.25	—	4.75	—
Malaysia	Zorro East	—	—	—	—	—	—	—	—	—	—	—	—	—	—	68.75	—	—	—	—
Maldives	Addoo	9.3	—	—	—	—	—	—	—	—	—	—	—	—	—	—	—	—	—	—
Maldives	Angaga Housereef Northeast	—															1.25			
Maldives	Angaga Housereef Southwest	—															1.0			
Maldives	Aquarium	1.25	—	—	—	—	—	—	—	—	—	—	—	—	—	—	—	—	—	—
Maldives	Banana Reef	—	—	—	—	—	—	—	—	—	—	—	—	—	0.25	—	—	—	0.25	—
Maldives	Banyan Tree House Reef	—	—	—	—	—	—	—	—	—	—	—	—	—	—	—	—	1.25	—	—
Maldives	Baros House Reef	—													1.0	2.25	2.0	—	0.63	—
Maldives	Bathalaa Maagaa	—														—	—	12.25	—	—
Maldives	Bathalaa Maagaa Kanthila	—														1.75	—	—	—	—
Maldives	Bathalaa Maagaa South	—														4.5	—	—	—	—
Maldives	Biyadhoo House Reef	—													—	—	5.75	—	—	—
Maldives	Bodu Giri	7.75	—	—	—	—	—	—	—	—	—	—	—	—	—	—	—	—	—	—
Maldives	Dega Giri	—	—	—	—	—	—	—	—	—	—	—	11.5	—	—	—	—	—	—	—
Maldives	Dega Giri, Ari Atoll	—	—	—	—	—	—	—	—	—	—	—	1.25	—	—	—	—	—	—	—
Maldives	Dega Thila	—	—	—	—	—	—	—	—	—	—	—	—	—	0.5	—	—	—	—	—

Continued

337

Table A4 (Continued) Global distribution of giant clams (Reef Check)

Country	Reef Site	Monitoring years (total clam density = number of individuals per 100 m²)																		
		1997	1998	1999	2000	2001	2002	2003	2004	2005	2006	2007	2008	2009	2010	2011	2012	2013	2014	2015
Maldives	Deh Giri	5.75	—	—	—	—	—	—	—	—	—	—	—	—	—	—	—	—	1.0	—
Maldives	Dhigga Thila	—	—	—	—	—	—	—	—	—	—	—	—	—	—	—	—	5.5	—	—
Maldives	Digga Thila	—	—	—	—	—	—	—	—	—	—	—	—	—	—	2.25	—	—	—	—
Maldives	Ellaidhoo Giri Nord	0.75	—	—	—	—	—	—	—	—	—	—	—	—	—	—	—	—	—	—
Maldives	Ellaidhoo House Reef	1.5	—	—	—	—	—	—	—	—	—	—	—	—	—	—	2.25	—	—	—
Maldives	Embudhoo	—	—	—	—	—	—	—	—	—	—	—	—	—	—	—	2.75	—	—	—
Maldives	Fan Reef	—	—	—	—	3.0	—	—	—	—	—	—	—	—	—	—	—	—	—	—
Maldives	Flat Reef	2.0	—	—	—	—	—	—	—	—	—	—	—	—	—	—	—	—	—	—
Maldives	Gangehi Island House Reef	—	—	—	—	—	—	—	—	—	—	—	—	—	—	8.0	—	—	—	—
Maldives	Gangehi North backreef	—	—	—	—	—	—	—	—	—	—	—	—	—	—	0.75	—	—	—	—
Maldives	Hembadhoo Hohola	3.0	—	—	—	—	—	—	—	—	—	—	—	—	—	—	—	—	—	—
Maldives	Holiday Thila North	—	—	—	—	—	—	—	—	—	—	—	—	—	—	4.5	—	—	—	—
Maldives	Holiday Thila South	—	—	—	—	—	—	—	—	—	—	—	—	—	—	2.0	—	—	—	—
Maldives	Honkey's	—	—	—	—	7.5	—	—	—	—	—	—	—	—	—	—	—	—	—	—
Maldives	House reef Angaga	0.92	—	—	—	—	—	—	—	—	—	—	—	—	—	—	—	—	—	—
Maldives	HP Reef	—	—	—	—	—	—	—	—	—	0.25	0.5	—	—	—	—	—	—	—	—
Maldives	Hufi Faru	1.75	—	—	—	—	—	—	—	—	—	—	—	—	—	—	—	—	0.5	—
Maldives	Hurasdhoo	1.0	—	—	—	—	—	—	—	—	—	—	—	—	—	—	—	—	—	—
Maldives	Kahanbu Thila	—	—	—	—	—	—	—	—	1.5	—	—	—	—	—	—	—	—	—	—
Maldives	Kahanbu Thila Fahru	—	—	—	—	—	—	—	—	—	3.5	—	—	—	—	—	—	—	—	—

Continued

Table A4 (Continued) Global distribution of giant clams (Reef Check)

Country	Reef Site	Monitoring years (total clam density = number of individuals per 100 m^2)																		
		1997	1998	1999	2000	2001	2002	2003	2004	2005	2006	2007	2008	2009	2010	2011	2012	2013	2014	2015
Maldives	Kuda Falhu	—	—	—	—	—	—	—	—	—	—	—	—	—	—	—	—	2.0	—	—
Maldives	Kuda Faru	9.25	—	—	—	—	—	—	—	—	—	—	—	—	—	—	—	—	2.0	—
Maldives	Kudafalu	—	—	—	—	—	—	—	—	—	—	—	—	—	—	3.0	—	—	—	—
Maldives	Kuramathi, Rasdhoo	—	—	—	—	—	—	—	—	—	—	—	—	—	—	1.0	—	—	—	—
Maldives	Kuredu House Reef	7.25	—	—	—	—	—	—	—	—	—	—	—	—	—	—	—	—	—	—
Maldives	Kuredu Zafari	1.75	—	—	—	—	—	—	—	—	—	—	—	—	—	—	—	—	—	—
Maldives	Laguna Beyru House Reef	—	—	—	—	—	—	—	—	—	—	—	—	—	—	11.25	—	—	—	—
Maldives	LGT1	—	—	—	—	—	—	—	—	—	—	12.92	8.0	3.25	—	—	—	—	—	—
Maldives	LGT2	—	—	—	—	—	—	—	—	—	—	4.5	4.08	0.5	—	—	—	—	—	—
Maldives	LGT3	—	—	—	—	—	—	—	—	—	—	6.33	5.33	1.0	—	—	—	—	—	—
Maldives	LGT4	—	—	—	—	—	—	—	—	—	—	1.33	1.42	1.25	—	—	—	—	—	—
Maldives	Lohifushi 1	—	—	—	—	1.5	—	—	—	—	—	—	—	—	—	—	—	—	—	—
Maldives	Lohifushi 2	—	—	—	—	4.0	—	—	—	—	—	—	—	—	—	—	—	—	—	—
Maldives	Maamigili	—	—	—	—	—	—	—	—	—	—	—	—	—	—	0.25	—	—	—	—
Maldives	Maaya Thila	—	—	—	—	—	—	—	—	—	1.25	—	—	—	—	—	—	—	—	—
Maldives	Madi Gaa	1.0	—	—	—	—	—	—	—	—	—	—	—	—	—	—	—	—	—	—
Maldives	Maduvaree Island Reef	5.5	—	—	—	—	—	—	—	—	—	—	—	—	—	—	—	—	0.38	—
Maldives	Meddu Faru Nord	1.5	—	—	—	—	—	—	—	—	—	—	—	—	—	—	—	—	—	—
Maldives	Musa	—	—	—	—	1.0	—	—	—	—	—	—	—	—	—	—	—	—	—	—
Maldives	Niumath Thilla	—	—	—	—	—	—	—	—	—	1.0	—	—	—	—	—	—	—	—	—
Maldives	Orimas Faru Nord	3.5	—	—	—	—	—	—	—	—	—	—	—	—	—	—	—	—	—	—
Maldives	Orimas Faru	—	—	—	—	—	—	—	—	—	—	—	—	—	—	—	0.5	—	—	—
Maldives	Panettone Reef	—	—	—	—	—	—	—	—	—	1.25	—	—	—	—	—	—	—	—	—

Continued

Table A4 (Continued) Global distribution of giant clams (Reef Check)

Country	Reef Site	Monitoring years (total clam density = number of individuals per 100 m^2)																		
		1997	1998	1999	2000	2001	2002	2003	2004	2005	2006	2007	2008	2009	2010	2011	2012	2013	2014	2015
Maldives	Rasdhoo Madivaru	—	—	—	—	—	—	—	—	—	—	—	—	—	—	5.75	—	—	—	—
Maldives	Rasdhoo Madivaru Beyru	—	—	—	—	—	—	—	—	—	—	—	—	—	—	—	—	7.5	—	—
Maldives	Rasdhoo North Ari	—	—	—	—	—	—	—	—	10.75	—	6.5	7.75	—	—	—	—	—	—	—
Maldives	Rashdoo Madivaru	—	—	—	—	—	—	—	—	—	2.62	—	—	—	—	—	—	—	—	—
Maldives	Reethi Faru	6.75	—	—	—	—	—	—	—	—	—	—	—	—	—	—	—	—	—	—
Maldives	Remas Faru	0.75	—	—	—	—	—	—	—	—	—	—	—	—	—	—	—	—	0.75	—
Maldives	Salomon Isle 1	—	—	—	—	249.5	—	—	—	—	—	—	—	—	—	—	—	—	—	—
Maldives	Sultans	—	—	—	—	0.5	—	—	—	—	—	—	—	—	—	—	—	—	—	—
Maldives	Tasdhoo Madivaru West	—	—	—	—	—	—	—	—	—	—	—	—	—	—	1.75	—	—	—	—
Maldives	Thuvaru Island Reef	3.25	—	—	—	—	—	—	—	—	—	—	—	—	—	—	—	—	—	—
Maldives	Vilm05/ Vilamendhoo southwest	—	—	—	—	—	—	—	—	—	—	—	—	—	—	—	4.25	—	—	—
Maldives	Weng Gaa	6.75	—	—	—	—	—	—	—	—	—	—	—	—	—	—	—	—	0.25	—
Marshall Islands	Ajejen	—	—	—	—	—	3.0	—	—	—	—	—	—	—	—	—	—	—	—	—
Marshall Islands	Ajejen 2	—	—	—	—	—	10.75	—	—	—	—	—	—	—	—	—	—	—	—	—
Marshall Islands	Enijet Bar	—	—	—	—	—	16.0	—	—	—	—	—	—	—	—	—	—	—	—	—
Mauritius	Chaland, Passe Armand (10m)	—	—	—	—	1.25	—	—	—	—	—	—	—	—	—	—	—	—	—	—

Continued

Table A4 (Continued) Global distribution of giant clams (Reef Check)

Country	Reef Site	Monitoring years (total clam density = number of individuals per 100 m²)																		
		1997	1998	1999	2000	2001	2002	2003	2004	2005	2006	2007	2008	2009	2010	2011	2012	2013	2014	2015
Mauritius	Chaland, Passe Armand (3m)	—	—	—	—	0.75	—	—	—	—	—	—	—	—	—	—	—	—	—	—
Mauritius	Island Reef, Anse La Raie	—	—	6.5	0.25	—	—	—	—	—	—	—	—	—	—	—	—	—	—	—
Mauritius	Passe Armand	—	—	1.75	—	—	0.5	0.25	—	—	—	—	—	—	—	—	—	—	—	—
Mauritius	Patte Cappor	—	—	2.75	—	—	—	1.5	—	—	—	—	—	—	—	—	—	—	—	—
Mauritius	Petit Brisane	—	—	0.25	—	—	—	—	—	—	—	—	—	—	—	—	—	—	—	—
Mayotte	Boa Sadia Reef	—	—	—	—	—	—	—	—	4.25	1.25	—	—	1.75	1.25	—	—	—	—	—
Mayotte	Boueni Village	—	—	—	—	—	—	—	—	—	—	—	—	—	0.25	—	—	—	—	—
Mayotte	Longoni Reef	—	—	—	—	—	—	1.75	2.25	0.75	—	1.25	—	1.75	2.75	—	—	—	1.0	—
Mayotte	Passe Boueni	—	—	—	—	—	—	—	—	—	—	—	—	—	0.75	—	—	—	—	—
Mayotte	Passe en S- Bouee 2	—	—	—	—	—	—	—	8.5	7.75	—	5.75	—	1.25	—	—	—	—	2.75	—
Mayotte	Passe en S, bouee 11	—	—	—	—	—	—	4.0	2.5	—	7.25	9.0	—	10.0	—	—	—	—	3.5	—
Mayotte	Reserve Naturelle de Mbouzi	—	—	—	—	—	—	—	—	—	—	—	—	—	2.0	—	—	—	—	—
Mayotte	Sakouli	—	—	—	—	—	—	—	—	—	—	—	—	—	1.5	—	—	—	—	—
Mayotte	Tanaraki	—	—	—	—	—	—	—	—	—	—	—	—	—	—	—	—	—	1.0	—
Mayotte	Ngouja	—	—	—	—	0.25	—	—	—	—	—	—	—	—	—	—	—	—	8.5	—
Mozambique	Baixo Vadiazi	—	—	—	—	—	8.25	—	—	—	—	—	—	—	—	—	—	—	—	—
Mozambique	Cabo Pequeve	—	—	—	—	—	0.75	—	—	—	—	—	—	—	—	—	—	—	—	—
Mozambique	Doodles Reef	—	—	—	—	0.25	—	—	—	—	—	—	—	—	—	—	—	—	—	—
Mozambique	Ilha Matemo	—	—	—	—	—	2.5	—	—	—	—	—	—	—	—	—	—	—	—	—
Mozambique	Ilha Medjumbi	—	—	—	—	—	3.25	—	—	—	—	—	—	—	—	—	—	—	—	—
Mozambique	Ilha Quilaluia	—	—	—	—	—	0.25	—	—	—	—	—	—	—	—	—	—	—	—	—
Mozambique	Ilha Quissanga	—	—	—	—	—	6.0	—	—	—	—	—	—	—	—	—	—	—	—	—
Mozambique	Ilha Rongui	—	—	—	—	—	1.25	—	—	—	—	—	—	—	—	—	—	—	—	—

Continued

Table A4 (Continued) Global distribution of giant clams (Reef Check)

Country	Reef Site	Monitoring years (total clam density = number of individuals per 100 m^2)																		
		1997	1998	1999	2000	2001	2002	2003	2004	2005	2006	2007	2008	2009	2010	2011	2012	2013	2014	2015
Mozambique	Ilha Tecomangi	—	—	—	—	—	0.5	—	—	—	—	—	—	—	—	—	—	—	—	—
Mozambique	Ilha Vamizi	—	—	—	—	—	1.0	—	—	—	—	—	—	—	—	—	—	—	—	—
Mozambique	Malongane	—	—	—	0.25	—	—	—	—	—	—	—	—	—	—	—	—	—	—	—
Mozambique	Quirimba Outer Reef	1.0	—	—	—	—	—	—	—	—	—	—	—	—	—	—	—	—	—	—
Myanmar	Bo Yar Nunt/Poni Island	—	—	—	—	—	—	—	—	—	—	—	—	—	—	—	—	0.5	—	—
Myanmar	Island 115	—	—	—	—	—	—	—	—	—	—	—	—	—	—	—	—	0.25	—	—
Myanmar	Kunn Thee Island	—	—	—	—	4.0	—	—	—	—	—	—	—	—	—	—	—	—	—	—
Myanmar	Kya Haing Island, W-Beach	—	—	—	—	—	—	—	—	3.5	—	—	—	—	—	—	—	—	—	—
Myanmar	Kyunn Me Gyee	—	—	—	—	—	—	—	0.75	—	—	—	—	—	—	—	—	—	—	—
Myanmar	Kyunn Phi Lar/Pi La Kyun/Great Swinton Island	—	—	—	—	—	—	1.75	—	—	—	—	—	—	—	—	—	0.25	—	—
Myanmar	Kyunn Thone Lon	—	—	—	—	0.5	—	1.75	—	—	—	—	—	—	—	—	—	—	—	—
Myanmar	Lampi Island	—	—	—	—	—	—	0.75	6.0	—	—	—	—	—	—	—	—	—	—	—
Myanmar	Lampi Island North	—	—	—	—	—	—	0.5	—	—	—	—	—	—	—	—	—	—	—	—
Myanmar	McLeod Island (Kho Yinn Khwa)	—	—	—	—	—	—	0.25	—	—	—	—	—	—	—	—	—	0.25	—	—
Myanmar	Say Tan Island	—	—	—	—	—	—	—	—	3.5	—	—	—	—	—	—	—	—	—	—
Myanmar	St. Paul's Island	—	—	—	—	0.25	—	—	—	—	—	—	—	—	—	—	—	—	—	—
Myanmar	Tar Yar Island	—	—	—	—	7.5	—	1.25	1.5	—	—	—	—	—	—	—	—	—	—	—

Continued

342

Table A4 (Continued) Global distribution of giant clams (Reef Check)

Country	Reef Site	Monitoring years (total clam density = number of individuals per 100 m²)																		
		1997	1998	1999	2000	2001	2002	2003	2004	2005	2006	2007	2008	2009	2010	2011	2012	2013	2014	2015
Myanmar	Than Yoke (Potter) Island	—	—	—	—	—	—	—	—	—	—	—	—	—	—	—	—	0.25	—	—
New Caledonia	Abore Reef	1.5	—	—	—	—	—	—	—	—	—	—	—	—	—	—	—	—	—	—
New Caledonia	Akaia	—	—	—	—	—	—	—	—	—	0.5	—	0.25	—	—	—	—	—	—	—
New Caledonia	Beco	—	—	—	—	—	—	4.5	9.0	9.75	—	11.5	12.25	—	11.5	16.0	—	—	—	—
New Caledonia	Bonne Anse	—	0.25	—	—	—	—	0.25	0.75	0.75	—	0.25	0.62	—	2.0	2.5	—	—	—	—
New Caledonia	Casy	—	0.25	—	—	—	—	—	0.75	—	—	—	0.25	—	0.25	—	—	—	—	—
New Caledonia	Donga Hienga	—	—	—	—	—	—	2.75	2.75	3.5	3.0	3.25	—	3.5	2.5	2.0	—	—	—	—
New Caledonia	Ever Prosperity	—	—	—	—	—	—	0.25	1.25	—	—	0.25	—	—	—	—	—	—	—	—
New Caledonia	Fausse Passe	—	1.0	—	—	—	—	—	—	—	—	—	—	—	—	—	—	—	—	—
New Caledonia	Fausse Passe Pouembout	—	—	—	—	—	—	1.25	1.0	2.0	1.75	2.75	1.5	—	3.5	7.25	—	—	—	—
New Caledonia	Goro	—	0.25	—	—	—	—	—	—	—	—	—	—	—	—	—	—	—	—	—
New Caledonia	Grand Recif Thio	—	—	—	—	—	—	2.75	3.25	4.25	1.5	—	2.25	4.25	—	9.0	—	—	—	—
New Caledonia	Hiengabat	—	—	—	—	—	—	6.5	14.25	11.5	16.0	11.5	—	9.5	10.25	7.0	—	—	—	—
New Caledonia	Hnapalu/Qanono	—	—	—	—	—	—	2.25	—	1.12	2.0	—	—	—	2.75	3.5	—	—	—	—
New Caledonia	Ile Verte	—	—	—	—	—	—	8.0	7.75	8.0	5.75	—	5.25	—	4.75	7.25	—	—	—	—

Continued

Table A4 (Continued) Global distribution of giant clams (Reef Check)

Country	Reef Site	Monitoring years (total clam density = number of individuals per 100 m²)																		
		1997	1998	1999	2000	2001	2002	2003	2004	2005	2006	2007	2008	2009	2010	2011	2012	2013	2014	2015
New Caledonia	Jinek	—	—	—	—	—	—	3.0	1.0	1.25	—	1.25	0.25	—	7.25	6.25	—	—	—	—
New Caledonia	Koniene	—	—	—	—	—	—	7.75	10.5	13.5	9.0	8.5	6.75	—	10.0	10.75	—	—	—	—
New Caledonia	Koulnoue	—	—	—	—	—	—	—	—	0.25	0.25	0.25	—	—	—	0.25	—	—	—	—
New Caledonia	Luecilla 2	—	—	—	—	—	—	—	—	0.25	—	—	—	—	—	—	—	—	—	—
New Caledonia	Luengoni 1	—	—	—	—	—	—	—	—	—	—	—	—	—	0.5	0.25	—	—	—	—
New Caledonia	Luengoni 2	—	—	—	—	—	—	0.75	2.0	1.0	—	—	—	—	1.0	—	—	—	—	—
New Caledonia	M'Bere Reef	—	0.75	—	—	—	—	—	—	—	—	—	—	—	—	—	—	—	—	—
New Caledonia	Maitre	—	—	—	—	0.5	—	0.5	—	—	—	0.5	—	0.25	—	—	—	—	—	—
New Caledonia	Mbere	—	—	—	—	—	—	3.75	1.5	—	1.5	2.25	—	2.75	—	—	—	—	—	—
New Caledonia	Moara	—	—	—	—	—	—	0.75	1.25	—	0.5	—	1.0	0.5	—	—	—	—	—	—
New Caledonia	Nouville	—	—	—	—	—	—	—	0.25	—	—	—	—	1.5	—	—	—	—	—	—
New Caledonia	Pindai	—	—	—	—	—	—	0.75	0.75	1.0	—	1.0	0.25	—	0.75	1.0	—	—	—	—
New Caledonia	Pinjien	—	—	—	—	—	—	0.25	0.25	0.25	0.25	—	—	—	0.25	—	—	—	—	—
New Caledonia	Recif interieur Thio	—	—	—	—	—	—	1.0	0.5	0.75	0.5	—	0.25	—	0.5	0.75	—	—	—	—
New Caledonia	Ricaudy	0.25	—	—	—	—	—	—	—	—	0.25	0.25	—	0.75	—	—	—	—	—	—

Continued

Table A4 (Continued) Global distribution of giant clams (Reef Check)

Country	Reef Site	Monitoring years (total clam density = number of individuals per 100 m^2)																		
		1997	1998	1999	2000	2001	2002	2003	2004	2005	2006	2007	2008	2009	2010	2011	2012	2013	2014	2015
New Caledonia	Sable	—	0.25	—	—	—	—	—	—	—	—	—	—	—	—	—	—	—	—	—
New Caledonia	Santal 1	—	—	—	—	—	—	1.25	1.0	1.5	—	1.0	1.0	—	2.5	0.5	—	—	—	—
New Caledonia	Santal 2	—	—	—	—	—	—	0.75	0.5	1.5	—	0.25	0.5	—	2.75	1.0	—	—	—	—
New Caledonia	Siande	—	2.0	—	—	—	—	3.0	9.5	9.75	6.75	—	5.75	—	10.5	8.25	—	—	—	—
New Caledonia	Signal	—	—	—	—	—	—	0.25	0.25	—	0.75	0.75	—	1.25	—	—	—	—	—	—
New Caledonia	Tabou	—	0.75	—	—	—	—	—	—	—	—	—	—	—	—	—	—	—	—	—
New Caledonia	Thio Barrier Reef	—	0.75	—	—	—	—	—	—	—	—	—	—	—	—	—	—	—	—	—
New Caledonia	We Port	—	—	—	—	—	—	0.25	—	0.5	—	—	—	—	0.75	0.5	—	—	—	—
Palau	Cemetery Reef	—	—	—	—	—	6.5	—	—	—	—	—	—	—	—	—	—	—	—	—
Palau	Ngederak Reef	5.75	—	—	7.25	2.0	0.75	1.25	—	—	1.25	—	—	—	—	—	—	—	—	—
Palau	Short Drop Off	3.75	—	—	3.38	1.38	3.75	1.62	—	—	1.0	—	—	—	—	—	—	—	—	—
Papua New Guinea	Anemone Patch	—	—	—	—	—	0.5	—	—	—	—	—	—	—	—	—	—	—	—	—
Papua New Guinea	Annsophie's Reef	—	—	0.5	—	—	—	—	—	—	—	—	—	—	—	—	—	—	—	—
Papua New Guinea	B25 Bomber	—	—	0.25	—	—	—	—	—	—	—	—	—	—	—	—	—	—	—	—
Papua New Guinea	Cape Hewsner	—	—	2.5	—	—	—	—	—	—	—	—	—	—	—	—	—	—	—	—
Papua New Guinea	Chermain's Reef	—	—	1.5	—	—	—	—	—	—	—	—	—	—	—	—	—	—	—	—

Continued

Table A4 (Continued) Global distribution of giant clams (Reef Check)

Country	Reef Site	Monitoring years (total clam density = number of individuals per 100 m²)																		
		1997	1998	1999	2000	2001	2002	2003	2004	2005	2006	2007	2008	2009	2010	2011	2012	2013	2014	2015
Papua New Guinea	Cyclone Reef	—	1.0	1.25	0.25	—	—	—	—	—	—	—	—	—	—	—	—	—	—	—
Papua New Guinea	First Reef	—	—	0.25	—	—	—	—	—	—	—	—	—	—	—	—	—	—	—	—
Papua New Guinea	Jais Aben alpha	—	—	—	—	—	0.12	—	—	—	—	—	—	—	—	—	—	—	—	—
Papua New Guinea	Jais Aben bravo	—	—	—	—	—	0.25	—	—	—	—	—	—	—	—	—	—	—	—	—
Papua New Guinea	Kaleu 1	—	—	—	—	—	—	—	—	—	—	—	—	15.25	—	—	—	—	—	—
Papua New Guinea	Kaleu 2	—	—	—	—	—	—	—	—	—	—	—	—	1.25	—	—	—	—	—	—
Papua New Guinea	Kaleu 3	—	—	—	—	—	—	—	—	—	—	—	—	9.0	—	—	—	—	—	—
Papua New Guinea	Keng MPA 1	—	—	—	—	—	—	—	—	—	—	—	—	0.75	—	—	—	—	—	—
Papua New Guinea	Lumu Reef	—	—	0.5	—	—	—	—	—	—	—	—	—	—	—	—	—	—	—	—
Papua New Guinea	Maclaren's Reef	—	—	—	0.75	—	—	—	—	—	—	—	—	—	—	—	—	—	—	—
Papua New Guinea	Magic Passage	—	—	1.25	—	—	—	—	—	—	—	—	—	—	—	—	—	—	—	—
Papua New Guinea	Marangis Reef 1	—	—	—	—	—	—	—	—	—	—	—	0.75	0.12	—	—	—	—	—	—
Papua New Guinea	Marangus MPA 2	—	—	—	—	—	—	—	—	—	—	—	—	2.0	—	—	—	—	—	—
Papua New Guinea	Marangus MPA 3	—	—	—	—	—	—	—	—	—	—	—	—	0.25	—	—	—	—	—	—
Papua New Guinea	Mata–Limut Reef	—	—	—	—	—	—	—	—	—	—	—	0.25	1.0	—	—	—	—	—	—

Continued

346

Table A4 (Continued) Global distribution of giant clams (Reef Check)

Country	Reef Site	1997	1998	1999	2000	2001	2002	2003	2004	2005	2006	2007	2008	2009	2010	2011	2012	2013	2014	2015
							Monitoring years (total clam density = number of individuals per 100 m²)													
Papua New Guinea	Midway Reef	—	—	—	—	—	0.5	—	—	—	—	—	—	—	—	—	—	—	—	—
Papua New Guinea	Motupore Island	—	—	—	—	—	0.25	—	—	—	—	—	—	—	—	—	—	—	—	—
Papua New Guinea	Nago 2	—	—	—	—	—	—	—	—	—	—	—	—	3.25	—	—	—	—	—	—
Papua New Guinea	Nago Island Reef 1	—	—	—	—	—	—	—	—	—	—	—	0.75	1.5	—	—	—	—	—	—
Papua New Guinea	Nago Island Reef Site 2	—	—	—	—	—	—	—	—	—	—	—	1.0	—	—	—	—	—	—	—
Papua New Guinea	Nonovaul Island No Take Area Reef	—	—	—	—	—	—	—	—	—	—	—	1.75	1.75	—	—	—	—	—	—
Papua New Guinea	Nusa Island Reef	—	—	—	—	—	—	—	—	—	—	—	0.5	0.25	—	—	—	—	—	—
Papua New Guinea	Nusa Lik Reef	—	—	—	—	—	—	—	—	—	—	—	0.5	0.75	—	—	—	—	—	—
Papua New Guinea	Oinari Point	—	—	—	1.0	—	—	—	—	—	—	—	—	—	—	—	—	—	—	—
Papua New Guinea	Pig Island Drop Off	—	—	0.75	—	—	—	—	—	—	—	—	—	—	—	—	—	—	—	—
Papua New Guinea	Pig Island Passage	—	—	1.5	—	—	—	—	—	—	—	—	—	—	—	—	—	—	—	—
Papua New Guinea	Sinub Island Northside	—	—	—	—	—	0.42	—	—	—	—	—	—	—	—	—	—	—	—	—
Papua New Guinea	Tabat Exposed	—	—	—	—	—	0.5	—	—	—	—	—	—	—	—	—	—	—	—	—
Papua New Guinea	Tufi Harbour Point	—	—	—	0.25	—	—	—	—	—	—	—	—	—	—	—	—	—	—	—

Continued

Table A4 (Continued) Global distribution of giant clams (Reef Check)

Country	Reef Site	Monitoring years (total clam density = number of individuals per 100 m²)																		
		1997	1998	1999	2000	2001	2002	2003	2004	2005	2006	2007	2008	2009	2010	2011	2012	2013	2014	2015
Papua New Guinea	Usen	—	—	—	—	—	—	—	0.5	—	—	—	—	—	—	—	—	—	—	—
Philippines	3rd Plateau/ Coral Garden	—	—	—	—	—	—	—	—	—	—	—	—	—	—	—	0.25	0.75	1.75	—
Philippines	7th Commando Outside	—	—	—	—	—	—	—	—	—	—	—	—	—	—	—	—	0.75	0.75	—
Philippines	Abdeen's Rock	—	—	—	—	—	—	—	—	—	—	—	—	—	—	—	—	4.5	1.13	—
Philippines	Acacia Resort and Dive Center	—	—	—	—	—	—	—	—	—	—	—	—	—	0.75	—	1.0	—	1.0	—
Philippines	Albaguen Island	—	1.12	—	—	—	—	—	—	—	—	—	—	—	—	—	—	—	—	—
Philippines	Alegre Beach Resort 2	—	—	—	—	—	—	—	0.75	—	—	—	—	—	—	—	—	—	—	—
Philippines	Alegre Beach Resort 5	—	—	—	—	—	—	—	0.5	—	—	—	—	—	—	—	—	—	—	—
Philippines	Alegre Beach Resort 6	—	—	—	—	—	—	—	0.75	—	—	—	—	—	—	—	—	—	—	—
Philippines	AMPO- AM01	—	—	—	—	—	—	—	—	—	—	0.12	—	—	—	—	—	—	—	—
Philippines	AMPO- AM02	—	—	—	—	—	—	—	—	—	—	0.25	—	—	—	—	—	—	—	—
Philippines	AMPO- AM05	—	—	—	—	—	—	—	—	—	—	0.06	—	—	—	—	—	—	—	—
Philippines	AMPO- AM06	—	—	—	—	—	—	—	—	—	0.12	—	—	—	—	—	—	—	—	—
Philippines	AMPO- AM07	—	—	—	—	—	—	—	—	—	0.25	0.08	—	—	—	—	—	—	—	—
Philippines	AMPO- AM08	—	—	—	—	—	—	—	—	—	0.08	—	—	—	—	—	—	—	—	—
Philippines	AMPO- AM09	—	—	—	—	—	—	—	—	—	0.12	—	0.25	—	—	—	—	—	—	—
Philippines	AMPO- AM10	—	—	—	—	—	—	—	—	—	—	0.25	—	—	—	—	—	—	—	—
Philippines	AMPO- AM13	—	—	—	—	—	—	—	—	—	—	0.08	—	—	—	—	—	—	—	—
Philippines	Apid Marine Sanctuary	—	—	—	—	—	—	—	—	—	0.25	—	—	—	—	—	—	—	—	—
Philippines	Apid MPA	—	—	—	—	—	—	—	—	—	0.25	0.25	0.5	—	—	—	—	—	—	—

Continued

Table A4 (Continued) Global distribution of giant clams (Reef Check)

Country	Reef Site	Monitoring years (total clam density = number of individuals per 100 m²)																		
		1997	1998	1999	2000	2001	2002	2003	2004	2005	2006	2007	2008	2009	2010	2011	2012	2013	2014	2015
Philippines	Apo Island Marine Reserve	—	—	—	0.12	0.12	—	—	—	—	—	—	—	—	—	—	—	—	—	—
Philippines	Apo Reef 1	—	—	—	—	—	—	—	—	0.5	—	—	—	—	—	—	—	—	—	—
Philippines	Apo Reef 3	—	—	—	—	—	—	—	—	1.0	—	—	—	—	—	—	—	—	—	—
Philippines	Apo Reef 4	—	—	—	—	—	—	—	—	0.25	—	—	—	—	—	—	—	—	—	—
Philippines	Arraceife Island	—	0.25	—	—	—	—	—	—	—	—	—	—	—	—	—	—	—	—	—
Philippines	Aslom Island	—	—	—	—	—	—	—	—	—	—	—	—	—	—	—	—	0.25	—	—
Philippines	Atop-Atop	—	—	—	—	—	—	—	—	—	1.5	—	—	—	—	—	—	—	—	—
Philippines	Balabag Reef	—	—	—	—	—	—	—	—	—	—	—	—	—	—	—	—	1.0	—	—
Philippines	Balabagon	—	—	—	—	—	—	—	—	3.0	—	—	—	—	—	—	—	—	—	—
Philippines	Balangingi Eastside	—	—	—	—	—	—	—	—	—	0.12	—	—	—	—	—	—	—	—	—
Philippines	Balangingi Westside	—	—	—	—	—	—	—	—	—	0.38	—	—	—	—	—	—	—	—	—
Philippines	Balatasan MPA North	—	—	—	—	—	—	—	—	—	—	—	—	—	—	—	—	0.25	—	—
Philippines	Balatasan MPA South	—	—	—	—	—	—	—	—	—	—	—	—	—	—	—	—	4.0	—	—
Philippines	Balicasag	—	—	—	—	—	—	—	—	0.25	—	—	—	—	—	—	—	—	—	—
Philippines	Balicasag MPA	—	—	—	—	—	—	—	—	—	0.25	0.25	—	—	—	—	—	—	—	—
Philippines	Bancoro	—	—	—	—	—	—	—	—	—	—	—	—	—	—	—	—	1.5	—	—
Philippines	Banlot Tongo Basdiot	—	—	—	—	—	—	—	—	0.5	—	—	—	—	—	—	—	—	—	—
Philippines	Barangay Talima 2	—	—	—	—	—	0.75	0.25	—	—	—	—	—	—	—	—	—	—	—	—
Philippines	Barge Centro/ Roberto	—	—	—	—	—	—	—	—	—	—	—	—	—	—	—	—	0.75	0.5	—
Philippines	Barge Laot/ Sabino	—	—	—	—	—	—	—	—	—	—	—	—	—	—	—	—	0.5	0.25	—

Continued

349

Table A4 (Continued) Global distribution of giant clams (Reef Check)

Country	Reef Site	Monitoring years (total clam density = number of individuals per 100 m²)																		
		1997	1998	1999	2000	2001	2002	2003	2004	2005	2006	2007	2008	2009	2010	2011	2012	2013	2014	2015
Philippines	Barge Tandol	—	—	—	—	—	—	—	—	—	—	—	—	—	—	—	—	0.75	—	—
Philippines	Baring	—	—	—	—	—	—	—	0.25	0.25	—	—	—	—	—	—	—	—	—	—
Philippines	Basdiot (North)	—	—	—	—	—	—	—	—	—	0.25	—	—	—	—	—	—	—	—	—
Philippines	Big Apple	—	—	—	—	—	—	—	—	—	—	—	—	—	—	—	0.25	—	—	—
Philippines	Big Manta Rock	—	—	0.5	—	—	—	—	—	—	—	—	—	—	—	—	—	—	—	—
Philippines	Binubusan Shoal	—	—	—	—	—	—	—	—	0.25	—	—	—	—	—	—	—	—	—	—
Philippines	Bitayan	—	—	—	—	—	—	—	0.25	—	—	0.25	—	—	—	—	—	—	—	—
Philippines	Bitoon	—	—	—	—	—	—	—	—	—	—	—	0.38	—	—	—	—	—	—	—
Philippines	Black Rock	—	—	—	—	—	—	—	—	—	—	—	—	—	—	—	51.75	—	—	—
Philippines	Blue Water	—	—	—	—	—	—	0.25	—	—	—	—	—	—	—	—	—	—	—	—
Philippines	BRGY, POOC	—	—	—	—	—	—	—	—	—	0.5	—	—	—	—	—	—	—	—	—
Philippines	Bugor MPA	—	—	—	—	—	—	—	—	—	—	—	—	—	—	—	—	1.88	2.0	—
Philippines	Bukal	—	—	—	—	—	—	—	—	—	—	—	—	—	—	—	—	0.75	2.0	—
Philippines	Bulalakaw (Ulogan Bay) Transect 1	—	—	—	—	—	—	—	—	—	—	—	9.5	—	—	—	—	—	—	—
Philippines	Bulalakaw (Ulogan Bay) Transect 2	—	—	—	—	—	—	—	—	—	—	—	12.25	—	—	—	—	—	—	—
Philippines	Buyayao Island	—	—	—	—	—	—	—	—	—	—	—	—	—	—	—	—	2.5	—	—
Philippines	Caalan MPA A	—	—	—	—	—	—	—	—	—	—	—	—	—	—	—	—	0.75	3.75	—
Philippines	Caalan MPA B	—	—	—	—	—	—	—	—	—	—	—	—	—	—	—	—	1.75	—	—
Philippines	Cagdanao Island	—	—	3.5	0.75	—	—	—	—	—	—	—	—	—	—	—	—	—	—	—
Philippines	Calanggaman	—	—	—	—	—	—	—	—	—	1.75	0.5	—	—	—	—	—	—	—	—
Philippines	Campomanes Bay	1.0	—	—	—	—	—	—	—	—	—	—	—	—	—	—	—	—	—	—
Philippines	Capitancillo	—	—	—	—	—	—	—	—	0.12	0.5	0.25	—	—	—	—	—	—	—	—
Philippines	Capitancillo Transect 2	—	—	—	—	—	—	—	—	0.25	—	—	—	—	—	—	—	—	—	—

Continued

350

Table A4 (Continued) Global distribution of giant clams (Reef Check)

Country	Reef Site	Monitoring years (total clam density = number of individuals per 100 m^2)																		
		1997	1998	1999	2000	2001	2002	2003	2004	2005	2006	2007	2008	2009	2010	2011	2012	2013	2014	2015
Philippines	Centro Site 1(shallow)	—	—	—	—	—	—	—	—	0.25	—	—	—	—	—	—	—	—	—	—
Philippines	Centro Site 2 (deep)	—	—	—	—	—	—	—	0.25	—	—	—	—	—	—	—	—	—	—	—
Philippines	Centro Site 2 (shallow)	—	—	—	—	—	—	—	—	0.25	—	—	—	—	—	—	—	—	—	—
Philippines	Coral Gardens	—	—	—	—	—	—	—	0.5	—	—	—	—	—	—	—	—	—	0.25	—
Philippines	Costabella	—	—	—	—	—	0.25	—	—	—	—	—	—	—	—	—	—	—	—	—
Philippines	Cueva Calintaan	—	—	—	—	—	—	—	—	—	—	—	—	—	—	—	—	0.5	—	—
Philippines	CYC East Reef	—	—	—	—	—	—	—	—	—	—	—	—	—	—	—	—	1.0	1.75	—
Philippines	CYC West Reef	—	—	—	—	—	—	—	—	—	—	—	—	—	—	—	—	0.75	0.5	—
Philippines	Dakit Dakit (Logon)	—	—	—	—	—	—	—	—	—	—	1.5	—	—	—	—	—	—	—	—
Philippines	Danjugan Island	—	—	0.25	—	0.25	—	—	—	—	—	—	—	—	—	—	—	—	—	—
Philippines	Dive and Trek Marine Sanctuary	—	—	—	—	—	—	—	—	—	—	—	—	—	—	—	—	—	7.0	—
Philippines	Dungon Bay	—	—	—	—	—	—	—	—	—	—	—	—	—	—	—	0.25	—	0.25	—
Philippines	East Outside Proposed MPA, Caubian Dako	—	—	—	—	—	—	—	0.5	—	—	—	—	—	—	—	—	—	—	—
Philippines	East Sangat Japanese Gunboat	—	—	—	—	—	—	—	1.75	—	—	—	—	—	—	—	—	—	—	—
Philippines	Fondeado Island	—	1.5	—	—	—	—	—	—	—	—	—	—	—	—	—	—	—	—	—
Philippines	Fusiliro Sombrero	—	—	—	—	—	—	—	—	0.5	—	—	—	—	—	—	—	—	—	—
Philippines	Grande	—	—	—	—	—	—	—	—	0.25	—	—	—	—	—	—	—	—	—	—
Philippines	Giant Clam	—	—	—	—	—	—	—	—	—	—	—	—	—	—	—	—	—	0.5	—

Continued

Table A4 (Continued) Global distribution of giant clams (Reef Check)

Country	Reef Site	Monitoring years (total clam density = number of individuals per 100 m²)																		
		1997	1998	1999	2000	2001	2002	2003	2004	2005	2006	2007	2008	2009	2010	2011	2012	2013	2014	2015
Philippines	Helens Reef	—	0.5	—	—	—	—	—	—	—	—	—	—	—	—	—	—	—	—	—
Philippines	Hidden Beach (Shallow)	—	—	—	—	—	—	—	—	—	0.5	—	—	—	—	—	—	—	—	—
Philippines	Hilantagaan (Outside MPA)	—	—	—	—	—	—	—	—	—	—	—	0.25	—	—	—	—	—	—	—
Philippines	Hilantagaan Diyot	—	—	—	—	—	—	—	—	—	—	—	0.75	—	—	—	—	—	—	—
Philippines	Hilantagaan Diyot MPA	—	—	—	—	—	—	—	—	—	—	—	0.5	—	—	—	—	—	—	—
Philippines	Himokilan	—	—	—	—	—	—	—	—	—	0.25	—	—	—	—	—	—	—	—	—
Philippines	Himokilan (outside MPA)	—	—	—	—	—	—	—	—	—	—	—	—	—	—	—	—	—	—	—
Philippines	Himokilan Marine Sanctuary	—	—	—	—	—	—	—	—	—	0.75	—	0.38	—	—	—	—	—	—	—
Philippines	Inside Talima MPA Site 1	—	—	—	—	—	—	—	—	0.5	—	—	—	—	—	—	—	—	—	—
Philippines	Inside Talima MPA Site 2	—	—	—	—	—	—	—	0.25	0.5	—	—	—	—	—	—	—	—	—	—
Philippines	Ipil MPA (Inside) Brgy. Buena Suerte	—	—	—	—	—	—	—	—	—	—	—	—	—	—	—	—	1.75	—	—
Philippines	Isla Rita Transect 1	—	—	—	—	—	—	—	—	—	—	—	26.25	—	—	—	—	—	—	—
Philippines	Isla Rita Transect 2	—	—	—	—	—	—	—	—	—	—	—	34.75	—	—	—	—	—	—	—
Philippines	Jahikan	—	—	—	—	—	—	—	—	—	0.25	3.25	—	—	—	—	—	—	—	—

Continued

352

Table A4 (Continued)   Global distribution of giant clams (Reef Check)

Country	Reef Site	Monitoring years (total clam density = number of individuals per 100 m²)																		
		1997	1998	1999	2000	2001	2002	2003	2004	2005	2006	2007	2008	2009	2010	2011	2012	2013	2014	2015
Philippines	Jahikan Site 1—Hilantangaan Island	—	—	—	—	—	—	—	—	0.5	—	—	—	—	—	—	—	—	—	—
Philippines	Jahikan Site 2—Hilantangaan Island	—	—	—	—	—	—	—	—	1.0	—	—	—	—	—	—	—	—	—	—
Philippines	Jilatagaan Is, Bantayan, Outside MPA	—	—	—	—	—	—	—	0.75	—	—	—	—	—	—	—	—	—	—	—
Philippines	Jilatagaan MPA Site 2	—	—	—	—	—	—	—	3.0	1.25	1.5	0.25	—	—	—	—	—	—	—	—
Philippines	Juag Southeast	—	—	—	—	—	—	—	—	—	—	—	—	—	—	—	—	0.75	—	—
Philippines	Kakulasian	—	—	—	—	—	—	—	—	—	—	—	—	—	—	—	0.25	—	—	—
Philippines	Kalanggaman 1	—	—	—	—	—	—	—	1.5	—	—	—	—	—	—	—	—	—	—	—
Philippines	Kalanggaman 2	—	—	—	—	—	—	—	—	0.25	—	—	—	—	—	—	—	—	—	—
Philippines	Kalingaw Beach Resort (Barangay Marigondon)	—	—	—	—	—	—	0.75	—	—	—	—	—	—	—	—	—	—	—	—
Philippines	Kasabangan Eastside	—	—	—	—	—	—	—	—	—	3.0	—	—	—	—	—	—	—	—	—
Philippines	Kasabangan North	—	—	—	—	—	—	—	—	—	0.25	—	—	—	—	—	—	—	—	—
Philippines	Kasabangan South	—	—	—	—	—	—	—	—	—	1.5	—	—	—	—	—	—	—	—	—
Philippines	Kawayan	—	—	—	—	—	—	—	—	—	—	—	—	—	—	—	—	—	2.5	—
Philippines	Koala, Bagalangit	—	—	—	—	—	—	—	—	—	—	—	—	—	0.25	—	—	3.75	—	—

Continued

353

Table A4 (Continued) Global distribution of giant clams (Reef Check)

Country	Reef Site	Monitoring years (total clam density = number of individuals per 100 m²)																		
		1997	1998	1999	2000	2001	2002	2003	2004	2005	2006	2007	2008	2009	2010	2011	2012	2013	2014	2015
Philippines	Kontiki	—	—	—	—	—	—	—	—	—	—	0.25	—	—	—	—	—	—	—	—
Philippines	Labangtaytay 1	—	—	—	—	—	—	—	—	—	0.25	—	—	—	—	—	—	—	—	—
Philippines	Lapus Lapus MPA	—	—	—	—	—	—	—	—	1.0	0.25	0.25	—	—	—	—	—	—	—	—
Philippines	Layag Layag Lot 19 West	—	—	—	—	—	—	—	—	—	—	—	—	—	—	—	—	—	0.25	—
Philippines	Liloan Analao	—	—	—	—	—	—	—	—	—	—	—	—	—	2.25	—	—	—	—	—
Philippines	Liloan Reef	—	—	—	—	—	0.25	—	—	—	—	—	—	—	—	—	—	—	—	—
Philippines	Lutoban Reef	—	—	—	—	0.25	—	—	—	—	—	—	—	—	—	—	—	—	—	—
Philippines	Maalequenquen Island	—	—	2.75	—	—	—	—	—	—	—	—	—	—	—	—	—	—	—	—
Philippines	Maapdit	—	—	—	—	—	—	—	—	—	—	—	—	—	—	—	—	0.25	—	—
Philippines	Maasin Island, Bulalacao	—	—	—	—	—	—	—	—	—	—	—	—	—	—	—	—	1.5	—	—
Philippines	Maca Reef 1 VSS Dive 1	—	—	—	—	—	—	—	—	—	2.75	—	—	—	—	—	—	—	—	—
Philippines	Magransing	—	—	—	—	—	—	—	—	—	—	—	—	—	—	—	—	3.0	2.75	—
Philippines	Mahaba Marine Sanctuary	—	—	—	—	—	—	—	—	—	0.25	—	—	—	—	—	—	—	—	—
Philippines	Maitre MPA	—	—	—	—	—	—	—	—	—	—	—	—	—	—	—	—	0.5	0.5	—
Philippines	Maitre MPA (Outside)	—	—	—	—	—	—	—	—	—	—	—	—	—	—	—	—	1.0	—	—
Philippines	Malbago	—	—	—	—	—	—	—	—	—	0.75	—	—	—	—	—	—	—	—	—
Philippines	Maliit na Tapik	—	—	—	—	—	—	—	—	—	—	—	—	—	—	—	—	14.0	7.25	—
Philippines	Manalo MPA (Honda Bay) Transect 1	—	—	—	—	—	—	—	—	—	—	—	0.5	—	—	—	—	—	—	—

Continued

Table A4 (Continued) Global distribution of giant clams (Reef Check)

Country	Reef Site	Monitoring years (total clam density = number of individuals per 100 m^2)																		
		1997	1998	1999	2000	2001	2002	2003	2004	2005	2006	2007	2008	2009	2010	2011	2012	2013	2014	2015
Philippines	Manalo MPA (Honda Bay) Transect 2	—	—	—	—	—	—	—	—	—	—	—	0.75	—	—	—	—	—	—	—
Philippines	Mantaray Reef	—	1.5	—	—	—	—	—	—	—	—	—	—	—	—	—	—	—	—	—
Philippines	Marigondon	—	—	—	—	—	—	—	—	—	—	0.25	—	—	—	—	—	—	—	—
Philippines	Masigasig/Esteban Reef	—	—	—	—	—	—	—	—	—	—	—	—	—	—	—	—	0.5	0.25	—
Philippines	Medicare-MC05	—	—	—	—	—	—	—	—	—	0.25	—	—	—	—	—	—	—	—	—
Philippines	Medicare-MC06	—	—	—	—	—	—	—	—	—	0.08	—	—	—	—	—	—	—	—	—
Philippines	Nalusuan MPA Transect 2	—	—	—	—	—	—	—	—	—	—	—	1.5	—	—	—	—	—	—	—
Philippines	Napantao	—	—	—	—	—	—	—	—	—	—	—	—	—	0.15	—	—	—	—	—
Philippines	Napantao 12	—	—	—	—	—	—	—	—	—	—	—	—	—	0.62	—	—	—	—	—
Philippines	Napantao 9	—	—	—	—	—	—	—	—	—	—	—	—	—	0.75	—	—	—	—	—
Philippines	North Wall	—	—	1.25	—	0.25	—	—	—	—	—	—	—	—	—	—	—	—	—	—
Philippines	Outside Lapus—Lapus MPA Site 1	—	—	—	—	—	—	—	0.5	0.25	—	—	—	—	—	—	—	—	—	—
Philippines	Outside Lapus—Lapus MPA Site 2	—	—	—	—	—	—	—	0.75	0.25	—	—	—	—	—	—	—	—	—	—
Philippines	Panal Reef 1	—	—	—	—	—	—	—	—	—	0.25	—	—	—	—	—	—	—	—	—
Philippines	Panal Reef 2	—	—	—	—	—	—	—	—	—	0.25	—	—	—	—	—	—	—	—	—
Philippines	Pangan-an Islet	—	—	—	—	—	0.5	—	—	—	—	—	—	—	—	—	—	—	—	—
Philippines	Paraiso Reef	—	3.25	—	—	—	—	—	—	—	—	—	—	—	—	—	—	—	—	—

Continued

Table A4 (Continued) Global distribution of giant clams (Reef Check)

Country	Reef Site	Monitoring years (total clam density = number of individuals per 100 m²)																		
		1997	1998	1999	2000	2001	2002	2003	2004	2005	2006	2007	2008	2009	2010	2011	2012	2013	2014	2015
Philippines	Pinagbakahan Central, Barangay Pagkilatan	—	—	—	—	—	—	—	—	—	—	—	—	—	—	—	0.25	—	—	—
Philippines	Plantation Bay	—	—	—	—	—	—	—	—	—	—	—	0.25	—	—	—	—	—	—	—
Philippines	Poblacion	—	—	—	—	—	—	—	—	—	0.25	—	0.25	—	—	—	—	—	—	—
Philippines	Poblacion East	—	—	—	—	—	—	—	—	0.75	—	2.25	—	—	—	—	—	—	—	—
Philippines	Poblacion West	—	—	—	—	—	—	—	—	0.5	0.75	17.25	—	—	—	—	—	—	—	—
Philippines	Pooc MPA	—	—	—	—	—	—	—	—	—	—	—	1.25	—	—	—	—	—	—	—
Philippines	Portulano Marine Sanctuary	—	—	—	—	—	—	—	—	—	—	—	—	—	—	—	—	—	0.75	—
Philippines	Puntod Ilis	—	—	—	—	—	—	—	—	—	—	—	0.25	—	—	—	—	—	—	—
Philippines	Puting Buhangin	—	—	—	—	—	—	—	—	—	—	—	0.75	—	—	—	—	—	—	—
Philippines	Rakit-Rakit Reef	—	—	—	—	—	—	—	—	—	—	—	—	—	—	—	—	1.0	—	—
Philippines	Rawis	—	—	—	—	—	—	—	—	—	—	—	—	—	—	—	—	0.5	0.5	—
Philippines	Rizal Site 2	—	—	—	—	—	—	—	—	—	0.25	—	—	—	—	—	—	—	—	—
Philippines	Saavedra	—	—	—	—	—	—	—	—	—	0.25	0.25	—	—	—	—	—	—	—	—
Philippines	Saavedra Site 2	—	—	—	—	—	—	—	—	0.25	—	—	—	—	—	—	—	—	—	—
Philippines	San Diego North	—	—	—	—	—	—	—	—	0.75	—	—	—	—	—	—	—	—	—	—
Philippines	San Diego Station A	—	—	—	—	—	—	—	—	—	—	—	—	—	—	—	0.5	—	—	—
Philippines	San Isidro MPA	—	—	—	—	—	—	—	—	—	—	0.75	—	—	—	—	—	—	—	—
Philippines	San Isidro-Dao MPA	—	—	—	—	—	—	—	—	1.25	0.25	—	—	—	—	—	—	—	—	—
Philippines	San Miguel MPA	—	—	—	—	—	—	—	—	—	—	—	—	—	—	—	—	0.38	—	—
Philippines	Santelmo North	—	—	—	—	—	—	—	—	—	—	—	—	—	—	—	—	—	0.25	—

Continued

GIANT CLAMS (BIVALVIA: CARDIIDAE: TRIDACNINAE)

Table A4 (Continued) Global distribution of giant clams (Reef Check)

Country	Reef Site	Monitoring years (total clam density = number of individuals per 100 m²)																		
		1997	1998	1999	2000	2001	2002	2003	2004	2005	2006	2007	2008	2009	2010	2011	2012	2013	2014	2015
Philippines	Sawang Gamay, E side of proposed Caubian Dako MPA	—			—	—	—	—	1.25	—	—	—	—	—	—	—	—	—	—	—
Philippines	Secret Garden	—	—	—	—	—	—	0.5	—	—	—	—	—	—	—	—	—	—	—	—
Philippines	Shangri-La	—	—	—	—	—	—	0.25	—	—	—	—	—	—	—	—	—	—	—	—
Philippines	Siete Pecados Islands	—	—	—	—	—	—	—	1.0	—	—	—	—	—	—	—	—	—	—	—
Philippines	Silad Island	—	—	—	—	—	—	—	—	—	—	—	—	—	—	—	—	0.5	—	—
Philippines	Sillon	—	—	—	—	—	—	—	—	—	—	3.0	—	—	—	—	—	—	—	—
Philippines	Sillon Site 1	—	—	—	—	—	—	—	—	0.25	—	—	—	—	—	—	—	—	—	—
Philippines	Sillon Site 2	—	—	—	—	—	—	—	—	0.5	—	—	—	—	—	—	—	—	—	—
Philippines	Silonay Island 1A	—	—	—	—	—	—	—	—	—	0.5	—	—	—	—	—	—	—	—	—
Philippines	Sigayan 02	—	—	—	—	—	—	—	—	—	—	—	—	—	—	—	—	—	0.25	—
Philippines	Sitio Pinagbakahan, Barangay Pagkilatan	—	—	—	—	—	—	—	—	—	—	—	—	—	0.75	—	—	—	—	—
Philippines	Susan's Reef, Bantayan	—	—	—	—	—	—	—	0.5	0.75	0.25	0.75	—	—	—	—	—	—	—	—
Philippines	Tabalong	—	—	—	—	—	—	—	—	—	—	0.25	—	—	—	—	—	—	—	—
Philippines	Tagbac Sanctuary	—	—	—	—	—	—	1.5	—	—	—	—	—	—	—	—	—	—	—	—
Philippines	Talangnan Site 2	—	—	—	—	—	—	—	—	0.25	—	—	—	—	—	—	—	—	—	—
Philippines	Talim Outer Reef	—	—	—	—	—	—	—	—	—	—	—	—	—	—	1.25	—	—	1.0	—
Philippines	Talima D1	—	—	—	—	—	—	—	0.25	—	—	0.25	—	—	—	—	—	—	—	—

Continued

357

Table A4 (Continued) Global distribution of giant clams (Reef Check)

Country	Reef Site																			
		1997	1998	1999	2000	2001	2002	2003	2004	2005	2006	2007	2008	2009	2010	2011	2012	2013	2014	2015
Philippines	Talisay Tree, Cabilao Island	—	—	—	—	—	—	—	—	—	—	—	—	—	—	—	0.25	—	0.25	—
Philippines	Tambuli	—	—	—	—	—	0.12	0.25	—	—	—	—	—	—	—	—	—	—	—	—
Philippines	Tandol Reef	—	—	—	—	—	—	—	—	—	—	—	—	—	—	—	—	4.0	7.0	—
Philippines	Tanglaw	—	—	—	—	—	—	—	—	—	—	—	—	—	—	—	—	13.75	7.25	—
Philippines	Tapik Centro	—	—	—	—	—	—	—	—	—	—	—	—	—	—	—	—	0.25	0.5	—
Philippines	Tingo (shallow)	—	—	—	—	—	—	—	—	0.25	—	—	—	—	—	—	—	—	—	—
Philippines	Tomonoy	—	—	—	—	—	—	—	—	3.0	0.5	4.25	—	—	—	—	—	—	—	—
Philippines	Tongo	—	—	—	—	—	—	—	—	—	0.5	—	—	—	—	—	—	—	—	—
Philippines	Tony's Reef	—	—	—	—	—	—	—	—	—	—	10.25	—	—	—	—	—	—	—	—
Philippines	Tony's Reef, Jilatagaan Is, Bantayan	—	—	—	—	—	—	—	0.5	3.25	9.5	—	—	—	—	—	—	—	—	—
Philippines	Tres Marias	—	—	—	—	—	—	—	—	—	—	—	—	—	—	—	—	—	1.0	—
Philippines	Twinpeaks	—	—	—	—	—	—	—	—	0.5	—	—	—	—	—	—	—	—	—	—
Philippines	Very West Caubian Dako	—	—	—	—	—	—	—	0.5	—	—	—	—	—	—	—	—	—	—	—
Philippines	White Beach, Puerto Princesa Bay	—	—	—	—	—	—	—	—	—	—	—	—	—	—	—	—	0.25	—	—
Philippines	White Sand Island Sanctuary	—	—	—	—	—	—	—	—	—	—	—	—	—	—	—	—	0.5	0.5	—
Saudi Arabia	Allith	—	—	—	—	—	—	—	—	—	—	—	—	0.13	—	—	—	—	—	—
Saudi Arabia	Amaq-Hali	—	—	—	—	—	—	—	—	—	—	—	—	1.5	—	—	—	—	—	—
Saudi Arabia	Channel Slope	—	—	0.25	—	—	—	—	—	—	—	—	—	—	—	—	—	—	—	—
Saudi Arabia	Chornich	—	—	—	—	—	—	—	—	—	—	—	4.5	—	—	—	—	—	—	—
Saudi Arabia	Duba (Cement Tabouk)	—	—	—	—	—	—	—	—	—	—	—	5.75	—	—	—	—	—	—	—

Continued

Table A4 (Continued)   Global distribution of giant clams (Reef Check)

Country	Reef Site	\multicolumn{19}{c}{Monitoring years (total clam density = number of individuals per 100 m²)}																	
		1997	1998	1999	2000	2001	2002	2003	2004	2005	2006	2007	2008	2009	2010	2011	2012	2013	2014	2015
Saudi Arabia	Farasan-Zfaf	—	—	—	—	—	—	—	—	—	—	—	—	0.5	—	—	—	—	—	—
Saudi Arabia	Hagal (Dora)	—	—	—	—	—	—	—	—	—	—	—	1.25	—	—	—	—	—	—	—
Saudi Arabia	Inner patch, J. Umm Rumah	—	—	66.25	—	—	—	—	—	—	—	—	—	—	—	—	—	—	—	—
Saudi Arabia	J. Qumma'an fringing, E	—	—	1.75	—	—	—	—	—	—	—	—	—	—	—	—	—	—	—	—
Saudi Arabia	J. Qumma'an fringing, SW	—	—	1.75	—	—	—	—	—	—	—	—	—	—	—	—	—	—	—	—
Saudi Arabia	J. Qumma'an fringing, W	—	—	0.25	—	—	—	—	—	—	—	—	—	—	—	—	—	—	—	—
Saudi Arabia	J. Shaybara barrier	—	—	2.75	—	—	—	—	—	—	—	—	—	—	—	—	—	—	—	—
Saudi Arabia	J. Shaybara inner	—	—	21.63	—	—	—	—	—	—	—	—	—	—	—	—	—	—	—	—
Saudi Arabia	Magna	—	—	—	—	—	—	—	—	—	—	—	1.38	—	—	—	—	—	—	—
Saudi Arabia	Masturah	—	—	—	—	—	—	—	—	—	—	—	1.63	—	—	—	—	—	—	—
Saudi Arabia	Mid-Bank Patch	—	—	1.5	—	—	—	—	—	—	—	—	—	—	—	—	—	—	—	—
Saudi Arabia	Mid-Bank slope, outside channel	—	—	27.0	—	—	—	—	—	—	—	—	—	—	—	—	—	—	—	—
Saudi Arabia	Outer barrier, central bank	—	—	1.75	—	—	—	—	—	—	—	—	—	—	—	—	—	—	—	—
Saudi Arabia	Outer barrier, J. Mizab	—	—	0.75	—	—	—	—	—	—	—	—	—	—	—	—	—	—	—	—
Saudi Arabia	Outer patch SW of J. Jusur Shurayrat	—	—	1.0	—	—	—	—	—	—	—	—	—	—	—	—	—	—	—	—
Saudi Arabia	Outer-Bank Patch	—	—	3.63	—	—	—	—	—	—	—	—	—	—	—	—	—	—	—	—
Saudi Arabia	Patch NE J. Qumma'an	—	—	0.5	—	—	—	—	—	—	—	—	—	—	—	—	—	—	—	—

Continued

Table A4 (Continued) Global distribution of giant clams (Reef Check)

Country	Reef Site	Monitoring years (total clam density = number of individuals per 100 m^2)																		
		1997	1998	1999	2000	2001	2002	2003	2004	2005	2006	2007	2008	2009	2010	2011	2012	2013	2014	2015
Saudi Arabia	Umlajj	—	—	—	—	—	—	—	—	—	—	—	34.0	—	—	—	—	—	—	—
Seychelles	Baie Ternay	0.25	—	—	—	—	—	—	—	—	—	—	—	—	—	—	—	—	—	—
Seychelles	Big Sister Island	0.25	—	—	—	—	—	—	—	—	—	—	—	—	—	—	—	—	—	—
Seychelles	Corsair Reef	0.25	—	—	—	—	—	—	—	—	—	—	—	—	—	—	—	—	—	—
Seychelles	Danzil Reef	0.25	—	—	—	—	—	—	—	—	—	—	—	—	—	—	—	—	—	—
Seychelles	Farquar	—	—	—	—	2.5	—	—	—	—	—	—	—	—	—	—	—	—	—	—
Seychelles	Turtle Reef	1.0	—	—	—	—	—	—	—	—	—	—	—	—	—	—	—	—	—	—
Solomon Islands	Field Station 1	—	—	—	—	—	—	—	—	—	—	—	0.75	3.0	—	—	—	—	—	—
Solomon Islands	Inside 5 yr MPA shallow	—	—	—	—	—	—	—	—	—	—	—	—	—	—	0.62	—	—	—	—
Solomon Islands	Inside Perm MPA shallow	—	—	—	—	—	—	—	—	—	—	—	—	—	—	0.25	—	—	—	—
Solomon Islands	Mbo	—	—	—	—	—	—	—	—	—	—	—	0.25	0.5	—	—	—	—	—	—
Solomon Islands	MLPP01	—	—	—	—	—	—	—	—	—	—	—	—	—	—	—	0.75	—	—	—
Solomon Islands	MLPP02	—	—	—	—	—	—	—	—	—	—	—	—	—	—	—	0.25	—	—	—
Solomon Islands	MLPT01	—	—	—	—	—	—	—	—	—	—	—	—	—	—	—	0.25	—	—	—
Solomon Islands	MLPT02	—	—	—	—	—	—	—	—	—	—	—	—	—	—	—	0.5	—	—	—
Solomon Islands	MLPT04	—	—	—	—	—	—	—	—	—	—	—	—	—	—	—	0.75	—	—	—
Solomon Islands	MLPT05	—	—	—	—	—	—	—	—	—	—	—	—	—	—	—	1.5	—	—	—
Solomon Islands	MLPT06	—	—	—	—	—	—	—	—	—	—	—	—	—	—	—	0.25	—	—	—

Continued

Table A4 (Continued)   Global distribution of giant clams (Reef Check)

Country	Reef Site	Monitoring years (total clam density = number of individuals per 100 m²)																		
		1997	1998	1999	2000	2001	2002	2003	2004	2005	2006	2007	2008	2009	2010	2011	2012	2013	2014	2015
Solomon Islands	No. 5, Site 1	—	—	—	—	—	—	—	—	0.75	—	1.25	1.0	1.5	—	—	—	—	—	—
Solomon Islands	No. 6	—	—	—	—	—	—	—	—	1.5	—	1.25	—	1.25	—	—	—	—	—	—
Solomon Islands	Number 4	—	—	—	—	—	—	—	—	—	—	—	—	0.12	—	—	—	—	—	—
Solomon Islands	Perm MPA shallow inside	—	—	—	—	—	—	—	—	—	—	—	—	—	—	0.25	—	—	—	—
Solomon Islands	Plantation	—	—	—	—	—	—	—	—	0.25	—	0.25	—	—	—	—	—	—	—	—
Solomon Islands	Sanbis Reef	—	—	—	—	—	—	—	—	2.25	—	—	—	—	—	—	—	—	—	—
Solomon Islands	Singi 1	—	—	—	—	—	—	—	—	—	0.5	0.25	—	0.25	—	—	—	—	—	—
Solomon Islands	Singi 3	—	—	—	—	—	—	—	—	3.25	2.5	—	—	—	—	—	—	—	—	—
Solomon Islands	Soe	—	—	—	—	—	—	—	—	2.5	—	—	—	—	—	—	—	—	—	—
Solomon Islands	Station 1	—	—	—	—	—	—	—	—	1.25	0.25	—	—	—	—	—	—	—	—	—
Solomon Islands	Station 2	—	—	—	—	—	—	—	—	1.75	0.62	—	2.5	0.62	—	—	—	—	—	—
Solomon Islands	Tehakatu'u	—	—	—	—	—	—	—	—	—	—	—	—	15.5	—	—	—	—	—	—
Solomon Islands	Tuo Village Reef	—	—	—	—	—	—	—	—	—	—	—	—	—	0.38	—	—	—	—	—
Solomon Islands	TUOO01	—	—	—	—	—	—	—	—	—	—	—	—	—	—	—	1.08	—	—	—
Solomon Islands	TUOO02	—	—	—	—	—	—	—	—	—	—	—	—	—	—	—	0.42	—	—	—

Continued

Table A4 (Continued) Global distribution of giant clams (Reef Check)

Country	Reef Site	Monitoring years (total clam density = number of individuals per 100 m²)																		
		1997	1998	1999	2000	2001	2002	2003	2004	2005	2006	2007	2008	2009	2010	2011	2012	2013	2014	2015
Solomon Islands	TUOO3	—	—	—	—	—	—	—	—	—	—	—	—	—	—	—	1.0	—	—	—
Solomon Islands	TUOO4	—	—	—	—	—	—	—	—	—	—	—	—	—	—	—	0.33	—	—	—
Solomon Islands	TUOP01	—	—	—	—	—	—	—	—	—	—	—	—	—	—	—	1.25	—	—	—
Solomon Islands	TUOP02	—	—	—	—	—	—	—	—	—	—	—	—	—	—	—	0.75	—	—	—
Solomon Islands	TUOP03	—	—	—	—	—	—	—	—	—	—	—	—	—	—	—	2.67	—	—	—
Solomon Islands	TUOT01	—	—	—	—	—	—	—	—	—	—	—	—	—	—	—	0.83	—	—	—
Solomon Islands	TUOT02	—	—	—	—	—	—	—	—	—	—	—	—	—	—	—	1.08	—	—	—
Solomon Islands	TUOT03	—	—	—	—	—	—	—	—	—	—	—	—	—	—	—	2.08	—	—	—
South Africa	2 Mile Reef	—	—	—	—	0.25	—	—	—	—	—	—	—	—	—	—				
South Africa	4 Buoy Reef	—	—	—	—	0.25	—	—	—	—	—	—	—	—	—	—				
South Africa	Alliwal Shoal	—	—	—	1.0	—	—	—	—	—	—	—	—	—	—	—				
South Africa	Central Two-Mile Reef	—	—	—	—	—	—	—	—	1.0	—	—	—	—	—	—				
South Africa	Inner Central Two-Mile Reef	—	—	—	—	—	—	—	—	1.5	—	—	—	—	—	—				
South Africa	Limestone Reef	—	—	—	—	—	0.25	—	—	—	—	—	—	—	—	—				
South Africa	Raggie Cave	—	—	—	—	0.25	—	—	—	—	—	—	—	—	—	—				
Sri Lanka	Coral Island	—	—	—	—	—	—	0.25	—	—	—	—	—	—	—	—				
Sri Lanka	Pigeon Island	—	—	—	—	—	—	1.75	—	—	—	—	—	—	—	—				
Sudan	Abu Hashish	—	—	—	—	—	—	—	—	—	—	—	—	2.25	—	—				
Sudan	Arkiyai	—	—	—	—	—	—	—	—	—	—	—	—	6.88	—	—				

Continued

Table A4 (Continued) Global distribution of giant clams (Reef Check)

Country	Reef Site	Monitoring years (total clam density = number of individuals per 100 m²)																		
		1997	1998	1999	2000	2001	2002	2003	2004	2005	2006	2007	2008	2009	2010	2011	2012	2013	2014	2015
Sudan	Bashayer Marine Terminal	—	—	—	—	—	—	—	2.25	—	—	—	—	—	—	—	—	—	—	—
Sudan	Damadma Fringing Reef	—	—	—	—	—	—	—	0.33	—	—	—	—	—	—	—	—	—	—	—
Sudan	Falamingo Fringing North	—	—	—	—	—	—	—	1.5	—	—	—	—	—	—	—	—	—	—	—
Sudan	Gota Wingate North	—	—	—	—	—	—	—	3.5	—	—	—	—	—	—	—	—	—	—	—
Sudan	O'seif Bay	—	—	—	—	—	—	—	—	—	—	—	—	3.38	—	—	—	—	—	—
Sudan	Sanganeb South-west	—	—	—	—	—	—	—	1.25	—	—	—	—	—	—	—	—	—	—	—
Sudan	Suakin	—	—	—	—	—	—	—	—	—	—	—	—	0.5	—	—	—	—	—	—
Sudan	Wingate Reef-Police Station	—	—	—	—	—	—	—	4.5	—	—	—	—	—	—	—	—	—	—	—
Sudan	Winget Barrier Reef	—	—	—	—	—	—	—	14.63	—	—	—	—	—	—	—	—	—	—	—
Taiwan	Beauty Cave	—	—	—	—	—	—	—	—	—	—	—	—	1.0	1.0	—	—	—	—	—
Taiwan	Centre Sanyuan	—	—	—	—	—	—	—	—	—	—	—	—	—	—	—	—	—	—	—
Taiwan	ChaiKou	—	—	—	—	—	—	—	—	—	—	—	—	1.0	1.25	—	—	—	—	—
Taiwan	Fanzaiao	—	—	—	—	—	—	—	—	—	—	—	—	—	0.25	—	—	—	—	—
Taiwan	Gateway Rock	—	—	—	—	—	—	—	—	—	—	—	1.5	—	—	—	—	—	—	—
Taiwan	Geban Bay	—	—	—	—	—	—	—	—	—	—	—	—	1.25	0.5	—	—	—	—	—
Taiwan	GeeChang	—	5.0	—	—	—	—	—	—	—	—	—	—	—	—	—	—	—	—	—
Taiwan	General Rock	—	—	—	—	—	—	—	—	—	—	—	—	—	1.0	—	—	—	—	—
Taiwan	Gong-guan	—	—	—	—	—	—	—	—	—	—	—	—	—	4.5	—	—	—	—	—
Taiwan	Haishen Flats	—	—	—	—	—	—	—	—	—	—	—	2.25	—	—	—	—	—	—	—
Taiwan	Hongtoe	—	2.25	—	—	—	—	—	—	—	—	—	—	—	—	—	—	—	—	—
Taiwan	Houshi	—	—	—	—	—	—	—	—	—	—	—	—	—	0.25	—	—	—	—	—

Continued

Table A4 (Continued)   Global distribution of giant clams (Reef Check)

Country	Reef Site	Monitoring years (total clam density = number of individuals per 100 m²)																		
		1997	1998	1999	2000	2001	2002	2003	2004	2005	2006	2007	2008	2009	2010	2011	2012	2013	2014	2015
Taiwan	Houshi Fringing Reef	—	—	—	—	—	—	—	—	—	—	—	—	0.25	—	—	—	—	—	—
Taiwan	Kungkuan	—	—	—	—	—	—	—	—	—	—	—	—	4.75	—	—	—	—	—	—
Taiwan	Lion Couple Rock	—	—	—	—	—	—	—	—	—	—	—	—	—	2.0	—	—	—	—	—
Taiwan	MeiRenDong	—	0.5	—	—	—	—	—	—	—	—	—	—	—	—	—	—	—	—	—
Taiwan	Nanliao	—	0.75	—	—	—	—	—	—	—	—	—	—	—	—	—	—	—	—	—
Taiwan	Reef Outside Airport	—	—	—	—	—	—	—	—	—	—	—	—	1.5	—	—	—	—	—	—
Taiwan	South Dongyuping	—	—	—	—	—	—	—	—	—	—	—	—	0.25	—	—	—	—	—	—
Taiwan	South Shanyuan Bay	—	—	—	—	—	—	—	—	—	—	—	—	—	0.25	—	—	—	—	—
Taiwan	Tsaikou	—	0.75	—	—	—	—	—	—	—	—	—	—	—	—	—	—	—	—	—
Taiwan	Tudigong Temple	—	—	—	—	—	—	—	—	—	—	—	—	0.75	—	—	—	—	—	—
Taiwan	Virgin Rock West	—	—	—	—	—	—	—	—	—	—	—	—	—	1.75	—	—	—	—	—
Taiwan	West Dongyuping	—	—	—	—	—	—	—	—	—	—	—	—	0.25	—	—	—	—	—	—
Taiwan	Yeyou	—	—	—	—	—	—	—	—	—	—	—	—	0.75	—	—	—	—	—	—
Taiwan	Yie-yin Village	—	—	—	—	—	—	—	—	—	—	—	—	—	0.75	—	—	—	—	—
Tanzania	Chumba Cha Chumbo	—	2.75	—	—	—	—	—	—	—	—	—	—	—	—	—	—	—	—	—
Tanzania	Chumbe Island Coral Park	—	—	—	—	—	—	—	2.08	—	—	—	—	—	—	—	—	—	—	—
Tanzania	Fungu Zinga Reef North	—	—	—	—	—	—	0.5	0.25	—	0.5	—	0.25	—	—	—	—	—	—	—
Tanzania	Fungu Zinga Reef South	—	—	—	—	—	—	—	—	0.25	1.25	—	0.25	—	—	—	—	—	—	—

Continued

364

Table A4 (Continued) Global distribution of giant clams (Reef Check)

Country	Reef Site	Monitoring years (total clam density = number of individuals per 100 m²)																		
		1997	1998	1999	2000	2001	2002	2003	2004	2005	2006	2007	2008	2009	2010	2011	2012	2013	2014	2015
Tanzania	Mazivwe Reef North	—	—	—	—	—	—	—	1.5	2.0	1.5	1.75	0.25	—	—	—	—	—	—	—
Tanzania	Mazivwe Reef South	—	—	—	—	—	—	—	—	0.25	—	0.5	0.25	—	—	—	—	—	—	—
Tanzania	Mwan wa Mwana, near Tumbatu Island	—	—	—	—	—	—	—	1.5	—	—	—	—	—	—	—	—	—	—	—
Tanzania	Ras Msimbati	3.25	—	—	—	—	—	—	—	—	—	—	—	—	—	—	—	—	—	—
Tanzania	Ravula	—	0.75	—	—	—	—	—	—	—	—	—	—	—	—	—	—	—	—	—
Tanzania	The Gap	—	0.5	—	—	—	—	—	—	—	—	—	—	—	—	—	—	—	—	—
Thailand	Ao Chong Kaad	—	—	—	—	—	—	—	—	6.25	—	—	—	—	—	—	—	—	—	—
Thailand	Ao Luek South Site	—	—	7.25	—	—	—	—	—	—	—	—	—	—	—	—	—	—	—	—
Thailand	Ao Mae Yai	—	—	—	—	—	—	—	—	3.0	—	—	—	—	—	—	—	—	—	—
Thailand	Ao Tao 2	—	—	—	—	—	—	—	—	1.5	—	—	—	—	—	—	—	—	—	—
Thailand	Aow Keuk, Koh Tao	—	—	—	—	—	—	—	—	—	—	—	7.75	—	—	—	—	—	—	—
Thailand	Aow Leuk	—	—	—	—	—	—	—	—	—	—	—	3.0	—	—	4.25	—	—	—	—
Thailand	Bida Nog Island	—	—	—	—	—	—	—	—	—	2.5	—	—	—	—	—	—	—	3.5	—
Thailand	Boulder City	—	—	0.25	—	—	—	—	—	—	—	—	—	—	—	—	—	—	—	—
Thailand	Hin Kong	—	—	—	—	—	—	—	—	1.5	—	—	—	—	—	—	—	—	—	—
Thailand	Japanese Gardens, Koh Tao	—	—	—	—	—	—	—	—	—	—	—	—	3.5	—	—	—	—	—	—
Thailand	Ko Bai Dang	—	—	—	—	—	—	—	—	—	1.5	—	—	—	—	—	—	—	—	—
Thailand	Ko Bai Dang North	—	—	—	—	—	—	—	—	—	1.25	—	—	—	—	—	—	—	—	—
Thailand	Ko Khang	—	—	—	—	—	—	—	—	0.25	—	—	—	—	—	—	—	—	—	—
Thailand	Ko Khlum	—	—	—	—	—	—	—	—	—	0.25	—	—	—	—	—	—	—	—	—

Continued

Table A4 (Continued) Global distribution of giant clams (Reef Check)

Country	Reef Site	Monitoring years (total clam density = number of individuals per 100 m^2)																		
		1997	1998	1999	2000	2001	2002	2003	2004	2005	2006	2007	2008	2009	2010	2011	2012	2013	2014	2015
Thailand	Ko Torinla	—	—	—	—	—	—	—	—	0.25	—	—	—	—	—	—	—	—	—	—
Thailand	Koh Butang (East)	—	—	—	—	—	—	—	—	—	13.25	—	—	—	—	—	—	—	—	—
Thailand	Koh Butang (South)	—	—	—	—	—	—	—	—	—	20.25	—	—	—	—	—	—	—	—	—
Thailand	Koh Door East	—	—	—	—	—	—	—	—	—	0.25	—	—	—	—	—	—	—	—	—
Thailand	Koh Jorakeh	—	—	—	—	—	—	—	—	—	1.5	—	—	—	—	—	—	—	—	—
Thailand	Koh Jorakeh East	—	—	—	—	—	—	—	—	—	2.0	—	—	—	—	—	—	—	—	—
Thailand	Koh Joung (lower)	—	—	—	—	2.5	—	—	—	—	—	—	—	—	—	—	—	—	—	—
Thailand	Koh Joung (upper)	—	—	—	—	0.5	—	—	—	—	—	—	—	—	—	—	—	—	—	—
Thailand	Koh Kata (South)	—	—	—	—	—	—	—	—	—	18.0	—	—	—	—	—	—	—	—	—
Thailand	Koh Khai Nok	—	—	—	1.75	—	—	—	—	—	—	—	—	—	—	—	—	—	—	—
Thailand	Koh Kood	—	—	—	—	—	—	—	—	—	1.25	—	—	—	—	—	—	—	—	—
Thailand	Koh Kra	—	—	—	—	—	—	—	—	—	14.25	—	—	—	—	—	—	—	—	—
Thailand	Koh Lan	—	—	—	—	—	—	—	—	—	—	—	—	0.08	—	—	—	—	—	—
Thailand	Koh Lom	—	—	—	—	—	—	—	—	—	2.0	—	—	—	—	—	—	—	—	—
Thailand	Koh Man Wichai	—	—	—	0.75	—	—	—	—	—	—	—	—	—	—	—	—	—	—	—
Thailand	Koh Mapring	—	—	—	—	—	—	—	—	—	2.5	—	—	—	—	—	—	—	—	—
Thailand	Koh Mattra	—	—	—	—	—	—	—	—	—	1.25	—	—	—	—	—	—	—	—	—
Thailand	Koh Nangyuan Stretch	—	—	—	—	—	—	—	9.0	—	—	—	—	—	—	—	—	—	—	—
Thailand	Koh Ngam Yai	—	—	—	—	—	—	—	—	—	6.25	—	—	—	—	—	—	—	—	—
Thailand	Koh Payang	—	—	—	—	1.0	—	—	—	—	—	—	—	—	—	—	—	—	—	—
Thailand	Koh Payu	—	—	—	—	0.75	—	—	—	—	—	—	—	—	—	—	—	—	—	—

Continued

Table A4 (Continued) Global distribution of giant clams (Reef Check)

Country	Reef Site	Monitoring years (total clam density = number of individuals per 100 m²)																		
		1997	1998	1999	2000	2001	2002	2003	2004	2005	2006	2007	2008	2009	2010	2011	2012	2013	2014	2015
Thailand	Koh Payu- North East (lower)	—	—	—	—	1.0	—	—	—	—	—	—	—	—	—	—	—	—	—	—
Thailand	Koh Pu	—	—	—	3.25	—	—	—	—	—	—	—	—	—	—	—	—	—	—	—
Thailand	Koh Raya Yai	—	0.75	—	—	—	—	—	—	—	—	—	—	—	—	—	—	—	—	—
Thailand	Koh Raya Yai- Staghorn Reef	—	—	—	2.0	—	—	—	—	—	—	—	—	—	—	—	—	—	—	—
Thailand	Koh Thong Lang	—	—	—	—	—	—	—	—	—	3.25	7.5	—	—	—	—	—	—	—	—
Thailand	Koh Tien/Koh Thain West	—	—	—	—	—	—	—	—	—	—	5.0	4.5	—	—	—	—	—	—	—
Thailand	Koh Yak	—	—	—	—	—	—	—	—	—	—	2.0	—	—	0.75	—	—	—	—	—
Thailand	Koh Yak Lek	—	—	—	—	—	—	—	—	—	1.25	—	—	—	—	—	—	—	—	—
Thailand	Koh Yak Yai	—	—	—	—	—	—	—	—	—	1.5	—	—	—	—	—	—	—	—	—
Thailand	Koh-Huyong	—	—	—	—	0.25	—	—	—	—	—	—	—	—	—	—	—	—	—	—
Thailand	KuekBay (inner)	—	—	—	—	0.75	—	—	—	—	—	—	—	—	—	—	—	—	—	—
Thailand	Leum Island (Ko Luam)	—	—	—	—	—	—	0.5	—	—	—	—	—	—	—	—	—	—	—	—
Thailand	Lighthouse Bay	—	—	—	—	—	—	—	—	—	—	—	5.0	—	5.25	—	—	—	5.0	—
Thailand	Loh Samah Bay	—	—	11.0	—	—	—	—	—	—	—	—	—	—	—	—	—	—	—	—
Thailand	Mae Haad Reef	—	—	—	—	—	—	—	—	—	—	—	2.5	—	2.08	—	—	0.62	1.63	—
Thailand	Maeyai Bay (02)	—	—	—	—	2.0	—	—	—	—	—	—	—	—	—	—	—	—	—	—
Thailand	Maikhao	—	—	—	—	—	—	—	—	—	—	—	—	—	—	—	—	—	—	—
Thailand	Mango Bay	—	—	—	—	—	—	—	—	—	—	3.25	3.17	2.0	2.44	1.5	—	1.5	1.0	—
Thailand	Mango Bay (Aow Mamuang)	—	—	2.75	—	—	—	—	—	—	—	—	—	—	—	—	—	—	—	—
Thailand	Middle Ao-Leuk	—	—	—	—	2.75	—	—	—	—	—	—	—	—	—	—	—	—	—	—
Thailand	Moskito Island	—	—	—	—	—	—	—	—	—	1.25	—	—	—	—	—	—	—	—	—
Thailand	North Ao-Leuk	—	—	—	—	1.75	—	—	—	—	—	—	—	—	—	—	—	—	—	—

Continued

Table A4 (Continued) Global distribution of giant clams (Reef Check)

Country	Reef Site	Monitoring years (total clam density = number of individuals per 100 m^2)																		
		1997	1998	1999	2000	2001	2002	2003	2004	2005	2006	2007	2008	2009	2010	2011	2012	2013	2014	2015
Thailand	North East Similan (01)	—	—	—	—	0.25	—	—	—	—	—	—	—	—	—	—	—	—	—	—
Thailand	North East Similan (02)	—	—	—	—	0.25	—	—	—	—	—	—	—	—	—	—	—	—	—	—
Thailand	North Maeyai (outer)	—	—	—	—	2.5	—	—	—	—	—	—	—	—	—	—	—	—	—	—
Thailand	North Maeyai Bay (inner)	—	—	—	—	16.75	—	—	—	—	—	—	—	—	—	—	—	—	—	—
Thailand	North Maeyai Bay (outer)	—	—	—	—	4.5	—	—	—	—	—	—	—	—	—	—	—	—	—	—
Thailand	North Patong-shallow	—	—	—	—	0.25	—	—	—	—	—	—	—	—	—	—	—	—	—	—
Thailand	North Sai Ree Beach	—	—	—	—	—	—	—	3.0	—	—	—	—	—	—	—	—	—	—	—
Thailand	North-Koh-Yawasam	—	—	—	—	0.25	—	—	—	—	—	—	—	—	—	—	—	—	—	—
Thailand	Patong-south	—	—	—	—	0.75	—	—	—	—	—	—	—	—	—	—	—	—	—	—
Thailand	Racha Yai, Bungalow Bay	—	—	—	0.75	—	—	—	—	—	—	—	—	—	—	—	—	—	—	—
Thailand	Racha-Yai-East (01)	—	—	—	—	10.0	—	—	—	—	—	—	—	—	—	—	—	—	—	—
Thailand	Racha-Yai-East (02)	—	—	—	—	1.5	—	—	—	—	—	—	—	—	—	—	—	—	—	—
Thailand	Racha-Yai-North	—	—	—	—	1.0	—	—	—	—	—	—	—	—	—	—	—	—	—	—
Thailand	Saien Bay	—	—	—	—	1.25	—	—	—	—	—	—	—	—	—	—	—	—	—	—
Thailand	Sairee, Koh Tao	—	—	—	—	—	—	—	—	—	—	—	—	8.0	—	3.0	—	—	—	—
Thailand	Scubacat Bay/ Racha Yai East #1	—	—	—	—	—	—	—	—	—	66.5	64.75	94.5	—	104.25	229.5	—	—	—	—

Continued

368

Table A4 (Continued) Global distribution of giant clams (Reef Check)

Country	Reef Site	Monitoring years (total clam density = number of individuals per 100 m²)																		
		1997	1998	1999	2000	2001	2002	2003	2004	2005	2006	2007	2008	2009	2010	2011	2012	2013	2014	2015
Thailand	Shark Island	—	—	2.75	—	—	—	—	—	—	—	—	—	—	—	—	—	—	—	—
Thailand	South Ao Leuk	—	—	—	—	2.75	—	—	—	—	—	—	—	—	—	—	—	—	—	—
Thailand	South East South Surin	—	—	—	—	2.5	—	—	—	—	—	—	—	—	—	—	—	—	—	—
Thailand	South Maeyai Bay	—	—	—	—	4.5	—	—	—	—	—	—	—	—	—	—	—	—	—	—
Thailand	Suthep Bay	—	—	—	—	0.5	—	—	—	—	—	—	—	—	—	—	—	—	—	—
Thailand	Tao Bay	—	—	—	—	8.25	—	—	—	—	—	—	—	—	—	—	—	—	—	—
Thailand	Tanote Bay	—	—	—	—	—	—	—	—	—	—	—	—	—	—	—	—	—	9.5	—
Thailand	Tonsai Bay West	—	—	—	—	—	—	—	—	6.5	6.75	—	—	—	—	—	—	—	—	—
Thailand	Torinla	—	—	—	—	0.75	—	—	—	—	—	—	—	—	—	—	—	—	—	—
Thailand	Twin Peaks	—	—	—	—	—	—	—	—	—	—	—	—	—	—	—	0.5	2.75	—	1.04
Thailand	Twins, Koh Tao	—	—	—	—	—	—	—	—	—	—	6.25	2.25	6.12	1.4	2.12	—	—	—	—
Thailand	Viking Bay	—	—	—	—	—	—	—	—	—	—	—	—	—	—	—	—	44.0	29.75	—
Thailand	West-Koh-see	—	—	—	—	0.5	—	—	—	—	—	—	—	—	—	—	—	—	—	—
Thailand	Yawasam Southwest	—	—	—	—	—	—	—	—	—	2.5	—	—	—	—	—	—	—	—	—
Tonga	Kito si'l	—	—	—	—	—	1.0	—	—	—	—	—	—	—	—	—	—	—	—	—
Tonga	Pangaimotu Reef Reserve (North)	—	—	—	—	—	0.25	—	—	—	—	—	—	—	—	—	—	—	—	—
Tonga	The Coral Gardens	—	—	—	—	—	—	—	—	—	—	—	—	—	—	—	—	0.25	—	—
Vanuatu	Asanvari	—	—	—	—	—	—	—	—	—	—	—	9.25	—	—	—	—	—	—	—
Vanuatu	Asanvari North	—	—	—	—	—	—	—	—	—	—	—	1.75	—	—	—	—	—	—	—

Continued

Table A4 (Continued) Global distribution of giant clams (Reef Check)

Country	Reef Site	Monitoring years (total clam density = number of individuals per 100 m²)																		
		1997	1998	1999	2000	2001	2002	2003	2004	2005	2006	2007	2008	2009	2010	2011	2012	2013	2014	2015
Vanuatu	Devil's Point (Kawene Region) Location 2, Site 3	—	—	—	—	—	—	—	0.25	—	—	—	—	—	—	—	—	—	—	—
Vanuatu	Hat Island- NW Coast	—	—	—	—	—	—	—	0.5	—	—	—	—	—	—	—	—	—	—	—
Vanuatu	Laone	—	—	—	—	—	—	—	—	—	—	—	0.75	—	—	—	—	—	—	—
Vanuatu	Netjanavigacas	—	—	—	—	—	—	—	—	—	—	—	—	—	—	8.0	—	—	—	—
Vanuatu	Netjanisiecen	—	—	—	—	—	—	—	—	—	—	—	—	—	—	0.25	—	—	—	—
Vanuatu	Netjanliluhu	—	—	—	—	—	—	—	—	—	—	—	—	—	—	0.75	—	—	—	—
Vanuatu	Nijcanauan	—	—	—	—	—	—	—	—	—	—	—	—	—	—	3.5	—	—	—	—
Vanuatu	Nuosinehei	—	—	—	—	—	—	—	—	—	—	—	—	—	—	0.25	—	—	—	—
Vanuatu	Sakao Island	—	—	—	—	—	—	—	—	—	—	—	—	—	—	—	0.5	—	—	—
Vanuatu	Suvu Bay	—	—	—	—	—	—	—	—	—	—	—	—	—	—	2.25	—	—	—	—
Vanuatu	Takara	—	—	—	—	—	—	—	0.5	—	—	—	—	—	—	—	—	—	—	—
Vanuatu	Vejel Reef	—	—	—	—	—	—	—	—	—	—	—	—	—	—	4.0	—	—	—	—
Viet Nam	Bai Bac	—	—	—	—	—	3.25	—	—	—	—	—	—	—	—	—	—	—	—	—
Viet Nam	Bai Dau Tai	—	—	—	—	—	2.5	—	—	—	—	—	—	—	—	—	—	—	—	—
Viet Nam	Bai Duong	—	4.25	—	—	—	—	—	—	—	—	—	—	—	—	—	—	—	—	—
Viet Nam	Bai Nhat	—	—	—	—	—	0.25	—	—	—	—	—	—	—	—	—	—	—	—	—
Viet Nam	Bai Ong Cuong	—	—	—	20.0	—	107.5	—	—	—	—	—	—	—	—	—	—	—	—	—
Viet Nam	Ben Dam	—	401.0	70.75	71.5	45.75	50.5	—	—	—	—	—	—	—	—	—	—	—	—	—
Viet Nam	Bong Lan	—	32.0	—	18.25	—	42.5	—	—	—	—	—	—	—	—	—	—	—	—	—
Viet Nam	CAN06	—	—	—	—	—	—	—	0.25	0.25	—	—	—	—	—	—	—	—	—	—
Viet Nam	CDA01	—	—	—	—	—	—	—	32.5	—	—	—	—	—	—	—	—	—	—	—
Viet Nam	CDA02	—	—	—	—	—	—	—	20.5	—	—	—	—	—	—	—	—	—	—	—
Viet Nam	CDA03	—	—	—	—	—	—	—	76.25	—	—	—	—	—	—	—	—	—	—	—

Continued

Table A4 (Continued) Global distribution of giant clams (Reef Check)

Country	Reef Site	Monitoring years (total clam density = number of individuals per 100 m^2)																		
		1997	1998	1999	2000	2001	2002	2003	2004	2005	2006	2007	2008	2009	2010	2011	2012	2013	2014	2015
Viet Nam	CDA04	—	—	—	—	—	—	—	75.75	—	—	—	—	—	—	—	—	—	—	—
Viet Nam	CDA05	—	—	—	—	—	—	—	68.75	—	—	—	—	—	—	—	—	—	—	—
Viet Nam	CDA06	—	—	—	—	—	—	—	21.75	—	—	—	—	—	—	—	—	—	—	—
Viet Nam	CDA07	—	—	—	—	—	—	—	22.0	—	—	—	—	—	—	—	—	—	—	—
Viet Nam	Chim Chim	—	—	20.0	41.0	61.0	40.5	—	—	—	—	—	—	—	—	—	—	—	—	—
Viet Nam	CLC05	—	—	—	—	—	—	—	0.75	—	—	—	—	—	—	—	—	—	—	—
Viet Nam	CLC06	—	—	—	—	—	—	—	0.25	—	—	—	—	—	—	—	—	—	—	—
Viet Nam	CLC08	—	—	—	—	—	—	—	0.5	—	—	—	—	—	—	—	—	—	—	—
Viet Nam	CLC09	—	—	—	—	—	—	—	1.25	—	—	—	—	—	—	—	—	—	—	—
Viet Nam	CLC10	—	—	—	—	—	—	—	0.25	—	—	—	—	—	—	—	—	—	—	—
Viet Nam	CLC11	—	—	—	—	—	—	—	0.5	—	—	—	—	—	—	—	—	—	—	—
Viet Nam	CLC12	—	—	—	—	—	—	—	1.5	—	—	—	—	—	—	—	—	—	—	—
Viet Nam	CLC13	—	—	—	—	—	—	—	0.75	—	—	—	—	—	—	—	—	—	—	—
Viet Nam	CLC15	—	—	—	—	—	—	—	0.25	—	—	—	—	—	—	—	—	—	—	—
Viet Nam	Con Chin	—	32.0	—	—	—	—	—	—	—	—	—	—	—	—	—	—	—	—	—
Viet Nam	Da Trang	—	—	107.0	135.25	81.75	106.75	—	—	—	—	—	—	—	—	—	—	—	—	—
Viet Nam	Dat Doc	—	—	169.5	108.25	84.75	17.25	—	—	—	—	—	—	—	—	—	—	—	—	—
Viet Nam	Dat Trang	—	122.0	—	—	—	—	—	—	—	—	—	—	—	—	—	—	—	—	—
Viet Nam	DNA09	—	—	—	—	—	—	—	—	0.25	—	—	—	—	—	—	—	—	—	—
Viet Nam	Dong Bac Hon Mun (site 3)	—	—	—	0.25	—	—	—	—	—	—	—	—	—	—	—	—	—	—	—
Viet Nam	Hang Rai/Ninh Thuan Site 5	—	—	—	—	—	0.5	—	—	—	—	—	—	—	—	—	—	—	—	—
Viet Nam	Hon Bo Tra	—	—	—	—	—	0.25	—	—	—	—	—	—	—	—	—	—	—	—	—
Viet Nam	Hon Cau	—	—	—	—	0.5	0.25	—	—	—	—	—	—	—	—	—	—	—	—	—
Viet Nam	Hon Dam	—	—	—	—	—	0.25	—	—	—	—	—	—	—	—	—	—	—	—	—

Continued

Table A4 (Continued) Global distribution of giant clams (Reef Check)

Country	Reef Site	Monitoring years (total clam density = number of individuals per 100 m^2)																		
		1997	1998	1999	2000	2001	2002	2003	2004	2005	2006	2007	2008	2009	2010	2011	2012	2013	2014	2015
Viet Nam	Hon Dam Ngang/Phu Quoc Site 8	—	—	—	—	—	0.75	—	—	—	—	—	—	—	—	—	—	—	—	—
Viet Nam	Hon Giai	—	—	—	—	—	1.25	—	—	—	—	—	—	—	—	—	—	—	—	—
Viet Nam	Hon Hoa Lu	—	—	—	—	—	0.5	—	—	—	—	—	—	—	—	—	—	—	—	—
Viet Nam	Hon Mau	—	—	—	—	—	0.25	—	—	—	—	—	—	—	—	—	—	—	—	—
Viet Nam	Hon Mun Site 1	—	—	—	—	—	0.5	—	—	—	—	—	—	—	—	—	—	—	—	—
Viet Nam	Hon Mun Site 3	—	1.0	2.0	—	—	—	—	—	—	—	—	—	—	—	—	—	—	—	—
Viet Nam	Hon Roi/Phu Quoc Site 2	—	—	—	—	—	1.5	1.5	—	—	—	—	—	—	—	—	—	—	—	—
Viet Nam	Hon Rua	—	—	0.75	—	—	—	—	—	—	—	—	—	—	—	—	—	—	—	—
Viet Nam	Hon Tai	—	—	—	—	23.5	1.75	—	—	—	—	—	—	—	—	—	—	—	—	—
Viet Nam	Hon Tai 2	—	—	—	—	—	80.75	—	—	—	—	—	—	—	—	—	—	—	—	—
Viet Nam	Hon Thom/Phu Quoc Site 3	—	—	—	—	—	0.25	0.75	—	—	—	—	—	—	—	—	—	—	—	—
Viet Nam	Hon Tu	—	—	—	—	—	0.5	—	—	—	—	—	—	—	—	—	—	—	—	—
Viet Nam	Luoi Dang/Ninh Thuan Site 3	—	—	—	—	—	0.25	—	—	—	—	—	—	—	—	—	—	—	—	—
Viet Nam	May Rut Trong/PQO15/Phu Quoc Site 5	—	—	—	—	—	0.5	0.25	—	—	1.75	—	—	—	—	—	—	—	—	—
Viet Nam	Mong Tay/Phu Quoc Site 6	—	—	—	—	—	—	0.5	—	—	—	—	—	—	—	—	—	—	—	—
Viet Nam	Mui Thi/Ninh Thuan Site 7	—	—	—	—	—	0.5	—	—	—	—	—	—	—	—	—	—	—	—	—
Viet Nam	My Hoa/Ninh Thuan Site 8	—	—	—	—	—	0.75	—	—	—	—	—	—	—	—	—	—	—	—	—

Continued

Table A4 (Continued) Global distribution of giant clams (Reef Check)

Country	Reef Site	Monitoring years (total clam density = number of individuals per 100 m²)																		
		1997	1998	1999	2000	2001	2002	2003	2004	2005	2006	2007	2008	2009	2010	2011	2012	2013	2014	2015
Viet Nam	NTA01/ Nhatrang Site 15	—	—	—	—	—	—	—	3.25	1.25	1.0	—	—	—	—	—	—	—	—	—
Viet Nam	NTA02/ Nhatrang Site 14	—	—	—	—	—	—	—	0.75	0.25	0.25	—	—	—	—	—	—	—	—	—
Viet Nam	NTA03/ Nhatrang Site 6	—	—	—	—	—	—	—	—	0.75	—	—	—	—	—	—	—	—	—	—
Viet Nam	NTA04/ Nhatrang Site 5	—	—	—	—	—	—	—	4.75	1.5	3.25	—	—	—	—	—	—	—	—	—
Viet Nam	NTA06	—	—	—	—	—	—	—	—	0.25	1.25	—	—	—	—	—	—	—	—	—
Viet Nam	NTA07	—	—	—	—	—	—	—	0.5	0.5	—	—	—	—	—	—	—	—	—	—
Viet Nam	NTA07	—	—	—	—	—	—	—	—	—	0.5	—	—	—	—	—	—	—	—	—
Viet Nam	NTA08	—	—	—	—	—	—	—	0.25	0.25	—	—	—	—	—	—	—	—	—	—
Viet Nam	NTA09/	—	—	—	—	—	—	—	0.25	—	—	—	—	—	—	—	—	—	—	—
Viet Nam	NTA09/ Nhatrang Site 8	—	—	—	—	—	—	—	—	—	0.5	—	—	—	—	—	—	—	—	—
Viet Nam	NTA09/ Nhatrang Site 8	—	—	—	—	—	—	—	—	—	—	—	—	—	—	—	—	—	—	—
Viet Nam	NTH01	—	—	—	—	—	—	—	—	1.75	—	—	—	—	—	—	—	—	—	—
Viet Nam	NTH03	—	—	—	—	—	—	—	—	0.25	—	—	—	—	—	—	—	—	—	—
Viet Nam	NTH05	—	—	—	—	—	—	—	—	0.25	—	—	—	—	—	—	—	—	—	—
Viet Nam	NTH07	—	—	—	—	—	—	—	—	0.25	—	—	—	—	—	—	—	—	—	—
Viet Nam	PQO02	—	—	—	—	—	—	—	—	—	0.5	—	—	—	—	—	—	—	—	—
Viet Nam	PQO03	—	—	—	—	—	—	—	—	—	0.5	—	—	—	—	—	—	—	—	—
Viet Nam	PQO04	—	—	—	—	—	—	—	—	—	1.75	—	—	—	—	—	—	—	—	—
Viet Nam	PQO05	—	—	—	—	—	—	—	—	—	1.75	—	—	—	—	—	—	—	—	—
Viet Nam	PQO06/Phu Quoc Site 1	—	—	—	—	—	—	—	—	—	0.5	—	—	—	—	—	—	—	—	—
Viet Nam	PQO09	—	—	—	—	—	—	—	—	—	0.5	—	—	—	—	—	—	—	—	—

Continued

Table A4 (Continued) Global distribution of giant clams (Reef Check)

Country	Reef Site	1997	1998	1999	2000	2001	2002	2003	2004	2005	2006	2007	2008	2009	2010	2011	2012	2013	2014	2015
											Monitoring years (total clam density = number of individuals per 100 m²)									
Viet Nam	PQO10	—	—	—	—	—	—	—	—	—	3.0	—	—	—	—	—	—	—	—	—
Viet Nam	PQO11	—	—	—	—	—	—	—	—	—	0.75	—	—	—	—	—	—	—	—	—
Viet Nam	PQO13	—	—	—	—	—	—	—	—	—	0.75	—	—	—	—	—	—	—	—	—
Viet Nam	PQO14	—	—	—	—	—	—	—	—	—	1.75	—	—	—	—	—	—	—	—	—
Viet Nam	PQO16/Phu Quoc Site 7	—	—	—	—	—	—	—	—	—	1.0	—	—	—	—	—	—	—	—	—
Viet Nam	PQO17	—	—	—	—	—	—	—	—	—	4.25	—	—	—	—	—	—	—	—	—
Viet Nam	PQO18	—	—	—	—	—	—	—	—	—	1.25	—	—	—	—	—	—	—	—	—
Viet Nam	PQO19	—	—	—	—	—	—	—	—	—	1.0	—	—	—	—	—	—	—	—	—
Viet Nam	PQO20/Phu Quoc Site 10	—	—	—	—	—	—	—	—	—	1.0	—	—	—	—	—	—	—	—	—
Viet Nam	Thai An	—	—	—	—	0.25	1.0	—	—	—	—	—	—	—	—	—	—	—	—	—
Viet Nam	VPO06	—	—	—	—	—	—	—	—	—	0.5	—	—	—	—	—	—	—	—	—
Viet Nam	VPO07	—	—	—	—	—	—	—	0.25	0.25	0.25	—	—	—	—	—	—	—	—	—
Viet Nam	VPO09	—	—	—	—	—	—	—	0.25	0.5	0.5	—	—	—	—	—	—	—	—	—
Viet Nam	VPO10	—	—	—	—	—	—	—	—	0.25	—	—	—	—	—	—	—	—	—	—
Viet Nam	CAN04	—	—	—	—	—	—	—	—	—	1.25	—	—	—	—	—	—	—	—	—
Viet Nam	CAN07	—	—	—	—	—	—	—	—	—	0.25	—	—	—	—	—	—	—	—	—
Yemen	Alamah	—	—	—	—	—	—	—	—	—	—	—	0.25	—	—	—	—	—	—	—
Yemen	Di Hamri	—	2.25	—	—	1.0	—	—	—	—	—	—	—	—	—	—	—	—	—	—
Yemen	Dihamri	—	—	—	—	—	—	—	—	—	—	—	0.13	—	—	—	—	—	—	—
Yemen	Hawlaf	—	—	—	—	2.13	—	—	—	—	—	—	—	—	—	—	—	—	—	—

Notes: Data extracted from Global Reef Tracker (Reef Check Worldwide)
Reef Check Survey Area = 400 m²

Appendix B: Full list of literature reviewed

Refer to Tables A1–A3 for list of localities and species, respectively.

Abbott, R.T. 1950. The molluscan Fauna of the Cocos-Keeling Islands, Indian Ocean. *Bulletin of Raffles Musuem* **22**, 68–98.

Accordi, G., Brilli, M., Carbone, F. & Voltaggio, M. 2010. The raised coral reef complex of the Kenyan coast: *Tridacna gigas* U-series dates and geological implications. *Journal of African Earth Sciences* **58**, 97–114.

Agombar, J.S., Dugdale, H.L. & Hawkswell, N.J. 2003. Species list and relative abundance of marine molluscs collected on Aride Island beach between March 2001 and February 2002. *Phelsuma* **11**, 29–38.

Al-Horani, F.A., Al-Rousan, S.A., Al-Zibdeh, M. & Khalaf, M.A. 2006. The status of coral reefs on the Jordanian coast of the Gulf of Aqaba, Red Sea. *Zoology in the Middle East* **38**, 99–110.

Alcala, A.C. 1986. Distribution and abundance of giant clams (Family Tridacnidae) in the South-Central Philippines. *Silliman Journal* **33**, 1–9.

Alder, J. & Braley, R.D. 1989. Serious mortality in populations of giant clams on reefs surrounding Lizard Island, Great Barrier Reef. *Australian Journal of Marine and Freshwater Research* **40**, 205–213.

Anam, R. & Mostarda, E. 2012. *Field Identification Guide to the Living Marine Resources of Kenya*. FAO Species Identification Guide for Fishery Purposes. Rome, Food and Agriculture Organization of the United Nations (FAO).

Andréfouët, S., Friedman, K., Gilbert, A. & Remoissenet, G. 2009. A comparison of two surveys of invertebrates at Pacific Ocean Islands: the giant clam at Raivavae Island, Australes Archipelago, French Polynesia. *ICES Journal of Marine Science* **66**, 1825–1836.

Andréfouët, S., Gilbert, A., Yan, L., Remoissenet, G., Payri, C. & Chancerelle, Y. 2005. The remarkable population size of the endangered clam *Tridacna maxima* assessed in Fangatau atoll (Eastern Tuamotu, French Polynesia) using *in situ* and remote sensing data. *ICES Journal of Marine Science* **62**, 1037–1048.

Andréfouët, S., Menou, J-L. & Naeem, S. 2012. Macro-invertebrate communities of Baa Atoll, Republic of Maldives. In *Biodiversity, Resources and Conservation of Baa Atoll (Republic of Maldives): A UNESCO Man and Biosphere Reserve*, S. Andréfouët (ed.). Atoll Research Bulletin No. 590, 125–142.

Andréfouët, S., Van Wynsberge, S., Fauvelot, C., Bruckner, A.W. & Remoissenet, G. 2014. Significance of new records of *Tridacna squamosa* Lamarck, 1819, in the Tuamotu and Gambier Archipelagos (French Polynesia). *Molluscan Research* **34**, 277–284.

Andréfouët, S., Van Wynsberge, S., Gaertner-Mazouni, N., Menkes, C., Gilbert, A. & Remoissenet, G. 2013. Climate variability and massive mortalities challenge giant clam conservation and management efforts in French Polynesia atolls. *Biological Conservation* **160**, 190–199.

Andrews, C.W., Smith, E.A., Bernard, H.M., Kirkpatrick, R. & Chapman, F.C. 1900. On the marine fauna of Christmas Island (Indian Ocean). *Proceedings of the Zoological Society of London* **69**, 115–140.

Anonymous 1994. Hima (Giant Clams). *Sport Fish & Wildlife Restoration*.

Apte, D. & Dutta, S. 2010. Ecological determinants and stochastic fluctuations of *Tridacna maxima* survival rate in Lakshadweep Archipelago. *Systematics and Biodiversity* **8**, 461–469.

Apte, D., Dutta, S. & Babu, I. 2010. Monitoring densities of the giant clam *Tridacna maxima* in Lakshadweep Archipelago. *Marine Biodiversity Records* **3**, e78 (9 pages).

Aubert, A., Lazareth, C.E., Cabioch, G., Boucher, H., Yamada, T., Iryu, Y. & Farman, R. 2009. The tropical giant clam *Hippopus hippopus* shell, a new archive of environmental conditions as revealed by sclerochronological and δ18O profiles. *Coral Reefs* **28**, 989–998.

Australian Government 2005. Status of the coral reefs at the Cocos (Keeling) Islands: a report on the status of the marine community at Cocos (Keeling) Islands, East Indian Ocean, 1997–2005. Canberra: Department of the Environment and Heritage. Online. http://www.environment.gov.au/resource/status-coral-reefs-cocos-keeling-islands-indian-ocean (accessed 13 April 2017).

Barnes, D.K.A. & Rawlinson, K.A. 2009. Traditional coastal invertebrate fisheries in south-western Madagascar. *Journal of the Marine Biological Association of the United Kingdom* **89**, 1589–1596.

Barnes, D.K.A., Corrie, A., Whittington, M., Carvalho, M.A. & Gell, F. 1998. Coastal shellfish resource use in the Quirimba Archipelago, Mozambique. *Journal of Shellfish Research* **17**, 51–58.

Barott, K.L., Caselle, J.E., Dinsdale, E.A., Friedlander, A.M., Maragos, J.E., Obura, D., Rohwer, F.L., Sandin, S.A., Smith, J.E. & Zgliczynski, B. 2010. The lagoon at Caroline/Millennium Atoll, Republic of Kiribati: natural history of a nearly pristine ecosystem. *PLoS ONE* **5**, e10950. doi:10.1371/journal.pone.0010950

Basker, J.R. 1991. Giant Clams in the Maldives – A stock assessment and study of their potential for culture. Madras, India: Bay of Bengal Programme, Reef Fish Research & Resources Survey. Online. http://www.fao.org/3/a-ae451e.pdf (accessed 13 April 2017).

Beger, M. & Pinca, S. 2003. Coral reef biodiversity community-based assessment and conservation planning in the Marshall Islands: Baseline surveys, capacity building and natural protection and management of coral reefs of the atolls of Rongelap and Mili. Final Report: Project 2002-0317-008. Natural Resources Assessment Surveys Team and Majuro, Republic of the Marshall Islands: College of the Marshall Islands. Online. http://www.nras-conservation.org/nraslibrary/NFWF_finalreport2003.pdf (accessed 13 April 2017).

Beger, M., Jacobson, D., Pinca, S., Richards, Z.T., Hess, D., Harris, F., Page, C., Peterson, E.L. & Baker, N. 2008. The state of coral reef ecosystems of the Republic of the Marshall Islands. In *The State of Coral Reef Ecosystems of the United States and Pacific Freely Associated States*, J.E. Waddell & A. Clarke (eds). NOAA: Silver Spring, Maryland, USA, 330–361.

Bell, L.A.J. & Amos, M.J. 1993. Republic of Vanuatu Fisheries Resources Profiles. FFA Report 93/49. Honiara, Solomon Islands: Pacific Islands Forum Fisheries Agency. Online. www.spc.int/DigitalLibrary/Doc/FAME/FFA/Reports/FFA_1993_049.pdf (accessed 13 April 2017).

Bell, L.A.J. 1993. Giant Clam Project American Samoa. FFA Report 93/06. Honiara, Solomon Islands: Pacific Islands Forum Fisheries Agency. Online. www.spc.int/DigitalLibrary/Doc/FAME/FFA/Reports/FFA_1993_006.pdf (accessed 13 April 2017).

Bellchambers, L.M. & Evans, S.N. 2013. A Summary of the Department of Fisheries, Western Australia Invertebrate Research at Cocos (Keeling) Islands 2006–2011. Fisheries Research Report No. 239. North Beach, Western Australia: Department of Fisheries, Western Australia. Online. http://www.fish.wa.gov.au/Documents/research_reports/frr239.pdf (accessed 13 April 2017).

Bernard, F.R., Cai, Y-Y. & Morton, B. 1993. *Catalogue of the Living Marine Bivalve Molluscs of China*. Hong Kong: Hong Kong University Press.

Berzunza-Sanzhez, M.M., Cabrera, M.C.C. & Pandolfi, J.M. 2013. Historical patterns of resource exploitation and the status of Papua New Guinea coral reefs. *Pacific Science* **67**, 425–440.

Bigot, L., Charpy, L., Maharavo, J., Abdou Rabi, F., Paupiah, N., Aumeeruddy, R., Villedieu, C. & Lieutaud, A. 2000. 5. Status of coral reefs of the Southern Indian Ocean: The Indian Ocean Commission node for Comoros, Madagascar, Mauritius, Reunion and Seychelles. In *Status of Coral Reefs of the World: 2000*, C. Wilkinson (ed.). Cape Ferguson, Queensland: Australian Institute of Marine Science, 77–93.

Bijukumar, A., Ravinesh, R., Arathi, A.R. & Idreesbabu, K.K. 2015. On the molluscan fauna of Lakshadweep included in various schedules of wildlife (protection) act of India. *Journal of Threatened Taxa* **7**, 7253–7268.

Black, R., Johnson, M.S., Prince, J., Brearley, A. & Bond, T. 2011. Evidence of large, local variations in recruitment and mortality in the small giant clam, *Tridacna maxima*, at Ningaloo Marine Park, Western Australia. *Marine and Freshwater Research* **62**, 1318–1326.

Bodoy, A. 1984. Assessment of human impact on giant clams (*Tridacna maxima*) near Jeddah, Saudi Arabia. *Proceedings of the Symposium on Coral Reef Environment of the Red Sea, Jeddah* **1984**, 472–490.

Borsa, P., Fauvelot, C., Tiavouane, J., Grulois, D., Wabnitz, C., Abdon Naguit, M.R. & Andréfouët, S. 2015. Distribution of Noah's giant clam, *Tridacna noae*. *Marine Biodiversity* **45**, 339–344.

Bouchet, P., Heros, V., Le Goff, A., Lozouet, P. & Maestrati, P. 2001. *Atelier biodiversité Lifou 2000, grottes et récifs coralliens*. Rapport de mission, IRD, Noumea, New Caledonia.

Braley, R.D. 1987a. Distribution and abundance of the giant clams *Tridacna gigas* and *T. derasa* on the Great Barrier Reef. *Micronesica* **20**, 215–223.

Braley, R.D. 1987b. Spatial distribution and population parameters of *Tridacna gigas* and *T. derasa*. *Micronesica* **20**, 225–246.

Braley, R.D. 1988. *The Status of Giant Clams Stocks and Potential for Clam Mariculture in Tuvalu*. Suva, Fiji: South Pacific Aquaculture Development Project, Food and Agriculture Organization of the United Nations.

Braley, R.D. 1989. *A Giant Clam Stock Survey and Preliminary Investigation of Pearl Oyster Resources in the Tokelau Islands.* Suva, Fiji: South Pacific Aquaculture Development Project, Food and Agriculture Organization of the United Nations.

Brown, J.H. & Muskanofola, M.R. 1985. An investigation of stocks of giant clams (family Tridacnidae) in Java and of their utilization and potential. *Aquaculture and Fisheries Management* **1**, 25–39.

Bryan, P.G. & McConnell, D.B. 1976. Status of giant clam stocks (Tridacnidae) on Helen Reef, Palau, Western Caroline Islands, April 1975. *Marine Fisheries Review* **38**, 15–18.

Calumpong, H.P. & Cadiz, P. 1993. Observations on the distribution of giant clams in protected areas. *Silliman Journal* **36**, 107–116.

Calumpong, H.P. & Macansantos, A.D. 2008. Distribution and abundance of giant clams in the Spratlys, South China Sea. In *Proceedings of the Conference on the Results of the Philippines-Vietnam Joint Oceanographic and Marine Scientific Research Expedition in the South China Sea (JOMSRE-SCS I to IV), 26–29 March 2008, Ha Long City, Vietnam*, A.C. Alcala (ed.). Pasay City, Republic of the Philippines. Technical Cooperation Council of the Philippines of the Department of Foreign Affairs, 55–59.

Calumpong, H.P., Apao, A.B., Lucañas, J.R. & Estacion, J.S. 2002. Community-based giant clam restocking – hope for biodiversity conservation. *Proceedings 9th International Coral Reef Symposium, Bali, Indonesia 23–27 October 2000, Volume 2*, Moosa et al. (eds). Jakarta: Indonesian Institute of Sciences, Jakarta: Ministry of Environment, Honolulu, Hawaii: International Society for Reef Studies pp. 101–110.

Chambers, C.N.L. 2007. *Pasua (Tridacna maxima)* size and abundance in Tongareva Lagoon, Cook Islands. *SPC Trochus Information Bulletin* **13**, 7–12.

Chantrapornsyl, S., Kittiwattanawong, K. & Adulyanukosol, K. 1996. Distribution and abundance of giant clam around Lee-Pae Island, the Andaman Sea, Thailand. *Phuket Marine Biological Center Special Publication* **16**, 195–200.

Chesher, R.H. 1993. Giant clam sanctuaries in the Kingdom of Tonga. Marine Studies of the University of the South Pacific Technical Report Series 95/2. Suva, Fiji: University of the South Pacific. Online. http://www.tellusconsultants.com/chesher-1993-Giant%20Clam%20Sanctuaries%20in%20the%20 Kingdom%20of%20Tonga.pdf (accessed 19 December 2016).

Chin, A., Lison De Loma, T., Reytar, K., Planes, S., Gerhardt, K., Clua, E., Burke, L. & Wilkinson, C. 2011. *Status of Coral Reefs of the Pacific and Outlook: 2011.* Global Coral Reef Monitoring Network. Online. http://www.icriforum.org/sites/default/files/Pacific-Coral-Reefs-2011.pdf (accessed 12 April 2017).

Conales, S.F., Bundal, N.A. & Dolorosa, R.G. 2015. High densities of *Tridacna crocea* in exposed massive corals proximate the Ranger Station of Tubbataha Reefs Natural Park, Cagayancillo, Palawan, Philippines. *The Palawan Scientist* **7**, 36–39.

Craig, P. (ed.) 2009. *Natural History Guide to American Samoa.* Pago Pago, American Samoa: National Park of American Samoa, 3rd edition. Online. https://www.nps.gov/npsa/learn/education/upload/ NatHistGuideAS09.pdf (accessed 12 April 2017).

Dalzell, P., Lindsay, S.R. & Patiale, H. 1993. Fisheries resources survey of the island of Niue. Inshore Fisheries Research Project, Technical Document No. 3. Noumea, New Caledonia: South Pacific Commission. Online. http://www.spc.int/DigitalLibrary/Doc/FAME/Reports/Dalzell_93_Niue.pdf (accessed 12 April 2017).

Daniels, C. 2004. Marine Science Report – Update report for DFMR, October 2004. Chumbe Island, Tanzania: Chumbe Island Coral Park Pte Ltd.

Dolorosa, R.G. & Jontila, J.B.S. 2012. Notes on common macrobenthic reef invertebrates of Tubbataha Reefs Natural Park, Philippines. *Science Diliman* **24**, 1–11.

Dolorosa, R.G. & Schoppe, S. 2005. Focal benthic mollusks (Mollusca: Bivalvia and Gastropoda) of selected sites in Tubbataha Reef National Marine Park, Palawan, Philippines. *Science Diliman* **17**, 1–10.

Dolorosa, R.G. 2010. Conservation status and trends of reef invertebrates in Tubbataha Reefs with emphasis on molluscs and sea cucumbers. Unpublished Technical Report. Online. http://tubbatahareef.org/downloads/research_reports/conservation_status_and_trends_of_reef_invertebrates_in_tubbataha_reefs_ with_emphasis_on_molluscs_and_sea_cucumbers.pdf (accessed 12 April 2017).

Dolorosa, R.G., Conales, S.F. & Bundal, N.A. 2014. Shell dimension-live weight relationships, growth and survival of *Hippopus porcellanus* in Tubbataha Reefs Natural Park, Philippines. *Atoll Research Bulletin* **604**, 1–9.

Dolorosa, R.G., Picardal, R.M. & Conales, S.F., Jr 2015. Bivalves and gastropods of Tubbataha Reefs Natural Park, Philippines. *Check List* **11**, 1506. doi:10.15560/11.1.1506

Dumas, P., Fauvelot, C., Andréfouët, S. & Gilbert, A. 2011. Les benitiers en Nouvelle-Caledonie: Statut des populations, impacts de l'exploitation & connectivitié. Rapport final d'opération, Programme ZONECO, Avril 2011. Noumea, New Caledonia: Institut de Recherche pour le Développement (Nouvelle-Calédonie); Saint-Denis, Réunion: Université de la Réunion

Dumas, P., Jimenez, H., Peignon, C., Wantiez, L. & Adjeroud, M. 2013. Small-scale habitat structure modulates the effects of no-take marine reserves for coral reef macroinvetebrates. *PLoS ONE* **8**, e58998. doi:10.1371/journal.pone.0058998

Eliata, A., Zahida, F., Wibowo, N.J. & Panggabean, L.M.G. 2003. Abundance of giant clam in coral reef ecosystem at Pari Island: a population comparison of 2003's to 1984's data. *Biota* **8**, 149–152.

Evans, S.M., Knowles, G., Pye-Smith, C. & Scott, R. 1977. Conserving shells in Kenya. *Oryx* **13**, 480–485.

Evans, S.N., Konzewitsch, N. & Bellchambers, L.M. 2016. *An update of the Department of Fisheries, Western Australia, Invertebrate and Reef Health Research and Monitoring at Cocos (Keeling) Islands.* Fisheries Research Report No. 272, Department of Fisheries, Western Australia.

Fiege, D., Neumann, V. & Li, J. 1994. Observations on coral reefs of Hainan Island, South China Sea. *Marine Pollution Bulletin* **29**, 84–89.

Fijiwara, S., Shibuno, T., Mito, K., Nakai, T., Sasaki, Y., Dai, C-F. & Chen, G. 2000. 8. Status of coral reefs of East and North Asia: China, Japan and Taiwan. In *Status of Coral Reefs of the World: 2000*, C. Wilkinson (ed.). Cape Ferguson, Queensland: Australian Institute of Marine Science, 131–140.

George, K.C., Thomas, P.A., Appukuttan, K.K. & Gopakumar, G. 1986. Ancillary living marine resources of Lakshadweep. Marine Fisheries Information Service: Special Issue on Lakshadweep, No. 68, 46–50.

Gerlach, G. & Gerlach, R. 2004. Species list of marine molluscs on Silhouette Island. *Phelsuma* **12**, 12–23.

Gilbert, A., Andréfouët, S., Yan, L. & Remoissenet, G. 2006. The giant clam *Tridacna maxima* communities of three French Polynesia islands: comparison of their population sizes and structures at early stages of their exploitation. *ICES Journal of Marine Science* **63**, 1573–1589.

Gilbert, A., Planes, S., Andréfouët, S., Friedman, K. & Remoissenet, G. 2007. First observation of the giant clam *Tridacna squamosa* in French Polynesia: a species range extension. *Coral Reefs* **26**, 229 only.

Gilbert, A., Remoissenet, G., Yan, L. & Andréfouët, S. 2006. Special traits and promises of the giant clam (*Tridacna maxima*) in French Polynesia. *SPC Fisheries Newsletter* **No. 118**, 44–52.

Gilbert, A., Yan, L., Remoissenet, G., Andréfouët, S., Payri, C. & Chancerelle, Y. 2005. Extraordinarily high giant clam density under protection in Tatakoto Atoll (eastern Tuamotu Archipelago, French Polynesia). *Coral Reefs* **24**, 495 only.

Gilligan, J., Hender, J., Hobbs, J.P., Neilson, J. & McDonald, C. 2008. Coral reef surveys and stock size estimates of shallow water (0–20m) marine resources at Christmas Island, Indian Ocean. Unpublished Report to Parks Australia North (Technical Report).

Gomez, E.D. & Alcala, A.C. 1988. Giant clams in the Philippines. In *Giant Clams in Asia and the Pacific*, J.W. Copland & J.S. Lucas (eds). Canberra: Australian Centre for International Agricultural Research, 51–53.

Gonzales, B.J., Becira, J.G., Galon, W.M. & Gonzales, M.M.G. 2014a. Protected versus unprotected area with reference to fishes, corals, marine invertebrates, and CPUE in Honda Bay, Palawan. *The Palawan Scientist* **6**, 42–59.

Gonzales, B.J., Dolorosa, R.G., Pagliawan, H.B. & Gonzales, M.M.G. 2014b. Marine resource assessment for sustainable management of Apulit Island, West Sulu Sea, Palawan, Philippines. *IJFAS* **2**, 130–136.

Gössling, S., Kunkel, T., Schumacher, K. & Zilger, M. 2004. Use of molluscs, fish, and other marine taxa by tourism in Zanzibar, Tanzania. *Biodiversity and Conservation* **13**, 2623–2639.

Govan, H., Nichols, P.V. & Tafea, H. 1988. Giant clam resource investigations in Solomon Islands. In *Giant Clams in Asia and the Pacific*, J.W. Copland & J.S. Lucas (eds). Canberra: Australian Centre for International Agricultural Research, 54–57.

Green, A. & Craig, P. 1999. Population size and structure of giant clams at Rose Atoll, an important refuge in the Samoan Archipelago. *Coral Reefs* **18**, 205–211.

Guest, J.R., Todd, P.A., Goh, E., Sivalonganathan, B.S. & Reddy, K.P. 2008. Can giant clam (*Tridacna squamosa*) populations be restored on Singapore's heavily impacted coral reefs? *Aquatic Conservation: Marine and Freshwater Ecosystems* **18**, 570–579.

Hamner, W.M. & Jones, M.S. 1976. Distribution, burrowing, and growth rates of the clam *Tridacna crocea* on interior reef flats. *Oecologia* **24**, 207–227.

Harding, S. & Randriamanantsoa, B. 2008. Coral reef monitoring in marine reserves of Northern Madagascar. In *Ten Years After Bleaching – Facing the Consequences of Climate Change in the Indian Ocean*, D.O. Obura et al. (eds). CORDIO Status Report 2008. Mombasa, Kenya: Coastal Oceans Research and Development in the Indian Ocean/Sida-SAREC, 93–106.

Harding, S., Randriamanantsoa, B., Hardy, T. & Curd, A. 2006. Coral reef monitoring and biodiversity assessment to support the planning of a proposed MPA at Andavadoaka. Unpublished Technical Report.

Hardy, J.T. & Hardy, S.A. 1969. Ecology of *Tridacna* in Palau. *Pacific Science* **23**, 467–472.

Hender, J., McDonald, C.A. & Gilligan, J.J. 2001. Baseline surveys of the marine environments and stock size estimates of marine resources of the south Cocos (Keeling) Atoll (0–15m), eastern Indian Ocean. Unpublished report to the Fisheries Resources Research Fund, Barton, Australia.

Hensley, R.A. & Sherwood, T.S. 1993. An overview of Guam's inshore fisheries. *Marine Fisheries Review* **55**, 129–138.

Hernawan, U.E. 2010. Study on giant clams (Cardiidae) population in Kei Kecil waters, Southeast-Maluku. *Widyariset* **13**, 101–108.

Hester, F.J. & Jones, E.C. 1974. A survey of giant clams, Tridacnidae, on Helen Reef, a Western Pacific atoll. *Marine Fisheries Review* **36**, 17–22.

Hirase, S. 1954. *An Illustrated Handbook of Shells in Natural Colours from the Japanese Islands and adjacent territory*. Tokyo: Maruzen Co. Ltd.

Hirschberger, W. 1980. Tridacnid clam stocks on Helen Reef, Palau, Western Caroline Islands. *Marine Fisheries Review* **42**, 8–15.

Hopkins, A. 2009. *Marine invertebrates as indicators of reef health: a study of the reefs in the region of Andavadoaka, South West Madagascar*. MSc Dissertation, Imperial College London, UK.

Hourston, M. 2010. Review of exploitation of marine resources of the Australian Indian Ocean Territories: the implications of biogeographic isolation for tropical island fisheries. Fisheries Research Report No. 208. Perth: Department of Fisheries, Western Australia.

Huang, C.W., Hsiung, T.W., Lin, S.M. & Wu, W.L. 2013. Molluscan fauna of Gueishan Island, Taiwan. *ZooKeys* **261**, 1–13.

Huber, M. & Eschner, A. 2011. *Tridacna* (*Chametrachea*) *costata* Roa-Quiaoit, Kochzius, Jantzen, Al-Zibdah and Richter from the Red Sea, a junior synonym of *Tridacna squamosina* Sturany, 1899 (Bivalvia, Tridacnidae). *Annalen des Naturhistorischen Museums in Wien B* **112**, 153–162.

Huber, M. 2010. *Compendium of Bivalves. A Full-Color Guide to 3,300 of the World's Marine Bivalves. A Status on Bivalvia after 250 years of Research*. Hackenheim: Conchbooks.

Hughes, R.N. 1977. The biota of reef-flats and limestone cliffs near Jeddah, Saudi Arabia. *Journal of Natural History* **11**, 77–96.

Irving, R. & Dawson, T. 2013. 22 Coral reefs of the Pitcairn Islands. In *Coral Reefs of the United Kingdom Overseas Territories*, C.R.C. Sheppard (ed.). Dordrecht, Netherlands: Springer, 299–318.

Jacob, P. 2000. The Status of Marine Resources and Coral Reefs of Nauru. Unpublished Status Report, Global Coral Reef Monitoring Network, 1–10.

Jaubert, J. 1977. Light, metabolism, and the distribution of *Tridacna maxima* in a South Pacific atoll: Takapoto (French Polynesia). *Proceedings 3rd International Coral Reef Symposium, Rosenstiel School of Marine and Atmospheric Science, University of Miami, Miami, Florida, USA, May, 1977*, 489–494.

Job, S. & Ceccarelli, D. 2012. Tuvalu Marine Life: an Alofa Tuvalu Project with the Tuvalu Fisheries Department and Funafuti, Nanumea, Nukulaelae Kaupules. Scientific Report, December 2012. Paris: Alofa Tuvalu. Online. http://alofatuvalu.tv/US/05_a_tuvalu/05_page_tml/livret4light.pdf (accessed 12 April 2017).

Johnson, M.S., Prince, J., Brearley, A., Rosser, N.L. & Black, R. 2016. Is *Tridacna maxima* (Bivalvia: Tridacnidae) at Ningaloo Reef, Western Australia? *Molluscan Research*, doi:10.1080/13235818.2016.1181141

Juinio, M.A.R., Meñez, L.A.B., Villanoy, C.L. & Gomez, E.D. 1989. Status of giant clam resources of the Philippines. *Journal of Molluscan Studies* **55**, 431–440.

Junchompoo, C., Sinrapasan, N., Penpain, C. & Patsorn, P. 2013. Changing seawater temperature effects on giant clams bleaching, Mannai Island, Rayong province, Thailand. *Kurenai* **2013-03**, 71–76.

Kanno, K., Kotaki, Y. & Yasumoto, T. 1976. Distribution of toxins in molluscs associated with coral reefs. *Bulletin of the Japanese Society of Scientific Fisheries* **42**, 1395–1398.

Kay, E.A. 1970. The littoral marine mollusks of Fanning Island. In *Fanning Island Expedition, January 1970*, K.E. Chave (ed.). Honolulu: Hawaii Institute of Geophysics, University of Hawaii, 111–133.

Kepler, A.K. & Kepler, C.B. 1994. Part I. History, physiography, botany and isle descriptions. *Atoll Research Bulletin* **397**, 1–225.

Kilada, R., Zakaria, S. & Farghalli, M.E. 1998. Distribution and abundance of the giant clam *Tridacna maxima* (Bivalvia: Tridacnidae) in the Northern Red Sea. *Bulletin of the National Institute of Oceanography and Fisheries* **24**, 221–240.

Kinch, J. 2001. Clam harvesting, the Convention on the International Trade in Endangered Species (CITES) and conservation in Milne Bay Province, Papua New Guinea. *SPC Fisheries Newsletter* **99**, 24–36.

Kinch, J. 2002. Giant clams: their status and trade in Milne Bay Province, Papua New Guinea. *TRAFFIC Bulletin* **19**, 1–9.

Kittiwattanawong, K. 1997. Genetic structure of giant clam, *Tridacna maxima* in the Andaman Sea, Thailand. *Phuket Marine Biological Center Special Publication* **17**, 109–114.

Kittiwattanawong, K. 2001. Records of extinct *Tridacna gigas* in Thailand. *Phuket Marine Biological Center Special Publication* **25**, 461–463.

Kittiwattanawong, K., Nugranad, J. & Srisawat, T. 2001. High genetic divergence of *Tridacna squamosa* living at the west and the east coasts of Thailand. *Phuket Marine Biological Center Special Publication* **25**, 343–347.

Koh, L.L., Tun, K.P.P. & Chou, L.M. 2003. The status of coral reefs of Surin Islands, Thailand based on surveys in December 2003. REST Technical Report No. 5. Singapore: National University of Singapore.

Kronen, M., Fisk, D., Pinca, S., Magron, F., Friedman, K., Boblin, P., Awira, R. & Chapman, L. 2008. Niue country report: profile and results from in-country survey work (May to June 2005). Noumea, New Caledonia: Pacific Regional Oceanic and Coastal Fisheries Development Programme. Online. http://www.spc.int/DigitalLibrary/Doc/FAME/Reports/PROCFish/PROCFish_2008_NiueReport.pdf (accessed 12 April 2017).

Kubo, H. & Iwai, K. 2007. On two sympatric species within *Tridacna* "*maxima*". *Annual Report Okinawa Fisheries Oceanography Research Centre* **68**, 205–210.

Kusnadi, A., Triandiza, T. & Hernawan, U.E. 2008. Inventarisasi Jenis dan Potensi Moluska Padang Lamun di Kepulauan Kei Kecil, Maluku Tenggara. *Biodiversitas* **9**, 30–34.

Langi, V. & Hesitoni 'Aloua 1988. Status of giant clams in Tonga. In *Giant Clams in Asia and the Pacific*, J.W. Copland & J.S. Lucas (eds). Canberra: Australian Centre for International Agricultural Research, 58–59.

Langi, V. 1990. Marine resource survey of Nanumea and Nui Islands, Tuvalu: (giant clam, commercial species, bêche-de-mer, pearl oysters and trochus). Canberra: Australian Centre for International Agricultural Research, and Suva, Fiji: Ministry of Primary Industries.

Larrue, S. 2006. Giant clam fishing on the islands of Tubuai, Austral Islands group: between local portrayals, economic necessity and ecological realities. *SPC Traditional Marine Resource Management and Knowledge Information Bulletin* **19**, 3–10.

Lasola, N. & Hoang, X.B. 2008. Assessment of commercially important macro-invertebrates in the Spratly Group of Islands. In *Proceedings of the Conference on the Results of the Philippines-Vietnam Joint Oceanographic and Marine Scientific Research Expedition in the South China Sea (JOMSRE-SCS I to IV), 26–29 March 2008, Ha Long City, Vietnam*, A.C. Alcala (ed.). Pasay City, Republic of the Philippines: Technical Cooperation Council of the Philippines of the Department of Foreign Affairs, 51–54.

Latypov, Y.Y. & Selin, N.I. 2011. Current status of coral reefs of islands in the Gulf of Siam and Southern Vietnam. *Russian Journal of Marine Biology* **37**, 255–262.

Latypov, Y.Y. & Selin, N.I. 2012a. Changes of reef community near Ku Lao Cham Islands (South China Sea) after Sangshen Typhoon. *American Journal of Climate Change* **1**, 41–47.

Latypov, Y.Y. & Selin, N.I. 2012b. The composition and structure of a protected coral reef in Cam Ranh Bay in the South China Sea. *Russian Journal of Marine Biology* **38**, 112–121.

Latypov, Y.Y. & Selin, N.I. 2013. Some data on spatio-temporal stability and variability of coral reefs in Khanh Hoa Province (Vietnam). *Environment, Ecology & Management* **2**, 1–16.

Latypov, Y.Y. 2000. Macrobenthos communities on reefs of the An Thoi Archipelago of the South China Sea. *Russian Journal of Marine Biology* **26**, 18–26.

Latypov, Y.Y. 2001. Communities of coral reefs of central Vietnam. *Russian Journal of Marine Biology* **27**, 197–200.

Latypov, Y.Y. 2006. Changes in the composition and structure of coral communities of Mju and Moon Islands, Nha Trang Bay, South China Sea. *Russian Journal of Marine Biology* **32**, 269–275.

Latypov, Y.Y. 2013. Barrier and platform reefs of the Vietnamese coast of the South China Sea. *International Journal of Marine Science* **3**, 23–32.

Laurent, V. 2001. Etude de stocks, relations biométriques et structure des populations de bénitiers, Tridacna maxima, dans trois lagons de Polynésie francaise (Moorea, Takapoto et Anaa). Report. Rennes: École nationale supérieure agronomique de Rennes.

Ledua, E., Manu, N. & Braley, R. 1993. Distribution, habitat and culture of the recently described giant clam *Tridacna tevoroa* in Fiji and Tonga. In *The Biology and Mariculture of Giant Clams: A Workshop Held in Conjunction with the 7th International Coral Reef Symposium 21–26 June 1992, Guam, USA*, W.K. Fitt (ed.). ACIAR Proceedings No. 47. Canberra: Australian Centre for International Agricultural Research, 147–153.

Lewis, A.D. & Ledua, E. 1988. A possible new species of *Tridacna* (Tridacnidae: Mollusca) from Fiji. In *Giant Clams in Asia and the Pacific*, J.W. Copland & J.S. Lucas (eds). Canberra: Australian Centre for International Agricultural Research, 82–84.

Lewis, A.D., Adams, T.J.H. & Ledua, E. 1988. Fiji's giant clam stocks – A review of their distribution, abundance, exploitation and management. In *Giant Clams in Asia and the Pacific*, J.W. Copland & J.S. Lucas (eds). Canberra: Australian Centre for International Agricultural Research, 66–72.

Liu, J.Y. 2013. Status of marine biodiversity of the China Seas. *PLoS ONE* **8**, e50719. doi:10.1371/journal.pone.0050719

Loh, T.L., Chaipichit, S., Songploy, S. & Chou, L.M. 2004. The status of coral reefs of Surin Islands, Thailand, based on surveys in December 2004. REST Technical Report No. 7. Singapore: National University of Singapore.

Long, N.V. & Vo, T.S. 2013. Degradation trend of coral reefs in the coastal waters of Vietnam. *Galaxea Special Issue* **15**, 79–83. doi:10.3755/galaxea.15.79

Lovell, E., Sykes, H., Deiye, M., Wantiez, L., Garrigue, C., Virly, S., Samuelu, J., Solofa, A., Poulasi, T., Pakoa, K., Sabetian, A., Afzal, D., Hughes, A. & Sulu, R. 2004. 12. Status of coral reefs in the south west Pacific: Fiji, Nauru, New Caledonia, Samoa, Solomon Islands, Tuvalu and Vanuatu. In *Status of Coral Reefs of the World: 2004, Volume 2*, C. Wilkinson (ed.). Townsville, Queensland: Australian Institute of Marine Sciences, 337–361.

Lucas, J.S., Ledua, E. & Braley, R.D. 1991. *Tridacna tevoroa* Lucas, Ledua and Braley: a recently described species of giant clam (Bivalvia: Tridacnidae) from Fiji and Tonga. *Nautilus* **105**, 92–103.

Maes, V.O. 1967. The littoral marine mollusks of Cocos-Keeling Islands (Indian Ocean). *Proceedings of the Academy of Natural Sciences of Philadelphia* **119**, 93–217.

McKenna S.A., Baillon N., Blaffart H., & Abrusci G. 2006. Une évaluation rapide de la biodiversité marine des récifs coralliens du Mont Panié, Province Nord, Nouvelle Calédonie. Bulletin PER d'évaluation biologique N°42.

McMichael, D.F. 1974. Growth rate, population size and mantle coloration in the small giant clam *Tridacna maxima* (Röding), at One Tree Island, Capricorn Group, Queensland. *Proceedings 2nd International Coral Reef Symposium, Brisbane, October 1974*. Brisbane: The Great Barrier Reef Committee, 241–254.

Mekawy, M.S. & Madkour, H.A. 2012. Studies on the Indo-Pacific Tridacnidae (*Tridacna maxima*) from the Northern Red Sea, Egypt. *International Journal of Geosciences* **3**, 1089–1095.

Mekawy, M.S. 2014. Environmental factors controlling the distribution patterns and abundance of sclerobionts on the shells of *Tridacna maxima* from the Egyptian Red Sea coast. *Arabian Journal of Geosciences* **7**, 3085–3092.

Michel, C. 1985. *Marine molluscs of Mauritius*. Gland, Switzerland: WWF and IUCN.

Militz, T.A., Kinch, J. & Southgate, P.C. 2015. Population demographics of *Tridacna noae* (Röding, 1798) in New Ireland, Papua New Guinea. *Journal of Shellfish Research* **34**, 329–335.

Miller, I. & Sweatman, H. 2004. 11. Status of coral reefs in Australia and Papua New Guinea in 2004. In *Status of Coral Reefs of the World: 2004, Volume 2*, C. Wilkinson (ed.). Townsville, Queensland: Australian Institute of Marine Sciences, 303–335.

Mohamed-Pauzi, A., Mohd. Adib, H., Ahmad, A. & Abdul-Aziz, Y. 1994. A preliminary survey of giant clams in Malaysia. *Proceedings of Fisheries Research Conference, Department of Fisheries, Malaysia* **IV**, 487–493.

Monsecour, K. 2016. A new species of giant clam (Bivalvia: Cardiidae) from the Western Indian Ocean. *Conchylia* **46**, 69–77.

Montagne, A., Naim, O., Tourrand, C., Pierson, B. & Menier, D. 2013. Status of coral reef communities on two carbonate platforms (Tun Sakaran Marine Park, East Sabah, Malaysia). *Journal of Ecosystems* **2013**, 1–15.

Morton, B. & Morton, J.E. 1983. *The Sea Shore Ecology of Hong Kong*. Hong Kong: Hong Kong University Press.

Munro, J.L. 1988. *Status of Giant Clam Stocks in the Central Gilbert Islands Group, Republic of Kiribati*. Workshop on Pacific inshore fishery resources, Noumea, New Caledonia, 14–25 March 1988. SPC/ Inshore Fish Res/BP54. Noumea, New Caledonia: South Pacific Commission.

Munro, J.L. 1989. 24 Fisheries for giant clams (Tridacnidae: Bivalvia) and prospects for stock enhancement. In *Marine Invertebrate Fisheries: Their Assessment and Management*, J.F. Caddy (ed.). New York: Wiley, 541–558.

Nadon, M.O., Griffiths, D., Doherty, E. & Harris, A. 2007. The status of coral reefs in the remote region of Andavadoaka, Southwest Madagascar. *Western Indian Ocean Journal of Marine Science* **6**, 207–218.

Nagaoka, L. 1993. Chapter 13 Faunal assemblages from the To'aga site. In *The To'aga Site. Three Millennia of Polynesian Occupation in the Manu'a Islands, American Samoa*, P.V. Kirch & T.L. Hunt (eds). Contributions of the University of California Archaeological Research Facility, Berkeley, Number 51. Online. http://digitalassets.lib.berkeley.edu/anthpubs/ucb/text/arf051-014.pdf (accessed 12 April 2017).

Naguit, M.R.A., Tisera, W.L. & Calumpong, H.P. 2012. Ecology and genetic structure of giant clams around Savu Sea, East Nusa Tenggara Province, Indonesia. *Asian Journal of Biodiversity* **3**, 174–194.

Nakamura, Y. 2013. Coastal resource use and management on Kilwa Island, southern Swahili Coast, Tanzania. *AWER Procedia Advances in Applied Sciences* **2013**, 364–370.

Namboodiri, P.N. & Sivadas, P. 1979. Zonation of molluscan assemblage at Kavaratti Atoll (Laccadives). *Mahasagar-Bulletin of the National Institute of Oceanography* **12**, 239–246.

Neo, M.L. & Todd, P.A. 2012a. Population density and genetic structure of the giant clams *Tridacna crocea* and *T. squamosa* on Singapore's reefs. *Aquatic Biology* **14**, 265–275.

Neo, M.L. & Todd, P.A. 2012b. Giant clams (Mollusca: Bivalvia: Tridacninae) in Singapore: history, research and conservation. *Raffles Bulletin of Zoology* **25**, 67–78.

Neo, M.L. & Todd, P.A. 2013. Conservation status reassessment of giant clams (Mollusca: Bivalvia: Tridacninae) in Singapore. *Nature in Singapore* **6**, 125–133.

Newman, W. & Gomez, E. 2007. The significance of the giant clam *Tridacna squamosa* at Tubuai, Austral Islands, French Polynesia. *Coral Reefs* **26**, 909.

Nimoho, G., Seko, A., Iinuma, M., Nishiyama, K. & Wakisaka, T. 2013. A baseline survey of coastal villages in Vanuatu. *SPC Traditional Marine Resource Management and Knowledge Information Bulletin* **32**, 3–84.

Okada, H. 1997. Market survey of aquarium giant clams in Japan. South Pacific Aquaculture Development Project (Phase II). FAO Fisheries and Aquaculture Department Field Document No. 8. Rome: Food and Agriculture Organization of the United Nations. Online. http://www.fao.org/docrep/005/ac892e/ AC892E00.htm (accessed 19 December 2016).

Oliver, P.G., Holmes, A.M., Killeen, I.J., Light, J.M. & Wood, H. 2004. Annotated checklist of the marine Bivalvia of Rodrigues. *Journal of Natural History* **38**, 3229–3272.

Pan, H-Z. & Lan, X. 1998. Molluscs from Xisha Islands. *Acta Palaeontologica Sinica* **37**, 121–132.

Panggabean, L.M.G. 2007. Karakteristik Pertumbuhan Kima Pasir, *Hippopus hippopus* yang dibesarkan di Pulau Pari. *Oseanologi dan Limnologi di Indonesia* **33**, 469–480.

Pasaribu, B.P. 1988. Status of giant clams in Indonesia. In *Giant Clams in Asia and the Pacific*, J.W. Copland & J.S. Lucas (eds). Canberra: Australian Centre for International Agricultural Research, 44–46.

Paulay, G. 1987. Biology of Cook Islands' bivalves, Part I. Heterodont families. Atoll Bulletin Research No. 298. Washington, DC: The Smithsonian Institution. doi:10.5479/si.00775630.298.1

Paulay, G. 1989. Marine invertebrates of the Pitcairn Islands: Species composition and biogeography of corals, molluscs, and echinoderms. Atoll Research Bulletin No. 326. Washington, DC: The Smithsonian Institution. doi:10.5479/si.00775630.326.1

Paulay, G. 2003. Marine bivalvia (Mollusca) of Guam. *Micronesia* **35–36**, 218–243.

Pearson, R.G. & Munro, J.L. 1991. Growth, mortality and recruitment rates of giant clams, *Tridacna gigas* and *T. derasa*, at Michaelmas Reef, central Great Barrier Reef, Australia. *Australian Journal of Marine and Freshwater Research* **42**, 241–262.

Penny, S.S. & Willan, R.C. 2014. Description of a new species of giant clam (Bivalvia: Tridacnidae) from Ningaloo Reef, Western Australia. *Molluscan Research* **34**, 201–211.

PERSGA 2010. The Status of Coral Reefs in the Red Sea and Gulf of Aden: 2009. PERSGA Technical Series Number 16, Jeddah. Saudi Arabia: The Regional Organization for the Conservation of the Environment in the Red Sea and Gulf of Aden.

Pilcher, N. & Alsuhaibany, A. 2000. 2. Regional status of coral reefs in the Red Sea and the Gulf of Aden. In *Status of Coral Reefs of the World: 2000*, C. Wilkinson (ed.). Townsville, Queensland: Australian Institute of Marine Sciences, 35–54.

Pilcher, N.J. & Djama, N. 2000. Status of coral reefs in Djibouti–2000. PERSGA Technical Series Report, Jeddah, Saudi Arabia: The Regional Organization for the Conservation of the Environment in the Red Sea and Gulf of Aden.

Pinca, S. & Beger, M. (eds) 2002. Coral reef biodiversity community-based assessment and conservation planning in the Marshall Islands: Baseline surveys, capacity building and natural protection and management of coral reefs of the atoll of Rongelap. Majuro, Marshall Islands: College of the Marshall Islands. Online. http://www.nras-conservation.org/nraslibrary/ReportRong2002full.pdf (accessed 12 April 2017).

Planes, S., Chauvet, C., Baldwin, J., Bonvallot, J., Fontaine-Vernaudon, Y., Gabrie, C., Holthus, P., Payri, C. & Galzin, R. 1993. Impact of tourism-related fishing on *Tridacna maxima* (Mollusca, Bivalvia) stocks in Bora-Bora Lagoon (French Polynesia). Atoll Research Bulletin No. 385. Washington, DC: The Smithsonian Institution. doi:10.5479/si.00775630.385.1

Pollock, N.J. 1992. Giant clams in Wallis: Prospects for development. In *Giant Clams in the Sustainable Development of the South Pacific: Socioeconomic Issues in Mariculture and Conservation*, C. Tisdell (ed.). ACIAR Monograph No. 18, 65–79.

Price, C.M. & Fagolimul, J.O. 1988. Reintroduction of giant clams to Yap State, Federated States of Micronesia. In *Giant Clams in Asia and the Pacific*, J.W. Copland & J.S. Lucas (eds). Canberra: Australian Centre for International Agricultural Research, 41–43.

Pringgenies, D., Suprihatin, J. & Lazo, L. 1995. Spatial and size distribution of giant clams in the Karumunjawa Islands, Indonesia. *Phuket Marine Biological Center Special Publication* **15**, 133–135.

Purcell, S.W., Gossuin, H. & Agudo, N.S. 2009. Status and management of the sea cucumber fishery of La Grande Terre, New Caledonia. Final report for ZoNéCo project, 2006–2008. Penang, Malaysia: WorldFish Center.

Qi, Z. (ed.) 2004. *Seashells of China*. Beijing: China Ocean Press.

Radtke, R. 1985. *Population dynamics of the giant clam,* Tridacna maxima, *at Rose Atoll*. Honolulu: Hawaii Institute of Marine Biology, University of Hawaii.

Ramadoss, K. 1983. Giant clam (*Tridacna*) resources. *CMFRI Bulletin* **34**, 79–81.

Ramohia, P. 2006. Fisheries resources: Commercially important macroinvertebrates. In *Solomon Islands Marine Assessment: Technical report of survey conducted May 13 to June 17, 2004*, A. Green et al. (eds). TNC Pacific Island Countries Report No. 1/06. South Brisbane, Queensland: The Nature Conservancy, 330–400.

Rees, M., Colquhoun, J., Smith, L. & Heyward, A. 2003. *Surveys of Trochus, Holothuria, Giant Clams and the Coral Communities at Ashmore Reef, Cartier Reef and Mermaid Reef, Northwestern Australia: 2003*. Unpublished report. Townsville, Queensland: Australian Institute of Marine Science.

Richard, G. 1977. Quantitative balance and production of *Tridacna maxima* in the Takapoto Lagoon (French Polynesia). *Proceedings of the 3rd International Coral Reef Symposium, Rosenstiel School of Marine and Atmospheric Science, University of Miami, Miami, Florida, USA, May 1977*, 599–605.

Richter, C., Roa-Quiaoit, H., Jantzen, C., Al-Zibdah, M. & Kochzius, M. 2008. Collapse of a new living species of giant clam in the Red Sea. *Current Biology* **18**, 1349–1354.

Roa-Quiaoit, H.A.F. 2005. *The ecology and culture of giant clams (Tridacnidae) in the Jordanian sector of Gulf of Aqaba, Red Sea*. PhD Dissertation, University of Bremen, Germany.

Rosewater, J. 1965. The family Tridacnidae in the Indo-Pacific. *Indo-Pacific Mollusca* **1**, 347–396.

Rosewater, J. 1982. A new species of *Hippopus* (Bivalvia: Tridacnidae). *The Nautilus* **96**, 3–6.

Sahari, A., Ilias, Z., Sulong, N. & Ibrahim, K. 2002. Giant clam species and distribution at Pulau Layang Layang, Sabah. *Marine Biodiversity of Pulau Layang Layang Malaysia*, 25–28.

Salvat, B. 2000. 11. Status of southeast and central Pacific coral reefs in 'Polynesia Mana Node': Cook Islands, French Polynesia, Kiribati, Niue, Tokelau, Wallis and Futuna. In *Status of Coral Reefs of the World: 2000*, C. Wilkinson (ed.). Townsville, Queensland: Australian Institute of Marine Sciences, 181–198.

Sandin, S.A., Smith, J.E., DeMartini, E.E., Dinsdale, E.A., Donner, S.D., Friedlander, A.M., Konotchick, T., Malay, M., Maragos, J.E., Obura, D., Pantos, O., Paulay, G., Richie, M., Rohwer, F., Schroeder, R.E., Walsh, S., Jackson, J.B.C., Knowlton, N. & Sala, E. 2008. Baselines and degradation of coral reefs in Northern Line Islands. *PLoS ONE* **3**, e1548. doi:10.1371./journal.pone.0001548

Sauni, S., Kronen, M., Pinca, S., Sauni, L., Friedman, K., Chapman, L. & Magron, F. 2008. Tuvalu country report: Profiles and results from survey work at Funafuti, Nukufetau, Vaitupu and Niutao (October–November 2004 and March–April 2005). Noumea, New Caledonia: Pacific Regional Oceanic and Coastal Fisheries Development Programme.

Savage, J.M., Osborne, P.E. & Hudson, M.D. 2013. Abundance and diversity of marine flora and fauna of protected and unprotected reefs of the Koh Rong Archipelago, Cambodia. *Cambodian Journal of Natural History* **2013**, 83–94.

Schwartzmann, C., Durrieu, G., Sow, M., Ciret, P., Lazareth, C.E. & Massabuau, J.-C. 2011. *In situ* giant clam growth rate behaviour in relation to temperature: a one-year coupled study of high-frequency noninvasive valvometry and sclerochronology. *Limnology and Oceanography* **56**, 1940–1951.

Seeto, J., Nunn, P.D. & Sanjana, S. 2012. Human-mediated prehistoric marine extinction in the tropical Pacific? Understanding the presence of *Hippopus hippopus* (Linn. 1758) in ancient shell middens on the Rove Peninsula, Southwest Viti Levu Island, Fiji. *Geoarchaeology, An International Journal* **27**, 2–17.

Selin, N.I. & Latypov, Y.Y. 2011. The size and age structure of *Tridacna crocea* Lamarck, 1819 (Bivalvia: Tridacnidae) in the coastal area of islands of the Cön Dao Archipelago in the South China Sea. *Russian Journal of Marine Biology* **37**, 376–383.

Selin, N.I., Latypov, Y.Y., Malyutin, A.N. & Bolshakova, L.N. 1992. Chapter 4: Species composition and abundance of corals and other invertebrates on the reefs of the Seychelles Islands. Atoll Research Bulletin No. 368. Washington, DC: The Smithsonian Institution. doi:10.5479/si.00775630.368.1

Sheppard, A.L.S. 1984. The molluscan fauna of Chagos (Indian Ocean) and an analysis of its broad distribution patterns. *Coral Reefs* **3**, 43–50.

Siaosi, F., Sapatu, M., Lalavanua, W., Pakoa, K., Yeeting, B., Magron, F., Moore, B., Bertram, I. & Chapman, L. 2012. *Climate Change Baseline Assessment: Funafuti Atoll, Tuvalu July–August 2011*. Noumea, New Caledonia: Coastal Fisheries Science and Management Section, Secretariat of the Pacific Community. Online. http://www.spc.int/DigitalLibrary/Doc/FAME/Reports/Siaosi_12_Tuvalu_Climate_Change_Baseline_Monitoring_Report.pdf (accessed 15 March 2017).

Sims, N.A. & Howard NT-A-K. 1988. Indigeneous tridacnid clam populations and the introduction of *Tridacna derasa* in the Cook Islands. In *Giant Clams in Asia and the Pacific*, J.W. Copland & J.S. Lucas (eds). Canberra: Australian Centre for International Agricultural Research, 34–40.

Sirenko, B.I. & Scarlato, O.A. 1991. *Tridacna rosewateri* sp. n. A new species of giant clam from Indian Ocean. *La Conchiglia* **22**, 4–9.

Smith, A.J. 1992. Federated States of Micronesia Marine Resources Profiles. FFA Report 92/17. Honiara, Solomon Islands: Pacific Islands Forum Fisheries Agency. Online. www.spc.int/DigitalLibrary/Doc/FAME/FFA/Reports/FFA_1992_017.pdf (accessed 13 April 2017).

Smith, S.D.A. 2011. Growth and population dynamics of the giant clam *Tridacna maxima* (Röding) at its southern limit of distribution in coastal, subtropical eastern Australia. *Molluscan Research* **31**, 37–41.

Sommer, C., Schneider, W. & Poutiers, J.M. 1996. *FAO Species Identification Field Guide for Fishery Purposes. The Living Marine Resources of Somalia*. Rome: Food and Agriculture Organization.

Sone, S. & Loto'ahea, T. 1995. Ocean culture of giant clam in Tonga. Joint FFA/SPC workshop on the management of South Pacific Inshore Fisheries, Noumea, New Caledonia, 26 June–7 July 1995. SPC/Inshore Fish Mgmt/BP8. Noumea, New Caledonia: South Pacific Commission.

South, R. & Skelton, P. 2000. 10. Status of coral reefs in the Southwest Pacific: Fiji, Nauru, New Caledonia, Samoa, Solomon Islands, Tuvalu and Vanuatu. In *Status of Coral Reefs of the World: 2000*, C. Wilkinson (ed.). Townsville, Queensland: Australian Institute of Marine Sciences, 159–180.

Stojkovich, J.O. 1977. Survey and species inventory of representative pristine marine communities of Guam. University of Guam Marine Laboratory, Technical Report No. 40, October 1977. Mangilao, Guam: University of Guam.

Strotz, L.C., Mamo, B.L., Topper, T.P. & Bagnato, C. 2010. The highest southern latitude record of a living *Tridacna gigas*. *Malacologia* **53**, 155–159.

Su, Y., Hung, J.-H., Kubo, H. & Liu, L.-L. 2014. *Tridacna noae* (Röding, 1798) – a valid giant clam species separated from *T. maxima* (Röding, 1798) by morphological and genetic data. *Raffles Bulletin of Zoology* **62**, 124–135.

Tabugo, S.R.M., Pattuinan, J.O., Sespene, N.J.J. & Jamasali, A.J. 2013. Some economically important bivalves and gastropods found in the Island of Hadji Panglima Tahil, in the province of Sulu, Philippines. *International Research Journal of Biological Sciences* **2**, 30–36.

Tacconi, L. & Tisdell, C. 1992. Domestic markets and demand for giant clam meat in the South Pacific Islands: Fiji, Tonga and Western Samoa. In *Giant Clams in the Sustainable Development of the South Pacific: Socioeconomic Issues in Mariculture and Conservation*, C. Tisdell (ed.). ACIAR Monograph No. 18. Canberra: Australian Centre for International Agricultural Research, 205–222.

Tadashi, K., Dai, C.F., Park, H-S., Huang, H. & Ang, P.O. 2008. 10. Status of coral reefs in East and North Asia (China, Hong Kong, Taiwan, South Korea and Japan). In *Status of Coral Reefs of the World: 2008*, C. Wilkinson (ed.). Townsville, Australia: Global Coral Reef Monitoring Network and Reef and Rainforest Research Centre, 145–158.

Tan, S.H. & Zulfigar, Y. 2001. Factors affecting the dispersal *Tridacna squamosa* larvae and gamete material in the Tioman Archipelago, The South China Sea. *Phuket Marine Biological Center Special Publication* **25**, 349–356.

Tan, S.-H. & Zulfigar, Y. 2003. Status of giant clam in Malaysia. *SPC Trochus Information Bulletin* **10**, 9–10.

Tan, K.S. & Kastoro, W.W. 2004. A small collection of gastropods and bivalves from the Anambas and Natuna Islands, South China Sea. *Raffles Bulletin of Zoology* **11**, 47–54.

Tan, S.H., Yasin, Z.B., Salleh, I.B. & Yusof, A.A. 1998. Status of giant clams in Pulau Tioman, Malaysia. *Malayan Nature Journal* **52**, 205–216.

Tan, S.-K. & Low, M.E.Y. 2014. Checklist of the Mollusca of Cocos (Keeling)/Christmas Island ecoregion. *Raffles Bulletin of Zoology* **30**, 313–375.

Tang, Y.C. 2005. *The systematic status of* Tridacna maxima *(Bivalvia: Tridacnidae) based on morphological and molecular evidence*. MSc Dissertation, National Taiwan Ocean University, Taiwan.

Taniera, T. 1988. Status of giant clams in Kiribati. In *Giant Clams in Asia and the Pacific*, J.W. Copland & J.S. Lucas (eds). Canberra: Australian Centre for International Agricultural Research, 47–48.

Taylor, J.D. & Reid, D.G. 1984. The abundance and trophic classification of molluscs upon coral reefs in the Sudanese Red Sea. *Journal of Natural History* **18**, 175–209.

Taylor, J.D. 1968. Coral reef and associated invertebrate communities (mainly molluscan) around Mahé, Seychelles. *Philosophical Transactions of the Royal Society of London, Series B, Biological Sciences* **254**, 129–206.

Thaman, R.R., Puia, T., Tongabaea, W., Namona, A. & Fong, T. 2011. Marine biodiversity and ethnobiodiversity of Bellona (Mungiki) Island, Solomon Islands. *Singapore Journal of Tropical Geography* **31**, 70–84.

Thomas, F.R. 2001. Mollusk habitats and fisheries in Kiribati: an assessment from the Gilbert Islands. *Pacific Science* **55**, 77–97.

Thomas, F.R. 2014. Shellfish gathering and conservation on low coral islands: Kiribati perspecitives. *The Journal of Island and Coastal Archaeology* **9**, 203–218.

Thorne, B.V., Mulligan, B., Mag Aoidh, R. & Longhurst, K. 2015. Current status of coral reef health around the Koh Rong Archipelago, Cambodia. *Cambodian Journal of Natural History* **2015**, 98–113.

Tiavouane, J. & Fauvelot, C. 2016. First record of the Devil Clam, *Tridacna mbalavuana* Ladd 1934, in New Caledonia. *Marine Biodiversity*, doi:10.1007/s12526-016-0506-1

Tiitii, U., Roebeck, U. & Gomez, R.G. 2014. Samoa aquaculture section team fully involved in giant clam farming. *Fisheries Newsletter* **145**, 29.

Tisdell, C. & Wittenberg, R. 1992. The market for giant clam meat in New Zealand: results of interviews with Pacific Island immigrants. In *Giant Clams in the Sustainable Development of the South Pacific: Socioeconomic Issues in Mariculture and Conservation*, C. Tisdell (ed.). ACIAR Monograph No. 18. Canberra: Australian Centre for International Agricultural Research, 258–274.

Tomlin, J.R. 1934. The marine mollusca of Christmas Island, Indian Ocean. *Bulletin of Raffles Museum* **9**, 74–84.

Tu'avao, T., Loto'ahea, T., Udagawa, K. & Sone, S. 1995. Results of the field surveys on giant clam stock in the Tongatapu Island Group. *Fisheries Research Bulletin of Tonga* **3**, 1–10.

Ullmann, J. 2013. Population status of giant clams (Mollusca: Tridacnidae) in the northern Red Sea, Egypt. *Zoology in the Middle East* **59**, 253–260.

Van Long, N., Hoang, P.K., Ben, H.X. & Stockwell, B. 2008. Status of the marine biodiversity in the Northern Spratly Islands, South China Sea. In *Proceedings of the Conference on the Results of the Philippines-Vietnam Joint Oceanographic and Marine Scientific Research Expedition in the South China Sea (JOMSRE-SCS I to IV), 26–29 March 2008, Ha Long City, Vietnam*, A.C. Alcala (ed.). Pasay City, Republic of the Philippines: Technical Cooperation Council of the Philippines of the Department of Foreign Affairs, 11–19.

Van Wynsberge, S., Andréfouët, S., Gilbert, A., Stein, A. & Remoissenet, G. 2013. Best management strategies for sustainable giant clam fishery in French Polynesia Islands: answers from a spatial modelling approach. *PLoS ONE* **8**, e64641. doi:10.1371/journal.pone.0064641

Vieux, C. 2009. Assessment of targeted invertebrate species of the northwestern lagoon of Grande-Terre (Poum to Koumac). In *A Rapid Marine Biodiversity Assessment of the Coral Reefs of the Northwest Lagoon, between Yandé and Koumac, Province Nord, New Caledonia*, S.A. McKenna & J. Spaggiari (eds). RAP Bulletin of Biological Assessment 53. Arlington, Virginia: Conservation International, 41–46.

Vieux, C., Aubanel, A., Axford, J., Chancerelle, Y., Fisk, D., Holland, P., Juncker, M., Kirata, T., Kronen, M., Osenberg, C., Pasisi, B., Power, M., Salvat, B., Shima, J. & Vavia, V. 2004. 13. A century of change in coral reef status in southeast and central Pacific: Polynesia Mana Node, Cook Islands, French Polynesia, Kiribati, Niue, Tokelau, Tonga, Wallis and Futuna. In *Status of Coral Reefs of the World: 2004, Volume 2*, C. Wilkinson (ed.). Townsville, Queensland: Australian Government and Australian Institute of Marine Sciences, 363–380.

Villanoy, C.L., Juinio, A.R. & Meñez, L.A. 1988. Fishing mortality rates of giant clams (family Tridacnidae) from the Sulu Archipelago and Southern Palawan, Philippines. *Coral Reefs* **7**, 1–5.

Virly, S. 2004. Etude préliminaire relative à la ressource en bénitier en Province Nord: Statut écologique et halieutique. Koohne, New Caledonia: Service de l'Environnement de la Province Nord.

Vuki, V., Tisdell, C. & Tacconi, L. 1992. Giant clams, socioeconomics and village life in the Lau group, Fiji: prospects for farming Tridacnids. In *Giant Clams in the Sustainable Development of the South Pacific: Socioeconomic Issues in Mariculture and Conservation*, C. Tisdell (ed.). ACIAR Monograph No. 18. Canberra: Australian Centre for International Agricultural Research, 17–37.

Wantiez, L., Bouilleret, F., Clément, G. & Virly, S. 2007a. Communautés biologiques et habitat corallien de la Corne Sud. Etat initial. Unpublished report. Noumea, New Caledonia: Province Sud de la Nouvelle-Calédonie, Université de la Nouvelle-Calédonie. doi:10.13140/RG.2.1.1871.9768

Wantiez, L., Bouilleret, F., Clément, G. & Virly, S. 2007b. Communautés biologiques et habitats coralliens de l'île des Pins. Etat initial. Unpublished report. Noumea, New Caledonia: Province Sud de la Nouvelle-Calédonie, Université de la Nouvelle-Calédonie. doi:10.13140/RG.2.1.3444.8400

Wantiez, L., Bouilleret, F., Clément, G. & Virly, S. 2007c. Communautés biologiques et habitats coralliens de Bourail. Etat inital. Unpublished report. Noumea, New Caledonia: Province Sud de la Nouvelle-Calédonie, Université de la Nouvelle-Calédonie. doi:10.13140/RG.2.1.2920.5522

Wantiez, L., Bouilleret, F., le Mouellic, S. & Virly, S. 2008a. Communautés biologiques et habitats coralliens du Grand Lagon Nord. Etat initial. Unpublished report. Noumea, New Caledonia: Province Nord de la Nouvelle-Calédonie, Aquarium des Lagons. doi:10.13140/RG.2.1.3313.7686

Wantiez, L., Sarramégna, S. & Virly, S. 2008b. Communautés biologiques et habitats coralliens de la réserve intégrale Merlet. Etat inital. Unpublished report. Noumea, New Caledonia: Province Sud de la Nouvelle-Calédonie, Aquarium des lagons. doi:10.13140/RG.2.1.4362.3441

Wells, F.E. & Kinch, J.P. 2003. Chapter 3 Molluscs of Milne Bay Province, Papua New Guinea. In *A Rapid Marine Biodiversity Assessment of Milne Bay Province, Papua New Guinea – Survey II (2000)*, G.R. Allen et al. (eds). RAP Bulletin of Biological Assessment 29. Washington, DC: Conservation International, 39–45.

Wells, F.E. & Slack-Smith, S.M. 2000. Molluscs of Christmas Island. In *Survey of the Marine Fauna of the Montebello Islands, Western Australia and Christmas Island, Indian Ocean*, P.F. Berry & F.E. Wells (eds). Records of the Western Australian Museum Supplement No. 59, 103–115. Online. http://museum.wa.gov.au/research/records-supplements/records/molluscs-christmas-island (accessed 13 April 2017).

Wells, F.E. 1994. *Chapter 12 Marine molluscs of the Cocos (Keeling) Islands.* Atoll Research Bulletin No. 410. Washington, DC: The Smithsonian Institution. doi:10.5479/si.00775630.410.1

Wells, F.E. 2001. Chapter 3 Molluscs of the Gulf of Tomini, Sulawesi, Indonesia. In *A Marine Rapid Assessment of the Togean and Banggai Islands, Sulawesi, Indonesia*, G.R. Allen & S.A. McKenna (eds). RAP Bulletin of Biological Assessment 20. Washington, DC: Conservation International, 38–43.

Wells, F.E. 2002. Chapter 2 Molluscs of Rajah Ampat Islands, Papua Province, Indonesia. In *A Marine Rapid Assessment of the Rajah Ampat Islands, Papua Province, Indonesia*, S.A. McKenna et al. (eds). RAP Bulletin of Biological Assessment 22. Washington, DC: Conservation International, 37–45.

Wells, S.M., Pyle, R.M. & Collins, N.M. 1983. *The IUCN Invertebrate Red Data Book.* Gland, Switzerland: International Union for Conservation of Nature.

Williams, G.J., Smith, J.E., Conklin, E.J., Gove, J.M., Sala, E. & Sandin, S.A. 2013. Benthic communities at two remote Pacific coral reefs: effects of reef habitat, depth, and wave energy gradients on spatial patterns. *PeerJ* 1, e81. doi:10.7717/peerj.81

Wu, W.-L. 1999. The list of Taiwan bivalve fauna. *Quarterly Journal of the Taiwan Musuem* 33, 55–208.

Yusuf, C., Ambariyanto & Hartati, R. 2009. Abundance of *Tridacna* (family Tridacnidae) at Seribu Islands and Manado waters, Indonesia. *Ilmu Kelautan* 14, 150–154.

Zann, L.P. & Ayling, A.M. 1988. Status of giant clams in Vanuatu. In *Giant Clams in Asia and the Pacific*, J.W. Copland & J.S. Lucas (eds). Canberra: Australian Centre for International Agricultural Research, 60–63.

Zann, L.P. 1989. A preliminary check list of the major species of fishes and other marine organisms in Western Samoa. FAO/UNDP SAM/89/002 Field Report No. 1.

Zann, L.P. 1991. The inshore resources of Upolu, Western Samoa: Coastal inventory and fisheries database. FAO/UNDP SAM/89/002 Field Report No. 5.

Zhuang, Q. 1978. The Tridacnids of the Xisha Islands, Guangdong Province, China. *Studia Marina Sinica* 12, 133–139.

Zulfigar, Y. & Tan, A.S.-H. 2000. Quantitative and qualitative effects of light on the distribution of giant clams at the Johore Islands in South China Sea. *Phuket Marine Biological Center Special Publication* 21, 113–118.

Zuschin, M. & Piller, W.E. 1997. Bivalve distribution on coral carpets in the Northern Bay of Safaga (Red Sea, Egypt) and its relation to environmental parameters. *Facies* 37, 183–194.

Zuschin, M. & Stachowitsch, M. 2007. The distribution of molluscan assemblages and their postmortem fate on coral reefs in the Gulf of Aqaba (northern Red Sea). *Marine Biology* 151, 2217–2230.

Oceanography and Marine Biology: An Annual Review, 2017, **55**, 389-420
© S. J. Hawkins, D. J. Hughes, I. P. Smith, A. C. Dale, L. B. Firth, and A. J. Evans, Editors
Taylor & Francis

HOW ANTHROPOGENIC ACTIVITIES AFFECT THE ESTABLISHMENT AND SPREAD OF NON-INDIGENOUS SPECIES POST-ARRIVAL

EMMA L. JOHNSTON[1,2]*, KATHERINE A. DAFFORN[1,2],
GRAEME F. CLARK[1], MARC RIUS[3,4] & OLIVER FLOERL[5]

[1]*Evolution & Ecology Research Centre, School of Biological, Earth and Environmental
Sciences, University of New South Wales, Sydney, New South Wales 2052, Australia*
[2]*Sydney Institute of Marine Science, Mosman, New South Wales 2088, Australia*
[3]*Ocean and Earth Science, University of Southampton, National Oceanography
Centre, European Way, Southampton, SO14 3ZH, United Kingdom*
[4]*Department of Zoology, University of Johannesburg,
Auckland Park, 2006, Johannesburg, South Africa*
[5]*Cawthron Institute, Nelson 7010, New Zealand*
Corresponding author: Emma L. Johnston
e-mail: e.johnston@unsw.edu.au, tel: +61 2 93851825

When humans transport a species to a location outside its native range, multiple biotic and abiotic factors influence its post-arrival establishment and spread. Abiotic factors such as disturbance and environmental conditions determine the suitability of the new environment for an invader, as well as influence resource availability and ecological succession. Biotic processes such as competition, facilitation, predation and disease can either limit or promote invasion, as can emergent community-level traits such as species diversity. Synergies arise when the abiotic and biotic factors controlling invasion success are themselves influenced by anthropogenic activities, such as those associated with coastal urbanization and industrialization. Here we present a review of the major anthropogenic activities that affect the success of non-indigenous species (NIS) post-arrival. We prioritize the factors in terms of their ecological and evolutionary importance, and present potential management actions to reduce NIS success post-arrival. Evidence-based management has the potential to mitigate anthropogenic activities that enhance invasion success. High priority management actions include: 1) the removal, or containment, of legacy contaminants and reduction of new inputs to reduce the competitive advantage that some invaders have in contaminated environments, 2) the redesign of artificial structures to reduce colonization by NIS through eco-engineering, selection of construction materials and the 'seeding' of structures with native species to provide a priority advantage, 3) the management of dominant regional transport pathways to ensure that the risk of transporting NIS via our increasingly complex transport networks is minimized and 4) the protection and maintenance of biotic resilience in the form of intact living habitats and endemic diversity. Further research is required to advance our understanding of the role of anthropogenic activities in driving post-arrival success of NIS. Such work is vital for developing responsive and mechanistic management plans and ultimately for reducing the impacts of marine invasive species.

Introduction

The invasion of natural ecosystems by non-indigenous species (NIS) is one of the greatest threats to native biodiversity (Wilcove et al. 1998, Butchart et al. 2010). Although only a small proportion of NIS artificially transported to new regions establish, spread and cause impacts (Williamson et al. 1986, Suarez et al. 2005, Blackburn et al. 2011), successful invasions have had a wide range of effects on native biota. The post-arrival establishment of NIS is strongly influenced by a number of biotic and abiotic factors (Theoharides & Dukes 2007, Forrest et al. 2009). For example, ecological interactions such as competition, facilitation, predation or disease and environmental factors such as temperature and salinity may produce synergies that allow ecological dominance of NIS (Castilla et al. 2004). Species traits, such as predatory avoidance or growth rate, can sometimes be linked to the success of NIS over natives (Van Kleunen et al. 2010, McKnight et al. 2016). Abiotic influences such as disturbance can regulate resource availability, which may in turn affect invasibility (Davis et al. 2000, Airoldi & Bulleri 2011). Therefore, understanding biotic and abiotic factors that govern the survival and success of NIS and their populations post-arrival is key. In addition, it is important to have a good understanding of major anthropogenic factors that interact with these factors and ultimately shape the success of NIS post-arrival. In particular, anthropogenic factors associated with urbanization and industrialization are key for improving our understanding of post-arrival NIS success (Figure 1). Anthropogenic activities on land and in the ocean change physico-chemical parameters of marine habitats, such as water and sediment quality, directly influencing NIS. However, many NIS have wide tolerances to environmental conditions (Dukes & Mooney 1999, Sorte et al. 2010, Zerebecki & Sorte 2011, Rius et al. 2014b) and to highly toxic chemicals

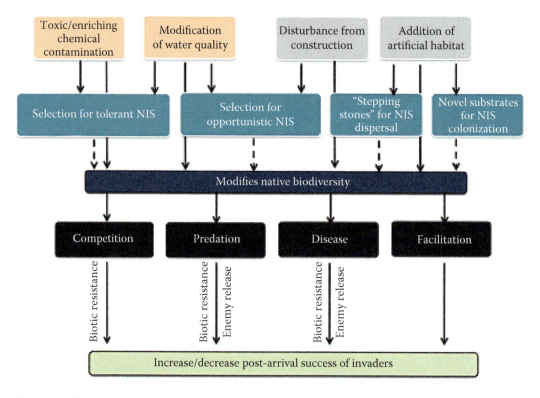

Figure 1 Chemical (orange) and physical (grey) disturbances from human activities that influence the post-arrival success of invaders. Effects of disturbance on NIS can be direct (dark green) or indirect (blue), and can affect associated biotic processes (black). Direct interactions are illustrated by an unbroken line; indirect interactions are illustrated by a broken line.

such as copper biocides (Prentis et al. 2008, Piola et al. 2009). Hence, anthropogenic activities that promote extreme environmental conditions might provide a post-arrival advantage for NIS.

The global increase in anthropogenic activities has resulted in previously-undisturbed marine and estuarine environments being transformed into habitats with artificial features such as pontoons, jetties, breakwaters, boating marinas and commercial ports (Dugan et al. 2011). The physical characteristics of artificial structures tend to differ markedly from that of natural systems (Airoldi et al. 2005, 2009, Airoldi & Beck 2007), creating environmental novelty and newly available artificial habitat (Glasby & Connell 1999). Urban sprawl into our waterways and the construction of vessel infrastructure also results in hydrological modifications that reduce flow and increase silt, nutrient and contaminant retention (Johnston et al. 2011, Rivero et al. 2013).

Human-assisted regional translocation of species can increase connectivity, overcoming barriers to natural dispersal and facilitating the post-establishment spread of NIS, with patterns and rates of spread being very different from those achieved via natural dispersal (Buchan & Padilla 1999, Ruiz & Carlton 2003). Intraregional transport increases propagule pressure of NIS (Zabin et al. 2014). Such transport patterns are likely to increase the frequency of propagule arrival, which is correlated with NIS success in both theoretical (Leung et al. 2004) and experimental studies (Clark & Johnston 2009, Hedge et al. 2012).

Vectors that initially transport a species beyond its native range have been the focus of NIS science and management for decades (Carlton 1985, Ruiz et al. 1997, Hewitt & Campbell 2008, Davidson et al. 2010). After a marine non-indigenous species has arrived, less attention and resources are allocated to its management as removal or control is automatically deemed too expensive or logistically impossible. If we pay more attention to the factors affecting NIS success post-arrival in a new region we can identify the biotic and abiotic conditions that will be important for the likelihood of a species' establishment and spread. For example, more information is needed to understand how human activities influence species traits that promote biological invasions. Such factors may be more amenable to management and more effective than attempts at direct eradication via physical removal or chemical/biological control.

In this paper, we explore the anthropogenic factors that influence the successful establishment and spread of introduced species in the marine environment, post-arrival (Figure 1). We first provide an overview of the major anthropogenic influences to marine environments and describe how they may affect NIS. We separate these factors into four major categories of change: chemical and physical changes to environments, changes to connectivity and changes to the biological aspects of recipient environments. Finally, we highlight areas in which there is potential for effective management of NIS post-arrival.

Chemical alteration of recipient environments

Contamination and changes to water quality

The intense and extensive development by humans across the planet has subjected much of the world's biological diversity to frequent chemical changes, which are often concentrated in urban and industrial areas (Grimm et al. 2008). Human activity is reliant on access to freshwater and trade such that it becomes concentrated around waterways. These activities inevitably release contaminants into water bodies and result in other modifications to physico-chemical conditions. As a consequence, estuaries in particular have been highly impacted by chemical change related to agriculture, industrialization and urbanization, with almost all estuaries suffering some degree of impact (Lotze et al. 2006). An example is fertilizer runoff into waterways; fertilizer use is already responsible for the eutrophication and formation of 'deadzones' in many of the world's coastal waterways (Rabalais et al. 2010); global nitrogen and phosphorous effluent is predicted to increase between 150–180% between the years 2000–2150 (Marchal et al. 2011, Alexandratos &

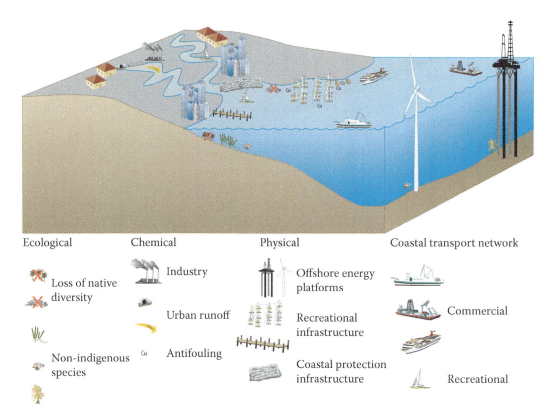

Ecological   Chemical   Physical   Coastal transport network

Loss of native diversity

Industry

Offshore energy platforms

Urban runoff

Recreational infrastructure

Commercial

Non-indigenous species

Antifouling

Coastal protection infrastructure

Recreational

Figure 2 Anthropogenic factors influencing the establishment and spread of NIS include chemical and physical alterations of habitat, which interact with ecological traits and processes, as well as coastal transport networks. The introduction of chemicals from industrial and urban runoff and vessel antifouling paint may facilitate metal-tolerant NIS and reduce native biodiversity. Increasing coastal development adds novel habitat such as recreational infrastructure (e.g. marinas and jetties) and coastal protection infrastructure (e.g. sea walls) for colonization by NIS, and creates stepping-stones for their spread. These structures are linked by busy coastal transport networks, such as commercial and recreational vessel movements or aquaculture operations, which can act as vectors for NIS for inter- and intraregional spread.

Bruinsma 2012). Important chemical stressors include toxic contaminants (e.g. metals) (Birch 2000, Rodríguez-Obeso et al. 2007, Burton & Johnston 2010) and enriching contaminants (e.g. nutrients) (Statham 2012) (Figure 2).

    Chemical stressors that are released into waterways impact the ecological composition and function of important habitats (Johnston & Roberts 2009, Burton & Johnston 2010). Metals are known to have toxic effects on aquatic taxa, including increased mortality (Trannum et al. 2004, Martínez-Lladó et al. 2007), reduced reproductive potential (Alquezar et al. 2006, Simpson & Spadaro 2011) and other sublethal effects (Fleeger et al. 2003). Nutrients such as dissolved nitrogen and phosphorus can also have community-wide effects, with high levels resulting in a community composed of very high densities of a few tolerant opportunistic species (Pearson & Rosenberg 1978). However, in contrast to toxic contaminants, such nutrients initially have an enriching effect, increasing the richness and abundance of primary producers with consequences at higher trophic levels (Tewfik et al. 2005, Smith et al. 2006, Elser et al. 2007, McKinley & Johnston 2010, Clark et al. 2015).

    Environmental suitability is now acknowledged as a strong predictor of invasion success, and consideration of the role of chemical parameters has generally emphasized natural environmental variables such as dissolved oxygen (DO), pH, salinity and temperature (Williamson et al. 1986, Blackburn & Duncan 2001). There has been less consideration of how anthropogenic modifications to environmental conditions may influence the establishment and spread of NIS. Locations where NIS are often introduced are low flow, high retention sites (naturally or by construction) (Floerl & Inglis 2003, Rivero et al. 2013) and therefore also locations of high contaminant retention (Dafforn et al. 2008, Johnston et al. 2011). Where the transport process for NIS is environmentally stressful (e.g. copper-coated vessel hull), it is likely that selection for environmentally tolerant (particularly to copper) NIS will take place (McKenzie et al 2012). This will lead to greater success of NIS in anthropogenically-modified habitats.

Toxic and enriching contaminants aid post-arrival success of NIS

The addition of toxic contaminants is perhaps the least intuitive reason for increasing NIS success, but is the area in which we have the strongest experimental and mensurative data to support the hypothesis. The mechanism is relatively simple and relies on selection for toxicant tolerance in NIS (Piola & Johnston 2009, McKenzie et al. 2012). Elevated concentrations of metals, for example, are highly toxic to many marine organisms (Hall et al. 1998), but some organisms have evolved effective detoxification and avoidance mechanisms and are considered to be superior in their ability to withstand exposures to these toxicants (Johnston 2011).

The majority of marine NIS are transported in ballast water or as hull-fouling organisms (Ruiz et al. 1997, 2000, Hewitt 2002, Godwin 2003, Clarke Murray et al. 2011) and both of these transport mechanisms are highly contaminated with metals (Alzieu et al. 1986, Claisse & Alzieu 1993, Schiff et al. 2004, Warnken et al. 2004, Piola et al. 2009, Dafforn et al. 2011). Hulls are contaminated because they are often coated in toxic antifouling paints containing metal-based biocides, and ballast water tanks are sometimes antifouled and often corroding internally (Tamburri et al. 2005). The transport process may therefore select for metal tolerance, and the major contaminants in ports and harbours are metals (Piola et al. 2009). Metal tolerance has now been observed in a wide range of marine organisms including polychaetes, bryozoans, algae, amphipods and barnacles (reviewed by Johnston 2011, Pineda et al. 2012). Hence, shipping selects for metal-tolerant species, then delivers them to metal-contaminated locations (Piola & Johnston 2008a). This gives NIS arrivals a competitive advantage over local native species that may not have developed tolerance. However, it should be noted that native species can also adapt or become tolerant and there is potential for toxicant tolerance to be a useful risk-identifier for predicting future NIS (Dafforn et al. 2009a). Some studies have observed that assemblages switch from native-dominated to NIS-dominated when exposed to a small strip of antifouling paint (Piola & Johnston 2008a) – an effect as relevant in small marinas as it is in large working ports (Dafforn et al. 2009a). Interestingly, large-scale surveys of selected NIS are now finding that these species possess higher tolerance to metals in more polluted environments than in pristine habitats (Clark et al., unpublished data), and it would appear that NIS have the ability to lose tolerance (which can be costly to fitness) when spreading from contaminated to clean systems (Piola & Johnston 2006).

The addition of nutrients to a system is an example of the addition of resources, and is therefore more readily understood as a mechanism for increasing the success of NIS (Davis et al. 2000). Many high-impact NIS are 'weedy' (reviewed by Sutherland 2004) with an 'r-type' life-history strategy (Ruiz & Hewitt 2002, Hänfling et al. 2011) so they are therefore capable of dominating in high-resource environments. This has been demonstrated for land-based weeds (Grime 1977, Dukes 2001) – more research is required for marine species, but disturbance that renews resources is certainly a facilitator (Clark & Johnston 2009, 2011, Airoldi & Bulleri 2011). Humans elevate nutrients in nearshore areas via run-off from agricultural and urbanized areas and through the release of sewage (Figure 2). Nutrients may be limiting in marine systems and hence anthropogenically-modified waterways may be more productive than natural systems (Nixon et al. 1986, 2001). Productivity increases may result from large changes in community composition (Duarte 1995). Productivity increases will occur up until thresholds are exceeded and excessive eutrophication takes place, reducing water quality and causing oxygen depletion as has happened in much of the Baltic Sea (Carstensen et al. 2014). Thresholds and tipping points will differ for each habitat and each NIS and are therefore difficult to predict without extensive monitoring. Up until such tipping points are reached, the system is increasingly susceptible to fast-growing weedy species that are able to rapidly take advantage of excess primary and secondary food sources (Clark et al. 2015). Non-indigenous species are often considered 'weedy' species and examples of this are the fast-growing invasive *Caulerpa* macroalgal species (Williams & Smith 2007) and harmful microalgae, which have a tendency to bloom in high-nutrient conditions (Hallegraeff & Gollasch 2006).

Changes in water quality and hydrological regimes

Due to a paucity of empirical studies, it is difficult to generalize the response of NIS to anthropogenic modifications of water quality *per se*. Where our activities push these parameters beyond natural realms of variability, we might expect that modifications will benefit species with wide environmental tolerances. Species that benefit are unlikely to be the resident native species that have evolved under historical conditions and will be disadvantaged by changed environmental or biological regimes. Water quality modifications may also be of concern if they represent a change in natural habitat or a uniformity of conditions that, as a result, drives biotic homogenization (McKinney & Lockwood 1999). Anthropogenic modifications of hydrological regimes will likely increase in the future as inland waterway transport is predicted to rise and there will be a greater need for expansion and new canal developments to support this trade (Galil et al. 2007, 2015). These modifications in hydrological regimes are likely to occur together with an increase in available suitable habitat for colonization by NIS. Increasing drought will drive water extractions from river sources, with associated impacts further downstream in estuaries. The interaction of water-usage practices and climate change anomalies has the potential to create invasion windows. For example, the co-occurrence of increased freshwater extraction and increased drought severity is thought to have created saline conditions in San Francisco estuary that benefitted a non-indigenous zooplankton species (Winder et al. 2011). Anthropogenic activities and associated stressors tend to be a common problem and may establish a particular set of conditions that are replicated in harbours around the globe (Halpern et al. 2008, Knights et al. 2013, Pearson et al. 2016). NIS are transported from multiple locations, but it is possible that these locations may have similar water quality conditions because they are busy ports or marinas, usually characterized by low flow, high turbidity, low DO and high nutrient conditions. These situations can create environments that suit a set of species representing 'harbour-tolerant' conditions as described in Floerl et al. (2009a).

Regime shifts associated with climate change may lead to the exacerbation of hydrological regime change and impacts on water quality (Delpla et al. 2009, Whitehead et al. 2009). It is inherently difficult to make predictions in complex ecological systems, but climate change will change the nature of basic chemical interactions. Increasing temperature and pH both have the potential to increase the availability of toxic contaminants (Schiedek et al. 2007, Sokolova & Lannig 2008, Nikinmaa 2013) as does increasing storm activity, which resuspends contaminated sediments (Eggleton & Thomas 2004). With temperature increases, we might also expect increases in primary productivity and an increased frequency of eutrophic events and hypoxia (Rabalais et al. 2009, Moss et al. 2011, O'Neil et al. 2012).

Physical alteration of recipient environments

Estuarine, coastal and offshore development

The estuarine environment faces increasing pressure from encroaching urban and industrial developments (Figure 2). Historically, the majority of human settlement has occurred within 100 km of the coast (Bulleri 2006, Firth et al. 2016) and, despite these areas being at most risk from climate change events, this trend continues (McGranahan et al. 2007). World population growth is projected to increase from 7.2 to 9.6 billion in 2050 (Gerland et al. 2014) and resource demand will result in increased exploitation of the marine environment. For example, some areas of Europe have lost 50–80% of coastal wetlands and seagrasses to development to support urban activities (reviewed by Airoldi & Beck 2007). Anthropogenic habitat modification also extends beyond the coastal zone as world population growth has driven the search for new energy sources off shore (Asif & Muneer 2007). The discovery of new oil and gas reserves, such as those in the Arctic region, will result in the continued construction of near and offshore production platforms (Asif & Muneer 2007). Over 7500 offshore oil and gas platforms had been constructed worldwide as of 2003 (Hamzah 2003, Parente et al. 2006). Similarly, renewable energy is moving off shore with the construction of marine wind farms (Kennedy 2005, Punt et al. 2009). To exploit renewable energy sources, several thousand turbines will be constructed in wind farm clusters along the European Atlantic coast (Kennedy 2005) and, presumably, other global locations (Firth et al. 2016). Offshore energy platforms may appear relatively isolated, but they are linked to coastal areas by vessel movements (e.g. maintenance) and therefore can act as sinks or sources of NIS propagules (Yeo et al. 2009, Sammarco et al. 2010, Adams et al. 2014). Furthermore, the decommissioning of offshore installations may remove the structures that support NIS or leave behind permanent structures for NIS that are no longer maintained or monitored (Schroeder & Love 2004, Page et al. 2006, Macreadie et al. 2011). Underwater pipelines have received less attention in relation to their potential impacts, but they connect offshore energy infrastructure with coastal zones and introduce other novel structures to the marine environment that may be colonized by NIS or facilitate their spread (Feary et al. 2011).

Habitat modification often involves the addition of structures that may increase or replace existing natural habitat (Glasby & Connell 1999). Common structures added to coastal zones include sea walls, break walls and groynes constructed to protect urban coastal zones and maritime vessels (Mineur et al. 2012). Marinas and ports are often protected by break walls and infrastructure within these areas includes pilings and pontoons to support vessel berthing. Furthermore, vessel transport is supported by hydrological modifications including the construction of canals and other waterways. Similarly, offshore energy platforms, while built above the waterline, require extensive underwater scaffolding (Wilson & Elliott 2009). Comparisons of artificial structures and natural habitats have revealed distinct differences in the assemblages able to colonize and persist on them (Connell & Glasby 1999, Glasby 1999a, Glasby & Connell 1999, Atilla et al. 2003, Chapman & Bulleri 2003, Bulleri & Chapman 2004, 2010, Firth et al. 2016). Differences between anthropogenic and natural hard-substratum habitats arise due to their physical characteristics, including substratum composition and microhabitats (Anderson & Underwood 1994, Glasby 2000, Chapman & Bulleri 2003, Chapman 2011, Firth et al. 2013, 2014, Browne & Chapman 2014), age (Perkol-Finkel et al. 2005, Pinn et al. 2005, Burt et al. 2011), orientation or incline (Connell 1999, Glasby & Connell 2001, Saunders & Connell 2001, Knott et al. 2004, Langhamer et al. 2009, Chapman & Underwood 2011, Firth et al. 2015), predation levels (Clynick et al. 2007, Nydam & Stachowicz 2007), illumination levels (Glasby 1999b, Shafer 1999, Marzinelli et al. 2011, Davies et al. 2014), disturbance levels (Airoldi & Bulleri 2011) and movement (Holloway & Connell 2002, Perkol-Finkel et al. 2008, Shenkar et al. 2008, Dafforn et al. 2009b). The increasing transformation of natural to urbanized coastlines has promoted the establishment and spread of NIS (Bulleri & Airoldi 2005, Bulleri et al. 2006, Airoldi et al. 2015).

Artificial structures aid establishment and dispersal of NIS

The addition of artificial structures in close proximity may provide 'stepping stones' for NIS (Glasby & Connell 1999, Coutts & Forrest 2007) (Figure 2), providing 'corridors' for their spread and dispersal (Bulleri & Airoldi 2005, Airoldi et al. 2015). Propagules released from one structure have a higher chance of making it to other structures where the presence of NIS is often higher than in surrounding natural habitats (often sedimentary, Airoldi et al. 2015). The large amount of artificial structures in ports and harbours provides suitable habitat in close proximity to key vectors, such as commercial and recreational vessels (Bulleri & Airoldi 2005) (Figure 2). Some invasive fouling NIS exhibit preferences for shallow floating artificial structures (Lambert & Lambert 1998, Glasby et al. 2007, Dafforn et al. 2009b), potentially because they present a similar surface to a vessel hull with respect to movement and depth (Neves et al. 2007). Such artificial habitats unprecedentedly provide downward-facing surfaces, which are uncommon in natural ecosystems (Miller & Etter 2008). Thus, certain species that were present in low abundance in nature are now thriving in these new environments. Research on larval phototaxis and geotaxis of ascidians (arguably one of the most important marine groups in terms of NIS) (Zhan et al. 2015), found that some globally-distributed species show settlement preference for downward surfaces (Svane & Dolmer 1995, Rius et al. 2010). In addition, studies have shown that recruitment of fouling species are enhanced within proximity to a pier (Hedge & Johnston 2012) and shading (Miller & Etter 2008). Moreover, the design of ports and marinas can disrupt tidal flushing and result in vastly local increased recruitment rates (Floerl & Inglis 2003, Johnston et al. 2011, Toh et al. 2016), but also increase regional connectivity by creating a network of substrata away from initial invader entry points (Knights et al., 2016). As a result, the characteristics of marinas increase retention of NIS propagules and provide a substratum for their establishment (Vaselli et al. 2008). This effect has been so strong that association with artificial structures has been used as a criterion for classifying species as non-indigenous (Chapman & Carlton 1991). Even temporary or removable infrastructure, such as slow-moving barges and drilling rigs, can cause changes to the local habitat. During their period of operation, these floating structures provide a hard substratum for colonization of NIS, often surrounded by soft-sediment habitats (Sheehy & Vik 2010). Invasive corals of the genus *Tubastraea* have, in recent years, colonized an ever-increasing proportion of oil-production infrastructure off the coasts of Brazil and southern USA (Sammarco et al. 2010, 2012, 2014, Costa et al. 2014). Hydrological modifications have also been implicated in the spread of NIS. For example, canals and waterways provide links between distant areas that would otherwise be isolated (Galil et al. 2007, Bishop et al. 2017). The increased addition of artificial structures into coastal areas and hydrological modifications that link isolated waterways provide NIS with 'stepping stones', networks and 'dispersal corridors', respectively (Glasby & Connell 1999, Bulleri & Airoldi 2005, Coutts & Forrest 2007).

The increasing intensity of storms and rising sea levels associated with climate change are likely to increase the need for artificial coastal defences to be constructed on a global scale (Nicholls & Mimura 1998, Moschella et al. 2005, FitzGerald et al. 2008). At the same time, climate change has created environmental conditions that have facilitated significant range expansion of various species (Barry et al. 1995, Hawkins et al. 2009, Ling et al. 2009, Mieszkowska et al. 2014, Rius et al. 2014a). Therefore, increased connectivity from networks of hard structures, together with climate change, may additively enhance the spread of NIS by providing habitat for colonization in areas opened up by warming temperatures (Ware et al. 2014, Firth et al. 2016). In addition to warming, there is a growing number of studies reporting more frequent extreme weather conditions (e.g. significantly larger differences between minimum and maximum seawater conditions, Wernberg et al. 2011, 2012, 2013), which facilitates the success of species with broader thermal ranges (Rius et al. 2014a).

In addition to increasing energy demands over the next decades, an increased demand for protein will result in a considerable expansion of the global aquaculture industry (FAO 2012). This will involve the construction of larger and denser aggregations of fin and shellfish farms in coastal regions that provide extensive artificial habitats to NIS (Fitridge et al. 2012). Another example is the Norwegian salmon farming industry that currently operates ~ 700 coastal farms, each comprising approximately 50,000 m^2 of artificial habitat (Bloecher et al. 2015), and is predicted to grow five-fold by 2050 (Olafsen 2012). Aquaculture facilities are a major vector of NIS (Voisin et al. 2005, Fitridge et al. 2012, Aldred & Clare 2014), not only because they intentionally introduce NIS to be

Figure 3 Extensive fouling of a salmon farm pontoon supports a diverse community of non-indigenous species, including the invasive ascidian, *Didemnum vexillum*. The development of dense aquaculture farming regions can facilitate the human-assisted spread of NIS and disease pathogens via the provision of stepping-stone habitats and a complex transport network. (Photo: Javier Atalah, Cawthron Institute)

farmed, but because they unintentionally transport associated organisms that may establish and spread in the new range (Rosa et al. 2013, Woodin et al. 2014, Grosholz et al. 2015) (Figure 3). Once the farmed or associated species grow, they release propagules that will settle both in the aquaculture facilities and elsewhere. After the initial introduction in the 1970s of the Mediterranean mussel, *Mytilus galloprovincialis*, for farming in South Africa, this species spread more than 2000 km, where it now dominates extensive sections of the rocky intertidal zone (Rius et al. 2011). In addition, the presence of aquaculture facilities has provided new artificial substrata where other NIS that coexist with the farmed NIS can thrive (Rius et al. 2011). Research has shown that such coexistence can allow persistence over long periods, which means that aquaculture facilities act as incubators for multiple NIS. Another problem associated with aquaculture facilities is the accidental release of the non-indigenous farmed stock (Schröder & De Leaniz 2011), such as the case of fish farms in Chile (Soto et al. 2001, 2006, Soto & Norambuena 2004). Finally, the development of dense aquaculture farming regions can facilitate the human-assisted spread of NIS and disease pathogens via the provision of stepping-stone habitats for natural dispersal and a complex transport network (Murray et al. 2002, Morrisey et al. 2011) (Figure 4).

Connectivity: coastal transport networks

Commercial and recreational vectors

Once established, the spread of marine NIS is often facilitated through the presence of extensive transport networks associated with coastal shipping, boating and aquaculture. Urbanized coastlines are characterized by the presence of commercial ports, boating marinas, ferry terminals and other infrastructure that are associated with a wide range of vectors, including merchant ships, cruise liners, naval vessels, car and passenger ferries, water taxis, recreational yachts, dredges, barges and others (Figure 2). Movements of these vectors occur at local (e.g. car ferries, water taxis, service vessels), regional or national scales (e.g. merchant or recreational vessels). Domestic transport networks can be complex. For example, New Zealand's recreational vessel network comprises > 500 distinct voyage routes among 36 of the country's main marina facilities, and involves > 8000 marina-to-marina voyages per year. In addition, there are ~ 7200 annual movements of large commercial vessels between New Zealand's commercial ports that occur via > 300 voyage routes (Floerl

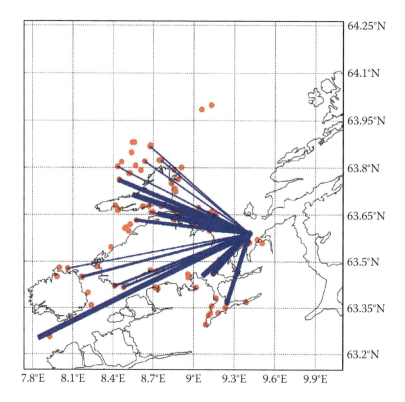

Figure 4 Annual 'connectivity' of a single Norwegian salmon farm in a network of 90 salmon farms on the coast of mid-Norway. The thickness of blue lines indicates the strength of connectivity. Vessel movements were tracked using automatic identification system (AIS) data (Floerl unpublished data).

et al. 2009a, Hayden et al. 2009). Many recreational, cruise, aquaculture and fishing vessels also regularly move between coastal centres and a wide range of relatively pristine natural coastal environments (Wasson et al. 2001, Gust et al. 2008, Zabin et al. 2014). For example, in 2013, a subset of 90 commercial salmon farms along the coast of Trøndelag, Norway, was visited by a total of 204 different vessels, including well-boats and feed, cleaning and service vessels (Figure 3). Individual farms received up to 400 visits from up to 57 different vessels, and farm-to-farm voyages of contractor vessels connected individual farms with up to 20 other farms (Floerl 2014). Human-assisted translocation can overcome barriers to natural dispersal and result in patterns and rates of spread very different from those achieved via natural dispersal (Buchan & Padilla 1999, Ruiz & Carlton 2003, Seebens et al. 2013) (Figure 4).

Coastal transport networks facilitate transport of NIS at local, regional and domestic scales

Transport of NIS via movements of vessels and other mobile submerged infrastructure predominantly occurs via biofouling on submerged surfaces or in internal ballast water (Drake & Lodge 2007, Hewitt & Campbell 2008). International conventions and guidelines to regulate shipping pathways are in development or already operational (IMO 2005, 2011b), but the domestic risk associated with both of these transportation modes remains largely unmanaged by most coastal nations. For example, Simkanin et al. (2009) established that approximately 27% (~ 6 million metric tons) of ballast water discharged at commercial ports on the west coast of the USA originates from other west coast ports, which can facilitate the translocation of organisms among regional ports. Similarly, domestic vessel movements in the USA and New Zealand (to name two examples) are likely to transport biofouling species between ports or from ports to remote natural environments (Floerl et al. 2009b, Zabin et al. 2014). Dispersal of biofouling species can be facilitated by all types of vessels and mobile infrastructure. The greatest risk is often attributed to vessels that spend extended periods (weeks to months) at their destinations, such as many recreational vessels, towed barges and inactive commercial vessels (Apte et al. 2000, Davidson et al. 2008, Floerl & Coutts 2009). However, Schimanski (2015) recently showed that the export of recruits from a local larval pool via vessel movements can occur following residency periods of a single day. Recreational vessels are implicated in the domestic spread of NIS in North America (Wasson et al. 2001, Davidson et al. 2008, Clarke Murray et al. 2011, Zabin et al. 2014), Europe (Fletcher & Farrell 1999, Dupont et al. 2010) and New Zealand (Goldstien et al. 2010). There is also evidence of commercial vessels acting as domestic transport vectors. For example, the translocation of a dumb barge from New Zealand's North Island to the South Island has facilitated the dispersal of the invasive ascidian, *Didemnum vexillum*, from its probable founder population into the heart of the country's aquaculture growing region 500 km further south, where it established highly prolific populations (Coutts & Forrest 2007, Forrest et al. 2013). Regional movements of leased marina pontoons and transfers of aquaculture stock between growing regions have also been identified as potential dispersal vectors of NIS (Forrest & Blakemore 2006, Gust et al. 2008).

The spread of NIS is determined by myriad factors that are not all well understood and that are likely to differ between species and environmental contexts. For example, there are known relationships between propagule pressure and colonization success (Hedge & Johnston 2012) and between aspects (size, gene pool, etc.) of founder populations and longer-term persistence (Simberloff & Gibbons 2004). The modes of anthropogenic transport described above can enhance the post-establishment success of marine NIS in several ways. First, they can facilitate the establishment of further regional satellite populations whose cumulative spread and impact can be greater and far more difficult to manage than that of a single invasive population (Moody & Mack 1988). Second, some elements of coastal transport networks, such as container vessel movements, involve repeat voyages or loops (Kaluza et al. 2010). These can facilitate recurring introductions of propagules to established NIS populations, which may enhance resilience and adaptive capacity of such populations to disturbance or environmental change (Carlton & Hodder 1995, Prentis et al. 2008, Hedge et al. 2012). Third, repeat introductions can also help small populations overcome Allee effects and become self-sustaining (Drake & Lodge 2006) thereby increasing invasion risk. Finally, the transport of biofouling species on vessels can select for individuals that are particularly robust (e.g. environmentally tolerant, Piola & Johnston 2008a). Vessels with particular voyage profiles, such as frequent short-distance voyages, may facilitate transport of recruits that are able to produce viable offspring for release in vessels' future destinations (Schimanski 2015).

There is mounting evidence that both recreational vessels and commercial shipping allow the translocation of genotypes around the world. Many studies show little genetic differentiation among distant populations found within a species' introduced range (Tepolt et al. 2009, Rius et al. 2012, Ordóñez et al. 2013), indicating the presence of population connectivity both at regional and global scales. Although genetic bottlenecks can have deleterious effects on recently-established introduced populations (Roman & Darling 2007), the majority of marine genetic studies support the idea that introduced populations have high levels of genetic diversity as a result of recurrent introductions from multiple and diverse sources (Rius et al. 2015). Human-assisted global reshuffling of genotypes may have evolutionary consequences for species and assemblages in both introduced and native ranges (Olden et al. 2004, Hudson et al. 2016). Human activities fundamentally alter evolutionary trajectories that have been shaped by millions of years. For example, both artificial transport of species and climate change facilitate contacts of previously isolated genotypes, which unprecedentedly increases hybridization rates (Rius & Darling 2014, Vallejo-Marin & Hiscock, 2016). However, more research is needed to understand how human activities are affecting species ranges of both native and NIS.

Ecological, physiological and genetic
alteration of recipient environments

When a NIS interacts with the receiving community, a gradient of possible outcomes can be expected. The first one is ecological dominance by the NIS, in which the absence of natural competitors and predators (the so-called 'enemy release hypothesis', Keane & Crawley 2002) facilitates NIS success. Another possibility is that a NIS becomes established in the new area but that the receiving community limits its success (i.e. biotic containment, Stachowicz et al. 2002a, Levine et al. 2004, Simkanin et al. 2013). Finally, biotic resistance or the ability of resident species to resist newcomers may prevent the establishment of NIS. Human activities have the potential to modify each of these processes and thereby influence the success of NIS post-arrival.

Loss of species diversity and changes in community interactions

One of the most studied but debated forms of community-level biotic resistance is that attributable to species diversity (Elton 1958). Species diversity is thought to affect biotic resistance through two main mechanisms: the sampling effect and species complementarity. The sampling effect refers to the probability that a community will contain one or more dominant species (e.g. superior competitors or predators) that are particularly effective in repelling invaders (Huston 1997). Dominant species may create habitat for many subordinate species, which increases species diversity and inhibits the establishment of new arrivals. Species complementarity, or resource partitioning, refers to differential resource use between species (Schoener 1974, Tilman 1997). A higher number of species can often utilize a larger proportion of the resource base, which reduces invasibility by leaving fewer unused resources available to invaders (Davis et al. 2000).

Experimental studies have shown that diverse communities can 'overyield', where they are more productive and use more resources than would be expected by the sum of component species (Hector et al. 2002). Species complementarity can also occur temporally, when diversity buffers the effects of temporary species loss (Levine 2000). This was observed in a marine system where the primary space-occupiers (colonial ascidians) underwent boom-and-bust cycles at different times of the year, and diverse communities tended to contain species in each boom-phase (Stachowicz et al. 1999). These experimental studies demonstrate effects of species diversity on invasibility at local scales, but there is debate over its importance at larger scales (Fridley et al. 2007, Clark & Johnston 2011, Clark et al. 2013).

Competition between species within trophic levels is an important process in many marine communities (Branch 1984), and represents a key form of biotic resistance. Classic ecological theory identifies three main types of competition (interference, exploitative and apparent), which include both direct and indirect mechanisms (Fellers 1987). Examples of these in hard-substratum marine systems are overgrowth interactions between neighbours (interference, Russ 1982) and competition for resources such as food and space (exploitative, Buss 1990), both of which can act to resist post-establishment invasion (Kimbro et al. 2013). Apparent competition is that mediated by a predator or herbivore and is more difficult to study, but has been implicated as a mechanism influencing the invasion success of some terrestrial plants (Dangremond et al. 2010, Combs et al. 2011).

The importance of competition shaping community composition is context-dependent (Firth et al. 2009, Klein et al. 2011) and is regulated by resources and stress levels (McQuaid et al. 2015), so is variable across space and time. It is less important in early successional or highly disturbed communities where resources are abundant, but becomes increasingly important as communities develop and resources become scarce (Parrish & Bazzaz 1982, Dohn et al. 2013). The degree of niche partitioning also influences the importance of competition, since divergent resource use between species diminishes the frequency and/or intensity of competitive interactions. There is some evidence for latitudinal trends in the intensity of competition (Barnes 2002), which may contribute to differences in the invasibility of latitudinal regions (Freestone & Osman 2011, Freestone et al. 2011, 2013).

Increased disturbance, for example by human activities that chemically or physically alter the environment, typically increases species turnover and the amount of available resources (Davis et al. 2000, Clark & Johnston 2005) (Figure 1). This reduces the importance of competition in structuring communities, and advantages species with r-selected traits and/or tolerance to disturbance. Many NIS are relatively successful in disturbed environments by their virtue of high dispersal (particularly in association with human transport vectors), rapid reproduction and wide environmental tolerance (Piola & Johnston 2008b) (Figure 1). Byers (2002) argued that anthropogenic disturbances create environmental conditions that favour NIS, removing the advantages of pre-adaptation that would normally be held by natives.

The loss of native species or a reduction in their abundance can compromise species interactions that would otherwise provide biotic resistance. Fewer species reduces the scope for species complementarity and the probability that communities will contain dominant taxa (i.e. the sampling effect). Stress that reduces the fitness of native species may weaken biotic resistance by decreasing the intensity of competition, or by altering the outcome of competitive interactions (Liancourt et al. 2005). Conversely, positive interactions (facilitation) between invaders can exacerbate their impacts, spread and subsequent invasions—a phenomenon known as 'invasional meltdown' (Simberloff & Von Holle 1999, Grosholz 2005) (Figure 1). An example in hard-substratum marine fauna is when habitat-forming invaders (e.g. the colonial bryozoan, *Watersipora subtorquata*) provide secondary substratum for other invaders (Floerl et al. 2004), sometimes on antifouling-painted surfaces that would otherwise be uninhabitable by non-tolerant invaders. Facilitative interactions can also occur between native and NIS. Most evidence of this comes from studies on terrestrial and freshwater ecosystems, but some marine examples exist (see review by Rodriguez 2006). Facilitative effects are often transitory (Holloway & Keough 2002) or dependent on environmental conditions (Maestre et al. 2009, Rius & McQuaid 2009, Holmgren & Scheffer 2010). Overall positive and negative ecological interactions affect levels of biodiversity, which ultimately influence the success of NIS.

Impacts of human activities on biotic resistance are spatially variable, as some habitats are dominated by taxa that are particularly susceptible to environmental change. For example, areas with more stable environments (e.g. subtidal reefs or deep-sea sediments) are more likely to contain species less able to adapt to or tolerate change, relative to areas with fluctuating conditions (e.g. tidal rock pools or shallow estuaries) (Levin & Lubchenco 2008). Change in biotic resistance may also be temporally variable, as species can approach their physiological limits during seasonal extremes (Durrant et al. 2013). These fluctuations in natural stress might interact with human stressors to create periods of heightened vulnerability to invasion.

Loss of top-down control or 'enemy release'

Predation or herbivory is a third type of ecological interaction that, in the context of biological invasions, is referred to as top-down control (McEnvoy & Coombs 1999). Non-indigenous marine invertebrates can be prey for some native fish or grazing invertebrates (e.g. echinoderms), and likewise non-indigenous fish can be prey for larger native fish or higher-order predators. The importance of this to marine bioinvasions is difficult to gauge since predation rates on many lower trophic levels are not often well understood. Some evidence exists from studies that describe predation of early life-history stages of marine epifaunal taxa (both native and/or NIS) as a key determinant of the development of benthic communities (Osman & Whitlatch 1995, 2004, Rius et al. 2014b). Soft-bodied marine invertebrates (e.g. solitary and colonial ascidians) may be more prone to predation than those with hard outer shells, so the susceptibility of invertebrate invaders to top-down control can be influenced by their morphology (Lavender et al. 2014), the natural predators present (e.g. specialist versus generalist) and other competitors (Russ 1980, 1982, Osman & Whitlatch 2004).

Chemical defences can also inhibit predation upon fish (Snyder & Burgess 2007) and marine invertebrates (Bakus 1981, Pawlik 1993, Teo & Ryland 1994), and herbivory upon algae (Steinberg 1986, Hay & Fenical 1988), providing some invaders with relative immunity to top-down control (Lagesa et al. 2006, Enge et al. Chapter 6 in this volume).

The 'enemy release hypothesis' refers to situations when the invader has partial or complete immunity from predation in its new range, and can lead to the proliferation and dominance of the invader (Keane & Crawley 2002) (Figure 1). Most examples of the enemy release hypothesis come from terrestrial studies (Colautti et al. 2004, Liu & Stiling 2006), but there is some evidence from the marine environment (Torchin et al. 2003, Blakeslee et al. 2009, 2012). A laboratory study compared the preference of a sea urchin for feeding upon native versus exotic ascidians, and found that the urchin preferred the native prey (Simoncini & Miller 2007). Another study found that even though Hawaiian herbivores grazed introduced algae in preference to native algae, the intensity of herbivory was lower there than in the invaders' native range (Vermeij et al. 2009). Field studies have highlighted that besides a relative immunity to predation, invaded communities dominated by NIS may benefit native predators by providing previously unavailable resources (Branch & Steffani 2004, Rius et al. 2009). Torchin et al. (2003) highlighted the importance of escape from parasites in the success of some exotic marine invertebrates, since parasites are known to reduce growth, survival and natality (Torchin et al. 2002). For example, infection of the native mussel, *Perna perna*, in South Africa by trematodes was found to reduce growth and adductor muscle strength, and increase water loss compared to the uninfected non-indigenous mussel, *Mytilus galloprovincialis* (Calvo-Ugarteburu & McQuaid 1998). While not top-down control, this appears to be another important form of enemy release in marine systems. Enemy release can accelerate invasion events that were initiated or facilitated by the anthropogenic factors mentioned in other sections of this chapter.

Managing anthropogenic factors to reduce the establishment and spread of NIS

Our existing insights of how humans can influence the post-arrival success of NIS by altering physical, chemical and biological parameters of the environment, or by facilitating dispersal, provide us with a wide range of options for reducing invasion risk. These are briefly discussed here and summarized in Table 1.

Ecological interactions, such as biotic resistance or containment, occur across and within trophic groups, as well as at multiple levels of biological organization and life-history stages (Kimbro et al. 2013). Human activities can impair biotic resistance to post-establishment spread of NIS by reducing the types, extent or magnitude of species diversity, community interactions and top-down control (Figure 1). At local or regional scales, for example, diversity loss can result from anthropogenic stressors such as contamination providing an advantage to non-indigenous species (Piola & Johnston 2008b). Physical modifications of habitats due to increasing coastal development can result in species removal and the loss of native species that might otherwise provide a barrier to invasion (Dafforn et al. 2015). Fishing practices that remove apex predators can reduce top-down control on marine communities and might also facilitate invasion at lower trophic levels (Baum & Worm 2009, reviewed by Johnson et al. 2011). At larger scales, human-induced climate change is modifying natural ranges of species (Ling et al. 2009) and may be increasing the rate of biotic homogenization (Stachowicz et al. 2002b, Olden et al. 2004).

The way we manage and conserve the diversity and integrity of native species assemblages will affect their ability to repel NIS now and in the future. Specifically, it will be important to conserve the native attributes of systems such that natural mechanisms of biotic resistance can operate most efficiently. For example, conserving native diversity will facilitate synergistic mechanisms of biotic resistance (e.g. species complementarity and indirect interactions) that would be virtually

Table 1 Summary of mechanisms for establishment and spread of NIS, likely impacts and suggestions for effective management

Establishment and/or dispersal vector	Likely impacts	Management suggestions
Reduced biotic resistance	Changes in species diversity, competitive interactions and top-down control	• Conserve native biodiversity (e.g. with marine sanctuaries) • Understand interactions between stressors • Protect natural predators for top-down control • Monitor key native species
Contamination, eutrophication and changes to water quality	Selection for tolerant species, selection for fast-growing species, freeing of resources for NIS	• Improve flushing in marinas to reduce water retention • Use of non-toxic antifouling paints • Remediate contaminated sediments to avoid resuspension of toxicants • Manage storm water runoff
Addition of artificial structures	Introduction of artificial hard substratum, invasion stepping stone	• Shift from hard defence structures to natural coastal protection • Design structures to conserve natural habitat complexity and reduced shading • Use fixed rather than floating structures
Commercial and recreational transport networks	Translocation of NIS via fouling on hulls or equipment, and in ballast water	• Pathway management • Domestic ballast water management • Improved ability for hull treatment

impossible to artificially engineer. Reducing the input of contaminants to receiving environments and removing historical legacies of toxicants would go some way to support native species resistance (Piola et al. 2009). For example, there has been evidence of macrofaunal recovery following the ban on tributyltin in antifouling paints (Smith et al. 2008, Langston et al. 2015). Similarly, the removal of organic contaminants associated with a fish farm resulted in positive changes to native ecological structure and function over time (Macleod et al. 2008). Broad conservation of processes that maintain strong ecological interactions will provide the most comprehensive protection against a wide range of possible scenarios.

The increased addition of artificial structures to coastal environments aids the establishment and dispersal of NIS. Where artificial structures are used for protection and defence (e.g. groynes, breakwaters, sea walls), the establishment of natural coastal protection would reduce these risks. For example, the addition of buffer zones for the landward extension of coastal vegetation (e.g. mangroves) would provide more natural protection from storms and storm surges (Hoang Tri et al. 1998, Kelly & Adger 2000, Costanza et al. 2008). Where coastal defence structures are needed, improving the design of artificial structures would go some way to reducing post-arrival success of NIS if structures matched the complexity of natural habitats (Atilla et al. 2005), shading was reduced to encourage native algal assemblage growth (Dafforn et al. 2012) and developments such as ports and marinas were designed to improve flushing and reduce retention rate for invasive propagules (Floerl & Inglis 2003, Vaselli et al. 2008). Future management strategies should take into account the potential for shallow moving structures to enhance invader dominance, and strongly consider using fixed structures to reduce opportunities for invaders (Dafforn et al. 2009b).

Physical changes that increase connectivity of habitats are a primary cause for invader spread. Canals have been implicated in invader spread (e.g. Suez and Panama canals) linking regions that would have otherwise been isolated by a natural barrier. The design of locks and weirs within such canals, combined with effective water or hull treatment technology, could help to reduce the transport of viable species between naturally isolated waterbodies (Galil et al. 2007, 2015). Upstream hydrological modifications should take into consideration the potential for changes to environmental conditions downstream that might enhance abiotic conditions for invaders (Winder et al. 2011).

A reduction in the risk of post-establishment spread via human transport mechanisms can also be achieved via the development and implementation of effective pathway management measures. Such initiatives may involve the setting of hygiene requirements or movement restrictions for vessels or infrastructure (e.g. aquaculture equipment) of particular types or origin to minimize their risk of translocating NIS (see discussions in Sinner et al. 2013, Inglis et al. 2014). One recent example of such measures is the Craft Risk Management Standard developed by the New Zealand government to limit the arrival and spread of non-indigenous biofouling species via overseas vessels (MPI 2014). Similar regional efforts are required to control domestic spread, and are being developed by several New Zealand regional jurisdictions (Sinner et al. 2013). To be effective and feasible, such measures need to be evidence-based and underpinned by effective prevention, inspection/surveillance and treatment technologies, industry codes of practice, incentivized schemes and educational measures (Floerl et al. 2016).

Longer-term international strategies for reducing risk include the implementation of widely-adopted, best-practice ballast water treatment and hull maintenance regimes that reduce biofouling (Hewitt & Campbell 2007, Tamelander et al. 2010) (Figure 5) and the development of more effective risk-based screening tools for border clearance. Genetic tools, including environmental DNA, are providing improved strategies for early detection, which can enable a faster and more effective detection and response to species invasions (Jerde et al. 2011, Zaiko et al. 2016). A sustained reduction in the overall per-vector risk of facilitating species transfers will be associated with long-term benefits for biosecurity (Drake and Lodge 2004). This is reflected in current international measures made for commercial ships (e.g. IMO 2011a) but needs to be better and more effectively implemented at domestic scales and across the range of anthropogenic transport mechanisms (Williams et al. 2013).

Multiple stressors (biological, chemical or physical) can impose additive or synergistic effects on ecosystems. Ameliorating as many stressors as possible will bolster the ability of ecosystems to deal with the remainder, and identifying important interactions between stressors may also help in prioritizing their management (Crain et al. 2008). Marine protected areas (MPAs), where certain human activities are restricted or forbidden, are one such tool with which to conserve the natural attributes of systems, and thereby biotic resistance. MPAs harbour natural predators that impose top-down control (Shears & Babcock 2002), and have the potential to conserve diversity and strong competitors within trophic levels to minimize excess resource availability (Baskett et al. 2007). Further understanding of species interactions most important to biotic resistance would be

Figure 5 Propeller of a domestic vessel fouled by the invasive ascidian, *Ciona robusta*. Longer-term strategies for reducing invasion risk include the implementation of widely-adopted, best-practice hull maintenance regimes that reduce fouling. (Photo: Javier Atalah, Cawthron Institute)

useful, allowing us to monitor key species and predict when and where biotic resistance is likely to be compromised.

Much research has focused on the vectors that initially transport a species beyond its native range (Carlton 1985, Ruiz et al. 1997, Hewitt & Campbell 2008, Davidson et al. 2010). However, it is clear that human activities have the potential to increase invader success through multiple stages of the invasion process (Williamson et al. 1986, Piola et al. 2009). Anthropogenic activities that increase the survival, establishment, proliferation and secondary spread of marine NIS post-arrival in a new region require greater management attention and research focus if we are to prevent the gradual homogenization of the world's coastal biota.

Acknowledgements

This research was supported by the Australian Research Council through an Australian Research Fellowship awarded to Johnston and a Linkage Grant (LP140100753) awarded to Dafforn and Johnston. This is SIMS publication number 195. Floerl's time was in part supported by SINTEF Fisheries and Aquaculture and by the National Institute of Water and Atmospheric Research under Coasts and Oceans Research Programme 6, Marine Biosecurity (2015/16 SCI). We thank a reviewer for constructive criticisms and Steve Hawkins, Louise Firth and Hanna Schuster for inviting us to submit this review.

References

Adams, T.P., Miller, R.G., Aleynik, D. & Burrows, M.T. 2014. Offshore marine renewable energy devices as stepping stones across biogeographical boundaries. *Journal of Applied Ecology* **51**, 330–338.

Airoldi, L., Abbiati, M., Beck, M.W., Hawkins, S.J., Jonsson, P.R., Martin, D., Moschella, P. S., Sundelof, A., Thompson, R.C. & Aberg, P. 2005. An ecological perspective on the deployment and design of low-crested and other hard coastal defence structures. *Coastal Engineering* **52**, 1073–1087.

Airoldi, L. & Beck, M.W. 2007. Loss, status and trends for coastal marine habitats of Europe. *Oceanography and Marine Biology: An Annual Review* **45**, 345–405.

Airoldi, L. & Bulleri, F. 2011. Anthropogenic disturbance can determine the magnitude of opportunistic species responses on marine urban infrastructures. *PLoS ONE* **6**, e22985.

Airoldi, L., Connell, S.D. & Beck, M.W. 2009. The loss of natural habitats and the addition of artificial substrata. In *Marine Hard Bottom Communities*, M. Wahl (ed.). Berlin: Springer, 269–280.

Airoldi, L., Turon, X., Perkol-Finkel, S. & Rius, M. 2015. Corridors for aliens but not for natives: effects of marine urban sprawl at a regional scale. *Diversity and Distributions* **21**, 755–768.

Aldred, N. & Clare, A.S. 2014. Mini-review: impact and dynamics of surface fouling by solitary and compound ascidians. *Biofouling* **30**, 259–270.

Alexandratos, N. & Bruinsma, J. 2012. World agriculture towards 2030/2050: the 2012 revision. ESA Working paper Rome, FAO.

Alquezar, R., Markich, S.J. & Booth, D.J. 2006. Effects of metals on condition and reproductive output of the smooth toadfish in Sydney estuaries, south-eastern Australia. *Environmental Pollution* **142**, 116–122.

Alzieu, C.L., Sanjuan, J., Deltreil, J.P. & Borel, M. 1986. Tin contamination in Arcachon Bay: effects on oyster shell anomalies. *Marine Pollution Bulletin* **17**, 494–498.

Anderson, M.J. & Underwood, A.J. 1994. Effects of substratum on the recruitment and development of an intertidal estuarine fouling assemblage. *Journal of Experimental Marine Biology and Ecology* **184**, 217–236.

Apte, S., Holland, B., Godwin, L. & Gardner, J. 2000. Jumping ship: a stepping stone event mediating transfer of a non-indigenous species via a potentially unsuitable environment. *Biological Invasions* **2**, 75–79.

Asif, M. & Muneer, T. 2007. Energy supply, its demand and security issues for developed and emerging economies. *Renewable and Sustainable Energy Reviews* **11**, 1388–1413.

Atilla, N., Fleeger, J.W. & Finelli, C.M. 2005. Effects of habitat complexity and hydrodynamics on the abundance and diversity of small invertebrates colonizing artificial substrates. *Journal of Marine Research* **63**, 1151–1172.

Atilla, N., Wetzel, M.A. & Fleeger, J.W. 2003. Abundance and colonization potential of artificial hard substrate-associated meiofauna. *Journal of Experimental Marine Biology and Ecology* **287**, 273–287.

Bakus, G.J. 1981. Chemical defense mechanisms on the Great Barrier reef, Australia. *Science* **211**, 497–499.

Barnes, D.K.A. 2002. Polarization of competition increases with latitude. *Proceedings of the Royal Society of London B: Biological Sciences* **269**, 2061–2069.

Barry, J.P., Baxter, C.H., Sagarin, R.D. & Gilman, S.E. 1995. Climate-related, long-term faunal changes in a California rocky intertidal community. *Science* **267**, 672–675.

Baskett, M.L., Micheli, F. & Levin, S.A. 2007. Designing marine reserves for interacting species: insights from theory. *Biological Conservation* **137**, 163–179.

Baum, J.K. & Worm, B. 2009. Cascading top-down effects of changing oceanic predator abundances. *Journal of Animal Ecology* **78**, 699–714.

Birch, G.F. 2000. Marine pollution in Australia, with special emphasis on central New South Wales estuaries and adjacent continental margin. *International Journal of Environment and Pollution* **13**, 573–607.

Bishop, M.J., Mayer-Pinto, M., Airoldi, L., Firth, L.B., Morris, R.L., Loke, L.H.L., Hawkins, S.J., Naylor, L.A., Coleman, R.A., Chee, S.Y. & Dafforn, K.A. 2017. Effects of ocean sprawl on ecological connectivity: impacts and solutions. *Journal of Experimental Marine Biology and Ecology*. https://doi.org/10.1016/j.jembe.2017.01.021

Blackburn, T.M. & Duncan, R.P. 2001. Determinants of establishment success in introduced birds. *Nature* **414**, 195–197.

Blackburn, T.M., Pyšek, P., Bacher, S., Carlton, J.T., Duncan, R.P., Jarosik, V., Wilson, J.R.U. & Richardson, D.M. 2011. A proposed unified framework for biological invasions. *Trends in Ecology & Evolution* **26**, 333–339.

Blakeslee, A.M., Altman, I., Miller, A.W., Byers, J.E., Hamer, C.E. & Ruiz, G.M. 2012. Parasites and invasions: a biogeographic examination of parasites and hosts in native and introduced ranges. *Journal of Biogeography* **39**, 609–622.

Blakeslee, A.M., Keogh, C.L., Byers, J.E., Lafferty, A.M.K.K.D. & Torchin, M.E. 2009. Differential escape from parasites by two competing introduced crabs. *Marine Ecology Progress Series* **393**, 83–96.

Bloecher, N., Floerl, O. & Sunde, L.M. 2015. Amplified recruitment pressure of biofouling organisms in commercial salmon farms: potential causes and implications for farm management. *Biofouling* **31**, 163–172.

Branch, G. 1984. Competition between marine organisms: ecological and evolutionary implications. *Oceanography and Marine Biology: An Annual Review* **22**, 429–593.

Branch, G.M. & Steffani, C.N. 2004. Can we predict the effects of alien species? A case-history of the invasion of South Africa by *Mytilus galloprovincialis* (Lamarck). *Journal of Experimental Marine Biology and Ecology* **300**, 189–215.

Browne, M. & Chapman, M. 2014. Mitigating against the loss of species by adding artificial intertidal pools to existing seawalls. *Marine Ecology Progress Series* **497**, 119–129.

Buchan, L.A.J. & Padilla, D.K. 1999. Estimating the probability of long-distance overland dispersal of invading aquatic species. *Ecological Applications* **9**, 254–265.

Bulleri, F. 2006. Is it time for urban ecology to include the marine realm? *Trends in Ecology & Evolution* **21**, 658–659.

Bulleri, F., Abbiati, M. & Airoldi, L. 2006. The colonisation of human-made structures by the invasive alga *Codium fragile* ssp. *tomentosoides* in the north Adriatic Sea (NE Mediterranean). *Hydrobiologia* **555**, 263–269.

Bulleri, F. & Airoldi, L. 2005. Artificial marine structures facilitate the spread of a non-indigenous green alga, *Codium fragile* ssp. *tomentosoides*, in the north Adriatic Sea. *Journal of Applied Ecology* **42**, 1063–1072.

Bulleri, F. & Chapman, M.G. 2004. Intertidal assemblages on artificial and natural habitats in marinas on the north-west coast of Italy. *Marine Biology* **145**, 381–391.

Bulleri, F. & Chapman, M.G. 2010. The introduction of coastal infrastructures as a driver of change in marine environments. *Journal of Applied Ecology* **47**, 26–35.

Burt, J., Bartholomew, A. & Sale, P.F. 2011. Benthic development on large-scale engineered reefs: a comparison of communities among breakwaters of different age and natural reefs. *Ecological Engineering* **37**, 191–198.

Burton, G.A. & Johnston, E.L. 2010. Assessing contaminated sediments in the context of multiple stressors. *Environmental Toxicology and Chemistry* **29**, 2625–2643.

Buss, L.W. 1990. Competition within and between encrusting clonal invertebrates. *Trends in Ecology & Evolution* **5**, 352–356.

Butchart, S.H.M., Walpole, M., Collen, B., van Strien, A., Scharlemann, J.P.W., Almond, R.E.A., Baillie, J.E.M., Bomhard, B., Brown, C., Bruno, J., Carpenter, K.E., Carr, G.M., Chanson, J., Chenery, A.M., Csirke, J., Davidson, N.C., Dentener, F., Foster, M., Galli, A., Galloway, J.N., Genovesi, P., Gregory, R.D., Hockings, M., Kapos, V., Lamarque, J.F., Leverington, F., Loh, J., McGeoch, M.A., McRae, L., Minasyan, A., Morcillo, M.H., Oldfield, T.E.E., Pauly, D., Quader, S., Revenga, C., Sauer, J.R., Skolnik, B., Spear, D., Stanwell-Smith, D., Stuart, S.N., Symes, A., Tierney, M., Tyrrell, T.D., Vie, J.C. & Watson, R. 2010. Global biodiversity: indicators of recent declines. *Science* **328**, 1164–1168.

Byers, J.E. 2002. Impact of non-indigenous species on natives enhanced by anthropogenic alteration of selection regimes. *Oikos* **97**, 449–458.

Calvo-Ugarteburu, G. & McQuaid, C. 1998. Parasitism and invasive species: effects of digenetic trematodes on mussels. *Marine Ecology Progress Series* **169**, 149–163.

Carlton, J.T. 1985. Transoceanic and interoceanic dispersal of coastal marine organisms: the biology of ballast water. *Oceanography and Marine Biology: An Annual Review* **23**, 313–373.

Carlton, J.T. & Hodder, J. 1995. Biogeography and dispersal of coastal marine organisms: experimental studies on a replica of a 16th-century sailing vessel. *Marine Biology* **121**, 721–730.

Carstensen, J., Conley, D., Bonsdorff, E., Gustafsson, B., Hietanen, S., Janas, U., Jilbert, T., Maximov, A., Norkko, A., Norkko, J., Reed, D., Slomp, C., Timmermann, K. & Voss, M. 2014. Hypoxia in the Baltic Sea: biogeochemical cycles, benthic fauna, and management. *Ambio* **43**, 26–36.

Castilla, J.C., Guiñez, R., Caro, A.U. & Ortiz, V. 2004. Invasion of a rocky intertidal shore by the tunicate *Pyura praeputialis* in the Bay of Antofagasta, Chile. *Proceedings of the National Academy of Sciences of the United States of America* **101**, 8517–8524.

Chapman, J.W. & Carlton, J.T. 1991. A test of criteria for introduced species: the global invasion by the isopod *Synidotea laevidorsalis* (Miers, 1881). *Journal of Crustacean Biology* **11**, 386–400.

Chapman, M.G. 2011. Restoring intertidal boulder-fields as habitat for "specialist" and "generalist" animals. *Restoration Ecology* **20**, 277–285.

Chapman, M.G. & Bulleri, F. 2003. Intertidal seawalls – new features of landscape in intertidal environments. *Landscape and Urban Planning* **62**, 159–172.

Chapman, M.G. & Underwood, A.J. 2011. Evaluation of ecological engineering of 'armoured' shorelines to improve their value as habitat. *Journal of Experimental Marine Biology and Ecology* **400**, 302–313.

Claisse, D. & Alzieu, C. 1993. Copper contamination as a result of antifouling paint regulations? *Marine Pollution Bulletin* **26,** 395–397.

Clark, G.F. & Johnston, E.L. 2005. Manipulating larval supply in the field: a controlled study of marine invasibility. *Marine Ecology Progress Series* **298**, 9–19.

Clark, G.F. & Johnston, E.L. 2009. Propagule pressure and disturbance interact to overcome biotic resistance of marine invertebrate communities. *Oikos* **118**, 1679–1686.

Clark, G.F. & Johnston, E.L. 2011. Temporal change in the diversity-invasibility relationship in the presence of a disturbance regime. *Ecology Letters* **14**, 52–57.

Clark, G.F., Johnston, E.L. & Leung, B. 2013. Intrinsic time dependence in the diversity-invasibility relationship. *Ecology* **94**, 25–31.

Clark, G.F., Kelaher, B.P., Dafforn, K.A., Coleman, M.A., Knott, N.A., Marzinelli, E.M. & Johnston, E.L. 2015. What does impacted look like? High diversity and abundance of epibiota in modified estuaries. *Environmental Pollution* **196**, 12–20.

Clarke Murray, C., Pakhomov, E.A. & Therriault, T.W. 2011. Recreational boating: a large unregulated vector transporting marine invasive species. *Diversity and Distributions* **17**, 1161–1172.

Clynick, B.G., Chapman, M.G. & Underwood, A.J. 2007. Effects of epibiota on assemblages of fish associated with urban structures. *Marine Ecology Progress Series* **332**, 201–210.

Colautti, R.I., Ricciardi, A., Grigorovich, I.A. & MacIsaac, H.J. 2004. Is invasion success explained by the enemy release hypothesis? *Ecology Letters* **7**, 721–733.

Combs, J.K., Reichard, S.H., Groom, M.J., Wilderman, D.L. & Camp, P.A. 2011. Invasive competitor and native seed predators contribute to rarity of the narrow endemic *Astragalus sinuatus* Piper. *Ecological Applications* **21**, 2498–2509.

Connell, S.D. 1999. Effects of surface orientation on the cover of epibiota. *Biofouling* **14**, 219–226.

Connell, S.D. & Glasby, T.M. 1999. Do urban structures influence local abundance and diversity of subtidal epibiota? A case study from Sydney Harbour, Australia. *Marine Environmental Research* **47**, 373–387.

Costa, T.J.F., Pinheiro, H.T., Teixeira, J.B., Mazzei, E.F., Bueno, L., Hora, M.S.C., Joyeux, J.-C., Carvalho-Filho, A., Amado-Filho, G., Sampaio, C.L.S. & Rocha, L.A. 2014. Expansion of an invasive coral species over Abrolhos Bank, Southwestern Atlantic. *Marine Pollution Bulletin* **85**, 252–253.

Costanza, R., Pérez-Maqueo, O., Martinez, M.L., Sutton, P., Anderson, S.J. & Mulder, K. 2008. The value of coastal wetlands for hurricane protection. *AMBIO: A Journal of the Human Environment* **37**, 241–248.

Coutts, A.D.M. & Forrest, B.M. 2007. Development and application of tools for incursion response: Lessons learned from the management of the fouling pest *Didemnum vexillum*. *Journal of Experimental Marine Biology and Ecology* **342**, 154–162.

Crain, C.M., Kroeker, K. & Halpern, B.S. 2008. Interactive and cumulative effects of multiple human stressors in marine systems. *Ecology Letters* **11**, 1304–1315.

Dafforn, K.A., Glasby, T.M., Airoldi, L., Rivero, N.K., Mayer-Pinto, M. & Johnston, E.L. 2015. Marine urbanisation: an ecological framework for designing multifunctional artificial structures. *Frontiers in Ecology and the Environment* **13**, 82–90.

Dafforn, K.A., Glasby, T.M. & Johnston, E.L. 2008. Differential effects of tributyltin and copper anti-foulants on recruitment of non-indigenous species. *Biofouling* **24**, 23–33.

Dafforn, K.A., Glasby, T.M. & Johnston, E.L. 2009a. Links between estuarine condition and spatial distributions of marine invaders. *Diversity and Distributions* **15**, 807–821.

Dafforn, K.A., Glasby, T.M. & Johnston, E.L. 2012. Comparing the invasibility of experimental "reefs" with field observations of natural reefs and artificial structures. *PLoS ONE* **7**, e38124.

Dafforn, K.A., Johnston, E.L. & Glasby, T.M. 2009b. Shallow moving structures promote marine invader dominance. *Biofouling* **25**, 277–287.

Dafforn, K.A., Lewis, J.A. & Johnston, E.L. 2011. Antifouling strategies: history and regulation, ecological impacts and mitigation. *Marine Pollution Bulletin* **62**, 453–465.

Dangremond, E.M., Pardini, E.A. & Knight, T.M. 2010. Apparent competition with an invasive plant hastens the extinction of an endangered lupine. *Ecology* **91**, 2261–2271.

Davidson, I.C., McCann, L.D., Fofonoff, P.W., Sytsma, M.D. & Ruiz, G.M. 2008. The potential for hull-mediated species transfers by obsolete ships on their final voyages. *Diversity and Distributions* **14**, 518–529.

Davidson, I.C., Zabin, C.J., Chang, A.L., Brown, C.W., Sytsma, M.D. & Ruiz, G.M. 2010. Recreational boats as potential vectors of marine organisms at an invasion hotspot. *Aquatic Biology* **11**, 179–191.

Davies, T.W., Duffy, J.P., Bennie, J. & Gaston, K.J. 2014. The nature, extent, and ecological implications of marine light pollution. *Frontiers in Ecology and the Environment* **12**, 347–355.

Davis, M.A., Grime, J.P. & Thompson, K. 2000. Fluctuating resources in plant communities: a general theory of invasibility. *Journal of Ecology* **88**, 528–534.

Delpla, I., Jung, A.-V., Baures, E., Clement, M. & Thomas, O. 2009. Impacts of climate change on surface water quality in relation to drinking water production. *Environment International* **35**:1225–1233.

Dohn, J., Dembélé, F., Karembé, M., Moustakas, A., Amévor, K.A. & Hanan, N.P. 2013. Tree effects on grass growth in savannas: competition, facilitation and the stress-gradient hypothesis. *Journal of Ecology* **101**, 202–209.

Drake, J.M. & Lodge, D.M. 2004. Global hot spots of biological invasions: evaluating options for ballast-water management. *Proceedings of the Royal Society of London B: Biological Sciences* **271**, 575–580.

Drake, J.M. & Lodge, D.M. 2006. Allee effects, propagule pressure and the probability of establishment: risk analysis for biological invasions. *Biological Invasions* **8**, 365–375.

Drake, J.M. & Lodge, D.M. 2007. Hull fouling is a risk factor for intercontinental species exchange in aquatic ecosystems. *Aquatic Invasions* **2**, 121–131.

Duarte, C.M. 1995. Submerged aquatic vegetation in relation to different nutrient regimes. *Ophelia* **41**, 87–112.

Dugan, J.E., Airoldi, L., Chapman, M.G., Walker, S.J. & Schlacher, T. 2011. Estuarine and coastal structures: environmental effects, a focus on shore and nearshore structures. In *Treatise on Estuarine and Coastal Science Vol 8*, E. Wolanski & D. McLusky (eds). Waltham: Academic Press, 17–41.

Dukes, J.S. 2001. Biodiversity and invasibility in grassland microcosms. *Oecologia* **126**, 563–568.

Dukes, J.S. & Mooney, H.M. 1999. Does global change increase the success of biological invaders? *Trends in Ecology & Evolution* **14**, 135–139.

Dupont, L., Viard, F., Davis, M.H., Nishikawa, T. & Bishop, J.D.D. 2010. Pathways of spread of the introduced ascidian *Styela clava* (Tunicata) in Northern Europe, as revealed by microsatellite markers. *Biological Invasions* **12**, 2707–2721.

Durrant, H.M.S., Clark, G.F., Dworjanyn, S.A., Byrne, M. & Johnston, E.L. 2013. Seasonal variation in the effects of ocean warming and acidification on a native bryozoan, *Celleporaria nodulosa*. *Marine Biology* **160**, 1903–1911.

Eggleton, J. & Thomas, K.V. 2004. A review of factors affecting the release and bioavailability of contaminants during sediment disturbance events. *Environment International* **30**, 973–980.

Elser, J.J., Bracken, M.E., Cleland, E.E., Gruner, D.S., Harpole, W.S., Hillebrand, H., Ngai, J.T., Seabloom, E.W., Shurin, J.B. & Smith, J.E. 2007. Global analysis of nitrogen and phosphorus limitation of primary producers in freshwater, marine and terrestrial ecosystems. *Ecology Letters* **10,** 1135–1142.

Elton, C.S. 1958. *The Ecology of Invasions by Animals and Plants*. London: Methuen.

FAO. 2012. The state of world fisheries and aquaculture. Food and Agriculture Organization of the United Nations, Rome.

Feary, D.A., Burt, J.A. & Bartholomew, A. 2011. Artificial marine habitats in the Arabian Gulf: review of current use, benefits and management implications. *Ocean & Coastal Management* **54**, 742–749.

Fellers, J.H. 1987. Interference and exploitation in a guild of woodland ants. *Ecology* **68**, 1466–1478.

Firth, L.B., Crowe, T.P., Moore, P., Thompson, R.C. & Hawkins, S.J. 2009. Predicting impacts of climate-induced range expansion: an experimental framework and a test involving key grazers on temperate rocky shores. *Global Change Biology* **15**, 1413–1422.

Firth, L., Knights, A., Thompson, R., Mieszkowska, N., Bridger, D., Evans, A., Moore, P., O'Connor, N., Sheehan, E. & Hawkins, S. 2016. Ocean sprawl: challenges and opportunities for biodiversity management in a changing world. *Oceanography and Marine Biology: An Annual Review* **54**, 193–269.

Firth, L.B., Schofield, M., White, F.J., Skov, M.W. & Hawkins, S.J. 2014. Biodiversity in intertidal rock pools: informing engineering criteria for artificial habitat enhancement in the built environment. *Marine Environmental Research* **102**, 122–130.

Firth, L.B., Thompson, R.C., White, F.J., Schofield, M., Skov, M.W., Hoggart, S.P.G., Jackson, J., Knights, A.M. & Hawkins, S.J. 2013. The importance of water-retaining features for biodiversity on artificial intertidal coastal defence structures. *Diversity and Distributions* **19**, 1275–1283.

Firth, L.B., White, F.J., Schofield, M., Hanley, M.E., Burrows, M.T., Thompson, R.C., Skov, M.W., Evans, A.J., Moore, P.J. & Hawkins, S.J. 2015. Facing the future: the importance of substratum features for ecological engineering of artificial habitats in the rocky intertidal. *Marine and Freshwater Research* **67**, 131–143.

Fitridge, I., Dempster, T., Guenther, J. & de Nys, R. 2012. The impact and control of biofouling in marine aquaculture: a review. *Biofouling* **28**, 649–669.

FitzGerald, D.M., Fenster, M.S., Argow, B.A. & Buynevich, I.V. 2008. Coastal impacts due to sea-level rise. *Annual Review of Earth and Planetary Sciences* **36**, 601–647.

Fleeger, J.W., Carman, K.R. & Nisbet, R.M. 2003. Indirect effects of contaminants in aquatic ecosystems. *Science of the Total Environment* **317**, 207–233.

Fletcher, R.L. & Farrell, P. 1999. Introduced brown algae in the North East Atlantic, with particular respect to *Undaria pinnatifida* (Harvey) Suringar. *Helgolander Meeresuntersuchungen* **52**, 259–275.

Floerl, O. 2014. Challenges and opportunities for understanding and managing biofouling in marine aquaculture. International Congress for Marine Corrosion and Fouling, Singapore.

Floerl, O. & Coutts, A. 2009. Potential ramifications of the global economic crisis on human-mediated dispersal of marine non-indigenous species. *Marine Pollution Bulletin* **58**, 1595–1598.

Floerl, O. & Inglis, G.J. 2003. Boat harbour design can exacerbate hull fouling. *Austral Ecology* **28**, 116–127.

Floerl, O., Inglis, G., Dey, K.L. & Smith, A. 2009a. The importance of transport hubs in stepping-stone invasions. *Journal of Applied Ecology* **46**, 37–45.

Floerl, O., Inglis, G.J. & Diettrich, J. 2016. Incorporating human behaviour into the risk-release relationship for invasion vectors: why targeting only the worst offenders can fail to reduce spread. *Journal of Applied Ecology* **53**, 742–750.

Floerl, O., Inglis, G.J. & Gordon, D.P. 2009b. Patterns of taxonomic diversity and relatedness among native and non-indigenous bryozoans. *Diversity and Distributions* **15**, 438–449.

Floerl, O., Pool, T.K. & Inglis, G.J. 2004. Positive interactions between nonidigenous species facilitate transport by human vectors. *Ecological Applications* **14**, 1724–1736.

Forrest, B. & Blakemore, K.A. 2006. Evaluation of treatments to reduce the spread of a marine plant pest with aquaculture transfers. *Aquaculture* **257**, 333–345.

Forrest, B.M., Fletcher, L.M., Atalah, J., Piola, R.F. & Hopkins, G.A. 2013. Predation limits spread of *Didemnum vexillum* into natural habitats from refuges on anthropogenic structures. *PLoS ONE* **8**, e82229.

Forrest, B.M., Gardner, J.P.A. & Taylor, M.D. 2009. Internal borders for managing invasive marine species. *Journal of Applied Ecology* **46**, 46–54.

Freestone, A.L. & Osman, R.W. 2011. Latitudinal variation in local interactions and regional enrichment shape patterns of marine community diversity. *Ecology* **92**, 208–217.

Freestone, A.L., Osman, R.W., Ruiz, G.M. & Torchin, M.E. 2011. Stronger predation in the tropics shapes species richness patterns in marine communities. *Ecology* **92**, 983–993.

Freestone, A.L., Ruiz, G.M. & Torchin, M.E. 2013. Stronger biotic resistance in tropics relative to temperate zone: effects of predation on marine invasion dynamics. *Ecology* **94**, 1370–1377.

Fridley, J.D., Stachowicz, J.J., Naeem, S., Sax, D.F., Seabloom, E.W., Smith, M.D., Stohlgren, T.J., Tilman, D. & Von Holle, B. 2007. The invasion paradox: reconciling pattern and process in species invasions. *Ecology* **88**, 3–17.

Galil, B., Boero, F., Campbell, M., Carlton, J., Cook, E., Fraschetti, S., Gollasch, S., Hewitt, C., Jelmert, A., Macpherson, E., Marchini, A., McKenzie, C., Minchin, D., Occhipinti-Ambrogi, A., Ojaveer, H., Olenin, S., Piraino, S. & Ruiz, G. 2015. 'Double trouble': the expansion of the Suez Canal and marine bioinvasions in the Mediterranean Sea. *Biological Invasions* **17**, 973–976.

Galil, B.S., Nehring, S. & Panov, V. 2007. Waterways as invasion highways – impact of climate change and globalization. In *Biological Invasions*, W. Nentwig (ed.). Berlin Heidelberg: Springer-Verlag, 59–74.

Gerland, P., Raftery, A.E., Ševčíková, H., Li, N., Gu, D., Spoorenberg, T., Alkema, L., Fosdick, B.K., Chunn, J. & Lalic, N. 2014. World population stabilization unlikely this century. *Science* **346**, 234–237.

Glasby, T.M. 1999a. Differences between subtidal epibiota on pier pilings and rocky reefs at marinas in Sydney, Australia. *Estuarine, Coastal and Shelf Science* **48**, 281–290.

Glasby, T.M. 1999b. Effects of shading on subtidal epibiotic assemblages. *Journal of Experimental Marine Biology and Ecology* **234**, 275–290.

Glasby, T.M. 2000. Surface composition and orientation interact to affect subtidal epibiota. *Journal of Experimental Marine Biology and Ecology* **248**, 177–190.

Glasby, T.M. & Connell, S.D. 1999. Urban structures as marine habitats. *Ambio* **28**, 595–598.

Glasby, T.M. & Connell, S.D. 2001. Orientation and position of substrata have large effects on epibiotic assemblages. *Marine Ecology Progress Series* **214**, 127–135.

Glasby, T.M., Connell, S.D., Holloway, M.G. & Hewitt, C.L. 2007. Nonindigenous biota on artificial structures: could habitat creation facilitate biological invasions? *Marine Biology* **151**, 887–895.

Godwin, L.S. 2003. Hull fouling of maritime vessels as a pathway for marine species invasions to the Hawaiian Islands. *Biofouling* **19**, 123–131.

Goldstien, S.J., Schiel, D.R. & Gemmell, N.J. 2010. Regional connectivity and coastal expansion: differentiating pre-border and post-border vectors for the invasive tunicate *Styela clava*. *Molecular Ecology* **19**, 874–885.

Grime, J.P. 1977. Evidence for existence of 3 primary strategies in plants and its relevance to ecological and evolutionary theory. *American Naturalist* **111**, 1169–1194.

Grimm, N.B., Foster, D., Groffman, P., Grove, J.M., Hopkinson, C.S., Nadelhoffer, K.J., Pataki, D.E. & Peters, D.P.C. 2008. The changing landscape: ecosystem responses to urbanization and pollution across climatic and societal gradients. *Frontiers in Ecology and the Environment* **6**, 264–272.

Grosholz, E.D. 2005. Recent biological invasion may hasten invasional meltdown by accelerating historical introductions. *Proceedings of the National Academy of Sciences of the United States of America* **102**, 1088–1091.

Grosholz, E.D., Crafton, R.E., Fontana, R.E., Pasari, J.R., Williams, S.L. & Zabin, C.J. 2015. Aquaculture as a vector for marine invasions in California. *Biological Invasions* **17**, 1471–1484.

Gust, N., Inglis, G., Floerl, O., Peacock, L., Denny, C. & Forrest, B. 2008. Assessment of population management options for *Styela clava*. NIWA, Christchurch.

Hall, L. W., Scott, M.C. & Killen, W.D. 1998. Ecological risk assessment of copper and cadmium in surface waters of Chesapeake Bay watershed. *Environmental Toxicology and Chemistry* **17**, 1172–1189.

Hallegraeff, G. & Gollasch, S. 2006. Anthropogenic introductions of microalgae. In *Ecology of Harmful Algae*, E. Granéli & J. Turner (eds). Berlin Heidelberg: Springer, 379–390.

Halpern, B.S., Walbridge, S., Selkoe, K.A., Kappel, C.V., Micheli, F., D'Agrosa, C., Bruno, J.F., Casey, K.S., Ebert, C., Fox, H.E., Fujita, R., Heinemann, D., Lenihan, H.S., Madin, E.M.P., Perry, M.T., Selig, E.R., Spalding, M., Steneck, R. & Watson, R. 2008. A global map of human impact on marine ecosystems. *Science* **319**, 948–952.

Hamzah, B.A. 2003. International rules on decommissioning of offshore installations: some observations. *Marine Policy* **27**, 339–348.

Hänfling, B., Edwards, F. & Gherardi, F. 2011. Invasive alien Crustacea: dispersal, establishment, impact and control. *BioControl* **56**, 573–595.

Hawkins, S., Sugden, H., Mieszkowska, N., Moore, P., Poloczanska, E., Leaper, R., Herbert, R. J., Genner, M., Moschella, P. & Thompson, R. 2009. Consequences of climate-driven biodiversity changes for ecosystem functioning of North European rocky shores. *Marine Ecology Progress Series* **396**, 245–259.

Hay, M.E. & Fenical, W. 1988. Marine plant-herbivore interactions: the ecology of chemical defense. *Annual Review of Ecology and Systematics* **19**, 111–145.

Hayden, B.J., Unwin, M., Roulston, H., Peacock, L., Floerl, O., Kospartov, M. & Seaward, K. 2009. Evaluation of vessel movements from the 24 ports and marinas surveyed through the port baseline survey programmes, ZBS2000–04 and ZBS2005–19 (ZBS2005–13). MPI Technical Paper No: 2014/04. Ministry of Primary Industries, Wellington.

Hector, A., Bazeley-White, E., Loreau, M., Otway, S. & Schmid, B. 2002. Overyielding in grassland communities: testing the sampling effect hypothesis with replicated biodiversity experiments. *Ecology Letters* **5**, 502–511.

Hedge, L.H. & Johnston, E.L. 2012. Propagule pressure determines recruitment from a commercial shipping pier. *Biofouling* **28**, 73–85.

Hedge, L.H., O'Connor, W.A. & Johnston, E.L. 2012. Manipulating the intrinsic parameters of propagule pressure: implications for bio-invasion. *Ecosphere* **3**, 1–13 Art48.

Hewitt, C. L. 2002. Distribution and biodiversity of Australian tropical marine bioinvasions. *Pacific Science* **56**, 213–222.

Hewitt, C.L. & Campbell, M.L. 2007. Mechanisms for the prevention of marine bioinvasions for better biosecurity. *Marine Pollution Bulletin* **55**, 395–401.

Hewitt, C.L. & Campbell, M.L. 2008. Assessment of relative contribution of vectors to the introduction and translocation of marine invasive species. Report for the Department of Agriculture, Fisheries and Forestry Australia, University of Tasmania

Hoang Tri, N., Adger, W.N. & Kelly, P.M. 1998. Natural resource management in mitigating climate impacts: the example of mangrove restoration in Vietnam. *Global Environmental Change* **8**, 49–61.

Holloway, M.G. & Connell, S.D. 2002. Why do floating structures create novel habitats for subtidal epibiota? *Marine Ecology Progress Series* **235**, 43–52.

Holloway, M.G. & Keough, M.J. 2002. An introduced polychaete affects recruitment and larval abundance of sessile invertebrates. *Ecological Applications* **12**, 1803–1823.

Holmgren, M. & Scheffer, M. 2010. Strong facilitation in mild environments: the stress gradient hypothesis revisited. *Journal of Ecology* **98**, 1269–1275.

Hudson, J., Viard, F., Roby, C. & Rius, M. 2016. Anthropogenic transport of species across native ranges: unpredictable genetic and evolutionary consequences. *Biology Letters* **12**, 20160620.

Huston, M.A. 1997. Hidden treatments in ecological experiments: re-evaluating the ecosystem function of biodiversity. *Oecologia* **108**, 449–460.

IMO. 2005. International convention on the control and management of ship's ballast water and sediments. International Maritime Organization, London.

IMO. 2011a. Guidelines for the control and management of ship's biofouling to minimize the transfer of invasive aquatic species (Annex 26, Resolution MEPC.207(62)) (http://www.imo.org/blast/blastDataHelper.asp?data_id=30766). International Maritime Organization, London.

IMO. 2011b. Guidelines for the control and management of ship's biofouling to minimize the transfer of invasive aquatic species (Annex 26, Resolution MEPC.207(62)) (http://www.imo.org/blast/blastDataHelper.asp?data_id=30766). International Maritime Organization, London.

Inglis, G., Morrisey, D., Woods, C., Sinner, J. & Newton, M. 2014. Managing the domestic spread of harmful marine organisms. Part A – Operational tools for management. Report prepared for New Zealand Ministry for Primary Industries. National Institute of Water and Atmospheric Research. Christchurch, 166 p.

Jerde, C.L., Mahon, A.R., Chadderton, W.L. & Lodge, D.M. 2011. "Sight-unseen" detection of rare aquatic species using environmental DNA. *Conservation Letters* **4**, 150–157.

Johnson, C.R., Banks, S.C., Barrett, N.S., Cazassus, F., Dunstan, P.K., Edgar, G.J., Frusher, S.D., Gardner, C., Haddon, M., Helidoniotis, F., Hill, K.L., Holbrook, N.J., Hosie, G.W., Last, P.R., Ling, S.D., Melbourne-Thomas, J., Miller, K., Pecl, G.T., Richardson, A.J., Ridgway, K.R., Rintoul, S.R., Ritz, D.A., Ross, D.J., Sanderson, J.C., Shepherd, S.A., Slotwinski, A., Swadling, K.M. & Taw, N. 2011. Climate change cascades: shifts in oceanography, species' ranges and subtidal marine community dynamics in eastern Tasmania. *Journal of Experimental Marine Biology and Ecology* **400**, 17–32.

Johnston, E. 2011. Tolerance to contaminants: evidence from chronically-exposed populations of aquatic organisms. In *Tolerance to Environmental Contaminants*, C. Amiard-Triquet et al. (eds). Boca Raton: CRC Press, Boca Raton, 25–46.

Johnston, E.L., Marzinelli, E.M., Wood, C.A., Speranza, D. & Bishop, J.D.D. 2011. Bearing the burden of boat harbours: heavy contaminant and fouling loads in a native habitat-forming alga. *Marine Pollution Bulletin* **62**, 2137–2144.

Johnston, E.L. & Roberts, D.A. 2009. Contaminants reduce the richness and evenness of marine communities: a review and meta-analysis. *Environmental Pollution* **157**, 1745–1752.

Kaluza, P., Kolzsch, A., Gastner, M.T. & Blasius, B. 2010. The complex network of global cargo ship movements. *Journal of the Royal Society Interface* **7**, 1093–1103.

Keane, R.M. & Crawley, M.J. 2002. Exotic plant invasions and the enemy release hypothesis. *Trends in Ecology & Evolution* **17**, 164–170.

Kelly, P.M. & Adger, W.N. 2000. Theory and practice in assessing vulnerability to climate change and facilitating adaptation. *Climatic Change* **47**, 325–352.

Kennedy, S. 2005. Wind power planning: assessing long-term costs and benefits. *Energy Policy* **33**, 1661–1675.

Kimbro, D.L., Cheng, B.S. & Grosholz, E.D. 2013. Biotic resistance in marine environments. *Ecology Letters* **16**, 821–833.

Klein, J.C., Underwood, A.J. & Chapman, M.G. 2011. Urban structures provide new insights into interactions among grazers and habitat. *Ecological Applications* **21**, 427–438.

Knights, A.M., Firth, L.B., Thompson, R.C., Yunnie, A.L.E., Hiscock, K. & Hawkins, S.J. 2016. Plymouth – a World Harbour through the ages. *Regional Studies in Marine Science* **8**, 297–307.

Knights, A.M., Koss, R.S. & Robinson, L.A. 2013. Identifying common pressure pathways from a complex network of human activities to support ecosystem-based management. *Ecological Applications* **23**, 755–765.

Knott, N.A., Underwood, A.J., Chapman, M.G. & Glasby, T.M. 2004. Epibiota on vertical and on horizontal surfaces on natural reefs and on artificial structures. *Journal of the Marine Biological Association of the United Kingdom* **84**, 1117–1130.

Lagesa, B.G., Fleurya, B.G., Ferreira, C.E.L. & Pereira, R.C. 2006. Chemical defense of an exotic coral as invasion strategy. *Journal of Experimental Marine Biology and Ecology* **328**, 127–135.

Lambert, C.C. & Lambert, G. 1998. Non-indigenous ascidians in southern California harbors and marinas. *Marine Biology* **130**, 675–688.

Langhamer, O., Wilhelmsson, D. & Engström, J. 2009. Artificial reef effect and fouling impacts on offshore wave power foundations and buoys – a pilot study. *Estuarine, Coastal and Shelf Science* **82**, 426–432.

Langston, W., Pope, N., Davey, M., Langston, K., O'Hara, S., Gibbs, P. & Pascoe, P. 2015. Recovery from TBT pollution in English Channel environments: a problem solved? *Marine Pollution Bulletin* **95**, 551–564.

Lavender, J.T., Dafforn, K.A. & Johnston, E.L. 2014. Meso-predators: a confounding variable in consumer exclusion studies. *Journal of Experimental Marine Biology and Ecology* **456**, 26–33.

Leung, B., Drake, J.M. & Lodge, D.M. 2004. Predicting invasions: Propagule pressure and the gravity of allee effects. *Ecology* **85**, 1651–1660.

Levin, S.A. & Lubchenco, J. 2008. Resilience, robustness, and marine ecosystem-based management. *Bioscience* **58**, 27–32.

Levine, J.M. 2000. Species diversity and biological invasions: relating local process to community pattern. *Science* **288**, 852–854.

Levine, J.M., Adler, P.B. & Yelenik, S.G. 2004. A meta-analysis of biotic resistance to exotic plant invasions. *Ecology Letters* **7**, 975–989.

Liancourt, P., Callaway, R.M. & Michalet, R. 2005. Stress tolerance and competitive-response ability determine the outcome of biotic interactions. *Ecology* **86**, 1611–1618.

Ling, S.D., Johnson, C.R., Ridgway, K., Hobday, A.J. & Haddon, M. 2009. Climate-driven range extension of a sea urchin: inferring future trends by analysis of recent population dynamics. *Global Change Biology* **15**, 719–731.

Liu, H. & Stiling, P. 2006. Testing the enemy release hypothesis: a review and meta-analysis. *Biological Invasions* **8**, 1535–1545.

Lotze, H.K., Lenihan, H.S., Bourque, B.J., Bradbury, R.H., Cooke, R.G., Kay, M.C., Kidwell, S.M., Kirby, M.X., Peterson, C.H. & Jackson, J.B.C. 2006. Depletion, degradation, and recovery potential of estuaries and coastal seas. *Science* **312**, 1806–1809.

Macleod, C.K., Moltschaniwskyj, N.A. & Crawford, C.M. 2008. Ecological and functional changes associated with long-term recovery from organic enrichment. *Marine Ecology Progress Series* **365**, 17–24.

Macreadie, P.I., Fowler, A.M. & Booth, D.J. 2011. Rigs-to-reefs: will the deep sea benefit from artificial habitat? *Frontiers in Ecology and the Environment* **9**, 455–461.

Maestre, F.T., Callaway, R.M., Valladares, F. & Lortie, C.J. 2009. Refining the stress-gradient hypothesis for competition and facilitation in plant communities. *Journal of Ecology* **97**, 199–205.

Marchal, V., Dellink, R., Van Vuuren, D., Clapp, C., Chateau, J., Magné, B. & van Vliet, J. 2011. OECD environmental outlook to 2050. Organization for Economic Co-operation and Development.

Martínez-Lladó, X., Gibert, O., Martí, V., Díez, S., Romo, J., Bayona, J.M. & de Pablo, J. 2007. Distribution of polycyclic aromatic hydrocarbons (PAHs) and tributyltin (TBT) in Barcelona harbour sediments and their impact on benthic communities. *Environmental Pollution* **149**, 104–113.

Marzinelli, E.M., Underwood, A.J. & Coleman, R.A. 2011. Modified habitats influence kelp epibiota via direct and indirect effects. *PLoS ONE* **6**, e21936.

McEnvoy, P.B. & Coombs, E.M. 1999. Biological control of plant invaders: regional patterns, field experiments, and structured population models. *Ecological Applications* **9**, 387–401.

McGranahan, G., Balk, D. & Anderson, B. 2007. The rising tide: assessing the risks of climate change and human settlements in low elevation coastal zones. *Environment and Urbanization* **19**, 17–37.

McKenzie, L.A., Johnston, E.L. & Brooks, R. 2012. Using clones and copper to resolve the genetic architecture of metal tolerance in a marine invader. *Ecology and Evolution* **2**, 1319–1329.

McKinley, A. & Johnston, E.L. 2010. Impacts of contaminant sources on marine fish abundance and species richness: a review and meta-analysis of evidence from the field. *Marine Ecology Progress Series* **420**, 175–191.

McKinney, M.L. & Lockwood, J.L. 1999. Biotic homogenization: a few winners replacing many losers in the next mass extinction. *Trends in Ecology & Evolution* **14**, 450–453.

McKnight, E., García-Berthou, E., Srean, P. & Rius, M. 2016. Global meta-analysis of native and nonindigenous trophic traits in aquatic ecosystems. *Global Change Biology*, doi:10.1111/gcb.13524

McQuaid, C.D., Porri, F., Nicastro, K. & Zardi, G. 2015. Simple, scale-dependent patterns emerge from very complex effects: an example from the intertidal mussels *Mytilus galloprovincialis* and *Perna perna*. *Oceanography and Marine Biology: An Annual Review* **53**, 127–156.

Mieszkowska, N., Sugden, H., Firth, L.B. & Hawkins, S.J. 2014. The role of sustained observations in tracking impacts of environmental change on marine biodiversity and ecosystems. *Philosophical Transactions of the Royal Society of London A: Mathematical, Physical and Engineering Sciences* **372**, 20130339.

Miller, R.J. & Etter, R.J. 2008. Shading facilitates sessile invertebrate dominance in the rocky subtidal Gulf of Maine. *Ecology* **89**, 452–462.

Mineur, F., Cook, E.J., Minchin, D., Bohn, K., MacLeod, A. & Maggs, C. 2012. Changing coasts: marine aliens and artificial structures. *Oceanography and Marine Biology: An Annual Review* **50**, 189–234.

Moody, M.E. & Mack, R.N. 1988. Controlling the spread of plant invasions: the importance of nascent foci. *Journal of Applied Ecology* **25**, 1009–1021.

Morrisey, D., Plew, D. & Seaward, K. 2011. Aquaculture Readiness Data: Phase II. *MAF Technical Paper* **58**.

Moschella, P.S., Abbiati, M., Åberg, P., Airoldi, L., Anderson, J.M., Bacchiocchi, F., Bulleri, F., Dinesen, G.E., Frost, M., Gacia, E., Granhag, L., Jonsson, P.R., Satta, M.P., Sundelöf, A., Thompson, R.C. & Hawkins, S.J. 2005. Low-crested coastal defence structures as artificial habitats for marine life: using ecological criteria in design. *Coastal Engineering* **52**, 1053–1071.

Moss, B., Kosten, S., Meerhof, M., Battarbee, R., Jeppesen, E., Mazzeo, N., Havens, K., Lacerot, G., Liu, Z. & De Meester, L. 2011. Allied attack: climate change and eutrophication. *Inland Waters* **1**, 101–105.

MPI. 2014. Craft risk management standard: biofouling on vessels arriving to New Zealand. Ministry for Primary Industries, Wellington, New Zealand.

Murray, A.G., Smith, R.J. & Stagg, R.M. 2002. Shipping and the spread of infectious salmon anemia in Scottish aquaculture. *Emerging Infectious Diseases* **8**, 1–5.

Neo, M.L. & Low, J.K.Y. 2017. First observations of *Tridacna noae* (Röding, 1798) (Bivalvia: Heterodonta: Cardiidae) in Christmas Island (Indian Ocean). *Marine Biodiversity*, doi:10.1007/s12526–017–0678–3

Neves, C.S., Rocha, R.M., Pitombo, F.B. & Roper, J.J. 2007. Use of artificial substrata by introduced and cryptogenic marine species in Paranagua Bay, southern Brazil. *Biofouling* **23**, 319–330.

Nicholls, R.J. & Mimura, N. 1998. Regional issues raised by sea-level rise and their policy implications. *Climate Research* **11**, 5–18.

Nikinmaa, M. 2013. Climate change and ocean acidification – interactions with aquatic toxicology. *Aquatic Toxicology* **126**, 365–372.

Nixon, S., Buckley, B., Granger, S. & Bintz, J. 2001. Responses of very shallow marine ecosystems to nutrient enrichment. *Human and Ecological Risk Assessment: An International Journal* **7**, 1457–1481.

Nixon, S., Oviatt, C., Frithsen, J. & Sullivan, B. 1986. Nutrients and the productivity of estuarine and coastal marine ecosystems. *Journal of the Limnological Society of Southern Africa* **12**, 43–71.

Nydam, M. & Stachowicz, J.J. 2007. Predator effects on fouling community development. *Marine Ecology Progress Series* **337**, 93–101.

O'Neil, J., Davis, T.W., Burford, M.A. & Gobler, C. 2012. The rise of harmful cyanobacteria blooms: the potential roles of eutrophication and climate change. *Harmful Algae* **14**, 313–334.

Olafsen, T. 2012. Value created from productive oceans in 2050. Report for the Royal Norwegian Society of Sciences and Letters (DKNVS) and the Norwegian Academy of Technological Sciences (NTVA), Trondheim.

Olden, J.D., Poff, N.L., Douglas, M.R., Douglas, M.E. & Fausch, K.D. 2004. Ecological and evolutionary consequences of biotic homogenization. *Trends in Ecology & Evolution* **19**, 18–24.

Ordóñez, V., Pascual, M., Rius, M. & Turon, X. 2013. Mixed but not admixed: a spatial analysis of genetic variation of an invasive ascidian on natural and artificial substrates. *Marine Biology* **160**, 1645–1660.

Osman, R. & Whitlatch, R. 1995. Predation on early ontogenetic life stages and its effect on recruitment into a marine epifaunal community. *Oceanographic Literature Review* **9**, 772.

Osman, R.W. & Whitlatch, R.B. 2004. The control of the development of a marine benthic community by predation on recruits. *Journal of Experimental Marine Biology and Ecology* **311**, 117–145.

Page, H.M., Dugan, J.E., Culver, C.S. & Hoesterey, J.C. 2006. Exotic invertebrate species on offshore oil platforms. *Marine Ecology Progress Series* **325**, 101–107.

Parente, V., Ferreira, D., dos Santos, E.M. & Luczynski, E. 2006. Offshore decommissioning issues: deductibility and transferability. *Energy Policy* **34**, 1992–2001.

Parrish, J.A.D. & Bazzaz, F.A. 1982. Responses of plants from three successional communities to a nutrient gradient. *Journal of Ecology* **70**, 233–248.

Pawlik, J.R. 1993. Marine invertebrate chemical defenses. *Chemical Reviews* **93**, 1911–1922.

Pearson, S., Windupranata, W., Pranowo, S.W., Putri, A., Ma, Y., Vila-Concejo, A., Fernández, E., Méndez, G., Banks, J., Knights, A.M. & Firth, L.B. 2016. Conflicts in some of the World Harbours: what needs to happen next? *Maritime Studies* **15**: 10–33.

Pearson, T.H. & Rosenberg, R. 1978. Macrobenthic succession in relation to organic enrichment and pollution of the marine environment. *Oceanography and Marine Biology: An Annual Review* **16**, 229–311.

Perkol-Finkel, S., Shashar, N., Barneah, O., Ben-David-Zaslow, R., Oren, U., Reichart, T., Yacobovich, T., Yahel, G., Yahel, R. & Benayahu, Y. 2005. Fouling reefal communities on artificial reefs: does age matter? *Biofouling* **21**, 127–140.

Perkol-Finkel, S., Zilman, G., Sella, I., Miloh, T. & Benayahu, Y. 2008. Floating and fixed artificial habitats: spatial and temporal patterns of benthic communities in a coral reef environment. *Estuarine, Coastal and Shelf Science* **77**, 491–500.

Pineda, M.C., McQuaid, C.D., Turon, X., López-Legentil, S., Ordóñez, V. & Rius, M. 2012. Tough adults, frail babies: an analysis of stress sensitivity across early life-history stages of widely introduced marine invertebrates. *PLoS ONE* **7**, e46672.

Pinn, E.H., Mitchell, K. & Corkill, J. 2005. The assemblages of groynes in relation to substratum age, aspect and microhabitat. *Estuarine, Coastal and Shelf Science* **62**, 271–282.

Piola, R.F., Dafforn, K.A. & Johnston, E.L. 2009. The influence of antifouling practices on marine invasions: a mini-review. *Biofouling* **25**, 633–644.

Piola, R.F. & Johnston, E.L. 2006. Differential tolerance to metals among populations of the introduced bryozoan *Bugula neritina*. *Marine Biology* **148**, 997–1010.

Piola, R.F. & Johnston, E.L. 2008a. Pollution reduces native diversity and increases invader dominance in marine hard-substrate communities. *Diversity and Distributions* **14**, 329–342.

Piola, R.F. & Johnston, E.L. 2008b. The potential for translocation of marine species via small-scale disruptions to antifouling surfaces. *Biofouling* **24**, 145–155.

Piola, R.F. & Johnston, E.L. 2009. Comparing differential tolerance of native and non-indigenous marine species to metal pollution using novel assay techniques. *Environmental Pollution* **157**, 2853–2864.

Prentis, P.J., Wilson, J.R.U., Dormontt, E.E., Richardson, D.M. & Lowe, A.J. 2008. Adaptive evolution in invasive species. *Trends in Plant Science* **13**, 288–294.

Punt, M.J., Groeneveld, R.A., van Ierland, E.C. & Stel, J.H. 2009. Spatial planning of offshore wind farms: a windfall to marine environmental protection? *Ecological Economics* **69**, 93–103.

Rabalais, N., Diaz, R., Levin, L., Turner, R., Gilbert, D. & Zhang, J. 2010. Dynamics and distribution of natural and human-caused hypoxia. *Biogeosciences* **7**, 585–619.

Rabalais, N.N., Turner, R.E., Díaz, R.J. & Justić, D. 2009. Global change and eutrophication of coastal waters. *ICES Journal of Marine Science: Journal du Conseil* **66**, 1528–1537.

Rius, M., Branch, G.M., Griffiths, C.L. & Turon, X. 2010. Larval settlement behaviour in six gregarious ascidians in relation to adult distribution. *Marine Ecology Progress Series* **418**, 151–163.

Rius, M., Clusella-Trullas, S., McQuaid, C.D., Navarro, R.A., Griffiths, C.L., Matthee, C.A., von der Heyden, S. & Turon, X. 2014a. Range expansions across ecoregions: interactions of climate change, physiology and genetic diversity. *Global Ecology and Biogeography* **23**, 76–88.

Rius, M. & Darling, J.A. 2014. How important is intraspecific genetic admixture to the success of colonising populations? *Trends in Ecology & Evolution* **29**, 233–242.

Rius, M., Heasman, K.G. & McQuaid, C.D. 2011. Long-term coexistence of non-indigenous species in aquaculture facilities. *Marine Pollution Bulletin* **62**, 2395–2403.

Rius, M. & McQuaid, C.D. 2009. Facilitation and competition between invasive and indigenous mussels over a gradient of physical stress. *Basic and Applied Ecology* **10**, 607–613.

Rius, M., Pineda, M.C. & Turon, X. 2009. Population dynamics and life cycle of the introduced ascidian *Microcosmus squamiger* in the Mediterranean Sea. *Biological Invasions* **11**, 2181–2194.

Rius, M., Potter, E.E., Aguirre, J.D. & Stachowicz, J.J. 2014b. Mechanisms of biotic resistance across complex life cycles. *Journal of Animal Ecology* **83**, 296–305.

Rius, M., Turon, X., Bernardi, G., Volckaert, F.A. & Viard, F. 2015. Marine invasion genetics: from spatio-temporal patterns to evolutionary outcomes. *Biological Invasions* **17**, 869–885.

Rius, M., Turon, X., Ordóñez, V. & Pascual, M. 2012. Tracking invasion histories in the sea: facing complex scenarios using multilocus data. *PLoS ONE* **7**, e35815.

Rivero, N.K., Dafforn, K.A., Coleman, M.A. & Johnston, E.L. 2013. Environmental and ecological changes associated with a marina. *Biofouling* **29**, 803–815.

Rodriguez, L.F. 2006. Can invasive species facilitate native species? Evidence of how, when, and why these impacts occur. *Biological Invasions* **8**, 927–939.

Rodríguez-Obeso, O., Alvarez-Guerra, M., Andrés, A., Viguri, J., DelValls, T., Riba, I. & Martín-Díaz, M. 2007. Monitoring and managing sediment quality and impact assessment in Spain in the past 10 years. *Trends in Analytical Chemistry* **26**, 252–260.

Roman, J. & Darling, J.A. 2007. Paradox lost: genetic diversity and the success of aquatic invasions. *Trends in Ecology & Evolution* **22**, 454–464.

Rosa, M., Holohan, B., Shumway, S., Bullard, S., Wikfors, G., Morton, S. & Getchis, T. 2013. Biofouling ascidians on aquaculture gear as potential vectors of harmful algal introductions. *Harmful Algae* **23**, 1–7.

Ruiz, G. & Carlton, J. 2003. *Invasive Species – Vectors and Management Strategies*. Washington, DC: Island Press.

Ruiz, G.M., Carlton, J.T., Grosholz, E.D. & Hines, A.H. 1997. Global invasions of marine and estuarine habitats by non-indigenous species: mechanisms, extent, and consequences. *American Zoologist* **37**, 621–632.

Ruiz, G.M., Fofonoff, P.W., Carlton, J.T., Wonham, M.J. & Hines, A.H. 2000. Invasion of coastal marine communities in North America: apparent patterns, processes and biases. *Annual Review of Ecology and Systematics* **31**, 481–531.

Ruiz, G.M. & Hewitt, C.L. 2002. Toward understanding patterns of coastal marine invasions: a prospectus. In *Invasive Aquatic Species of Europe. Distribution, Impacts and Management*, E. Leppäkoski et al. (eds). Dordrecht: Springer, 529–547.

Russ, G.R. 1980. Effects of predation by fishes, competition, and the structural complexity of the substratum on the establishment of a marine epifaunal community. *Journal of Experimental Marine Biology and Ecology* **42**, 55–69.

Russ, G.R. 1982. Overgrowth in a marine epifaunal community: competitive hierarchies and competitive networks. *Oecologia* **53**, 12–19.

Sammarco, P.W., Brazeau, D.A. & Sinclair, J. 2012. Genetic connectivity in scleractinian corals across the northern Gulf of Mexico: oil/gas platforms, and relationship to the Flower Garden Banks. *PLoS ONE* **7**, e30144.

Sammarco, P.W., Porter, S.A. & Cairns, S.D. 2010. A new coral species introduced into the Atlantic Ocean – *Tubastraea micranthus* (Ehrenberg 1834)(Cnidaria, Anthozoa, Scleractinia): an invasive threat. *Aquatic Invasions* **5**, 131–140.

Sammarco, P.W., Porter, S.A., Sinclair, J. & Genazzio, M. 2014. Population expansion of a new invasive coral species, *Tubastraea micranthus*, in the northern Gulf of Mexico. *Marine Ecology Progress Series* **495**, 161–173.

Saunders, R.J. & Connell, S.D. 2001. Interactive effects of shade and surface orientation on the recruitment of spirorbid polychaetes. *Austral Ecology* **26**, 109–115.

Schiedek, D., Sundelin, B., Readman, J.W. & Macdonald, R.W. 2007. Interactions between climate change and contaminants. *Marine Pollution Bulletin* **54**, 1845–1856.

Schiff, K., Diehl, D. & Valkirs, A. 2004. Copper emissions from antifouling paint on recreational vessels. *Marine Pollution Bulletin* **48**, 371–377.

Schimanski, K.B. 2015. *The importance of selective filters on vessel biofouling invasion processes*. PhD thesis, University of Canterbury, New Zealand.

Schoener, T.W. 1974. Resource partitioning in ecological communities. *Science* **185**, 27–39.

Schröder, V. & De Leaniz, C.G. 2011. Discrimination between farmed and free-living invasive salmonids in Chilean Patagonia using stable isotope analysis. *Biological Invasions* **13**, 203–213.

Schroeder, D.M. & Love, M.S. 2004. Ecological and political issues surrounding decommissioning of offshore oil facilities in the Southern California Bight. *Ocean & Coastal Management* **47**, 21–48.

Seebens, H., Gastner, M.T. & Blasius, B. 2013. The risk of marine bioinvasion caused by global shipping. *Ecology Letters* **16**, 782–790.

Shafer, D.J. 1999. The effects of dock shading on the seagrass *Halodule wrightii* in Perdido Bay, Alabama. *Estuaries* **22**, 936–943.

Shears, N.T. & Babcock, R.C. 2002. Marine reserves demonstrate top-down control of community structure on temperate reefs. *Oecologia* **132**, 131–142.

Sheehy, D.J. & Vik, S.F. 2010. The role of constructed reefs in non-indigenous species introductions and range expansions. *Ecological Engineering* **36**, 1–11.

Shenkar, N., Zeldman, Y. & Loya, Y. 2008. Ascidian recruitment patterns on an artificial reef in Eilat (Red Sea). *Biofouling* **24**, 119–128.

Simberloff, D. & Gibbons, L. 2004. Now you see them, now you don't – population crashes of established introduced species. *Biological Invasions* **6**, 161–172.

Simberloff, D. & Von Holle, B. 1999. Positive interactions of nonindigenous species: invasional meltdown? *Biological Invasions* **1**, 21–32.

Simkanin, C., Davidson, I., Falkner, M., Sytsma, M. & Ruiz, G. 2009. Intra-coastal ballast water flux and the potential for secondary spread of non-native species on the US West Coast. *Marine Pollution Bulletin* **58**, 366–374.

Simkanin, C., Dower, J.F., Filip, N., Jamieson, G. & Therriault, T.W. 2013. Biotic resistance to the infiltration of natural benthic habitats: examining the role of predation in the distribution of the invasive ascidian *Botrylloides violaceus*. *Journal of Experimental Marine Biology and Ecology* **439**, 76–83.

Simoncini, M. & Miller, R.J. 2007. Feeding preference of *Strongylocentrotus droebachiensis* (Echinoidea) for a dominant native ascidian, *Aplidium glabrum*, relative to the invasive ascidian *Botrylloides violaceus*. *Journal of Experimental Marine Biology and Ecology* **342**, 93–98.

Simpson, S.L. & Spadaro, D.A. 2011. Performance and sensitivity of rapid sublethal sediment toxicity tests with the amphipod *Melita plumulosa* and copepod *Nitocra spinipes*. *Environmental Toxicology and Chemistry* **30**, 2326–2334.

Sinner, J., Forrest, B.M., Newton, M., Hopkins, G.A., Inglis, G., Woods, C. & Morrisey, D. 2013. Managing the domestic spread of harmful marine organisms, Part B: statutory framework and analysis of options. Report prepared for New Zealand Ministry for Primary Industries. Cawthron Institute report No. 2442. Nelson, 73 p.

Smith, R., Bolam, S., Rees, H. & Mason, C. 2008. Macrofaunal recovery following TBT ban. *Environmental Monitoring and Assessment* **136**, 245–256.

Smith, V.H., Joye, S.B. & Howarth R.W. 2006. Eutrophication of freshwater and marine ecosystems. *Limnology and Oceanography* **51**, 351–355.

Snyder, D.B. & Burgess, G.H. 2007. The Indo-Pacific red lionfish, *Pterois volitans* (Pisces: Scorpaenidae), new to Bahamian ichthyofauna. *Coral Reefs* **26**, 175.

Sokolova, I.M. & Lannig, G. 2008. Interactive effects of metal pollution and temperature on metabolism in aquatic ectotherms: implications of global climate change. *Climate Research* **37**, 181–201.

Sorte, C.J.B., Williams, S.L. & Zerebecki, R.A. 2010. Ocean warming increases threat of invasive species in a marine fouling community. *Ecology* **91**, 2198–2204.

Soto, D., Arismendi, I., Gonzalez, J., Sanzana, J., Jara, F., Jara, C., Guzman, E. & Lara, A. 2006. Southern Chile, trout and salmon country: invasion patterns and threats for native species. *Revista Chilena de Historia Natural* **79**, 97–117.

Soto, D., Jara, F. & Moreno, C. 2001. Escaped salmon in the inner seas, southern Chile: facing ecological and social conflicts. *Ecological Applications* **11**, 1750–1762.

Soto, D. & Norambuena, F. 2004. Evaluation of salmon farming effects on marine systems in the inner seas of southern Chile: a large-scale mensurative experiment. *Journal of Applied Ichthyology* **20**, 493–501.

Stachowicz, J.J., Fried, H., Osman, R.W. & Whitlach, R.B. 2002a. Biodiversity, invasion resistance, and marine ecosystem function: reconciling pattern and process. *Ecology* **83**, 2575–2590.

Stachowicz, J.J., Terwin, J.R., Whitlatch, R.B. & Osman, R.W. 2002b. Linking climate change and biological invasions: ocean warming facilitates nonindigenous species invasions. *Proceedings of the National Academy of Sciences of the United States of America* **99**, 15497–15500.

Stachowicz, J.J., Whitlatch, R.B. & Osman, R.W. 1999. Species diversity and invasion resistance in a marine ecosystem. *Science* **286**, 1577–1579.

Statham, P.J. 2012. Nutrients in estuaries – an overview and the potential impacts of climate change. *Science of the Total Environment* **434**, 213–227.

Steinberg, P.D. 1986. Chemical defenses and the susceptibility of tropical marine brown algae to herbivores. *Oecologia* **69**, 628–630.

Suarez, A.V., Holway, D.A. & Ward, P.S. 2005. The role of opportunity in the unintentional introduction of nonnative ants. *Proceedings of the National Academy of Sciences of the United States of America* **102**, 17032–17035.

Sutherland, S. 2004. What makes a weed a weed: life history traits of native and exotic plants in the USA. *Oecologia* **141**, 24–39.

Svane, I. & Dolmer, P. 1995. Perception of light at settlement: a comparative study of two invertebrate larvae, a scyphozoan planula and a simple ascidian tadpole. *Journal of Experimental Marine Biology and Ecology* **187**, 51–61.

Tamburri, M.N., Ruiz, G.M., Apple, R., Altshuller, D., Fellbeck, H. & Hurley, W.L. 2005. Evaluations of a ballast water treatment to stop invasive species and tank corrosion. Discussion. *Transactions-Society of Naval Architects and Marine Engineers* **113**, 558–568.

Tamelander, J., Riddering, L., Haag, F., Matheickal, J. & No, G.M.S. 2010. *Guidelines for development of a national ballast water management strategy*, GloBallast Partnerships Project Coordination Unit, International Maritime Organization.

Teo, S.L.M. & Ryland, J.S. 1994. Toxicity and palatability of some British ascidians. *Marine Biology* **120**, 297–303.

Tepolt, C., Darling, J., Bagley, M., Geller, J., Blum, M. & Grosholz, E. 2009. European green crabs (*Carcinus maenas*) in the northeastern Pacific: genetic evidence for high population connectivity and current-mediated expansion from a single introduced source population. *Diversity and Distributions* **15**, 997–1009.

Tewfik, A., Rasmussen, J. & McCann, K.S. 2005. Anthropogenic enrichment alters a marine benthic foodweb. *Ecology* **86**, 2726–2736.

Theoharides, K.A. & Dukes, J.S. 2007. Plant invasion across space and time: factors affecting nonindigenous species success during four stages of invasion. *New Phytologist* **176**, 256–273.

Tilman, D. 1997. Community invasibility, recruitment limitation, and grassland biodiversity. *Ecology* **78**, 81–92.

Toh, K.B., Ng, C.S.L., Wu, B., Toh, T.C., Cheo, P.R., Tun, K. & Chou, L.M. 2016. Spatial variability of epibiotic assemblages on marina pontoons in Singapore. *Urban Ecosystems*, doi:10.1007/s11252–016–0589–2

Torchin, M.E., Lafferty, K.D., Dobson, A.P., McKenzie, V.J. & Kuris, A.M. 2003. Introduced species and their missing parasites. *Nature* **421**, 628–630.

Torchin, M., Lafferty, K. & Kuris, A. 2002. Parasites and marine invasions. *Parasitology* **124**, 137–151.

Trannum, H.C., Olsgard, F., Skei, J.M., Indrehus, J., Overas, S. & Eriksen, J. 2004. Effects of copper, cadmium and contaminated harbour sediments on recolonisation of soft-bottom communities. *Journal of Experimental Marine Biology and Ecology* **310**, 87–114.

Vallejo-Marín, M. & Hiscock, S.J. 2016. Hybridization and hybrid speciation under global change. *The New Phytologist* **211**, 1170–1187.

Van Kleunen, M., Weber, E. & Fischer, M. 2010. A meta-analysis of trait differences between invasive and non-invasive plant species. *Ecology Letters* **13**, 235–245.

Vaselli, S., Bulleri, F. & Benedetti-Cecchi, L. 2008. Hard coastal-defence structures as habitats for native and exotic rocky-bottom species. *Marine Environmental Research* **66**, 395–403.

Vermeij, M.J.A., Smith, T.B., Dailer, M.L. & Smith, C.M. 2009. Release from native herbivores facilitates the persistence of invasive marine algae: a biogeographical comparison of the relative contribution of nutrients and herbivory to invasion success. *Biological Invasions* **11**, 1463–1474.

Voisin, M., Engel, C.R. & Viard, F. 2005. Differential shuffling of native genetic diversity across introduced regions in a brown alga: aquaculture vs. maritime traffic effects. *Proceedings of the National Academy of Sciences of the United States of America* **102**, 5432–5437.

Ware, C., Berge, J., Sundet, J.H., Kirkpatrick, J.B., Coutts, A.D.M., Jelmert, A., Olsen, S.M., Floerl, O., Wisz, M.S. & Alsos, I.G. 2014. Climate change, non-indigenous species and shipping: assessing the risk of species introduction to a high-Arctic archipelago. *Diversity and Distributions* **20**, 10–19.

Warnken, J., Dunn, R.J.K. & Teasdale, P.R. 2004. Investigation of recreational boats as a source of copper at anchorage sites using time-integrated diffusive gradients in thin film and sediment measurements. *Marine Pollution Bulletin* **49**, 833–843.

Wasson, K., Zabin, C.J., Bedinger, L., Diaz, M.C. & Pearse, J.S. 2001. Biological invasions of estuaries without international shipping: the importance of intraregional transport. *Biological Conservation* **102**, 143–153.

Wernberg, T., Russell, B.D., Moore, P.J., Ling, S.D., Smale, D.A., Campbell, A., Coleman, M.A., Steinberg, P.D., Kendrick, G.A. & Connell, S.D. 2011. Impacts of climate change in a global hotspot for temperate marine biodiversity and ocean warming. *Journal of Experimental Marine Biology and Ecology* **400**, 7–16.

Wernberg, T., Smale, D.A. & Thomsen, M.S. 2012. A decade of climate change experiments on marine organisms: procedures, patterns and problems. *Global Change Biology* **18**, 1491–1498.

Wernberg, T., Smale, D.A., Tuya, F., Thomsen, M.S., Langlois, T.J., de Bettignies, T., Bennett, S. & Rousseaux, C.S. 2013. An extreme climatic event alters marine ecosystem structure in a global biodiversity hotspot. *Nature Climate Change* **3**, 78–82.

Whitehead, P., Wilby, R., Battarbee, R., Kernan, M. & Wade, A.J. 2009. A review of the potential impacts of climate change on surface water quality. *Hydrological Sciences Journal* **54**, 101–123.

Wilcove, D.S., Rothstein, D., Jason, D., Phillips, A. & Losos, E. 1998. Quantifying threats to imperiled species in the United States. *Bioscience* **48**, 607–615.

Williams, S.L., Davidson, I.C., Pasari, J.R., Ashton, G.V., Carlton, J.T., Crafton, R.E., Fontana, R.E., Grosholz, E.D., Miller, A.W., Ruiz, G.M. & Zabin, C.J. 2013. Managing multiple vectors for marine invasions in an increasingly connected world. *BioScience* **63**, 952–966.

Williams, S.L. & Smith, J.E. 2007. A global review of the distribution, taxonomy, and impacts of introduced seaweeds. *Annual Review of Ecology, Evolution, and Systematics* **38**, 327–359.

Williamson, M.H., Brown, K.C., Holdgate, M.W., Kornberg, H., Southwood, R. & Mollison, D. 1986. The analysis and modelling of British invasions. *Philosophical Transactions of the Royal Society of London* **314**, 505–522.

Wilson, J.C. & Elliott, M. 2009. The habitat-creation potential of offshore wind farms. *Wind Energy* **12**, 203–212.

Winder, M., Jassby, A.D. & Mac Nally, R. 2011. Synergies between climate anomalies and hydrological modifications facilitate estuarine biotic invasions. *Ecology Letters* **14**, 749–757.

Woodin, S.A., Wethey, D.S. & Dubois, S.F. 2014. Population structure and spread of the polychaete *Diopatra biscayensis* along the French Atlantic coast: human-assisted transport by-passes larval dispersal. *Marine Environmental Research* **102**, 110–121.

Yeo, D.C., Ahyong, S.T., Lodge, D.M., Ng, P.K., Naruse, T. & Lane, D.J. 2009. Semisubmersible oil platforms: understudied and potentially major vectors of biofouling-mediated invasions. *Biofouling* **26**, 179–186.

Zabin, C.J., Ashton, G.V., Brown, C.W., Davidson, I.C., Sytsma, M.D. & Ruiz, G.M. 2014. Small boats provide connectivity for nonindigenous marine species between a highly invaded international port and nearby coastal harbors *Management of Biological Invasions* **5**, 97–112.

Zaiko, A., Schimanski, K., Pochon, X., Hopkins, G.A., Goldstien, S., Floerl, O. & Wood, S.A. 2016. Metabarcoding improves detection of eukaryotes from early biofouling communities: implications for pest monitoring and pathway management. *Biofouling* **32**, 671–684.

Zerebecki, R.A. & Sorte, C.J.B. 2011. Temperature tolerance and stress proteins as mechanisms of invasive species success. *PloS one* **6**, e14806.

Zhan, A., Briski, E., Bock, D.G., Ghabooli, S. & MacIsaac, H.J. 2015. Ascidians as models for studying invasion success. *Marine Biology* **162**, 2449–2470.

Oceanography and Marine Biology: An Annual Review, 2017, **55**, 421-440
© S. J. Hawkins, D. J. Hughes, I. P. Smith, A. C. Dale, L. B. Firth, and A. J. Evans, Editors
Taylor & Francis

A REVIEW OF HERBIVORE EFFECTS
ON SEAWEED INVASIONS

SWANTJE ENGE[1,2], JOSEFIN SAGERMAN[3], SOFIA A. WIKSTRÖM[3] & HENRIK PAVIA[4]

*[1]Department of Plant and Environmental Sciences,
University of Copenhagen, 1871 Frederiksberg, Denmark
[2]Institute for Chemistry and Biology of the Marine Environment,
University of Oldenburg, 26382 Wilhelmshaven, Germany
[3]Baltic Sea Centre, Stockholm University, 106 91 Stockholm, Sweden
[4]Department of Marine Sciences-Tjärnö, University of Gothenburg, 452 96 Strömstad, Sweden
Corresponding author: Henrik Pavia
e-mail: henrik.pavia@marine.gu.se

Almost 300 non-native seaweeds are identified worldwide and an increasing number of these are classified as invasive with potential negative effects on the diversity and functioning of native ecosystems. Marine herbivores affect seaweed biomass and community structure in marine habitats across the globe. Consequently, herbivore-seaweed interactions are expected to be important for the establishment and invasion success of non-native seaweeds. To synthesize current knowledge of consumer effects on non-native seaweeds, we performed a meta-analysis on feeding preferences of native herbivores for non-native versus native seaweeds. Data were included from 35 studies, published from 1992–2015 and comprising 18 non-native seaweeds. Results showed that overall, native herbivores tended to prefer to feed on native rather than non-native seaweeds. Preferences were, however, variable across studies with significant differences between taxonomic and functional groups of seaweeds. In particular, filamentous red non-native seaweeds were of low palatability to native herbivores. No general feeding preferences were apparent between natives and non-natives for brown and green seaweeds, or for leathery and corticated seaweeds. In addition, we reviewed the existing studies on the effects of consumers on the performance of native and non-native seaweeds in invaded communities. This indicated that non-native seaweeds performed better than their native competitors in the presence of grazers, but in many cases had superior competitive abilities also in the absence of herbivory. To achieve a comprehensive evaluation of consumers' role in seaweed invasion success, future research should have a larger focus on manipulative community experiments, ideally on time scales that include seasonal changes and complete life cycles of the seaweeds.

Introduction

At any time, several thousand marine species are shuffled between biogeographical regions of the world's oceans (Johnson & Chapman 2007). Some of these species establish in their new communities and become widespread and abundant, being deemed invasive, with significant impacts on community composition and ecosystem properties (Williamson & Fitter 1996). The global number of non-indigenous seaweeds has reached more than 270 species (Williams & Smith 2007), many of them reported to negatively affect native seaweed communities in terms of cover, density and biodiversity (Schaffelke & Hewitt 2007, Engelen et al. 2015, Maggi et al. 2015). Identification of the factors that control establishment and invasion success of non-native species is a key challenge in invasion ecology and important for risk assessment and management of non-indigenous seaweeds.

Herbivores have large influence on the abundance and community structure of both terrestrial and marine primary producers across ecosystems (e.g. Lubchenco & Gaines 1981, Hawkins & Hartnoll 1983, Hawkins et al. 1992, Hay & Steinberg 1992, Burkepile & Hay 2008, Poore et al. 2012). Consequently, herbivore-plant interactions have long been suggested to be crucial also for the outcome of plant invasions. On the one hand, herbivores are recognized to contribute to biotic resistance against the establishment and proliferation of non-native plants (Elton 1958, Maron & Vila 2001). On the other hand, herbivore foraging has been suggested to drive invasions, when native herbivores preferably consume native over the non-native plant species. The latter argument forms the basis for the enemy release hypothesis (ERH), which states that non-native species become invasive since they escape the regulation of their co-evolved enemies in their native range and are less affected by enemies compared to the native competitors in their new range (Darwin 1859, Elton 1958, Keane & Crawley 2002). The reduced negative impact of herbivores is expected to give the non-native species a competitive advantage over the native plants in the community, leading to dominance and a biological invasion (Keane & Crawley 2002).

The concepts of biotic resistance and enemy release originate from terrestrial plant systems; the findings from these systems have been summarized in several reviews and meta-analyses (Colautti et al. 2004, Levine et al. 2004, Liu & Stiling 2006, Parker et al. 2006, Chun et al. 2010). Recently, Kimbro et al. (2013) published a meta-analysis on biotic resistance in marine environments, indicating negative effects of consumers on non-native marine primary producers. Seaweeds were only represented by seven studies in their analysis and seaweed data were analyzed together with data on a non-native salt marsh grass. Thus, a comprehensive quantitative synthesis of research on herbivore effects on non-native seaweeds is still lacking.

Seaweed communities are strongly dominated by generalist herbivores (Hawkins & Hartnoll 1983, Hawkins et al. 1992, Hay & Fenical 1992), which are more likely than specialists to include newly encountered species into their diet. Accordingly, generalist consumers have been suggested to counteract invasions instead of facilitating invasions through enemy release (Parker & Hay 2005, Parker et al. 2006). There are, however, mechanisms by which non-native plants can escape generalist herbivores in their new range, especially by means of chemical defences (Wikström et al. 2006, Verhoeven et al. 2009, Forslund et al. 2010, Schaffner et al. 2011, Enge et al. 2012, Nylund et al. 2012). Seaweeds are known to be rich in secondary metabolites (Hay & Fenical 1992) and there is an increasing number of examples where potent chemical defences against native herbivores have been demonstrated in non-native seaweeds (Lemee et al. 1996, Lyons et al. 2007, Nylund et al. 2011, Enge et al. 2012). It is not known, however, if low palatability to generalist herbivores is a common trait of non-native seaweeds or if this trait is important for invasion success in seaweed communities.

The aim of this study was to review and synthesize current findings of consumer effects on non-native seaweeds. As part of the predictions of the ERH, we specifically explored via a meta-analysis whether non-native seaweeds are less palatable than native species, and thus generally experience a reduced impact by herbivores compared to native competitors. We further examined if there are differences among 1) seaweed taxonomic groups, 2) seaweed functional groups, 3) herbivore groups or 4) regions. Furthermore, we reviewed and summarized the literature that assessed the effects of herbivory on the competitive outcome between non-native and native seaweeds in a community.

Material and methods

Literature search and data extraction

Relevant studies for the meta-analysis were identified in the online database ISI Web of Science in August 2015, with no restrictions on publication year, using the following combination of search terms: ((introduced OR invasive OR non-native OR exotic OR alien OR non-indigenous) AND (seaweed* OR alga* OR macroalga*) AND (herbivor* OR consum* OR *graz* OR enem* OR prefer*)). In order to retrieve studies that tested non-native seaweeds but without specifying the species as non-native in the title or abstract, we performed an additional search for all non-native seaweeds listed in Williamson & Smith (2007) using the search terms: (('algal species name') AND (enem* OR herbivor* OR consum* OR graz*)). To determine if the seaweeds in these studies were non-native, we compared the study region with the reported natural distribution of the species. We further included three as yet non-peer-reviewed datasets (S.A. Wikström unpublished, K. Hill unpublished, S. Jakobsson unpublished). All titles and abstracts of the search results were screened for studies assessing feeding preferences or herbivore damage on non-native compared to native seaweeds, as well as for studies examining seaweed performances and competitive relationships between the non-native and native seaweeds in the presence and absence of herbivory. To be included in the final dataset, the studies had to meet the following criteria: 1) the investigated seaweed was non-native to the study region while the seaweeds used for comparison and the herbivores were native to the study region; 2) the study assessed herbivore preference, damage or their effects on both non-native and native seaweed performance under laboratory or field conditions in two- or multiple-species experiments using living algal material; 3) the experimental design included proper controls and presented all necessary measures for calculating the effect size. We excluded data of epifaunal abundances on non-native compared to native seaweeds because abundance may reflect habitat choice rather than food preference (e.g. work on refuges from predation: Duffy & Hay 1991, Enge et al. 2013). We also excluded data from no-choice feeding experiments since consumption in a no-choice situation can be confounded by compensatory feeding (Cruz-Rivera & Hay 2000). Furthermore, studies using gut content analysis of herbivores collected in the field were excluded because it was not possible to relate gut content to the availability of seaweeds in the field. Finally, studies where herbivores were preconditioned on one of the experimental seaweeds by either being specifically collected from or fed with only that seaweed prior to the experiments were also excluded.

For the retained studies the following data were compiled (see Table 1): seaweed species name, taxonomic and functional group, the study region and its corresponding climate region, the origin of the non-native seaweeds, herbivore species identity and phylum, as well as the experiment type (i.e. two- or multiple-choice). Mean values and measures of dispersion were extracted from figures using the WebPlotDigitizer software (Rohatgi 2015) or directly from text, tables or original datasets. All retrieved studies on herbivore feeding preference reported consumption or relative growth of the seaweeds as the response variable, which were either presented already corrected for autogenic changes of the seaweeds or together with the means of controls for autogenic changes. In the latter case, the consumption or relative growth data were corrected for the autogenic changes before values were entered into the database. To obtain a reference value for the native seaweeds in multiple-choice experiments, we calculated the average consumption of all native seaweeds and used their pooled variance as a measure of variability, which assumes the means of the populations to differ but their variances to be the same.

Meta-analysis

Hedges' d standardized mean difference (Hedges & Olkin 1985) was used as the effect size measure and was calculated as the difference between non-native and native seaweeds: $d = [(\overline{X}_{NN} - \overline{X}_N)/s]$, where $\overline{X}_{NN}$ corresponds to the mean consumption of the non-native seaweed and $\overline{X}_N$ to the mean consumption of the native species, s designates their pooled standard deviations and J is the small-sample-size bias correction factor. Positive Hedges' d values thereby reflect herbivore preference for the non-native seaweeds, while negative values reflect herbivore preference for the native seaweeds.

Many publications reported data of several independently performed two- or multiple-choice experiments using different native seaweed species and/or different herbivores, which resulted in up to 16 data entries from one study and for one non-native seaweed. To balance the influence of studies in the analyses and to decrease possible non-independence of these multiple entries, we

Table 1 Summary of the data on herbivore preference used in the meta-analysis

Species	Functional group	Studies (number of data entries per study)	Studied regions	Climate region	Origin	Herbivore phylum	Herbivore species	Experiment type
Chlorophyta								
Caulerpa filiformis	Corticated	Davis et al. 2005 (1); Cummings & Williamson 2008 (1)	SW Pacific	Subtropical	Indo-Pacific	Mollusca	*Turbo undulatus*[4]; natural herbivore assemblage	Multiple-choice
Caulerpa racemosa var cylindracea[1]	Corticated	Tomas et al. 2011a (1); Tomas et al. 2011b (1)	Mediterranean Sea	Subtropical	SW Pacific	Chordata; Echinodermata;	*Paracentrotus lividus*; *Sarpa salpa*	Two-choice
Codium fragile ssp. *atlanticum*	Corticated	Trowbridge & Todd 1999 (1)	NE Atlantic	Temperate	NW Pacific	Mollusca	*Littorina littorea*	Two-choice
Codium fragile ssp. *tomentosoides*[2]	Corticated	Prince & Leblanc 1992 (1); Trowbridge 1995 (3); Cruz-Rivera & Hay 2001 (1); Scheibling & Anthony 2001 (1); Chavanich & Harris 2002 (1); Levin et al. 2002 (2); Sumi & Scheibling 2005 (1); Hill (unpublished data) (2); Lyons & Scheibling 2007 (1); Jakobsson (unpublished data) (1)	NW Atlantic; NE Atlantic; SW Pacific	Temperate; Subtropical	NW Pacific	Arthropoda; Echinodermata; Mollusca	*Ampithoe longimana*; *Cookia sulcata*; *Evechinus chloroticus*; *Idotea granulosa*; *Lacuna vincta*; *Strongylocentrotus droebachiensis*; *Turbo smaragdus*[5]	Two-choice; Multiple-choice

Continued

Table 1 (Continued) Summary of the data on herbivore preference used in the meta-analysis

Species	Functional group	Studies (number of data entries per study)	Studied regions	Climate region	Origin	Herbivore phylum	Herbivore species	Experiment type
Phaeophyceae								
Fucus evanescens	Leathery	Schaffelke et al. 1995 (1); Wikström et al. 2006 (2)	NE Atlantic	Temperate	N Circumpol	Arthropoda; Mollusca	*Idotea granulosa; Littorina obtusata*	Two-choice; Multiple-choice
Fucus serratus	Leathery	Wikström et al. 2006 (1); Steinarsdóttir et al. 2009 (3); Wikström unpublished data (2)	NE Atlantic	Subpolar	N Circumpol	Arthropoda; Mollusca	*Gammarus obtusatus*[6]; *Idotea granulosa; Littorina littorea; Littorina obtusata*	Two-choice
Sargassum muticum	Leathery	Britton-Simmons 2004 (1); Pedersen et al. 2005 (1); Hill (unpublished data) (2); Monteiro et al. 2009 (6); Strong et al. 2009 (1); Cacabelos et al. 2010 (4); Britton-Simmons et al. 2011 (1); Engelen et al. 2011 (3)	NE Atlantic; NE Pacific	Temperate; Subtropical	NW Pacific	Arthropoda; Echinodermata; Mollusca	*Aplysia punctata; Dexamine spinosa; Gammarus insensibilis; Gibbula spp.; Hydrobia ulvae*[7]; *Idotea granulosa; Lacuna vincta; Littorina littorea; Littorina obtusata; Paracentrotus lividus; Psammechinus miliaris; Stenosoma nadejda; Strongylocentrotus droebachiensis*	Two-choice; Multiple-choice

Continued

425

Table 1 (Continued)   Summary of the data on herbivore preference used in the meta-analysis

Species	Functional group	Studies (number of data entries per study)	Studied regions	Climate region	Origin	Herbivore phylum	Herbivore species	Experiment type
Undaria pinnatifda	Leathery	Thornber et al. 2004 (1)	NE Pacific	Subtropical	NW Pacific	Arthropoda	*Pugettia producta*	Two-choice
Rhodophyta								
Acrothamnion preissii	Filamentous	Tomas et al. 2011a (1)	Mediterranean Sea	Subtropical	Indo-Pacific	Echinodermata	*Paracentrotus lividus*	Two-choice
Bonnemaisonia hamifera	Filamentous	Hill (unpublished data) (2); Enge et al. 2012 (4)	NE Atlantic	Temperate	NW Pacific	Arthropoda; Mollusca	*Aplysia punctata*; *Gammarellus angulosus*; *Gammarus locusta*; *Idotea neglecta*; *Littorina littorea*	Two-choice
Gracilaria salicornia	Corticated	Smith et al. 2004 (4)	N Pacific	Tropical	Indo-Pacific	Chordata	*Acanthurus blochii*; *Acanthurus triostegus*; juvenile scarids; *Zebrasoma flavescens*	Two-choice
Gracilaria vermiculophylla	Corticated	Weinberger et al. 2008 (2); Nejrup et al. 2012 (3)	Baltic Sea; NE Atlantic	Temperate	NW Pacific	Arthropoda; Mollusca	*Gammarus locusta*; *Idotea balthica*; *Littorina littorea*; *Littorina* sp.	Two-choice

Continued

Table 1 (Continued) Summary of the data on herbivore preference used in the meta-analysis

Species	Functional group	Studies (number of data entries per study)	Studied regions	Climate region	Origin	Herbivore phylum	Herbivore species	Experiment type
Heterosiphonia japonica[3]	Filamentous	Low et al. 2015 (2); Sagerman et al. 2015 (2)	NE Atlantic; NW Atlantic	Temperate	NW Pacific	Arthropoda; Mollusca	*Gammarellus angulosus; Gammarus locusta; Idotea balthica; Idotea granulosa; Lacuna vincta*	Two-choice; Multiple-choice
Lophocladia lallemandii	Filamentous	Tomas et al. 2011a (1)	Mediterranean Sea	Subtropical	Indo-Pacific	Echinodermata	*Paracentrotus lividus*	Two-choice
Mastocarpus stellatus	Corticated	Yun & Molis 2012 (2)	NE Atlantic	Temperate	NE Atlantic	Arthropoda; Mollusca	*Idotea balthica; Littorina littorea*	Two-choice
Womersleyella setacea	Filamentous	Tomas et al. 2011a (1)	Mediterranean Sea	Subtropical	Circum-equatorial	Echinodermata	*Paracentrotus lividus*	Two-choice
Gracilaria salicornia + *Acanthophora spicifera* + *Kappaphycus* sp.	Mixed	Stimson et al. 2007 (1)	N Pacific	Tropical	Indo-Pacific; NW Atlantic	Echinodermata	*Tripneustes gratilla*	Multiple-choice

Note: The functional group division follows Littler & Littler (1984): filamentous = delicately branched, uniseriate to slightly corticated; corticated = coarsely branched, corticated; leathery = thick branched, heavily corticated or thick walled.

[1] *Caulerpa racemosa* var. *cylindracea* currently known as *Caulerpa cylindracea*; [2] *Codium fragile* ssp. *tomentosoides* currently known as *Codium fragile* ssp. *fragile*; [3] *Heterosiphonia japonica* currently known as *Dasysiphonia japonica*; [4] *Turbo undulatus* currently known as *Lunella undulata*; [5] *Turbo smaragdus* currently known as *Lunella smaragda*; [6] *Gammarus obtusatus* currently known as *Echinogammarus obtusatus*; [7] *Hydrobia ulvae* currently known as *Peringia ulvae*.

calculated a study-specific mean effect size for each herbivore species and non-native species across all experiments testing different combinations with native seaweeds in a study. In this way, we kept the resolution between the non-native seaweed and herbivore species, but reduced the problem of overweighing and consequent false precision estimates (Rothstein et al. 2013). We also calculated a study-specific mean effect size when experiments were repeatedly performed at different times during the year. In both cases, the study-specific mean effects were estimated using a fixed-effects model, which assumes the results of the different experiments in one study to vary only because of random sampling error. This procedure reduced the number of entries in the dataset with a factor of up to five for some publications and in total from 145 to 74 entries.

The meta-analysis on consumer preference for non-native versus native seaweeds was conducted using the metafor-package in R (Viechtbauer 2010) and the OpenMee software (Dietz et al. 2016). The weighted overall mean effect of herbivore preference for non-native or native seaweeds was calculated by a random-effects model using the restricted maximum-likelihood estimator for residual heterogeneity. Bootstrapped 95% confidence intervals were calculated for the overall mean effect size generated from 4999 iterations. To check the robustness of the meta-analysis outcome, we calculated the fail-safe number with the weighted method of Rosenberg (2005), which represents the number of additional studies with no effect needed to change the result of the meta-analysis from significant to non-significant. Publication bias was further examined with a funnel-plot and the rank correlation test for funnel plot asymmetry (Begg & Mazumdar 1994). The influence of outliers on the overall mean effects size was tested by evaluating the change of the overall effect when one study at a time was left out of the analysis. Since hypothesis-driven research tends to favour large effect sizes in support of the hypothesis in earlier publications, we examined temporal trends in the data with a cumulative meta-analysis sorted by publication year (Jennions & Møller 2002).

We used meta-regression with mixed-effects models and with a restricted maximum-likelihood estimator for residual heterogeneity to assess if the predefined covariables explained any of the observed heterogeneity and to explore their influence on consumer preference as well as differences between subgroups. Specifically, we tested how much of the observed heterogeneity the non-native seaweed itself accounted for, if there were differences in consumer preferences among phyla or functional groups of the non-native species, and if the climate region of the study site, herbivore phylum and experiment type influenced the study outcome. Origin and study region as covariables were omitted from these analyses since subgroups were often only represented by one or two species and few data entries in some of the subgroups, which were considered insufficient for a valid interpretation.

Only five studies that tested for consumer effects on the performance of non-native and native seaweeds on a community scale were identified, too few to perform a formal meta-analysis. However, findings of the few existing studies were summarized and discussed.

Results

The literature search identified 35 studies published from 1992–2015 that examined feeding preferences of native herbivores for non-native compared to native seaweeds. These studies addressed in total 18 non-native seaweed taxa: four green algae (Chlorophyta), four brown algae (Phaeophyceae) and ten red algae (Rhodophyta) (Table 1). This is comparable with the taxonomic composition of all registered non-native seaweeds reported by Williams & Smith (2007): of 276 taxa, 45 were green algae, 66 brown algae and 165 red algae. Accordingly, between 6–9% of all non-native seaweeds in each of the taxonomic groups were covered by our meta-analysis. All comparisons were made between non-native seaweeds and native seaweeds with the exception of one study with five data entries where the palatability of non-native seaweeds was compared to a seagrass species. By far the most intensively studied seaweeds in the context of feeding preference of herbivores were *Codium fragile* ssp. *tomentosoides* (currently accepted name *Codium fragile* ssp. *fragile*) and *Sargassum muticum*, which also resulted in the highest number of entries for these species in the meta-analysis

(Table 1). More than two thirds of the non-native seaweeds included in the meta-analysis have their original distribution in the north-west Pacific and Indo-Pacific, whereas the most intensively studied region of introduction was the north-east Atlantic (Table 1). Isopods and amphipods (Arthropoda), gastropods (Mollusca), sea urchins (Echinodermata) and herbivorous fish (Chordata) constituted the majority of the tested herbivores in the studies (Table 1).

Meta-analysis of consumer preference

The weighted overall mean effect was estimated to be –0.528 with a bootstrapped 95% CI = (–0.872, –0.187). This effect was significantly different from zero (p = 0.002, Figure 1) and showed

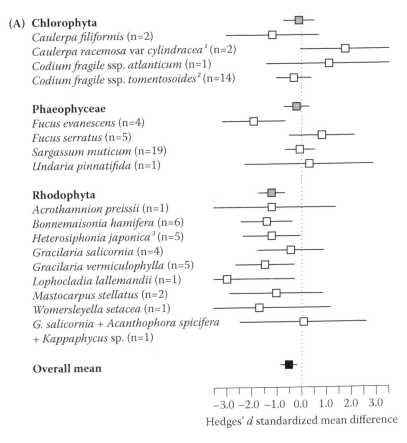

(B) Co-variable	Q_M	Q_E	I^2	R^2
Species	31.27*	466***	92.33	17.53
Phylum	9.34**	725***	92.95	11.72

Figure 1 (A) Forest plot of the estimated effect sizes (Hedges' d standardized mean difference) grouped by each non-native species (open squares) and by the phylum of the non-native species (grey squares). The black square indicates the overall mean effect size of feeding preferences. Error bars represent 95% confidence intervals. The numbers in brackets describe the number of dataset entries for each species. (B) The results of the meta-regression with the non-native seaweed or phylum of the non-native species as an explanatory variable in a random-effects model; * p < 0.5; ** p < 0.01; *** p < 0.001. [1]Currently accepted name *Caulerpa cylindracea*; [2]Currently accepted name *Codium fragile* ssp. *fragile*; [3]Currently accepted name *Dasysiphonia japonica*.

that non-native seaweeds were on average less preferred by native herbivores compared to native seaweeds. The results of the different studies were, however, highly heterogeneous (residual hetero-geneity among studies: $Q_{1.74} = 909.67$, $p < 0.001$, $I^2 = 93.8\%$). Including the predefined covariables in the model showed that the identity of the non-native seaweed, taxonomic and functional groups as well as the experiment type explained significant amounts of the observed heterogeneity (Figures 1 and 2). Residual heterogeneity was always high indicating that there may be other moderators not embraced by our analysis that influenced the outcome.

Grouping the non-native species by their taxonomic group revealed that only non-native red sea-weeds were of low palatability to native herbivores, while the non-native green and brown seaweeds did not differ significantly in palatability compared to native seaweeds (Figure 1). Grouping accord-ing to functional groups suggested that only non-native filamentous seaweeds were less preferred, whereas the palatability of corticated or leathery non-native seaweeds did not differ from native counterparts (Figure 2). Due to the dataset structure, functional and taxonomic group were highly confounded: all green algae were corticated, all brown algae were leathery macrophytes and all filamentous algae were red seaweeds (Table 1). However, when functional groups were separately

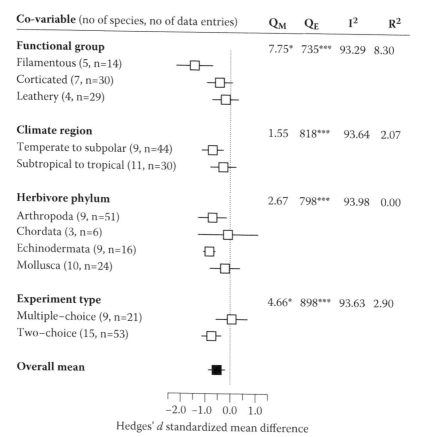

Figure 2 Forest plot of the estimated effect sizes (Hedges' d standardized mean difference) grouped by explanatory variables (open squares). The black square indicates the overall mean effect size of the dataset. Error bars represent 95% confidence intervals. The numbers in brackets describe the number of non-native species and the number of dataset entries represented by each subgroup. The statistics of the meta-regression of a random-effects model using functional group of the non-native seaweed, climate region and experiment type as an explanatory variable are displayed to the right; * $p < 0.5$; ** $p < 0.01$; *** $p < 0.001$.

tested for non-native red seaweeds, only filamentous seaweeds were again significantly less preferred ($Z_{filamentous} = -4.28$, p < 0.001; $Z_{corticated} = -1.80$, p = 0.07).

Analysis of the herbivore grouping revealed that arthropods and echinoderms found non-native seaweeds less palatable than native seaweeds, whereas molluscs and fish did not show any preference (Figure 2). Furthermore, herbivores from temperate to subpolar regions significantly preferred native seaweeds to non-native seaweeds, but herbivores from tropical or subtropical regions did not show a preference (Figure 2).

The experiment type had a significant influence on the effect size. In contrast to the two-choice experiments, the multiple-choice experiments, in which consumption of the non-native seaweed was compared to the averaged consumption of all native seaweeds, did not detect an overall significant feeding preference for native or non-native seaweeds (Figure 2).

Sensitivity analysis, publication bias and temporal trends

There was no indication that the results obtained from the meta-analysis lacked robustness. Exclusion of any data entry in the meta-analysis always resulted in similar overall mean effect size and confidence intervals (results not shown), which indicated that there were no serious outliers present. The cumulative meta-analysis by publication year showed that from the eighth data entry (2001), the overall mean effect size was constantly negative, oscillating between −0.628 and −0.302, though the 95% confidence interval included zero over some periods (Figure 3). Furthermore, Rosenberg's fail-safe number was sufficiently large (4219) to conclude that the observed outcome was a reliable estimate of the overall effect size. Additionally, the funnel plot and rank correlation test for funnel plot asymmetry gave no indication that publication bias affected the observed outcome (rank correlation test, Kendall's T = 0.098, p = 0.2196, Figure 4).

Review of community studies

Our literature search identified nine studies that examined adult performance of non-native seaweeds in the presence and absence of consumers. Only five reported effects on cover or biomass for both the non-native species and native seaweeds. These studies included one green, two brown and two red algal taxa and are summarized in Table 2. All taxa were also covered by the meta-analysis of feeding preference.

In the presence of herbivores, the filamentous red alga *Bonnemaisonia hamifera* reached higher cover and its biomass increased in short-term community experiments under laboratory conditions. In the absence of herbivores, *B. hamifera* was an inferior competitor compared to the native red seaweeds in the community and decreased in abundance (Enge et al. 2013, Sagerman et al. 2014). In contrast, the filamentous red alga *Heterosiphonia japonica* (currently accepted name *Dasysiphonia japonica*) dominated the community independent of herbivore presence due to its extreme growth rate (Sagerman et al. 2014). Compared to six native species, the leathery brown alga *Sargassum muticum* was the only seaweed that could maintain clear positive growth in the presence of herbivores in a short-term laboratory community experiment, but was also a superior competitor in the absence of herbivores (Engelen et al. 2011). Grazing had no effect on the cover of the corticated green alga *Codium fragile* ssp. *tomentosoides* (currently accepted name *Codium fragile* ssp. *fragile*) in a 13-week field experiment, while the native competitor *Laminaria longicruris* (currently accepted name *Saccharina longicruris*) could not persist and cover of turf algae strongly decreased under natural sea urchin densities. But again, *Codium fragile* ssp. *tomentosoides* had superior competitive abilities compared to *Laminaria longicruris* even in the absence of herbivores (Sumi & Scheibling 2005). On sea urchin barrens, grazing could not prevent canopy development of the leathery brown *Undaria pinnatifida* over a 30-month period, while the native canopy species did not exceed more than 0.7% cover (Valentine & Johnson 2005). In addition, the native canopy species showed inferior

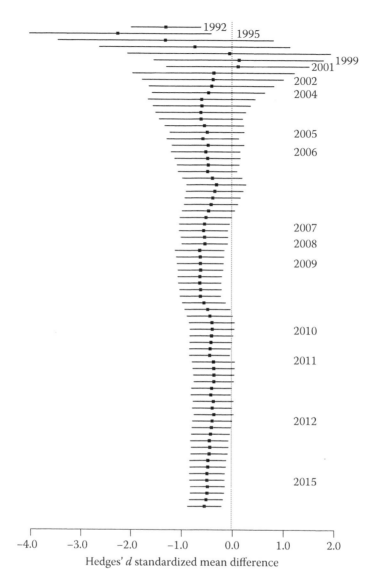

Figure 3 Forest plot of the cumulative meta-analysis of herbivore feeding preference between native and non-native seaweeds, ordered by publication year. Each data point represents the change of the estimated overall mean effect size (Hedges' *d* standardized mean difference) by adding the next newest entry into the meta-analysis. The dotted line indicates the absence of a significant effect. Error bars represent 95% confidence intervals.

competitive abilities compared to *U. pinnatifida* in the absence of herbivores (Valentine & Johnson 2005).

Discussion

Our meta-analysis showed that overall, non-native seaweeds tend to be less palatable than native seaweeds to herbivores in the new community. However, the meta-analysis also revealed considerable variability among the results of different studies. This variability could largely be explained by the identity of the seaweed taxon, indicating that the relative palatability of native compared

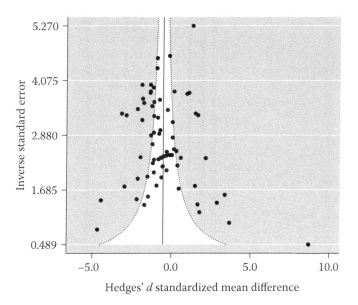

Figure 4 Funnel plot for the meta-analysis of feeding preferences using a random-effects model. Each data entry is represented by a circle showing the relation of the effect size (Hedges' *d* standardized mean difference) to its inversed standard error. The black line indicates the estimated overall mean effect and the white region represents the region in which 95% of the studies are expected to lie in the absence of biases and heterogeneity.

Table 2 Summary of community studies on herbivore effects on non-indigenous seaweeds

Species	Taxonomic group	Functional group	Studies	Method
Codium fragile ssp. *tomentosoides*[1]	Chlorophyta	Corticated	Sumi & Scheibling 2005	Field experiment
Sargassum muticum	Phaeophyceae	Leathery	Engelen et al. 2011	Laboratory experiment
Undaria pinnatifida	Phaeophyceae	Leathery	Valentine & Johnson 2005	Field experiment
Bonnemaisonia hamifera	Rhodophyta	Filamentous	Enge et al. 2013, Sagerman et al. 2014	Laboratory experiment
Heterosiphonia japonica[2]	Rhodophyta	Filamentous	Sagerman et al. 2014	Laboratory experiment

Note: The Functional group classification follows Littler & Littler (1984): filamentous = delicately branched, uniseriate to slightly corticated; corticated = coarsely branched, corticated; leathery = thick branched, heavily corticated or thick walled.

[1] currently accepted name *Codium fragile* ssp. *fragile*

[2] currently accepted name *Dasysiphonia japonica*

to non-native seaweeds differs between the groups of red, brown and green seaweeds. It was only for reds, but not for the brown and green seaweeds, that the meta-analysis demonstrated an overall difference in herbivore preference between native and non-native species. Furthermore, the meta-analysis revealed that filamentous non-native species, which were all red seaweeds in our analysis, tended to be of low palatability to native herbivores.

Feeding preference of herbivores is positively correlated to the nutritional quality and the shelter provided by the seaweed, and seaweeds can in turn deter herbivores by structural and chemical defences (Lubchenco & Gaines 1981, Hay & Fenical 1992). The low herbivore preference for non-native filamentous algae is an unexpected result because filamentous algae are commonly regarded to be palatable and highly susceptible to most consumers (Littler & Littler 1980, Steneck & Watling

1982, Littler et al. 1983). Consequently, this group could be expected to face a higher degree of consumptive biotic resistance in new regions, compared to non-native seaweeds from other functional groups. Our result matches recent findings suggesting that leathery and foliose algae are more susceptible to herbivores than filamentous or corticated algae (Poore et al. 2012), which indicates that structural traits, such as tissue toughness, is far from always a determining factor for food choice of herbivores and that feeding preferences are driven by other seaweed traits.

The presence of chemical defences can explain low palatability of certain seaweeds to marine herbivores (Hay & Fenical 1992, Pavia et al. 2012). Accordingly, it has been postulated that chemical defences can protect non-native seaweeds, as well as vascular plants, from being attacked by native herbivores in a new region, either by comparatively high defence concentrations or by molecular structures that are evolutionarily novel to the native herbivores (Cappuccino & Arnason 2006, Wikström et al. 2006, Verhoeven et al. 2009, Enge et al. 2012). Red seaweeds in particular produce an immense diversity and high quantities of often halogenated secondary metabolites, which have been frequently demonstrated to possess effective antimicrobial (Persson et al. 2011, Nylund et al. 2013), antifouling (Dworjanyn et al. 2006), allelopathic (Svensson et al. 2013) and antiherbivore activities (Kladi et al. 2005, Cabrita et al. 2010, Enge et al. 2012). The diversity of chemical defence compounds in red algae may explain why non-native red seaweeds showed especially low palatability in our meta-analysis.

The establishment of a specific chemical basis for a low preference of potential native consumers for an introduced organism is, however, a demanding task and marine examples are still rare. Evidence for chemical defences against native herbivores has so far only been provided for a few non-native seaweed species. The green algae, *Caulerpa taxifolia* and *C. racemosa*, produce caulerpenyne with effects on sea urchins (Amade & Lemée 1998, Dumay et al. 2002). *Codium fragile* ssp. *fragile* (=*Codium fragile* ssp. *tomentosoides*) possesses wound-activated defences involving dimethylsulfoniopropionate (DMSP), which deters native sea urchins (Lyons et al. 2007). The arctic brown alga, *Fucus evanescens**, contains significantly higher concentrations of phlorotannins (polyphenolic defence compounds) than native fucoids in its new range, deterring native isopods and molluscs (Wikström et al. 2006, Forslund et al. 2010). The highly invasive red alga, *Gracilaria vermiculophylla*, produces prostaglandins, hydroxylated fatty acids and arachidonic acid-derived lactones on wounding, which provides resistance against native isopods and molluscs (Nylund et al. 2011, Hammann et al. 2016). Another red seaweed, the filamentous *Bonnemaisonia hamifera*, produces volatile brominated compounds that provide defence against native isopods, gammarids and ophistobranch consumers (Enge et al. 2012). These examples show that chemical defence can explain the low palatability of some non-native seaweeds, but further studies are needed before it can be concluded that chemical defence is a common trait of low-preferred non-native seaweeds (especially of the filamentous red algae).

Notably, two of the species for which chemical defences have been characterized (*Caulerpa racemosa* and *Codium fragile* ssp. *tomentosoides*) were not consistently of low preference to native herbivores in our meta-analysis. In both cases, the chemical defence compound was only documented to be active against one herbivore species, while multiple herbivores were tested in the feeding preference experiments. A specific chemical defence is usually not effective against all herbivore species, since herbivores can adapt to and/or circumvent the effects of secondary metabolites (Sotka, 2005). Furthermore, concentrations and the effectiveness of the chemical defence compounds can vary between seaweed and herbivore populations (Pavia et al. 2003, Sotka 2005), which makes it

---

* *Fucus evanescens* may have expanded its range naturally, but available evidence suggests that the spread of this species to southern Scandinavia and the British Isles was aided by human transport. It exhibits a disjunct distribution with new occurrences that were discovered in harbours in the beginning of the 20th century, making introduction from shipping plausible. Thus, we chose to include *Fucus evanescens* in the definition of a non-native species that we used in the literature search, i.e. a species that has been translocated to a new range by humans.

important that the palatability of non-native species introduced into several regions are tested with a set of the native herbivores that are relevant in the new regions.

Herbivore preference can be a first indicator for plant performance and competitive ability, but cannot be directly translated into community composition and population dynamics under natural conditions. We found that studies using long-term community experiments assessing the effects of consumers on non-native seaweeds in interaction with native seaweeds are essentially lacking. The few existing studies included in our review showed that performance of non-native seaweeds can be increased, equal or reduced in the presence of herbivores. To date, the most rigorous example of a successful seaweed invasion based on chemical defence concerns the filamentous red alga *Bonnemaisonia hamifera* (Figure 5). The documented chemical defence (1,1,3,3-tetrabromo-2-heptanone) provides this relatively poor competitor (in the absence of native herbivores), with a strong competitive advantage in its new range in the presence of native herbivores (Enge et al. 2013, Sagerman et al. 2014). In addition, the same brominated compound inhibits the recruitment of native algal competitors (Svensson et al. 2013) and reduces bacterial load (Nylund et al. 2008). Thus, the multiple ecological benefits of this compound outweigh the cost of its production for the invader in the new range (Nylund et al. 2013). In some of the other studies the non-native seaweeds often performed better than their native competitors in the presence of herbivores, but the non-native species were superior competitors also in the absence of herbivores. Thus, the extent to which low herbivore preference contributes to invasion success of non-native seaweeds remains elusive. There is a need for more studies on the effects of herbivores on non-native seaweed populations, ideally experiments that include seasonal changes and complete life cycles of the seaweed and grazers. This is a challenging task in marine environments with species with complex life cycles.

In conclusion, the results of our meta-analysis show that low palatability does not seem to be a universal trait among non-native seaweeds and only certain seaweeds escape native herbivores in

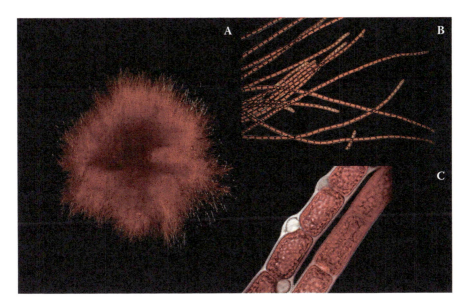

Figure 5 The invasive filamentous red seaweed *Bonnemaisonia hamifera*, which produces a potent chemical defence (1,1,3,3-tetrabromo-2-heptanone) that makes it unpalatable to native herbivores in the north Atlantic, thereby providing it with a strong competitive advantage over native seaweeds in its new range. (A) Tetrasporophytic phase, growing as small turfs, which consist of numerous sparsely branched filaments (B). Filaments are one cell-layer thick and have numerous gland cells, containing chemical defences, located between the vegetative cells (C). (From Nylund et al. 2008.)

their new range. Accordingly, the prediction of the enemy release hypothesis that introduced species are less attacked by herbivores than their native counterparts in the new range (Keane & Crawley 2002), does not hold for all non-native seaweeds. Interestingly, we found that non-native filamentous red seaweeds tend to be especially less palatable to herbivores. This is an important finding considering the majority of seaweed introductions are filamentous or corticated red algae (Williams & Smith 2007). Due to their morphology, these species can often be cryptic or less apparent compared to larger brown and green seaweeds. Probably, therefore, they are less frequently studied and often overlooked in their community impacts, even though effects on biodiversity and ecosystem processes have been proven (Schaffelke & Hewitt 2007, Sagerman et al. 2014). The low preference for many non-native filamentous red algae indicates that low impact of herbivores may contribute to invasion success in this group. However, to assess if consumers play a crucial role in seaweed invasions, future research should focus on examining consumer impacts on competitive interactions between non-native and native seaweeds.

Acknowledgements

This work was supported by the Swedish Research Council through grant no. 621–2011–5630 to H.P., by a grant from the Swedish Research Council Formas to S.A.W., and by the Linnaeus Centre for Marine Evolutionary Biology (http://www.cemeb.science.gu.se/). S.A.W. was partly financed by the Baltic Eye project. Stephen Hawkins and Ally Evans provided comments that improved the manuscript and Gunilla Toth and Göran Nylund helped with the editing.

References

Amade, P. & Lemée, R. 1998. Chemical defence of the mediterranean alga *Caulerpa taxifolia*: variations in caulerpenyne production. *Aquatic Toxicology* **43**, 287–300.

Begg, C.B. & Mazumdar, M. 1994. Operating characteristics of a rank correlation test for publication bias. *Biometrics* **50**, 1088–1101.

Britton-Simmons, K.H. 2004. Direct and indirect effects of the introduced alga *Sargassum muticum* on benthic, subtidal communities of Washington State, USA. *Marine Ecology Progress Series* **277**, 61–78.

Britton-Simmons, K.H., Pister, B., Sánchez, I. & Okamoto, D. 2011. Response of a native, herbivorous snail to the introduced seaweed *Sargassum muticum*. *Hydrobiologia* **661**, 187–196.

Burkepile, D.E. & Hay, M.E. 2008. Herbivore species richness and feeding complementarity affect community structure and function on a coral reef. *Proceedings of the National Academy of Science USA* **105**, 16201–16206.

Cabrita, M.T., Vale, C. & Rauter, A.P. 2010. Halogenated compounds from marine algae. *Marine Drugs* **8**, 2301–2317.

Cacabelos, E., Olabarria, C., Incera, M. & Troncoso, J.S. 2010. Do grazers prefer invasive seaweeds? *Journal of Experimental Marine Biology and Ecology* **393**, 182–187.

Cappuccino, N. & Arnason, J.T. 2006. Novel chemistry of invasive exotic plants. *Biology Letters* **2**, 189–193.

Chavanich, S. & Harris, L.G. 2002. The influence of macroalgae on seasonal abundance and feeding preference of a subtidal snail, *Launa vincta* (Montagu) (Littorinidae) in the Gulf of Maine. *Journal of Molluscan Studies* **68**, 73–78.

Chun, Y.J., van Kleunen, M. & Dawson, W. 2010. The role of enemy release, tolerance and resistance in plant invasions: linking damage to performance. *Ecology Letters* **13**, 937–946.

Colautti, R.I., Ricciardi, A., Grigorovich, I.A. & MacIsaac, H.J. 2004. Is invasion success explained by the enemy release hypothesis? *Ecology Letters* **7**, 721–733.

Cruz-Rivera, E. & Hay, M.E. 2000. Can quantity replace quality? Food choice, compensatory feeding, and fitness of marine mesograzers. *Ecology* **81**, 201–219.

Cruz-Rivera, E. & Hay, M. 2001. Macroalgal traits and the feeding and fitness of an herbivorous amphipod: the roles of selectivity, mixing, and compensation. *Marine Ecology Progress Series* **218**, 249–266.

Cummings, D.O. & Williamson, J.E. 2008. The role of herbivory and fouling on the invasive green alga *Caulerpa filiformis* in temperate Australian waters. *Marine and Freshwater Research* **59**, 279–290.

Darwin, C. 1859. *On the Origin of Species by Means of Natural Selection, or, the Preservation of Favoured Races in the Struggle for Life*. London: J. Murray.

Davis, A.R., Benkendorff, K. & Ward, D.W. 2005. Responses of common SE Australian herbivores to three suspected invasive *Caulerpa* spp.. *Marine Biology* **146**, 859–868.

Dietz, G., Dahabreh, I.J., Gurevitch J, Lajeunesse, M.J., Schmid, C.H., Trikalinos, T.A. & Wallace, B.C. 2016. OpenMEE: Software for Ecological and Evolutionary Meta-analysis (Computer program). Available at (http://www.cebm.brown.edu/open_mee)

Duffy, J.E. & Hay, M.E. 1991. Food and shelter as determinants of food choice by an herbivorous marine amphipod. *Ecology* **72**, 1286–1298.

Dumay, O., Pergent, G., Pergent-Martini, C. & Amade, P. 2002. Variations in caulerpenyne contents in *Caulerpa taxifolia* and *Caulerpa racemosa*. *Journal of Chemical Ecology* **28**, 343–352.

Dworjanyn, S.A., de Nys, R. & Steinberg, P.D. 2006. Chemically mediated antifouling in the red alga *Delisea pulchra*. *Marine Ecology Progress Series* **318,** 153–163.

Elton, C.S. 1958. *The Ecology of Invasions by Animals and Plants*. London: Springer.

Enge, S., Nylund, G.M., Harder, T. & Pavia, H. 2012. An exotic chemical weapon explains low herbivore damage in an invasive alga. *Ecology* **93**, 2736–2745.

Enge, S., Nylund, G.M & Pavia, H. 2013. Native generalist herbivores promote invasion of a chemically defended seaweed via refuge-mediated apparent competition. *Ecology Letters* **16**, 487–492.

Engelen, A.H., Henriques, N., Monteiro, C. & Santos, R. 2011. Mesograzers prefer mostly native seaweeds over the invasive brown seaweed *Sargassum muticum*. *Hydrobiologia* **669**, 157–165.

Engelen, A.H., Serebryakova, A., Ang, P., Britton-Simmons, K., Mineur, F., Pedersen, M.F., Arenas, F., Fernández, C., Steen, H., Svenson, R., Pavia, H., Toth, G., Viard, F. & Santos, R. 2015. Circumglobal invasion by the brown seaweed *Sargassum muticum*. *Oceanography and Marine Biology: An Annual Review* **53**, 81–126.

Forslund, H., Wikström, S. & Pavia, H. 2010. Higher resistance to herbivory in introduced compared to native populations of a seaweed. *Oecologia* **164**, 833–840.

Hammann, M., Rempt, M., Pohnert, G., Wang, G., Boo, S.M. & Weinberger, F. 2016. Increased potential for wound activated production of Prostaglandin E2 and related toxic compounds in non-native populations of *Gracilaria vermiculophylla*. *Harmful Algae* **51**, 81–88.

Hawkins, S.J. & Hartnoll, R.G. 1983. Grazing of intertidal algae by marine-invertebrates. *Oceanography and Marine Biology: An Annual Review* **21**, 195–282.

Hawkins, S.J., Hartnoll, R.G., Kain, J.M. & Norton, T.A. 1992. Plant-animal interactions on hard substrata in the north-east Atlantic. In *Plant-Animal Interactions in the Marine Benthos*, D.M. John et al. (eds). Oxford: Clarendon Press, 1–32.

Hay, M.E. & Fenical W. 1992. Chemical mediation of seaweed-herbivore interactions. In *Plant-Animal Interactions in the Marine Benthos*, D.M. John et al. (eds). Oxford: Clarendon Press, 319–338.

Hay, M.E. & Steinberg, P.D. 1992. The chemical ecology of plant-herbivore interactions in marine versus terrestrial communities. In *Herbivores: Their Interactions with Secondary Metabolites, Evolutionary and Ecological Processes,* G. Rosenthal & M. Berenbaum (eds). San Diego, USA: Academic Press, 371–413.

Hedges, L.V. & Olkin, I. 1985. Statistical methods for meta-analysis. Orlando: Academic Press.

Jennions, M.D. & Møller, A.P. 2002. Relationships fade with time: a meta-analysis of temporal trends in publication in ecology and evolution. *Proceedings of the Royal Society of London B: Biological Sciences* **269**, 43–48.

Johnson, C.R. & Chapman, A.R.O. 2007. Seaweed invasions: introduction and scope. *Botanica Marina* **50**, 321–325.

Keane, R. & Crawley, M.J. 2002. Exotic plant invasions and the enemy release hypothesis. *Trends in Ecology & Evolution* **17**, 164–170.

Kimbro, D.L., Cheng, B.S. & Grosholz, E.D. 2013. Biotic resistance in marine environments. *Ecology Letters* **16**, 821–833.

Kladi, M., Vagias, C. & Roussis, V. 2005. Volatile halogenated metabolites from marine red algae. *Phytochemistry Reviews* **3**, 337–366.

Lemee, R., Boudouresque, C., Gobert, J., Malestroit, P., Mari, X., Meinesz, A., Menager, V. & Ruitton, S. 1996. Feeding behaviour of *Paracentrotus lividus* in the presence of *Caulerpa taxifolia* introduced in the Mediterranean Sea. *Oceanologica Acta* **19**, 245–253.

Levin, P.S., Coyer, J.A., Petrik, R. & Good, T.P. 2002. Community-wide effects of noninigenous species on temperate rocky reefs. *Ecology* **83**, 3182–3193.

Levine, J.M., Adler, P.B. & Yelenik, S.G. 2004. A meta-analysis of biotic resistance to exotic plant invasions. *Ecology Letters* **7**, 975–989.

Littler, M.M. & Littler, D.S. 1980. The evolution of thallus form and survival strategies in benthic marine macroalgae: field and laboratory tests of a functional form model. *American Naturalist* **116**, 25–44.

Littler, M.M. & Littler, D.S. 1984. Relationships between macroalgal functional form groups and substrata stability in a subtropical rocky-intertidal system. *Journal of Experimental Marine Biology and Ecology* **74**, 13–34.

Littler, M.M., Taylor, P.R. & Littler, D.S. 1983. Algal resistance to herbivory on a Caribbean barrier reef. *Coral Reefs* **2**, 111–118.

Liu, H. & Stiling, P. 2006. Testing the enemy release hypothesis: a review and meta-analysis. *Biological Invasions* **8**, 1535–1545.

Low, N.H.N., Drouin, A., Marks, C.J. & Bracken, M.E.S. 2015. Invader traits and community context contribute to the recent invasion success of the macroalga *Heterosiphonia japonica* on New England rocky reefs. *Biological Invasions* **17**, 257–271.

Lubchenco, J. & Gaines, S.D. 1981. A unified approach to marine plant-herbivore interactions. I. Populations and communities. *Annual Review of Ecology and Systematics* **12**, 405–437.

Lyons, D.A. & Scheibling, R.E. 2007. Effect of dietary history and algal traits on feeding rate and food preference in the green sea urchin *Strongylocentrotus droebachiensis*. *Journal of Experimental Marine Biology and Ecology* **349**, 194–204.

Lyons, D.A., Van Alstyne, K.L. & Scheibling, R.E. 2007. Anti-grazing activity and seasonal variation of dimethylsulfoniopropionate-associated compounds in the invasive alga *Codium fragile* ssp. *tomentosoides*. *Marine Biology* **153**, 179–188.

Maggi, E., Benedetti-Cecchi, L., Castelli, A., Chatzinikolaou, E., Crowe, T.P., Ghedini, G., Kotta, J., Lyons, D.A., Ravaglioli, C., Rilov, G., Rindi, L. & Bulleri, F. 2015. Ecological impacts of invading seaweeds: a meta-analysis of their effects at different trophic levels. *Diversity and Distributions* **21**, 1–12.

Maron, J.L. & Vila, M. 2001. When do herbivores affect plant invasion? Evidence for the natural enemies and biotic resistance hypotheses. *Oikos* **95**, 361–373.

Monteiro, C.A., Engelen, A.H. & Santos, R.O.P. 2009. Macro- and mesoherbivores prefer native seaweeds over the invasive brown seaweed *Sargassum muticum*: a potential regulating role on invasions. *Marine Biology* **156**, 2505–2515.

Nejrup, L.B., Pedersen, M.F. & Vinzent, J. 2012. Grazer avoidance may explain the invasiveness of the red alga *Gracilaria vermiculophylla* in Scandinavian waters. *Marine Biology* **159**, 1703–1712.

Nylund, G.M., Cervin, G., Persson, F., Hermansson, M., Steinberg, P.D. & Pavia, H. 2008. Seaweed defence against bacteria: a poly-halogenated 2-heptanone from the red alga *Bonnemaisonia hamifera* inhibits bacterial colonisation at natural surface concentrations. *Marine Ecology Progress Series* **369**, 39–50.

Nylund, G.M., Enge, S. & Pavia, H. 2013. Cost and benefits of chemical defence in the red alga *Bonnemaisonia hamifera*. *PLoS ONE* **8**, e61291.

Nylund, G.M., Pereyra, R.T., Wood, H.L., Johannesson, K. & Pavia, H. 2012. Increased resistance towards generalist herbivory in the new range of a habitat-forming seaweed. *Ecosphere* **3**, 1–13 Art125.

Nylund, G.M., Weinberger, F., Rempt, M. & Pohnert, G. 2011 Metabolomic assessment of induced and activated chemical defence in the invasive red alga *Gracilaria vermiculophylla*. *PLoS ONE* **6**, e29359.

Parker, J.D., Burkepile, D.E. & Hay, M.E. 2006. Opposing effects of native and exotic herbivores on plant invasions. *Science* **311**, 1459–1461.

Parker, J.D. & Hay, M.E. 2005. Biotic resistance to plant invasions? Native herbivores prefer non-native plants. *Ecology Letters* **8**, 959–967.

Pavia, H., Baumgartner, F., Cervin, G., Enge, S., Kubanek, J., Nylund, G.M., Selander, E., Svensson, J.R. & Toth, G.B. 2012. Chemical defences against herbivores. In *Chemical Ecology in Aquatic Systems*, C. Brönmark & L.-A. Hansson (eds). Oxford: Oxford University Press, 210–235.

Pavia, H., Toth, G.B., Lindgren, A. & Åberg, P. 2003. Intraspecific variation in the phlorotannin content of the brown alga *Ascophyllum nodosum*. *Phycologia* **42**, 378–383.

Pedersen, M.F., Stæhr, P.A., Wernberg, T. & Thomsen, M.S. 2005. Biomass dynamics of exotic *Sargassum muticum* and native *Halidrys siliquosa* in Limfjorden, Denmark – –implications of species replacements on turnover rates. *Aquatic Botany* **83**, 31–47.

Persson, F., Svensson ,R., Nylund, G.M., Fredriksson, J., Pavia, H. & Hermansson, M. 2011. Ecological role of a seaweed secondary metabolite for a colonizing bacterial community. *Biofouling* **27**, 579–588.

Poore, A.G.B., Campbell, A.H., Coleman, R.A., Edgar, G.J., Jormalainen, V., Reynolds, P.L., Sotka, E.E., Stachowicz, J.J., Taylor, R.B., Vanderklift, M.A. & Duffy, J.E. 2012. Global patterns in the impact of marine herbivores on benthic primary producers. *Ecology Letters* **15**, 912–922.

Prince, J.S. & LeBlanc, W.G. 1992. Comparative feeding preference of *Strongylocentrotus droebachiensis* (Echinoidea) for the invasive seaweed *Codium fragile* ssp. *tomentosoides* (Chlorophyceae) and four other seaweeds. *Marine Biology* **113**, 159–163.

Rohatgi, A. 2015. *WebPlotDigitalizer. Version 3.10.* Austin, Texas. Available at (http://arohatgi.info/WebPlotDigitizer/app/).

Rosenberg, M.S. 2005. The file-drawer problem revisited: a general weighted method for calculating fail-safe numbers in meta-analysis. *Evolution* **59**, 464–468.

Rothstein, H.R., Lortie, C.J., Stewart, G.B., Koricheva, J. & Gurevitch, J. 2013. Quality standards for research synthesis. In *Handbook of Meta-Analysis in Ecology and Evolution.* J. Koricheva et al. (eds). Princeton: Princeton University Press, 323–338.

Sagerman, J., Enge, S., Pavia, H. & Wikström, S.A. 2014. Divergent ecological strategies determine different impacts on community production by two successful non-native seaweeds. *Oecologia* **175**, 937–946.

Sagerman, J., Enge, S., Pavia, H. & Wikström, S.A. 2015. Low feeding preference of native herbivores for the successful non-native seaweed *Heterosiphonia japonica*. *Marine Biology* **162**, 2471–2479.

Schaffelke, B. Evers, D. & Walhorn, A. 1995. Selective grazing of the isopod *Idotea baltica* between *Fucus evanescens* and *F. vesiculosus* from Kiel Fjord (western Baltic). *Marine Biology* **124**, 215–218.

Schaffelke, B. & Hewitt, C.L. 2007. Impacts of introduced seaweeds. *Botanica Marina* **50**, 397–417.

Schaffner, U., Ridenour, W.M., Wolf VC, Bassett, T., Muller, C., Muller-Scharer, H., Sutherland, S., Lortie, C.J. & Callaway, R.M. 2011. Plant invasions, generalist herbivores, and novel defense weapons. *Ecology* **92**, 829–835.

Scheibling, R. & Anthony, S. 2001. Feeding, growth and reproduction of sea urchins (*Strongylocentrotus droebachiensis*) on single and mixed diets of kelp (*Laminaria* spp.) and the invasive alga *Codium fragile* ssp. *tomentosoides*. *Marine Biology* **139**, 139–146.

Smith, J.E., Hunter, C.L., Conklin, E.J., Most, R., Sauvage, T., Squair, C. & Smith, C.M. 2004. Ecology of the invasive red alga *Gracilaria salicornia* (Rhodophyta) on O'ahu, Hawai'i. *Pacific Science* **58**, 325–343.

Sotka, E.E. 2005. Local adaptation in host use among marine invertebrates. *Ecology Letters* **8**, 448–459.

Steinarsdóttir, M.B., Ingólfsson, A. & Ólafsson, E. 2009. Trophic relationships on a fucoid shore in south-western Iceland as revealed by stable isotope analyses, laboratory experiments, field observations and gut analyses. *Journal of Sea Research* **61**, 206–215.

Steneck, R.S. & Watling, L. 1982. Feeding capabilities and limitation of herbivorous mollusks – a functional-group approach. *Marine Biology* **68,** 299–319.

Stimson, J., Cunha, T. & Philippoff, J. 2007. Food preferences and related behavior of the browsing sea urchin *Tripneustes gratilla* (Linnaeus) and its potential for use as a biological control agent. *Marine Biology* **151**, 1761–1772.

Strong, J.A., Maggs, C.A. & Johnson, M.R. 2009. The extent of grazing release from epiphytism for *Sargassum muticum* (Phaeophyceae) within the invaded range. *Journal of the Marine Biological Association of the United Kingdom* **89**, 303–314.

Sumi, C.B.T. & Scheibling, R.E. 2005. Role of grazing by sea urchins *Strongylocentrotus droebachiensis* in regulating the invasive alga *Codium fragile* ssp. *tomentosoides* in Nova Scotia. *Marine Ecology Progress Series* **292**, 203–212.

Svensson, J.R., Nylund, G.M., Cervin, G., Toth, G.B. & Pavia, H. 2013. Novel chemical weapon of an exotic macroalga inhibits recruitment of native competitors in the invaded range. *Journal of Ecology* **101**, 140–148.

Thornber, C.S., Kinlan, B.P., Graham, M.H. & Stachowicz, J.J. 2004. Population ecology of the invasive kelp *Undaria pinnatifida* in California: environmental and biological controls on demography. *Marine Ecology Progress Series* **268**, 69–80.

Tomas, F., Box, A. & Terrados, J. 2011a. Effects of invasive seaweeds on feeding preference and performance of a keystone Mediterranean herbivore. *Biological Invasions* **13**, 1559–1570.

Tomas, F., Cebrian, E. & Ballesteros, E. 2011b. Differential herbivory of invasive algae by native fish in the Mediterranean Sea. *Estuarine, Coastal and Shelf Science* **92**, 27–34.

Trowbridge, C.D. 1995. Establishment of the green alga *Codium fragile* ssp. *tomentosoides* on New Zealand rocky shores: current distribution and invertebrate grazers. *The Journal of Ecology* **83**, 949–965.

Trowbridge, C.D. & Todd, C.D. 1999. The familiar is exotic: I. *Codium fragile* ssp. *atlanticum* on Scottish rocky intertidal shores. *Botanical Journal of Scotland* **51**, 139–160.

Valentine, J.P. & Johnson, C.R. 2005. Persistence of the exotic kelp *Undaria pinnatifida* does not depend on sea urchin grazing. *Marine Ecology Progress Series* **285**, 43–55.

Verhoeven, K.J.F., Biere, A., Harvey, J.A. & van der Putten, W.H. 2009. Plant invaders and their novel natural enemies: who is naïve? *Ecology Letters* **12**, 107–117.

Viechtbauer, W. 2010. Conducting meta-analyses in R with the metafor package. *Journal of Statistical Software* **36**, 1–48.

Weinberger, F., Buchholz, B., Karez, R. & Wahl, M. 2008. The invasive red alga *Gracilaria vermiculophylla* in the Baltic Sea: adaptation to brackish water may compensate for light limitation. *Aquatic Biology* **3**, 251–264.

Wikström, S.A., Steinarsdóttir, M.B., Kautsky, L. & Pavia, H. 2006. Increased chemical resistance explains low herbivore colonization of introduced seaweed. *Oecologia* **148**, 593–601.

Williams, S.L. & Smith, J.E. 2007. A global review of the distribution, taxonomy, and impacts of introduced seaweeds. *Annual Review of Ecology, Evolution, and Systematics* **38**, 327–359.

Williamson, M. & Fitter, A. 1996. The varying success of invaders. *Ecology* **77**, 1661–1666.

Yun, H.Y. & Molis, M. 2012. Comparing the ability of a non-indigenous and a native seaweed to induce anti-herbivory defenses. *Marine Biology* **159**, 1475–1484.

AUTHOR INDEX

Page numbers in **boldface** denote complete articles.

SYSTEMATIC INDEX

SUBJECT INDEX

Page numbers in **boldface** denote tables.

#0003 - 281017 - C498 - 254/178/27 [29] - CB - 9781138197862